Gertz Likhtenshtein
Stilbenes

Further Reading

Ackermann, L. (ed.)

Modern Arylation Methods

2009
ISBN: 978-3-527-31937-4

Bandini, M., Umani-Ronchi, A. (eds.)

Catalytic Asymmetric Friedel-Crafts Alkylations

2009
ISBN: 978-3-527-32380-7

Albini, A., Fagnoni, M. (eds.)

Handbook of Synthetic Photochemistry

2010
ISBN: 978-3-527-32391-3

Mohr, F. (ed.)

Gold Chemistry

Applications and Future Directions in the Life Sciences

2009
ISBN: 978-3-527-32086-8

Gertz Likhtenshtein

Stilbenes

Applications in Chemistry, Life Sciences
and Materials Science

WILEY-VCH Verlag GmbH & Co. KGaA

The Author

Prof. Gertz Likhtenshtein
Ben-Gurion University
Department of Chemistry
653, P. O. Box
84105 Beer-Sheva 84105
Israel

All books published by Wiley-VCH are carefully produced. Nevertheless, authors, editors, and publisher do not warrant the information contained in these books, including this book, to be free of errors. Readers are advised to keep in mind that statements, data, illustrations, procedural details or other items may inadvertently be inaccurate.

Library of Congress Card No.: applied for

British Library Cataloguing-in-Publication Data
A catalogue record for this book is available from the British Library.

Bibliographic information published by the Deutsche Nationalbibliothek
The Deutsche Nationalbibliothek lists this publication in the Deutsche Nationalbibliografie; detailed bibliographic data are available on the Internet at http://dnb.d-nb.de.

© 2010 WILEY-VCH Verlag GmbH & Co. KGaA, Weinheim

All rights reserved (including those of translation into other languages). No part of this book may be reproduced in any form – by photoprinting, microfilm, or any other means – nor transmitted or translated into a machine language without written permission from the publishers. Registered names, trademarks, etc. used in this book, even when not specifically marked as such, are not to be considered unprotected by law

Composition   Thomson Digital, Noida, India
Printing   Strauss GmbH, Mörlenbach
Bookbinding   Litges & Dopf GmbH, Heppenheim
Cover Design   Schulz Grafik Design, Fußgönheim

Printed in the Federal Republic of Germany
Printed on acid-free paper

ISBN: 978-3-527-32388-3

Contents

Preface *XI*

1   **Stilbenes Preparation and Analysis** *1*
1.1   General *1*
1.2   Classical Methods and Their Development *2*
1.2.1   Aldol-Type Condensation *2*
1.2.2   Siegrist Method *3*
1.2.3   Wittig Reaction *4*
1.2.4   Heck Reaction *5*
1.2.5   Negishi–Stille Coupling *7*
1.2.6   Barton–Kellogg–Staudinger Reaction *8*
1.2.7   McMurry Reaction *8*
1.2.8   Perkin Reaction *9*
1.3   Miscellaneous Chemical Methods of Stilbene Synthesis *10*
1.3.1   Palladium-Catalyzed Reactions *10*
1.3.2   Horner–Wadsworth–Emmons and Wittig–Horner Olefination Reactions *13*
1.3.3   Other Synthetic Reactions *13*
1.4   Physically Promoted Reactions *19*
1.5   Synthesis of Stilbene Dendrimers *19*
1.6   Stilbene Cyclodextrin Derivatives *23*
1.7   Stilbenes on Templates *25*
1.8   Stilbenes Analysis *28*
1.8.1   Methods Using Liquid and Gas Chromatography *28*
1.8.2   Miscellaneous Analytical Methods *33*
    References *35*

2   **Stilbene Chemical Reactions** *43*
2.1   Halogenation of Stilbenes *43*
2.2   Oxidation of Stilbenes *46*

2.2.1	Epoxidation 46
2.2.2	Other Oxidation Reactions 48
2.3	Stilbene Reduction 51
2.4	Other Reactions 52
2.4.1	Vinyl Lithiation 52
2.4.2	Carbolization 52
2.4.3	Addition Reactions 53
2.4.4	Substituted Groups Reactions 54
2.5	Stilbenes in Polymerization 56
2.5.1	Radical Polymerization 56
2.5.2	Anionic Polymerization 57
2.6	Complexation 58
2.6.1	Complexation with Small Molecules 58
2.6.2	Complexation with Proteins 60
2.6.2.1	Tubulins 60
2.6.2.2	Antibodies 60
	References 62
3	**Stilbene Photophysics** 67
3.1	General 67
3.2	Stilbene Excited States 67
3.2.1	Excited Singlet State 67
3.2.2	Excited Triplet State 69
3.3	Absorption Spectra 71
3.3.1	Singlet–Singlet Absorption Spectra 71
3.3.2	Triplet–Triplet Absorption Spectra 71
3.3.3	Two-Photon Absorption and Fluorescence 72
3.4	Fluorescence from Excited Singlet States 76
3.4.1	Fluorescence Behavior and Hammett Relationships 76
3.4.1.1	Excitation to the Franck–Condon State 76
3.4.1.2	Radiative Deactivation 78
3.4.2	Miscellaneous Data on Stilbenes Fluorescence 80
3.4.2.1	Stilbenes in Solution 80
3.4.2.2	Stilbene Dendrimer Fluorescence 84
3.4.2.3	Stilbenes on Templates and in Proteins 86
3.5	Interactions Involving Triplet State and Phosphorescence 90
3.6	Fluorescence of Excimers and Exciplexes 92
3.7	Energy Transfer 93
3.8	Intramolecular Charge Transfer 95
	References 95
4	**Stilbene Photoisomerization** 99
4.1	General 99
4.2	Mechanisms of Photoisomerization 101

4.2.1	Ideas, Concepts, and Theoretical Calculations *101*
4.2.1.1	Through Double-Bond Twisting (Saltiel) Mechanism *102*
4.2.1.2	Single-Bond Twisting Mechanism *105*
4.2.1.3	Planar Intramolecular Charge Transfer Precursor Mechanism *105*
4.2.1.4	Double-Bond Twisting Mechanism in Linear Quinoid Structure *106*
4.2.1.5	A Volume-Conserving Mechanism *108*
4.2.1.6	Media "Melting" Mechanism: Photoisomerization in Rigid Surroundings *110*
4.2.1.7	Nonvertical Energy Transfer *110*
4.2.1.8	A Dual Thermal Bond Activation Mechanism *112*
4.2.2	Experimental and Theoretical Studies of the Photoisomerization Mechanisms *113*
4.3	Effect of Substituents and Polarity *117*
4.4	Viscosity Effect *119*
4.5	Miscellaneous Experimental Data on Photoisomerization *121*
4.5.1	Photoisomerization in Solutions *121*
4.5.1.1	Direct Photoisomerization *121*
4.5.1.2	Sensitized Photoisomerization *124*
4.5.1.3	Photoisomerization of Stilbenophanes *126*
4.5.1.4	Stilbene Photoisomerization in Dendrimers *126*
4.5.1.5	Stilbene Photoswitching Processes *128*
4.5.1.6	Stilbene Photoisomerization on Templates *129*
	References *131*
5	**Miscellaneous Stilbene Photochemical Reactions** *137*
5.1	Photocyclization *137*
5.2	Bimolecular Reactions *140*
5.2.1	Photodimerization *140*
5.2.2	Reactions with Alkenes and Dienes *143*
5.2.3	Reactions with Amines, Imines, Nitroso Oxide, and Protic Solvents *143*
5.3	Photoreactions in Stilbene Dendrimers *146*
5.4	Reactions in Polymers and Other Matrices *147*
5.5	Reaction Using Two-Photon Excitation *148*
5.6	Charge Transfer Ionization *150*
	References *156*
6	**Stilbene Materials** *159*
6.1	Stilbene Lasers *159*
6.1.1	Dye Lasers *159*
6.1.2	Stilbene Solid Lasers *160*
6.2	Electro-Optic Materials *161*
6.3	Electrophotographic Material *165*
6.4	Light-Emitting Diodes *167*

6.5	Materials for Nonlinear Optics	169
6.6	Light-Emitting Materials	175
6.7	Materials for Image-Forming Apparatuses	177
6.8	Radioluminescence Materials: Scintillators	180
6.9	Miscellaneous	183
	References	185
7	**Bioactive Stilbenes**	**189**
7.1	Resveratrol	189
7.1.1	General	189
7.1.2	Resveratrol Content in Biological Objects	190
7.1.3	Metabolism and Pharmacokinetics	192
7.1.4	Antioxidant Activity	195
7.1.5	Resveratrol and Apoptosis	197
7.1.6	Biochemical Effect	202
7.1.6.1	Enzymes	202
7.1.6.2	Cells and Animals	204
7.1.6.3	Effects on Metabolism of Estrogens	207
7.1.6.4	Signaling Pathway	207
7.1.7	Resveratrol in Genetics	208
7.1.8	Effect on Aging	209
7.1.9	Miscellaneous	210
7.2	Combretastatin and Its Analogues	213
7.2.1	Effect on Tubulin Polymerization	215
7.2.2	Miscellaneous	216
7.3	Pterostilbene	218
	References	219
8	**Preclinic Effects of Stilbenes**	**225**
8.1	Resveratrol	225
8.1.1	Cancer Protection in Animal	225
8.1.2	Cell Cancer Protection	227
8.1.3	Miscellaneous Effects	230
8.2	Combretastatin	233
8.2.1	Effects on Cancer Cells	233
8.2.2	Xenografts and Tumors	239
8.2.3	Animals	244
8.3	Pterostilbene	250
8.3.1	Cells	250
8.3.2	Animals	251
	References	256
9	**Stilbenes in Clinics**	**261**
9.1	General	261
9.2	*trans*-Resveratrol	262

9.3	Combretastatin 267
9.3.1	Vascular Damaging Agents 267
9.3.2	Pharmacometrics of Stilbenes: Seguing Toward the Clinic 268
9.4	Other Stilbenoids 273
	References 274

10	**Stilbenes as Molecular Probes** 277
10.1	General 277
10.2	Theoretical Grounds 278
10.2.1	Local Properties of Medium 278
10.2.1.1	Polarity 278
10.2.1.2	Molecular Dynamics 278
10.2.2	Excited Energy Transfer 279
10.2.2.1	Fluorescence Resonance Energy Transfer 279
10.3	Experimental Methods and Their Applications 280
10.3.1	Probing Based on Solvatochromism 280
10.3.2	Image and Structure Probing 283
10.3.3	Methods Based on Accessibility of Reactive Groups 287
10.3.4	Stilbene Probes Binding to Proteins 288
10.3.5	Depth of Immersion of a Stilbene Probe in Biomembranes 289
10.3.6	Fluorescence–Photochrome Method 290
10.3.6.1	General 290
10.3.6.2	Molecular Dynamics of Proteins and Biomembranes 290
10.3.6.3	Molecular Dynamics of anti-DNP Antibody Binding Site 292
10.3.7	Systems Immobilized on Quartz Slides 293
10.3.7.1	Sensoring for Surface Microviscosity and Ascorbic Acid 293
10.3.7.2	A Fluorescent–Photochrome Method for the Quantitative Characterization of Solid-Phase Antibody Orientation 296
10.3.8	Triplet-Photochrome Method 297
10.3.9	Cascade Spin-Triplet-Photochrome Methods 299
10.3.10	Fluorescence–Photochrome Immunoassay 301
10.3.11	Suppermolecules Containing Stilbene and Fluorescent Quenching Groups 303
	References 305

11	**Modern Methods of Stilbene Investigations** 309
11.1	General 309
11.2	Nanosecond Transient Absorption Spectroscopy 310
11.3	Femtosecond Broadband Pump-Probe Spectroscopy 312
11.4	Fluorescence Picosecond Time-Resolved Single Photon Counting 314
11.5	The Fluorescence Upconversion Spectroscopy 318
11.6	Femtosecond Time-Resolved Fluorescence Depletion Spectroscopy 321

11.7	High-Speed Asynchronous Optical Sampling *323*	
11.8	Multiphoton Excitation *323*	
11.9	Time-Resolved Vibrational Spectroscopy *328*	
	References *332*	

12 Conclusions *335*

Index *337*

Preface

Stilbenes and their derivatives, stilbenoids, form a multidisciplinary field that combines many important branches of chemistry, biology, and physics. This field is under active investigation now. The number of publications on stilbenes available on Internet has reached about 32 000. Classical and modern synthetic organic chemistry (more than 6300 articles), providing a whole arsenal of stilbenes of different structures, is paving the way for numerous fundamental research and practical applications. Unique features of this class of compounds, including a combination of fluorescence, phosphorescence, photochrome, photochemical, and photophysical properties, have long been a matter of great interest to researchers. Specifically, recent years have seen the accumulation of new facts and ideas in the following areas: (i) stilbenes have proved to be convenient models ("proving ground") for experimental and theoretical investigations of detailed mechanisms of photochemical and photophysical processes including photochromism and multiphonone phenomena; (ii) stilbenoids are naturally present in plants (e.g., resveratrol and pterostilbene) and make an important contribution to biochemical and physiological processes in plants and have been used as drugs and antitumor agents; (iii) stilbene derivatives have been used as molecular probes and labels for investigation of dynamic properties of proteins and biomembranes; (iv) stilbenes are used as dyes, brighteners, whiteners of paper and textile, and photobleachers; (v) stilbene and its combination with polymers and inorganic materials have been made a basis for numerous optical and measuring instruments, devices, and apparatuses for dye lasers, organic solid lasers, and scintillators, phosphorus, neutron, and radiation detectors, electrophotographic photoconductor and image-forming apparatus, and photochromic, light-transmitting, dichroic, electroluminescent nonlinear optic, and organic–inorganic hybrid multichromophoric optical limiting materials.

This book will concisely cover practically all aspects of stilbenes: the chemical synthesis, the modern methods of investigations, their chemical, photochemical, and photophysical properties, their biological role, their use as therapeutic agents, and the instrumentation and materials.

This book is not intended to provide an exhaustive survey of each topic but rather discusses their theoretical and experimental backgrounds, and recent developments.

The literature on stilbenes is so vast, and many scientists have made important contribution in the area, that it is impossible to give a representative set of references in the space allowed for this book. The author apologizes to those he has not been able to include. More than 1100 references are given that should provide a key to essential relevant literature.

Fundamentals, classical and new methods of preparation, and basic chemical properties of stilbenes form the main topics of Chapters 1 and 2. In Chapters 3 to 5, the general theoretical and experimental backgrounds and recent results are explained for photophysical and photochemical properties of stilbenes stressing a detailed mechanism of photoisomerization. Advances in traditional and new areas of construction and investigation of photophysical and photochemical materials on the basis of stilbenes form the subject of Chapter 6. Chapters 7 to 9 consider biochemical, biomedical, therapeutic, and clinical applications of stilbenes, that is, in areas that appear to be of great importance for human well-being. Chapter 10 discusses fundamentals and recent results of using stilbenes as probes for investigating molecular structure, dynamics, and functional activity of proteins, enzymes, and biomembranes and real-time analysis of biologically active compounds. Chapter 11 describes the recent advances in modern absorption, fluorescence, and vibration techniques and related areas that to a considerable extent were stimulated by the growing requirements of stilbene applications.

This monograph is intended for scientists working on chemistry, physics, and biology of stilbenes and related areas such as optical materials and devices production, molecular biophysics, plant biochemistry, biomedicine, and pharmacology. The book can also be used as a subsidiary manual for instructors and graduate and undergraduate students of university physics, biochemistry, and chemistry departments.

Gertz Likhtenshtein

1
Stilbenes Preparation and Analysis

1.1
General

The name for stilbene (1,2-diphenylethylene) was derived from the Greek word *stilbos*, which means shining. There are two isomeric forms of 1,2-diphenylethylene: (*E*)-stilbene (*trans*-stilbene), which is not sterically hindered, and (*Z*)-stilbene (*cis*-stilbene), which is sterically hindered and therefore less stable.

trans-stilbene   *cis*-stilbene

(*E*)-Stilbene has a melting point of about 125 °C, while the melting point of (*Z*)-stilbene is 6 °C. Stilbene is a relatively unreactive colorless compound practically insoluble in water [1]. *trans*-Stilbene isomerizes to *cis*-stilbene under the influence of light. The reverse path can be induced by heat or light. The stilbene feature is associated with intense absorption and fluorescence properties, which correspond to the excitation of π-electrons of the conjugated ethenediyl group into π^* orbitals, as well as some other dynamic processes. The excited singlet state behavior of *trans*-stilbene is governed by fluorescence from the S_1 state that effectively competes with isomerization. This phenomenon of photochromism, namely, *trans–cis* photoisomerization of stilbene derivatives, can be readily monitored by a single steady-state fluorescence technique. A necessary stage in the olefinic photoisomerization process, in the singlet or triplet excited state, involves twisting (about the former double bond) of stilbene fragments relative to one another. The chemistry and photochemistry of stilbenes have been extensively investigated for decades and have been reviewed [2–25].

Stilbene derivatives are synthesized relatively easily, are usually thermally and chemically stable, and possess absorption and fluorescence properties that are

Stilbenes. Applications in Chemistry, Life Sciences and Materials Science. Gertz Likhtenshtein
Copyright © 2010 WILEY-VCH Verlag GmbH & Co. KGaA, Weinheim
ISBN: 978-3-527-32388-3

convenient for monitoring by relevant optical techniques. Stilbenes are widely used in the manufacture of industrial dyes, dye lasers, optical brighteners, phosphor, scintillator, and other materials. They are playing an increasingly prominent role in the area of photophysical, photochemical, biophysical, and biomedical investigations.

Hydroxylated derivatives of stilbene (stilbenoids) are secondary products of heartwood formation in trees that can act as phytoalexins (antibiotics produced by plants).

Because of the chemical stability of phenyl moiety of 1,2-diphenylethylene, stilbene is not a suitable starting compound for synthesis of stilbene derivatives. In order to form more complex molecules, it is necessary to introduce more reactive functional groups.

1.2
Classical Methods and Their Development

Many synthetic routes to stilbene derivatives have been reported, and only most important methods, which were used in the total synthesis, will be considered in this section. Figure 1.1 summarizes the five most important methods for forming the C=C bond of the 1,2-ethenediyl unit in stilbenes.

The following classical methods will be described in this section:

1. Aldol-type condensation
2. Siegrist method
3. Wittig–Horner reaction
4. Heck reaction
5. Negishi–Stille reactions
6. Barton–Kellogg–Staudinger reaction
7. McMurry reaction
8. Perkin reaction

1.2.1
Aldol-Type Condensation

Aldol-type condensation of an aromatic aldehyde with activated methylarene or phenylacetic acid is a useful reaction for preparing stilbene derivatives. Starting from *para*-substituted toluenes or *para*-substituted aromatic aldehydes, one can obtain 4,4′-disubstituted stilbenes. This reaction is relatively simple but has low yield. As an example, condensation of 2,4-dinitrotoluene and 4-nitrophenylacetic acid with aromatic aldehyde was studied [26]. The reaction involves carbanion addition to the carbonyl group. The carbanion is formed by the extraction of proton from the active methylene group of 2,4-dinitrotoluene by the base (usually, piperidine). The carbanion then adds to carbon atoms of the carbonyl group of the aldehyde. The reaction will therefore be facilitated by the ease of both the formation of the

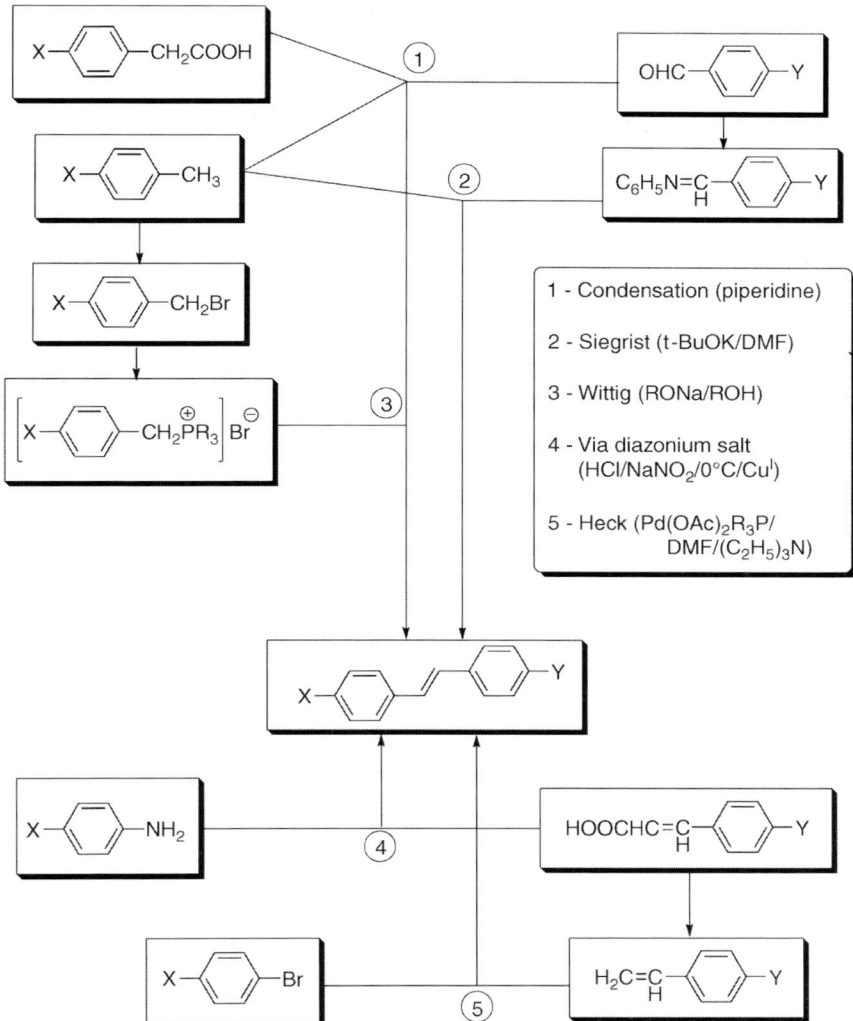

Figure 1.1 The most important synthetic routes to stilbene compounds. (Reproduced with permission from Ref. [24].)

relatively stable carbanion and the formation of the carbonium ion, which is obtained by the migration of p-electrons from carbonyl to the oxygen atom.

1.2.2
Siegrist Method

The total yield of the Siergist method (Figure 1.1) [27] is often inferior to those of the other four methods – its main advantage is its remarkably high selectivity.

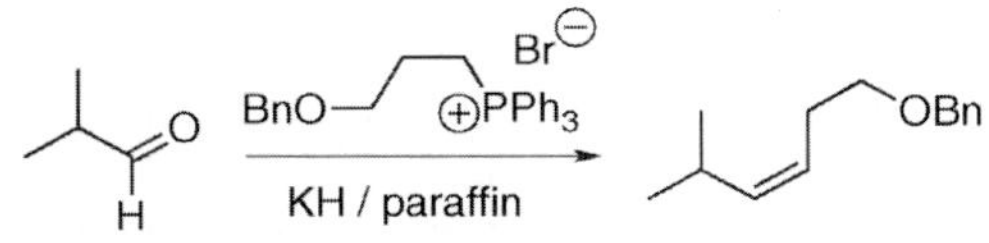

Figure 1.2 Wittig reaction using potassium hydride in paraffin. (Reproduced with permission from Ref. [30].)

For example, the synthesis of 4-methoxy-4-methyl stilbene showed that the selectivity of *cis*-configuration is 100 times more [28]. The decisive factor here in this reaction is the anti-elimination (E2) from the least energetic conformation.

1.2.3
Wittig Reaction

The Wittig reaction is the reaction of an aldehyde or ketone with a triphenyl phosphonium ylide to give an alkene and triphenylphosphine oxide. The Wittig reaction was discovered in 1954 by Georg Wittig and described in his pioneering publication titled "Über Triphenyl-phosphin-methylene als olefinbildende Reagenzien I" [29]. A recent example of the Wittig reaction is shown in Figure 1.2.

The Wittig reaction has proved to be quite versatile in the preparation of different substituted stilbenes [31–36]. This reaction is not sensitive to atmospheric oxygen, thus allowing simpler experimental procedures. It furnishes the *trans*-isomer in the steriospecific reaction. Moreover, the *trans*-isomer can be separated from the *cis*-isomer in the course of reaction because it is less soluble in the reaction solvent (usually, methanol, if sodium/lithium methoxide is used as a base) and precipitates on standing.

The Horner–Wadsworth–Emmons reaction (or HWE reaction) is the reaction of stabilized phosphonate carbanions with aldehydes (or ketones) to produce predominantly *E*-alkenes. In 1958, Horner published a modified Wittig reaction using phosphonate-stabilized carbanions [32]. Wadsworth and Emmons further defined the reaction [33]. Compared to phosphonium ylides used in the Wittig reaction, phosphonate-stabilized carbanions are more nucleophilic and more basic. Likewise, phosphonate-stabilized carbanions can be alkylated, unlike phosphonium ylides. The dialkylphosphate salt by-product is easily removed by aqueous extraction. A reliable and versatile synthesis of a stilbene derivative, 2,2-aryl-substituted cinnamic acid esters, using the Wittig reaction was reported [34–36] (Figure 1.3).

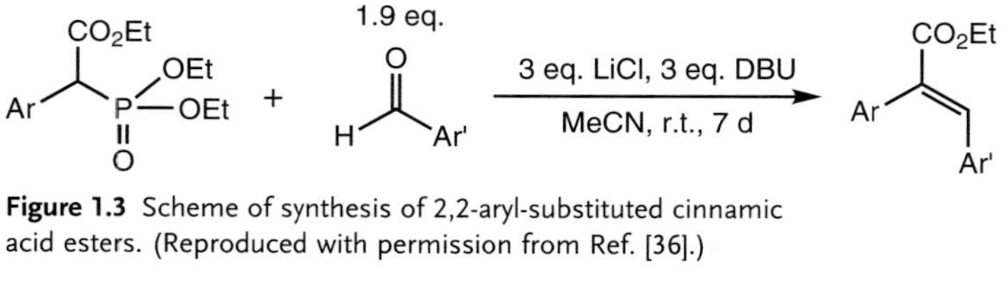

Figure 1.3 Scheme of synthesis of 2,2-aryl-substituted cinnamic acid esters. (Reproduced with permission from Ref. [36].)

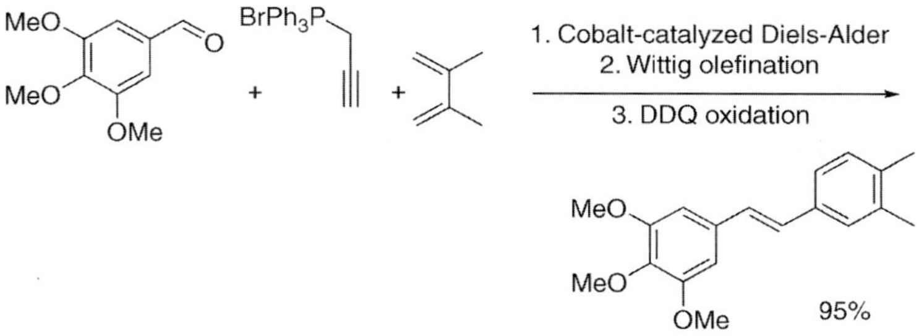

Figure 1.4 Scheme of cobalt-catalyzed Diels–Alder/Wittig olefination reaction. (Reproduced with permission from Ref. [37].)

A concise synthesis of substituted stilbenes from propargylic phosphonium salts by a cobalt-catalyzed Diels–Alder/Wittig olefination reaction has been described (Figure 1.4) [37]. It was shown that the cobalt(I)-catalyzed Diels–Alder reaction of propargylic phosphonium salts and alkyne-functionalized phosphonium salts with 1,3-dienes led to dihydroaromatic phosphonium salt intermediates that were directly used in a one-pot Wittig-type olefination reaction with aldehydes. Subsequent oxidation led to styrene- and stilbene-type products with the formation of three new carbon–carbon bonds. The reaction gives predominantly the E-configured products.

A convenient procedure to effect the Wittig and Horner–Wadsworth–Emmons reactions employing guanidine TBD and MTBD as base promoters was developed. Mild reaction conditions highly efficiently facilitated isolation of the final products (Figure 1.5) [38]. Further developments of the Wittig reaction have been reported [39, 40].

1.2.4
Heck Reaction

The Heck reaction (Mizoroki–Heck reaction) is the reaction of an unsaturated halide (or triflate) with an alkene and a strong base and palladium catalyst to form a substituted alkene [41, 42]. The reaction is performed in the presence of an organopalladium catalyst. The halide or triflate is an aryl, benzyl, or vinyl compound, and the alkene contains at least one proton and is often electron deficient, such as acrylate ester or an acrylonitrile. The catalyst can be tetrakis(triphenylphosphine)palladium

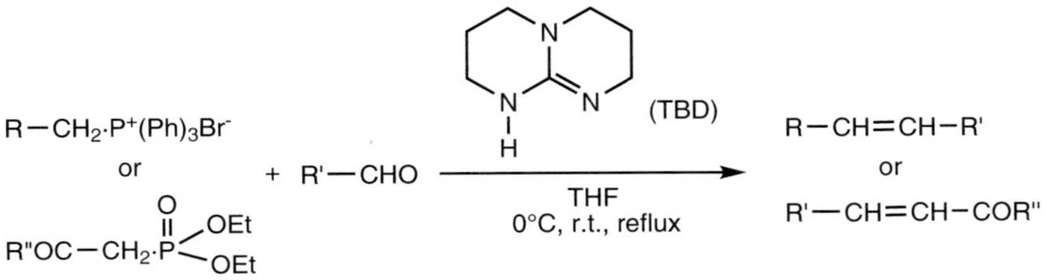

Figure 1.5 Horner–Wadsworth–Emmons reactions base-promoted with guanidine. (Reproduced with permission from Ref. [38].)

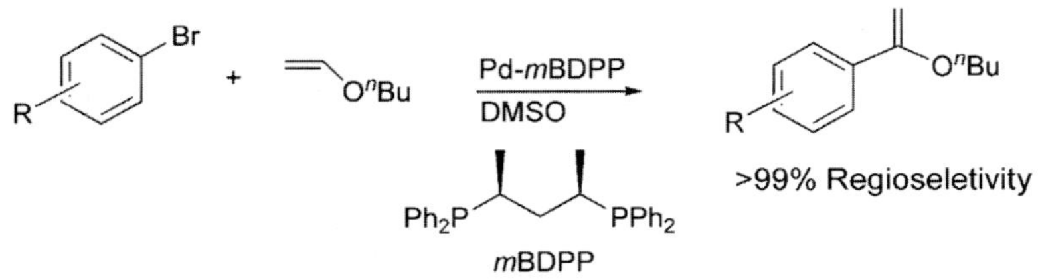

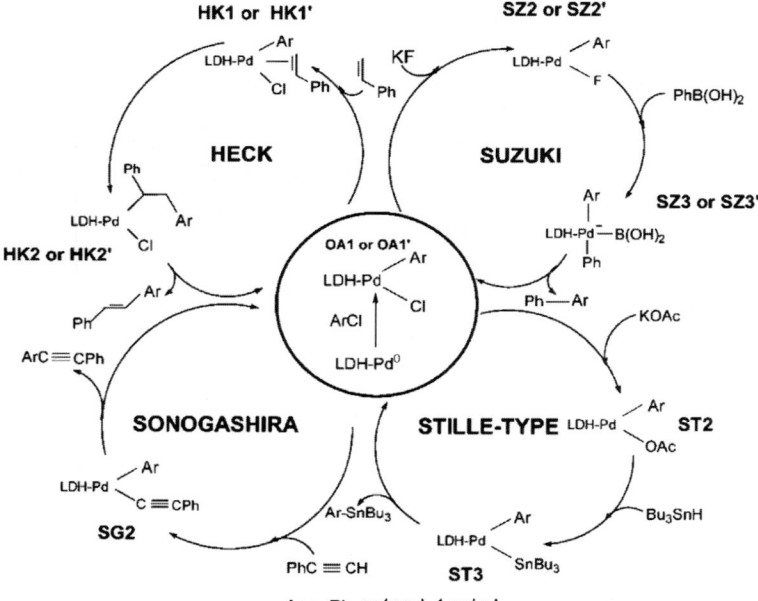

Figure 1.6 Pd-mBDPP-catalyzed regioselective internal arylation of electron-rich olefins by aryl halides. (Reproduced with permission from Ref. [43].)

(0), palladium chloride, or palladium(II) acetate. The ligand can be triphenylphosphine. The base is triethylamine, potassium carbonate, or sodium acetate.

An example of the Heck reaction is shown in Figure 1.6. Several reviews on this topic have been published [44–46].

A proposed mechanism of Heck reaction and other reactions using palladium compounds as catalysts and running via surface transient organometallic (STO) intermediates is presented in Figure 1.7. The formation of only one STO intermediate in all the Heck-, Suzuki-, Sonogashira-, and Stille-type coupling reactions during their reaction sequences was stressed.

The palladium(0) compound required in this cycle is generally prepared *in situ* from a palladium(II) precursor [48]. Further modifications of the Heck reaction are described in Ref. [49]. Among these modifications, the following can be considered. In the ionic

Figure 1.7 Proposed validated mechanistic cycle for the coupling reactions. (Reproduced with permission from Ref. [47].)

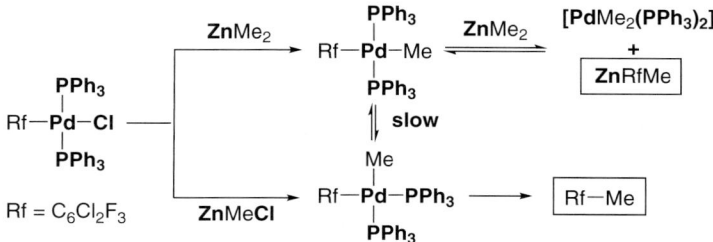

Figure 1.8 The Negishi coupling reaction and concentration/time data for the reaction 1 + ZnMe$_2$ obtained by ^{19}F NMR, in THF at 298 K. Starting conditions: [1] = 1.65 × 10^{-2} M. (Reproduced with permission from Ref. [52].)

liquid Heck reaction, palladium acetate and the ionic liquid (bmim)PF$_6$ were immobilized inside the cavities of reversed-phase silica gel. In this way, the reaction proceeded in water and the catalyst was reusable [50]. In the Heck oxyarylation modification, the palladium substituent in the *syn*-addition intermediate was displaced by a hydroxyl group, and the reaction product contained a tetrahydrofuran ring [49]. In the amino-Heck reaction, a nitrogen–carbon bond was formed. The catalyst used was tetrakis(triphenylphosphine)palladium(0) and the base was triethylamine [51].

1.2.5
Negishi–Stille Coupling

The Negishi coupling is a cross-coupling reaction between organozinc and alkenyl or aryl halide or triflate promoted by Pd catalyst (Figure 1.8).

The Stille coupling is the palladium-catalyzed cross-coupling between organotin and alkenyl or aryl halide or triflate [53–56].

In the Stille reaction mechanism (Figure 1.9), the first step in this catalytic cycle is the reduction of the palladium catalyst to the active Pd(0) species. The oxidative

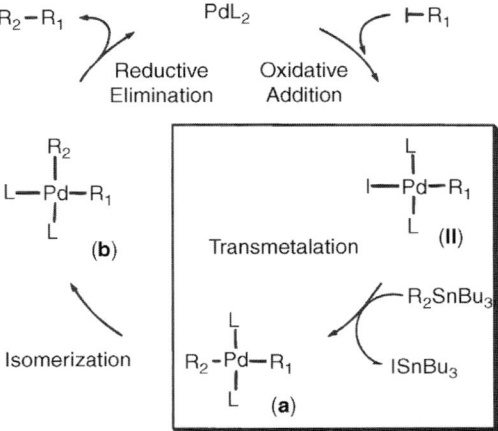

Figure 1.9 The Stille coupling mechanism [54].

addition of the organohalide gives a *cis* intermediate that rapidly isomerizes to the *trans* intermediate. Transmetalation with organostannane forms intermediate, which produces the desired product and the active Pd(0) species after reductive elimination. The oxidative addition and reductive elimination retain the stereochemical configuration of the respective reactants.

An interesting development in Stille coupling by using only catalytic amounts of tin was reported [56].

1.2.6
Barton–Kellogg–Staudinger Reaction

The Barton–Kellogg reaction (diazo-thioketone coupling) is a reaction between a ketone and a thioketone through a diazo intermediate forming an alkene [57–60]. In Ref. [59], the authors presented a new methodology to prepare sterically overcrowded alkenes by using the Barton–Kellog method (Figure 1.10).

As versatile synthetic intermediates with a tetramethylindanylindane (stiff-stilbene) core, the *cis* and *trans* isomers of 5,16-dibromo-2,2,13′,13′-tetramethylindanylindanes were synthesized by the Barton–Kellogg coupling [60].

1.2.7
McMurry Reaction

The McMurry reductive coupling reaction is an organic reaction in which two ketone or aldehyde groups are coupled to an alkene in the presence of titanium(III) chloride and a reducing agent [61, 62]. As an example, intramolecular reductive McMurry coupling reactions of bis(formylphenoxy)-substituted calix[4]arenediols mediated by titanium(IV) chloride and activated zinc followed by cyclocondensation of the diols with tetra- and penta(ethylene glycol) bistosylates provided stilbene- and crown ether-bridged calix[4]arenes. For the synthesis of stilbenes, some authors use the

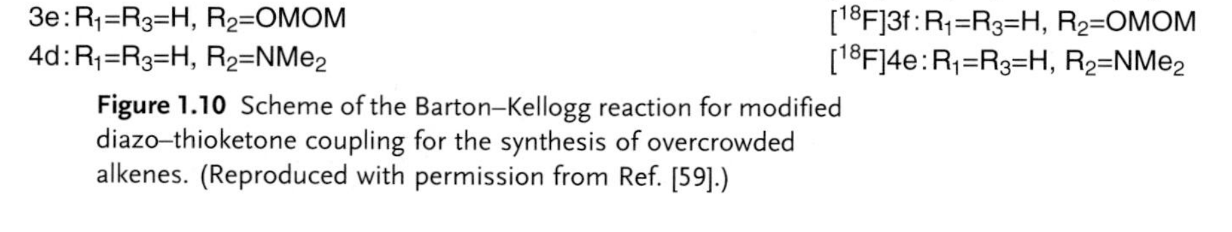

2e: $R_1=R_3$=OMOM, R_2=H
3e: $R_1=R_3$=H, R_2=OMOM
4d: $R_1=R_3$=H, R_2=NMe$_2$

[^{18}F]2f: $R_1=R_3$=OMOM, R_2=H
[^{18}F]3f: $R_1=R_3$=H, R_2=OMOM
[^{18}F]4e: $R_1=R_3$=H, R_2=NMe$_2$

Figure 1.10 Scheme of the Barton–Kellogg reaction for modified diazo–thioketone coupling for the synthesis of overcrowded alkenes. (Reproduced with permission from Ref. [59].)

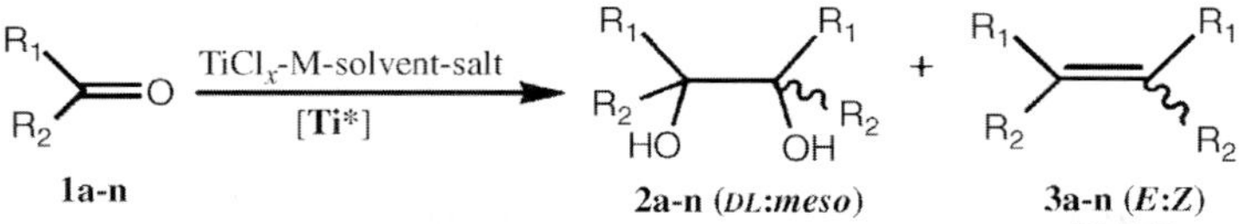

1.2 Classical Methods and Their Development

Figure 1.11 McMurry olefination. (Reprinted from [64].)

Reaction scheme:

$R_1R_2C{=}O$ (1a–n) $\xrightarrow{\text{TiCl}_x\text{-M-solvent-salt, [Ti*]}}$ $R_1R_2C(OH){-}C(OH)R_1R_2$ (2a–n, *DL:meso*) + $R_1R_2C{=}CR_1R_2$ (3a–n, *E:Z*) 30–85%

[Ti*] = "Activated" titanium

TiCl$_x$ x = 3, 4
M (reducing metal) = Li, Mg
Solvent = THF, DME
Salt = LiCl, LiI, KCl, CsCl, MgCl$_2$, ZnCl$_2$

Aromatic carbonyls
- a: R_1 = Ph, R_2 = Me
- b: R_1 = 2-naphthyl, R_2 = Me
- c: R_1 = 4-MeC$_6$H$_4$, R_2 = Me
- d: R_1 = 4-*iso*-PrC$_6$H$_4$, R_2 = Me
- e: R_1 = 4-*t*-BuC$_6$H$_4$, R_2 = Me
- f: R_1 = R_2 = Ph
- g: R_1 = Ph, R_2 = H
- h: R_1 = 4-ClC$_6$H$_4$, R_2 = H
- i: R_1 = 4-MeOC$_6$H$_3$, R_2 = H
- j: R_1 = (3,4-CH$_2$O$_2$)C$_6$H$_3$, R_2 = H

Aliphatic carbonyls
- k: R_1 = R_2 = *cyc*-Hex
- l: R_1 = R_2 = *cyc*-Pent
- m: R_1 = PhCH$_2$, R_2 = H
- n: R_1 = *n*-C$_6$H$_{13}$, R_2 = H

Suzuki–Miyaura coupling that is the reaction of an aryl- or vinylboronic acid with an aryl or vinyl halide catalyzed by a palladium(0) complex (Figure 1.1) [63].

A combination of alkali metal salts, particularly potassium chloride, with low-valent titanium reagents generated from titanium chlorides with lithium or magnesium in either THF or DME are effective reagents for stereoselective McMurry coupling reactions of aldehydes and ketones to substituted alkenes (Figure 1.11).

1.2.8
Perkin Reaction

The Perkin reaction is an organic reaction developed by William Henry Perkin that can be used to make cinnamic acids by the aldol condensation of aromatic aldehydes and acid anhydrides in the presence of an alkali salt of the acid [65, 66].

A mild and convenient one-pot two-step synthesis of hydroxystilbenes (*E*)-4-chloro-4′-hydroxy-3′-methoxystilbene from 4-hydroxy-3-methoxybenzaldehyde and 4-chlorophenylacetic acid with *trans* selectivity developed through a modified Perkin reaction between benzaldehydes and phenylacetic acids was recently reported (Figure 1.12) [67].

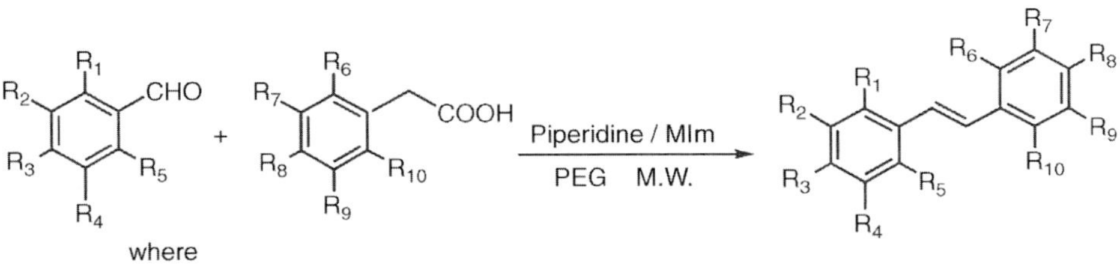

Reaction scheme: substituted benzaldehyde + substituted phenylacetic acid $\xrightarrow{\text{Piperidine / MIm, PEG M.W.}}$ substituted stilbene

where at least one of R_1 or R_3 or R_5 or R_6 or R_8 or R_{10} = OH
and R_1–R_{10} = H, OMe, OH, Cl, etc.

Figure 1.12 Scheme of the modified Perkin reaction. (Reprinted from [67].)

The observation of a simultaneous condensation–decarboxylation leading to the unusual formation of hydroxystilbenes in lieu of α-phenylcinnamic acid reveals an interesting facet of the classical Perkin reaction.

1.3
Miscellaneous Chemical Methods of Stilbene Synthesis

1.3.1
Palladium-Catalyzed Reactions

Pd-catalyzed cross-coupling reactions were studied in one-pot multicatalytic processes to synthesize disubstituted alkenes and alkanes from carbonyl derivatives [68]. The use of Cu-catalyzed methylenation reactions was the key starting reaction to produce terminal alkenes that are not isolated but submitted to further structure elongation (hydroboration followed by Suzuki cross-coupling) (Figure 1.13). These processes have been used to synthesize methoxylated (E)-stilbenoids (i.e., (E)-1,3-dimethoxy-5-(4-methoxystyryl)benzene).

Synthesis of symmetrical trans-stilbenes by a double Heck reaction of (arylazo) amines with vinyltriethoxysilane has been reported [69]. A detailed procedure of the method for the synthesis of trans-4,4'-dibromostilbenes was described. Bis-stilbene (I) was obtained using the Heck reaction with catalyst generated in situ from equimolar amount of Pd acetate and corresponding phosphine [70]. Due to its relatively low melting point, this compound formed a smectic mesophase and was readily soluble in dioxane, THF, CH_2Cl_2, and $CHCl_3$. Palladium-catalyzed stereoselective synthesis of (E)-stilbenes via organozinc reagents and carbonyl compounds has been reported [71]. In the presence of a catalytic amount of $PdCl_2(PPh_3)$ and a silylating agent, organozinc halides reacted with carbonyl compounds to give the corresponding (E)-stilbenes, in good to excellent yields under mild conditions. Two types of domino reactions from the same internal alkynes and hindered Grignard reagents based on carbopalladation, Pd-catalyzed cross-coupling reaction, and a C—H activation strategy were described [72]. The reaction was used in Pd $(OAc)_2$-catalyzed domino carbopalladation cross-coupling of PhC≡CPh and mesitylmagnesium bromide in the presence of $BrCH_2CH_2Br$ and PPh_3 (4 equiv) in refluxing THF to give 71% cis-stilbene II(Figure 1.14).

Synthesis of polyhydroxylated ester analogues of the stilbene resveratrol was accomplished using decarbonylative Heck couplings [73]. Levulinate- and

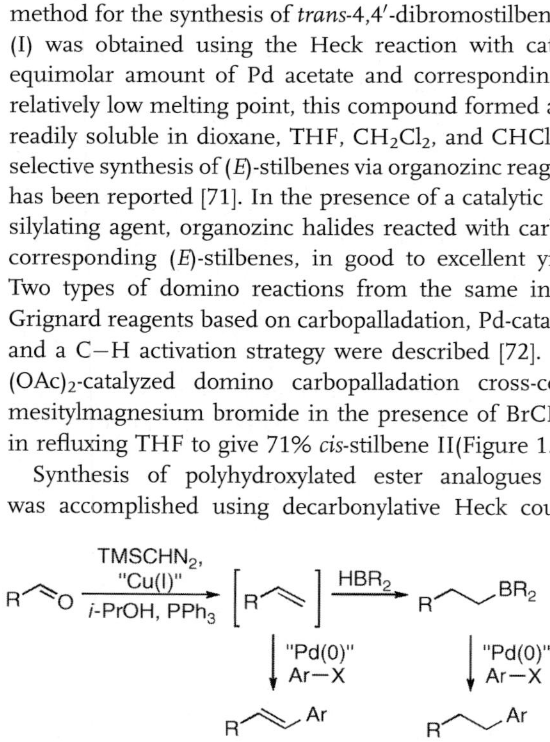

Figure 1.13 One-pot multicatalytic processes. (Reproduced with permission from Ref. [68].)

chloroacetate-protected 3,5-dihydroxybenzoyl chlorides were coupled with styrenes, $H_2C:CHC_6H_4X$-4 (X = OH, OAc, $OCOCH_2Cl$, $OCOCH_2CH_2COMe$, F), to give hydroxylated stilbenes, analogues of resveratrol I (X = Y = Z = OH).

Levulinate and chloroacetate protecting groups allowed selective production of mono- and diacetate variations under palladium-N-heterocyclic carbene (NHC) catalyzed decarbonylative coupling conditions. Fluorinated analogues, such as I (X = F, Y = Z = OH; X = Y = OH, Z = F; X = Y = F, Z = OH; X = Y = Z = F), were also produced using Heck conditions with bromofluorobenzenes. Luminescent stilbenoid chromophores with diethoxysilane end groups were prepared via Heck reactions [74]. Diethoxysilane-substituted styrenes were used as vinylic components, thus allowing the combined connection of the chromophore to the silane moiety with an extension of the π-system. Monodisperse oligo(phenylenevinylene)s of different conjugation lengths and bromine or iodine as reactive sites were used as coupling partners.

It was shown [75] that CuI-mediated substitution of 2-bromopyridine with sodium 4-bromophenylsulfinate in DMF followed by Pd-catalyzed coupling with (E)-2-(4-fluorophenyl)vinylboronic acid gave pyridinylsulfonyl stilbene. In the palladium-catalyzed Heck vinylation performed in nonaqueous ionic liquids, catalytic amounts of ligand-free $PdCl_2$ yielded stilbene from chlorobenzene and styrene in high yield [76]. The reaction occurred without the need for further promoting salt additives such as tetraphenylphosphonium chloride. The heterodinuclear Ru–Pd complex photocatalyst that catalyzes selective reduction of tolane to produce cis-stilbene without added H_2 was designed [77]. The photocatalyst contained a photoactive Ru(II) fragment as a light absorber, a $PdCl_2$ unit coordinated to the other end as a catalytic center, and a bridging unit connecting the two metal centers. It was suggested that intramolecular photoinduced electron transfer in the heterodinuclear complex facilitates the photocatalytic reactions.

Novel photoswitchable chiral compounds having an axis chiral 2,2′-dihydroxy-1,1′-binaphthyl (BINOL)-appended stiff-stilbene, trans-(R,R)-**1** and trans-(S,S)-**1**, were synthesized by palladium-catalyzed Suzuki–Miyaura coupling forming aryl–aryl bond) and low-valence titanium-catalyzed McMurry coupling as key steps [78]. The Suzuki–Miyaura reaction of aryl halides with trans-(2-phenylvinyl)boronic acid using a series of related in situ generated N-heterocyclic carbene palladium(II) complexes was studied [79]. The nature of the substituents of the carbene ligand was found to be critical. Specifically, the presence of alkyl groups on the ortho positions of the Ph substituents was a requisite for obtaining the most efficient catalyst systems. The synthesis by Heck and/or Wittig reactions and characterization of a new class of molecules based on 6b,10b-dihydrobenzo[j]cyclobut[a]acenaphthylene (DBCA) with potentially interesting optical and electronic properties were described [80]. The new compounds contain one or two DBCA units linked via a double bond to an aromatic system.

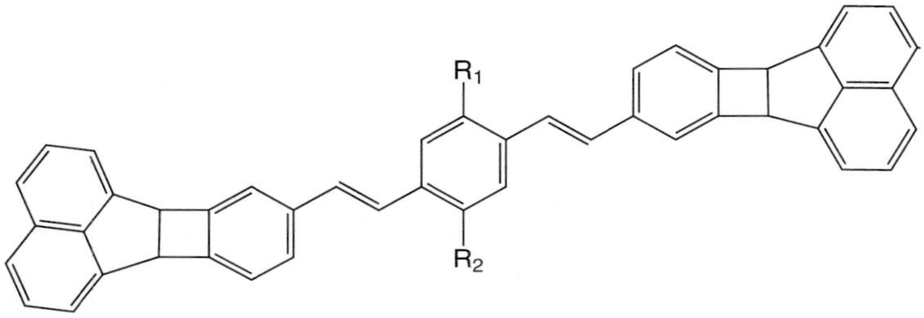

$R_1 = CH_3O$, $R_2 = CH_3(CH_2)_3CH(C_2H_5)CH_2O$

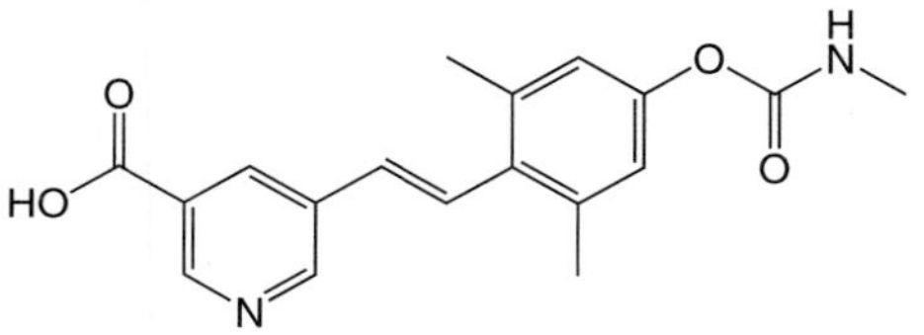

The synthesis of several aza-stilbene derivatives similar to was carried out (Figure 1.15) [81]. The compounds were tested for their c-RAF enzyme inhibition.

Convenient methods for highly stereoselective synthesis of unsymmetrical stilbenoids were accomplished [82]. Cross-metathesis of 4-chlorostyrene with (vinyl)

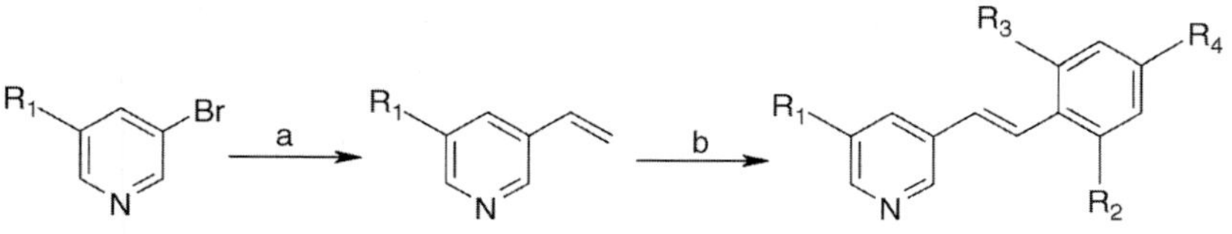

Figure 1.15 Synthesis of several aza-stilbene derivatives. Reagents and conditions: (a) tributylvinyl tin, LiCl, BHT, Pd(PPh$_3$)$_2$Cl$_2$, DMF, 70 °C; (b) aryl bromide(iodide), Pd$_2$dba$_3$, TEA, P(o-tol)$_3$, DMF, 95 °C [81]. (Reproduced with permission from Elsevier.)

silane derivatives in the presence of second generation of Grubbs catalyst [Cl$_2$(PCy$_3$)(IMesH$_2$)Ru(=CHPh)] or silylative coupling in the presence of [RuH(Cl)(CO)(PPh$_3$)$_3$] followed by palladium-catalyzed Hiyama coupling has been proved.

1.3.2
Horner–Wadsworth–Emmons and Wittig–Horner Olefination Reactions

The synthesis of ^{18}F-labeled stilbenes [^{18}F]2g, [^{18}F]3g, and [^{18}F]4e (E-isomers) by the Horner–Wadsworth–Emmons reaction was accomplished [83]. This carbonyl-olefination reaction was performed via a "multistep/one-pot" reaction by the coupling of benzylic phosphonic acid esters (3,5-bis-methoxymethoxybenzyl)-phosphonic acid diethyl ester, (4-methoxymethoxybenzyl)phosphonic acid diethyl ester, and (4-dimethylaminobenzyl)phosphonic acid diethyl ester with 4-[^{18}F]fluor-obenzaldehyde. The radiochemical yields ranged from 9 to 22%. Three new polyfluorinated compounds ([(E)-4-(4-bromostyryl)-2,3,5,6-tetrafluorobenzonitrile]$_x$·[(E)-4-(4-bromo-2,3,5,6-tetrafluorostyryl)benzonitrile]$_{1-x}$) were obtained by the Horner–Wadsworth–Emmons approach to study intermolecular interactions in the crystal state and the formation of cocrystals [84].

Three new coordination polymers, [Cd(SCN)$_2$L$_2$]$_n$, [CdHg(SCN)4L$_2$]$_n$, and [MnHg(SCN)$_4$L$_{2n}$, were prepared by the self-assembly of L with the corresponding metal salts and NaSCN (L is a new functional rigid imidazole ligand, trans-4-imidazolyl-4′-(N,N-diethylamino)stilbene) [85]. The crystal structures of the coordination polymers were detected by single-crystal X-ray diffraction.

A recent example of Horner–Wadsworth–Emmons reaction has been reported in Ref. [86]. A modified Wittig–Horner reaction and a rearrangement in the presence of t-BuOK in toluene under mild conditions have been developed for the synthesis of stilbenes bearing electron-withdrawing group(s) by using benzils and arylmethyldi-phenylphosphine oxides [87]. The authors suggested that this approach could be readily applied to a facile synthesis of biologically important natural products, resveratrol and its derivatives, such as (Z)- and (E)-trimethoxystilbenes. A reaction of α-selenylation and Wittig–Horner olefination of benzylphosphonates was developed [88]. The reaction between (EtO)$_2$P(O)CH$_2$Ph and PhSeCl gave (EtO)$_2$P(O)CH(SePh)Ph, which gave vinylselenides RCH:C(Ph)SePh (**6a–h**; R = Ph, 4-MeC$_6$H$_4$, PhCH:CH, 4-MeOC$_6$H$_4$, 4-ClC$_6$H$_4$, n-Pr, iPr, H) by reaction with aldehydes RCHO. Selenium–lithium exchange of **6a–d** through the reaction with n-BuLi followed by capture with several electrophiles (H$_2$O, PhCHO, iPrCHO, DMF) gave trans-stilbene, (Z)-allyl alcohols, and (E)-α-phenyl-α,β-unsaturated aldehydes. Seventeen derivatives of stilbenes, including resveratrol, were synthesized using a scheme.

1.3.3
Other Synthetic Reactions

Bichromophoric photochromes based on the photoinduced opening and thermal closing of a [1,3]oxazine ring were designed [89]. In particular, by incorporating fused 3H-indole and 4-nitrophenoxy fragments, the compound containing stilbenylvinyl

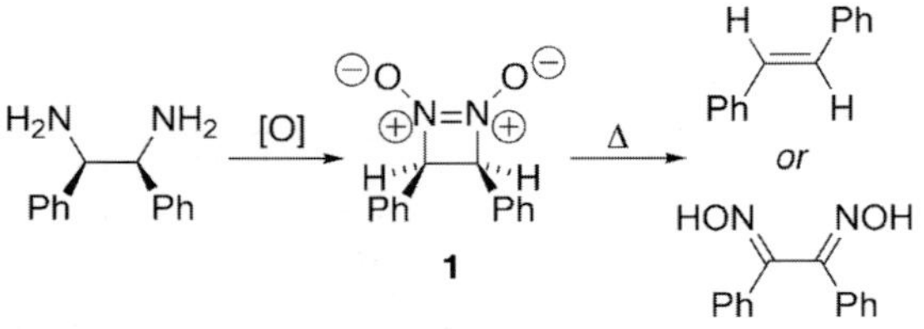

Figure 1.16 Possible products from the decomposition of diazetine dioxides **1**. (Reproduced with permission from Ref. [91].)

groups was prepared. The photoinduced process simultaneously generates a 4-nitrophenolate anion and a 3H-indolium cation. A series of liquid crystal stilbene derivatives containing 1,2-dienylalkoxy chains 1 ($n = 7$, 9, 11) have been synthesized [90]. The mesomorphic properties of stilbenes have been measured by polarizing optical microscopy, differential scanning calorimetry, and absorption spectroscopy. The effect of terminal alkoxy chain length and polymerizable function on the mesomorphic behavior was discussed. In the work [91], diazetine dioxide has been prepared in a single step via oxidation of *meso*-2,3-diphenyl-1,2-ethanediamine with dimethyldioxirane, albeit in low yield (7%). Thermal decomposition of 1,2-diazetine N,N'-dioxide afforded predominantly *trans*-stilbenes (Figure 1.16).

The reactions of [ReCl$_3$(MeCN)(PPh$_3$)$_2$] with benzil PhC(O)C(O)Ph and with a natural 1,2-naphthoquinone derivative, β-lapachone, resulted in oxidative addition with the formation of Re(V) complexes with stilbenediolate, [ReCl$_3$(PhC(O)=C(O)Ph)(PPh$_3$)] (**1**) [92]. General procedures for the preparation of thiol end-capped stilbenes and oligo(phenylenevinylene)s (OPVs) with *tert*-butyl- and acetyl-protected thiol termini have been developed (Figure 1.17) [93]. These reactions proceed via

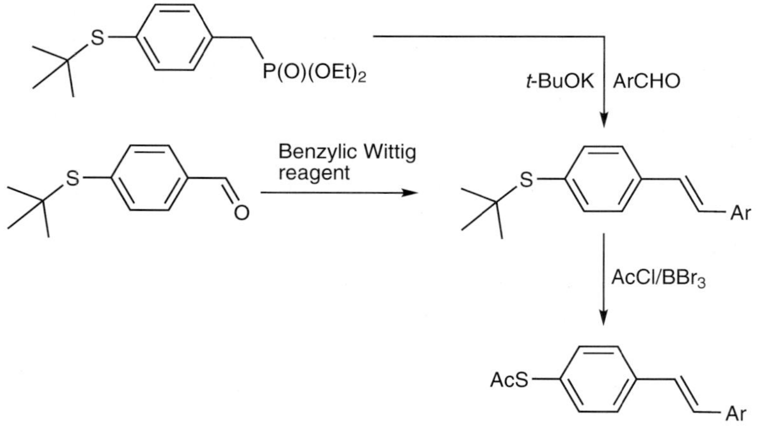

Figure 1.17 General procedures for the preparation of thiol end-capped stilbenes and oligo(phenylenevinylene)s. (Reproduced with permission from Ref. [93].)

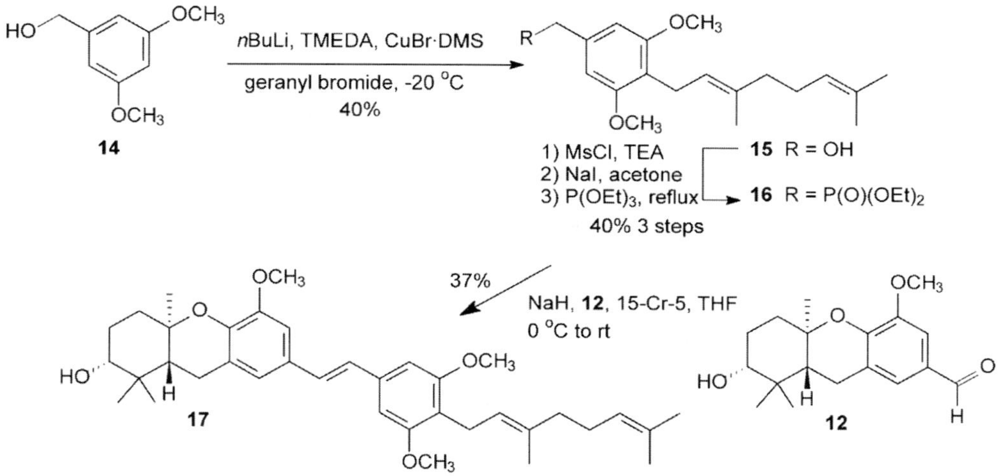

Figure 1.18 Synthesis of deoxyschweinfurthin. (Reproduced with permission from Ref. [96].)

Br/Li exchange, McMurry, and Wittig-type reactions. The reprotection of the thiol group is accomplished by means of acetyl chloride and boron tribromide.

New labeled stilbene derivatives, such as cis-3,5-dimethoxy-4'-[^{11}C]methoxystilbene, cis-3,4',5-trimethoxy-3'-[^{11}C]methoxystilbene, trans-3,5-dimethoxy-4'-[^{11}C]methoxystilbene, trans-3,4',5-trimethoxy-3'-[^{11}C]methoxystilbene, cis-3,5-dimethoxy-4'-[^{18}F]fluorostilbene, and trans-3,5-dimethoxy-4'-[^{18}F]fluorostilbene, were designed and synthesized [94]. The synthesis of (E)-tris-O-methylresveratrol and (E)-3,5-dimethoxystilbene via the Miyaura–Suzuki coupling was described [95]. This reaction has been carried out in air without solvent/substrate purification and in the absence of additional free ligand. Figure 1.18 shows a scheme of deoxyschweinfurthin synthesis accomplished in Ref. [96].

The synthesis, structural and spectroscopic characterization, and photophysical and photochemical properties of cyclic trans-stilbenes have been carried out [97].

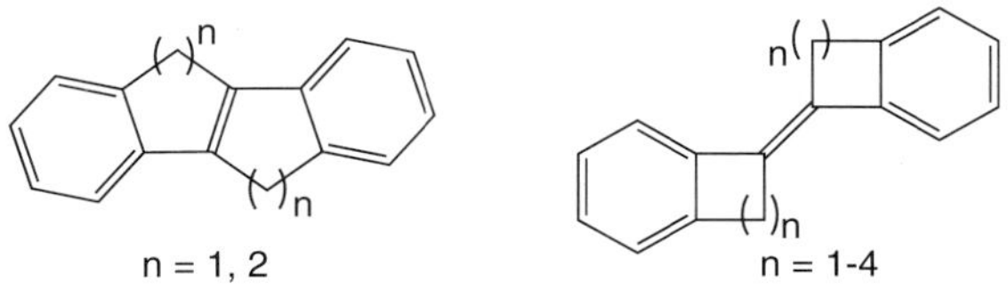

A method of stilbene synthesis via homocoupling of aryl aldehyde tosylhydrazones in the presence of lithium tert-butoxide and trimethyl borate under reflux in THF has been described.

Series of stilbenes were prepared in good yields via homocoupling of aryl aldehyde tosylhydrazones in the presence of lithium tert-butoxide and trimethyl borate under reflux in THF (Figure 1.19) [98].

Synthesis of trans- and cis-3,4',5-trihydroxystilbene from 3,5-dimethoxybenzaldehyde and 4-methoxyphenylacetonitrile via condensation reaction to form stilbene

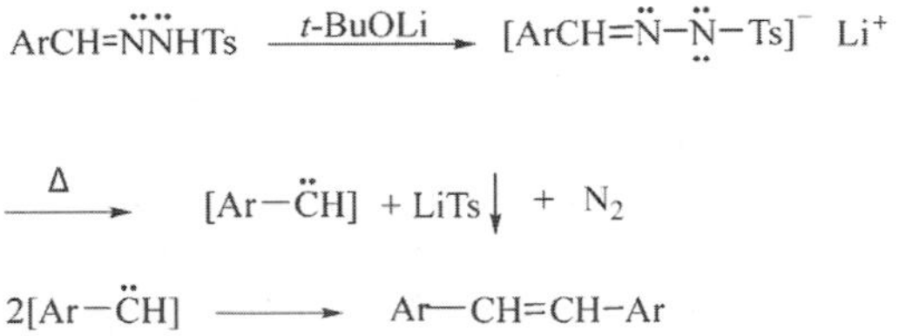

$$\text{ArCH=NNHTs} \xrightarrow{t\text{-BuOLi}} [\text{ArCH=}\ddot{\text{N}}\text{-}\ddot{\text{N}}\text{-Ts}]^- \text{Li}^+$$

$$\xrightarrow{\Delta} [\text{Ar}-\ddot{\text{C}}\text{H}] + \text{LiTs}\downarrow + \text{N}_2$$

$$2[\text{Ar}-\ddot{\text{C}}\text{H}] \longrightarrow \text{Ar}-\text{CH=CH}-\text{Ar}$$

Figure 1.19 Synthesis of homocoupling stilbenes [98]. (Reproduced with permission from Elsevier.)

skeleton, after hydrolysis, decarboxylation, and demethylation to obtain *trans*-3,4′,5-trihydroxystilbene was accomplished [99]. Highly functionalized (*E*)-stilbenes and 4-aryl-6-styrylpyran-2-ylideneacetonitriles were prepared and delineated through the ring transformation of 3,4-disubstituted 6-aryl-2*H*-pyran-2-ones with (*E*/*Z*)-4-phenyl-3-buten-2-one without using any catalyst [100]. Synthesis of 21 new (*E*)-4-[piperidino (4′-methylpiperidino-,morpholino-)*N*-alkoxy]stilbenes of chemical structure (where

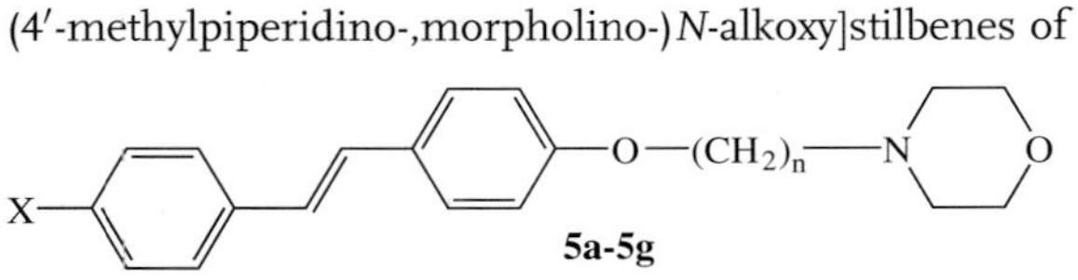

5a-5g

a X=H n=2;   e X=NO$_2$ n=3
b X=H n=3;   f X=NO$_2$ n=4
c X=H n=4;   g X=NO$_2$ n=5
d X=H n=5;

$n = 3, 4, R = X = H; n = 4, 5, R = H, X = NO_2$) and their antimicrobial activities was reported [101].

A series of stilbene derivatives of formula (R^1 = (substituted) NH$_2$, OH, alkoxy,

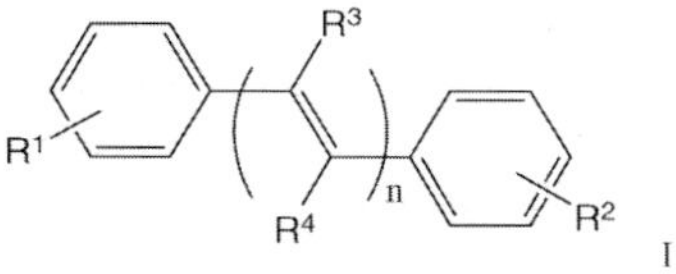

I

hydroxyalkyl; R^2 = (substituted) (OCH$_2$CH$_2$)$_q$-Z; $q = 1$–10; Z = halo, halobenzoyloxy, haloaryl, chelating group, and so on; R^3, R^4 = H, OH, NH$_2$, alkoxy, and so on; $n = 1$–6) were prepared [102].

A series of synthesis of stilbene derivatives was presented in Ref. [103]. In the work [104], another series of 4,4′-disubstituted organic–organometallic stilbenes were synthesized, that is, the 4′-substituted stilbenoid-NCN-pincer platinum(II) complexes [PtCl(NCN-R-4)] (NCN-R-4 = [C$_6$H$_2$(CH$_2$NMe$_2$)$_2$-2,6-R-4]– in which R = C$_2$H$_2$C$_6$H$_4$-R′-4′ with R′ = NMe$_2$, OMe, SiMe$_3$, H, I, CN, NO$_2$) (1–7). In these compounds, the PtCl grouping can be considered to be present as a donor substituent. Their synthesis involved a Horner–Wadsworth–Emmons reaction of [PtCl(NCN-CHO-4)] (9) with the appropriate phosphonate ester derivatives (8a–g). Under these reaction conditions, the C–Pt bond in aldehyde 9 was not affected, and the platinated stilbene products were obtained in 53–90% yield. The solid-state structures of complexes **1, 2** and **5–7** were detected by single-crystal X-ray diffraction, which revealed interesting bent conformations for **2, 5,** and **7**. Linear correlations were found between both the ^{13}C{^{1}H} (C ipso to Pt) and the ^{195}Pt{^{1}H} NMR chemical shifts and the Hammett σ_p value of the R' substituent; therefore, these NMR shifts can be used as a qualitative probe for the electronic properties of the delocalized π-system to which it is connected. Platinum–stilbene complexes were investigated for charge transfer (CT) properties in solvents of different polarities.

Three novel organic optical materials, 4′-(N,N-dihydroxyethylamino)-4-(pyridine-4-vinyl)stilbene, N-((4-N,N-dihydroxyethylamino)benzylidene)-4-(pyridine-4-vinyl) aniline, and 4′-(N,N-dihydroxyethylamino)-4-(pyridine-4-vinyl)azobenzene, were synthesized [105]. Tolunitriles reacted with donor-substituted aromatic aldehydes in high yielding reactions of the synthesis of donor–acceptor cyanostilbenes without the need of inert atmosphere. The keys to this reaction were the use of anhydride DMF solvent and the phase transfer agent – tris(3,6-dioxaheptyl)amine (TDA). High yields of stilbenes were also obtained with amino-substituted aromatic aldehydes [106].

According to Ref. [107], the addition of diphenylacetylene (110 °C, 3 h) to the hydrido acyl complex Ru(H){2-PPh$_2$C$_6$H$_4$C(O)}(CO)$_2$(PPh$_3$) afforded the novel complex Ru{2-PPh$_2$C$_6$H$_4$C(O)PhC:CHPh}(CO){PPh$_3$}, incorporating the newly assembled o-(diphenylphosphino)phenyl (E)-stilbenyl ketone ligand. The α,β-unsaturated ketone moiety of the latter was bound to the metal in an η^4 coordination mode involving both a side-on coordination of the carbonyl group and a classical η^2 linkage of the olefinic bond. A series of unsymmetrical *trans*-stilbenes have been prepared using the sequential coupling reactions of bromobenzenesulfonate with formylarylboronic acids, benzylphosphonates, and arylmagnesium bromides [108]. The nickel-catalyzed reactions of stilbene sulfonates with aryl Grignard reagents produced the corresponding stilbenes via the nucleophilic aromatic substitution of the neopentyloxysulfonyl group by aryl nucleophiles. The high chemoselectivity of the alkyloxysulfonyl group allowed the stepwise construction of unsymmetrical *trans*-stilbenes possessing terphenyl moieties. Total synthesis of stilbene artochamins F (I),

1a: R=Boc
1b: R=TBS

H (II), I (III), and J(IV) has been achieved through a flexible and expedient strategy that features a cascade sequence involving two concurrent [3,3] sigmatopic rearrangements and an unusual intramolecular formal [2 + 2] thermal cycloaddition reaction between an electron-rich stilbene and a prenyl group [109].

The photochemical coupling of various stilbenes (**S**) and chloranil (**Q**) was effected by the specific charge transfer activation of the precursor electron donor–acceptor (EDA) complex [**S**, **Q**] [110]. The [2 + 2] cycloaddition was established by X-ray structure elucidation of the crystalline *trans*-oxetanes formed selectively in high yields.

It was shown [111] that intramolecular reductive McMurry coupling reactions of bis(formylphenoxy)-substituted calix[4]arenediols mediated by titanium(IV) chloride and activated zinc followed by cyclocondensation of the diols with tetra- and penta(ethylene glycol) bistosylates provided the stilbene- and crown ether-bridged calix[4]arenes. Two synthesized analogues with two ethoxyethoxyethoxy substituents replacing the crown ether moiety and two calix[4]arene crown ethers with methoxy groups replacing the bridging stilbene moiety were tested for their extraction of alkali metal cations from aqueous solutions into chloroform. Stilbene-bridged calix[4] arenes with bridging tetra(ethylene glycol) ethers selectively extracted potassium ion over other alkali metal cations, while stilbene-bridged calix[4]arenes with bridging penta(ethylene glycol) ethers favor the complexation of cesium cation over other alkali metal cations.

To investigate the gelation ability of novel oxamide-based derivatives bearing a stilbene as a photoresponsive unit, oxamide-based derivatives, containing one or two oxamide moieties coupled to the 4- or 4,4′-positions of *cis*- and *trans*-stilbenes, have been synthesized [112]. *trans*-4-Me(CH$_2$)$_{11}$C$_6$H$_4$CH:CHC$_6$H$_4$NHCOCOR (I, R = OEt, NH-L-Leu-OMe) was found to act as efficient gelators of various organic solvents.

Several synthesis on the basis of stilbenes as starting materials were reported. The paper [113] presented experimental data regarding some azo dyes synthesized by coupling of the diazonium salt of 4,4′-diamino-stilbene-2,2′-disulfonic acid with different acetoacetarylides. Reaction products were purified and characterized by means of elemental analysis by UV-VIS, IR, ^{1}H-NMR, and ^{13}C-NMR spectroscopy. A series of stilbene and fluorene compounds were prepared [114]. Compounds of

formula where X′ is H, halo, C1–4 haloalkyl(amino), Sn(alkyl)$_3$, and so on; R^1, R^2, and R^3 are independently H, OH, halo, C1–4 alkyl, C1–4 alkoxy, CN, and so on; R^4 is C1–4 alkylthio, C1–4 alkylsulfonyl, OH, C1–4 alkoxy, NH$_2$ and derivatives, and so on; R^5 is H and C1–4 alkyl; and their pharmaceutically acceptable salts were prepared.

These synthesized compounds were found to be useful for rescuing cells from beta-amyloid toxicity and in the treatment of Alzheimer's disease.

1.4
Physically Promoted Reactions

The microwave-promoted Heck reaction of aryl iodides and bromides with terminal olefins using a $Pd(OAc)_2$ (0.05 mol%)/K_3PO_4 catalytic system under ligand-free and solvent-free conditions was described [115]. Microwave radiation was used for synthesis of a number stilbene derivatives [115–117]. The reaction mixture was placed inside the cavity of the microwave reactor and irradiated at 300 W for 25 min. After the reaction mixture was cooled to room temperature, the solid was extracted with ethyl acetate and the solvent was evaporated in a vacuum. The residue was purified by flash-column chromatography on silica gel using ethyl acetate/hexane (1: 20). *para*/*ortho*-Hydroxylated (*E*)-stilbenes have been synthesized by a metal-free protocol for decarboxylation of substituted α-phenylcinnamic acid derivatives in aqueous media [116]. The authors stressed a synergism between methylimidazole and aqueous $NaHCO_3$ in polyethylene glycol under microwave irradiation using discover© focused microwave (2450 MHz, 300 W).

A one-pot two-step synthesis of hydroxystilbenes with *trans* selectivity was developed through a modified Perkin reaction between benzaldehydes and phenylacetic acids bearing 4- or 2-hydroxy substitution at the aromatic ring [67]. The reaction was performed under mild conditions in the presence of piperidine–methylimidazole and polyethylene glycol under microwave irradiation. As a result, 71% yield of (*E*)-4-chloro-4′-hydroxy-3′-methoxystilbene from 4-hydroxy-3-methoxybenzaldehyde and 4-chlorophenylacetic acid was obtained. A microwave-induced one-pot process for the preparation of arylethenes has been patented [118]. For the preparation of a series of arylethenes (I; R^1–R^5 = H, OH, OMe, AcO, halo, NO_2; R^1, R^3, R^5 = OH, AcO; R = H, substituted aryl), reaction of 2- or 4-hydroxy substituted cinnamic acids or derivatives in the presence of a base, under reflux or microwave irradiation, has been used. For example, a mixture of α-phenyl-4-hydroxy-3-methoxycinnamic acid, $NaHCO_3$, methylimidazole, and polyethylene glycol was microwaved at 200 W and 180 °C for 10 min to give 96% 4-hydroxy-3-methoxystilbene.

Ultrasound-assisted synthesis of *Z* and *E*-stilbenes by Suzuki cross-coupling reactions of organotellurides with potassium organotrifluoroborate salts has been reported [119]. Palladium-catalyzed cross-coupling reactions between potassium aryl- or vinyltrifluoroborate salts and aryl or vinyl tellurides proceeded to give stilbenes containing a variety of functional groups. For example, a suspension of *Z*-(2-butyl-tellanyl-vinyl)-benzene, potassium phenyltrifluoroborate ($Pd(PPh_3)_4$), and silver(I) oxide of methanol was irradiated in a water bath of an ultrasonic cleaner for 40 min.

1.5
Synthesis of Stilbene Dendrimers

Dendrimers are repeatedly branched molecules. The name comes from the Greek "δενδρον"/*dendron*, which means tree. A series of stilbene dendrimers with a stilbene core and benzyl ether type dendrons has been synthesized and investigated

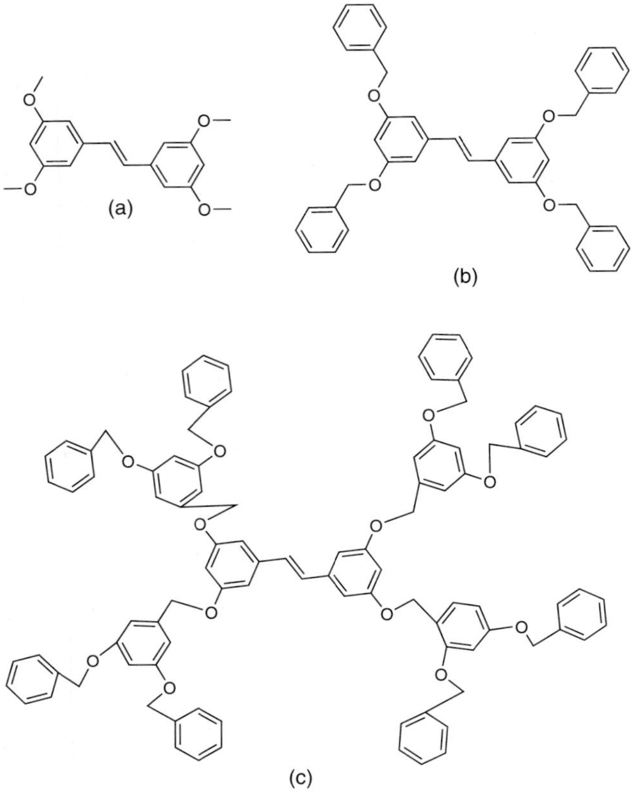

Figure 1.20 Dendrimers synthesized in the work [120]. (Reproduced with permission from Ref. [120].)

in an acetonitrile and 1,2-dichloroethane mixture (3:1) to elucidate the dendrimer effects (Figure 1.20) [120]. It was shown that the quantum yield of the formation of stilbene core radical cation during the 308-nm TPI was independent of the dendron generation of the dendrimers, whereas the generation dependence of the quantum yield of the radical cation was observed during the 266-nm TPI, where both the stilbene core and the benzyl ether-type dendron were ionized, suggesting that the subsequent hole transfer occurs from the dendron to the stilbene core and that the dendron acts as a hole-harvesting antenna.

Stilbenoid dendrons with various donor and acceptor groups on the focal unit were synthesized by a Wittig–Horner reaction, starting from an aldehyde-functionalized dendron and various substituted phosphonic acid esters [121]. It was shown that the target molecules were composed of *meta*-branched arms, two of them with extended conjugation (distyrylbenzene) and three flexible dodecyloxy chains; the focal group consists of a donor- or acceptor-substituted styryl unit. The synthesized stilbenoid dendrons were photosensitive, and degradation of the supramolecular order proceeds even in the glassy liquid crystal state.

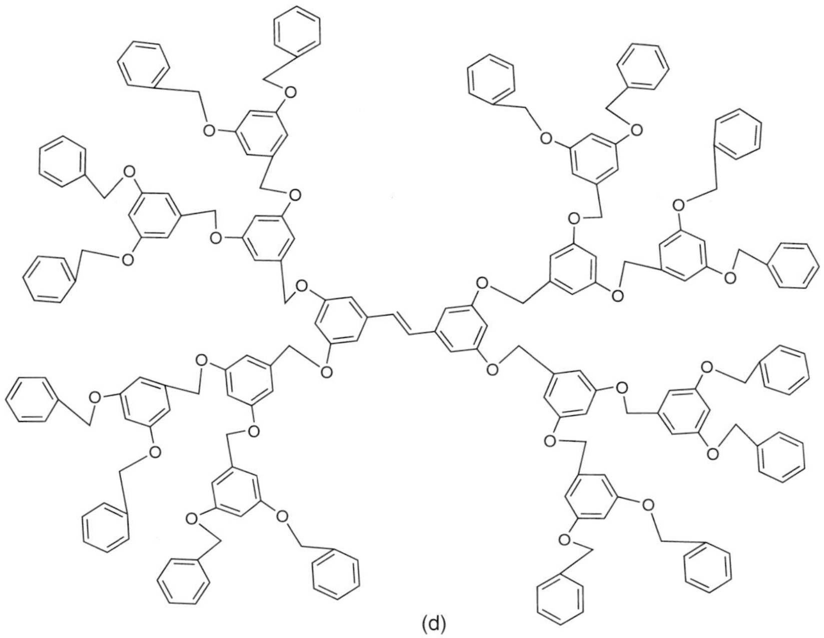

(d)

Figure 1.20 (Continued)

Polyamidoamine dendrimers, constructed on the surface of silica, were phosphonated by the Heck reaction using diphenylphosphinomethanol and complexed to form a palladium-dimethyl TMEDA [122]. This catalyst was found to be effective in the Heck reaction of aryl bromides with both butyl acrylate and styrene, affording coupling products in moderate to good yields. The heterogeneous palladium catalyst can also be recycled and reused with only moderate reduction in activity. Water-soluble self-assembly of amphiphilic pyrene-cored poly(aryl ether) dendrimers was prepared and its fluorescence properties were studied [123]. Dendron-conjugated branches of stilbene and 4-styrylstilbene groups have been attached to resorcinarene cores [124]. The optical properties of thin films were identical to those of the solutions indicating the absence of intermolecular interactions. Two generations of dendritic nanoparticles were prepared, which contain (E)-stilbene or (E,E)-1,4-distyrylbenzene chromophores in the 4 or 8 terminal positions of the propylene imine dendrons [125]. Two large π-conjugated dendrimers (G0 and G, molecular weight of 10 973 Da) employing the stilbenoid moiety as the bridge unit have been synthesized through the Suzuki and the Horner–Wadsworth–Emmons reactions [126]. Stilbene dendrimers were prepared by coupling 4,4'-dihydroxystilbene with first-, second-, third-, or fourth-generation benzyl ether-type dendrons [127]. All the generations of stilbene dendrimers underwent

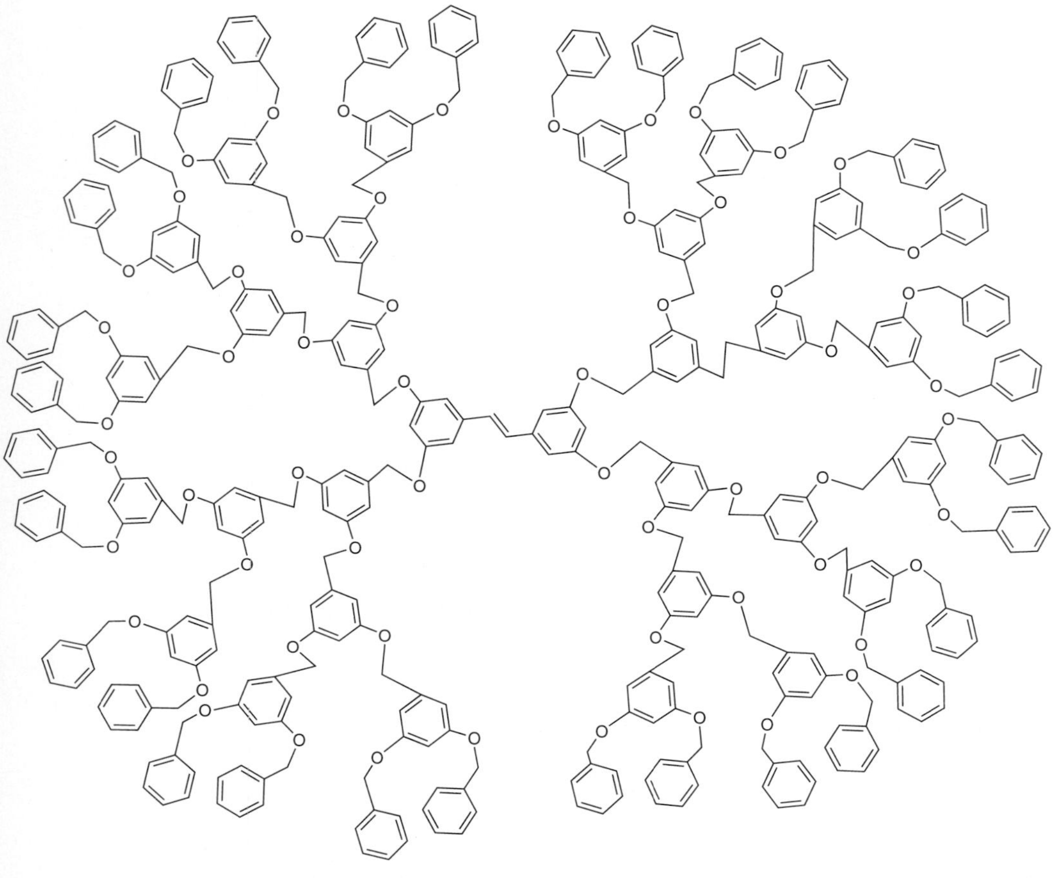

(e)

Figure 1.20 *(Continued)*

photoisomerization with the same efficiency as that of 4,4′-dimethoxystilbene. Dendrimers with terminal (*E*)-stilbene moieties based either on a hexamine core or on a benzenetricarboxylic acid core were synthesized [128]. Under irradiation, both types of dendrimers undergo photochemical reactions to yield cross-linked products lacking styryl moieties.

Photochemistry and mobility of the stilbenoid dendrimers [all-(*E*)-1,3,5-tris[2-(3,4,5-tridodecyloxyphenyl)ethenyl]benzene] and [all-(*E*)-1,3,5-tris(2-{3,5-bis[2-(3,4,5-tridodecyloxyphenyl)ethenyl]phenyl}ethenyl)benzene] in their neat phases were synthesized and investigated [129]. Selectively deuterated, dodecyloxy-substituted stilbenoid dendrimers of the first and second generations (Figure 1.21) were synthesized by a convergent synthesis, using the Wittig–Horner reaction. Molecules deuterated at the

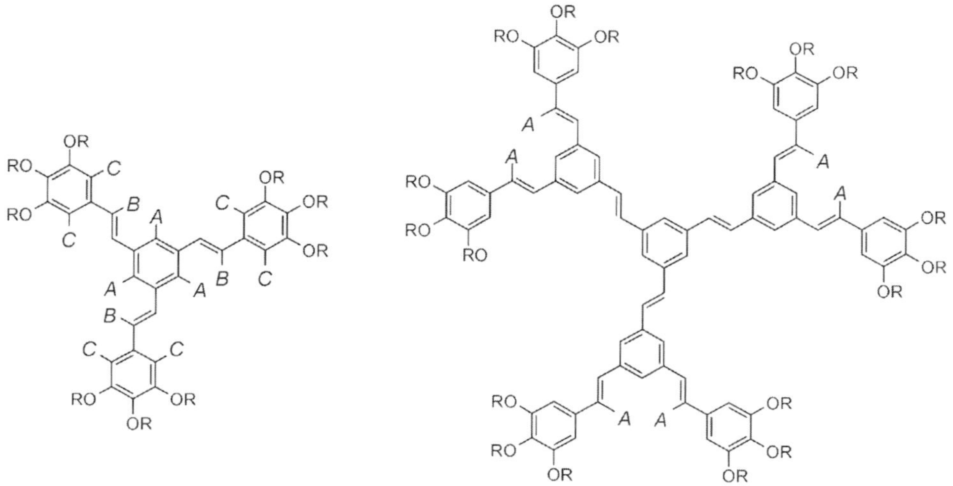

1a	A–C = ¹H, R = C₁₂H₂₅
1b	A = ²H, B, C = ¹H, R = C₁₂H₂₅
1c	B = ²H, A, C = ¹H, R = C₁₂H₂₅
1d	C = ²H, A, B = ¹H, R = C₁₂H₂₅
1e	A–C = ¹H, R = C²H₂C₁₁H₂₃
2a	A = ¹H, R = C₁₂H₂₅
2b	A = ²H, R = C₁₂H₂₅
2c	A = ¹H, R = C²H₂C₁₁H₂₃

Figure 1.21 Chemical structure of the stilbenoid dendrimers synthesized and investigated in Ref. [129]. (Reproduced with permission from Ref. [129].)

α-position of the alkoxy chains were used to study the photoreactions in the neat phases by ^{1}H NMR. No photoreactions occur in the crystal state. The mobility of the dendrimers was studied by means of ^{2}H solid-state NMR spectroscopy. The onset of the photochemistry for dendrimer 1 [all-(E)-1,3,5-tris[2-(3,4,5-tridodecyloxyphenyl)ethenyl]benzene] corresponds to the increasing mobility at the Cr/LC transition. The first-generation dendrimers still show large angle motion, whereas dendrimers of the second generation 2 [all-(E)-1,3,5-tris(2-{3,5-bis[2-(3,4,5-tridodecyloxyphenyl)ethenyl]phenyl}ethenyl)benzene] were restricted to librational motions. Photochemical conversion and fluorescence quenching for first- and second-generation dendrimers 1 and 2 were found to increase with increasing molecular motion and reach a maximum in the isotropic phase.

1.6
Stilbene Cyclodextrin Derivatives

Rotaxane is a mechanically interlocked molecular architecture consisting of a "dumbbell-shaped molecule" that is threaded through a "macrocycle." The name is derived from the Latin words for wheel (*rota*) and axle (*axis*). The synthesis of a novel rotaxane containing α-CD (α-cyclodextrin) as the macrocycle, stilbene as the "string," and the isophthalic acid as stopper via a palladium-catalyzed Suzuki coupling reaction in water at room temperature and supramolecular self-assembly was reported [130]. The molar ratio of α-CD to the "string" was 1:1 and the rotaxane was [2]rotaxane according to the results of ^{1}H NMR spectroscopy and mass spectrometry of rotaxane. The synthesis of homo- and hetero[3]rotaxanes with two π-system (dumbbell) components threaded through a single γ-cyclodextrin macrocycle has been accomplished [131]. The synthesis was carried out in two steps: first dumbbell was synthesized threaded through the macrocycle to give a [2]rotaxane and then a second dumbbell was prepared through the remaining cavity of the [2] rotaxane. A hetero[3]rotaxane with one stilbene and one cyanine dye threaded through γ-cyclodextrin has been synthesized. The stilbene [2]rotaxane intermediate in this synthesis was shown to have a high affinity for suitably shaped hydrophobic guests in aqueous solution, facilitating the synthesis of [3]rotaxanes and suggesting possible applications in sensors.

Two [2]rotaxanes (Figure 1.22), each comprising α-cyclodextrin as the rotor, stilbene as the axle, and 2,4,6-trinitrophenyl substituents as the capping groups, were prepared and their conformations were examined in solution and in solid state using ^{1}H NMR spectroscopy and X-ray crystallography, respectively [132]. In the solid state, the axles of rotaxanes form extended molecular fibers that are separated

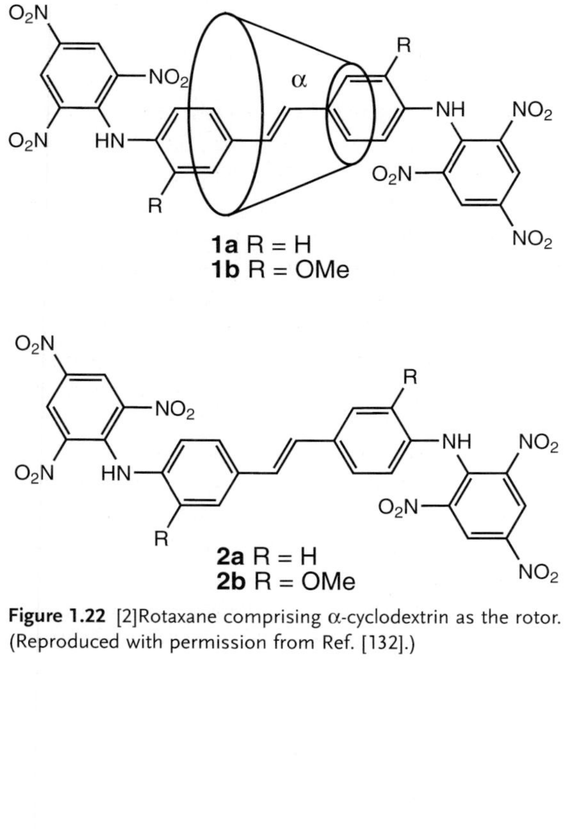

Figure 1.22 [2]Rotaxane comprising α-cyclodextrin as the rotor. (Reproduced with permission from Ref. [132].)

from each other and aligned along a single axis. The molecular fibers are strikingly similar to those formed by the axle component of one of the rotaxanes in the absence of the cyclodextrin, but in the latter case they are neither separated nor all aligned.

1.7
Stilbenes on Templates

A new catalyst system for Heck reaction, silica-supported poly-γ-aminopropylsiloxane palladium (Pd^{2+})-transition metal (Cu^{2+}) complex, has been designed [133]. The catalyst has been prepared from organic silica via immobilization on fumed silica, followed by treatment with $Cu(OAc)_2$ and $PdCl_2$ in ethanol. The catalyst was efficient for Heck arylation of aryl iodides with alkene. Stilbene-based azo dyes were synthesized and poly(vinyl alcohol) polarizing films were prepared [134]. Two series of combined liquid crystal polyphosphates bearing dual photoreactive mesogenic units (stilbene and azobenzene/α-methylstilbene and azobenzene) were synthesized by the solution polycondensation method [135]. Structures of the synthesized polymers were confirmed by various spectroscopic techniques. The photochemical response, photocross-linking reaction, and conversion of *trans* to *cis* form of azobenzene unit were investigated. The terminal substituents in the side chain affected the texture of liquid crystal phase for all the polymers. Poly[2-{bis (4-methoxyphenyl)amino}phenyl-5-(2-ethylhexyloxy)-1,4-phenylenevinylene] was prepared via the Gilch reaction of the *p*-bis(chloromethyl)benzene monomer, 2-{bis(4-methoxyphenyl)aminophenyl}-α,α'-dichloro-*p*-xylene [136]. 2,3-Bis[*N,N*-bis(4-methoxyphenyl)aminophenyl]stilbene 2 and its aminium diradical 2 + were also prepared as model dimer compounds.

A new urethane acrylic monomer with stilbene in its structure, *trans*-4-(2-methacryloyloxyethylcarbamoyloxymethyl)stilbene (SUM), was synthesized to be further free radically copolymered with methyl methacrylate (MMA) [137]. The structures of SUM and the resulting copolymer, *trans*-4-(2-methacryloyloxyethylcarbamoyloxymethyl)stilbene-*co*-methyl methacrylate (SUMMA), were characterized by a set of physicochemical methods. Morphological changes in the surface of the polymeric film during the photoisomerization were visualized by means of atomic force microscopy (AFM), and the newly formed cone-shaped structures from the irradiated surface were attributed to J-aggregates. In order to gain insight into the properties of the triplet excited states in platinum-acetylide polymers, four platinum complexes were synthesized in which the metal is linked to the *trans*-stilbene through acetylide bonds [138]. Comparison of the properties of these complexes provided information on the geometry of the π-conjugated acetylide ligands and the existence of the metal-to-ligand charge transfer (MLCT) state on the photophysics of these systems.

Three new coordination polymers, $[Cd(SCN)_2L_2]_n$ (Figure 1.23), $[CdHg(SCN)_4L_2]_n$, and $[MnHg(SCN)_4L_2]_n$, were synthesized by the self-assembly of L with the corresponding metal salts and NaSCN (L is a functional rigid imidazole ligand, *trans*-4-imidazolyl-4'-(*N,N*-diethylamino)stilbene) [139]. It was shown that adjacent Cd(II)

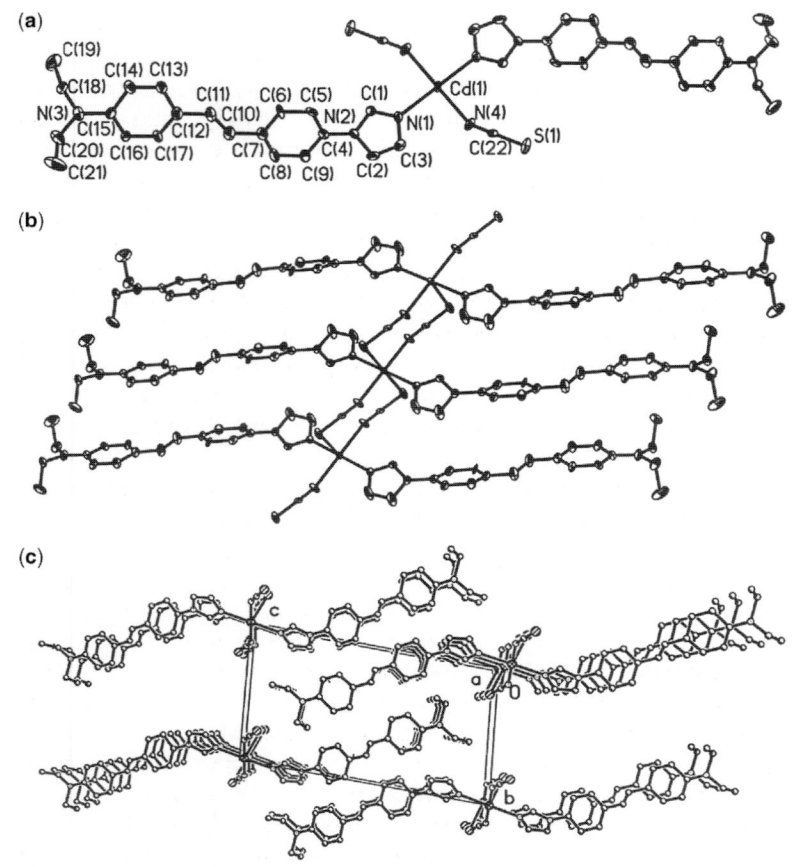

Figure 1.23 Structural unit of complex from coordination polymers [Cd(SCN)$_2$L$_2$]$_n$ [139]. (Reproduced with permission from Elsevier.)

ions are bridged by SCN− ligands to form infinite chains with the remaining two positions of six-coordinated Cd(II) ion occupied by two imidazole ligands. Cd(II) and Hg(II) centers were bridged by SCN− ligands to form two-dimensional framework sheets. Cd(II) and Hg(II) ions were coordinated by six N atoms and four thiocyanate S atoms, respectively. Mn(II) and Hg(II) ions were linked by bridging NCS− groups to form two-dimensional sheets. Figure 1.23 shows the structural unit of the complex from coordination polymers [Cd(SCN)$_2$L$_2$]$_n$.

A cross-linkable embossed film containing polyester stilbene has been prepared [140]. The substrate film comprised a copolyester consisting of repeating units derived from stilbene dicarboxylic acid (1–40 mol%), 1,4-cyclohexane dicarboxylic acid (60–99 mol%), and 1,4-cyclohexane dimethanol (50–100 mol%). UV-curable polyesters containing stilbene structural unit with good thermal stability have been synthesized [141]. The polymers consisted of (a) 1–40 mol% *trans* 3,3′ or *trans* 4,4′ of stilbene dicarboxylic acid, (b) 60–99 mol% *cis, trans* of 1,4-cyclohexane dicarboxylic acid, and (c) 50–100 mol% *cis, trans* of 1,4-cyclohexane dimethanol. A number of nanoparticle–monomer–receptor (NMR) sensors, where the nanoparticles were SiO_2 and ZnS: Mn/CdS core/shell quantum dots, the monomers were stilbene derivatives, and the receptors were isoquinoline and 3-aminohexafluoropropanol, have been synthesized [142]. Using these sensors, nerve gas analogues DCP and DMMP and acids such as HCl were detected with these NMR sensors by fluorescence change in the wavelength range 380–500 nm. Wang and Muralidharan [143] have synthesized 12 different stilbene-based monomers for the nanoparticle–monomer–nanomolecule–receptor (NMNR) and NMR sensors and investigated their efficacies for the detection of nerve gas analogues DCP and DMMP and acids such as HCl. Both NMNR sensors with $Eu(dppz)_3$ (dppz = dipyrido[3,2-a:2′,3′-c]phenazine) complex and NMR sensors have been obtained with the stilbene-based monomers.

The synthesis of poly(MMA/2,2,2-trifluoroethyl methacrylate (3FMA)/benzyl methacrylate = 52.0/42.0/6.0 (w/w/w)) and poly(MMA/3FMA = 85.0/15.0 (w/w)) containing 2.8 wt% of *trans*-stilbene that exhibited birefringence close to zero was reported [144]. Zigzag polymers consisting of dithia[3.3](2,6)pyridinophane units were prepared [145]. The resulting polymer complex exhibited a high catalytic activity for the Heck coupling reaction. A perdeuterated *trans*-stilbene grafted polystyrene has been synthesized [146]. The effects of chromophore concentration, solvent polarity, excitation energy, chromophore aggregation, and UV irradiation on photophysical properties of this photoactive material have been investigated.

The paper [147] has reported the preparation and characterization of pure Langmuir and Langmuir–Blodgett (LB) films of a stilbene derivative containing two alkyl chains, 4-dioctadecylamino-4′-nitrostilbene. Mixed films incorporating docosanoic acid and stilbene derivatives were also studied. Brewster angle microscopy (BAM) analysis revealed the existence of randomly oriented 3D aggregates, spontaneously formed immediately after the spreading process of the stilbene derivative onto the H_2O surface. It was shown that monolayers were transferred undisturbed onto solid substrates with AFM revealing that the one-layer LB films are constituted by a monolayer of the stilbene derivative together with some three-dimensional aggregates.

The synthesis and properties of DNA minihairpin conjugates possessing stilbene capping groups have been investigated for two hairpin base sequences with three stilbene capping groups [148]. It was found that the two hairpin sequences 5′-TTTCACCGAAA and 5′-ATTCACCGAAT differ in the orientation of the terminal base pair, the latter forming the more stable hairpin. Conjugation of these hairpins with a 5′-stilbenecarboxamide capping group significantly increased hairpin stability and reduced the difference in stability observed for the unmodified hairpins. The synthesis and properties of nicked dumbbell and dumbbell DNA conjugates having A-tract base pair domains connected by rod-like stilbenedicarboxamide linkers were reported [149]. Structures of the nicked dumbbells and dumbbells (Figure 1.24) have been investigated using a combination of CD spectroscopy and molecular modeling.

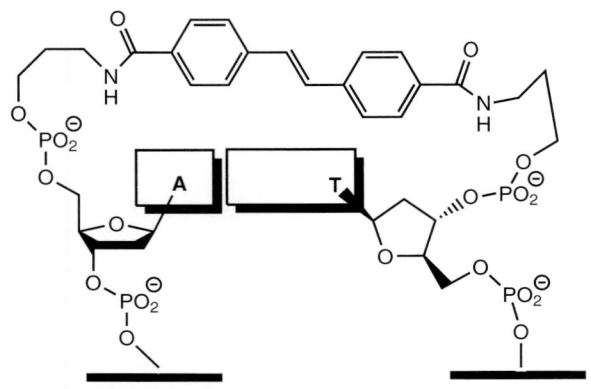

Figure 1.24 Structure of the hairpin loop region of synthetic hairpins having a stilbenedicarboxamide (SA) linker. (Reproduced with permission from Ref. [149].)

A synthesized dual stilbene–nitroxide probe was covalently immobilized onto the surface of a quartz plate as an eventual fiber-optic sensor (Figure 1.25) [150]. The immobilization procedure included a cyanogen bromide surface activation followed by smoothing with a protein tether. The rate of fluorescence change was monitored in aqueous glycerol solution of different viscosities and contents of ascorbic acid.

The attachment of *para*-NH$_2$-stilbene to the surface of the cowpea mosaic virus (CPMV) coat protein was performed with an indicating antibody–antigen interaction [151]. Antibody binding was subsequently blocked by the installation of polyethylene glycol chains. The authors claimed that these results typify the type of site-specific control that is available with CPMV and related virus building blocks. A preparation scheme of virus–stilbene conjugate is shown in Figure 1.26. The authors claimed that these results typify the type of site-specific control that is available with CPMV and related virus building blocks.

A complex antibody–donor–acceptor-substituted stilbene has been investigated (Figure 1.27) [152]. Photophysical and structural analyses indicated that antibody binding alters the excited-state behavior of stilbene. The authors suggested that such complexes may find *in vivo* application as fluorescent biosensors.

1.8
Stilbenes Analysis

1.8.1
Methods Using Liquid and Gas Chromatography

A method has been developed to detect residual stilbenes such as diethylstilbestrol (DES), dienestrol (DIS), and hexestrol (HS) in animal tissues using solid-phase extraction (SPE) and gas chromatography–mass spectrometry (GC–MS) [153]. The

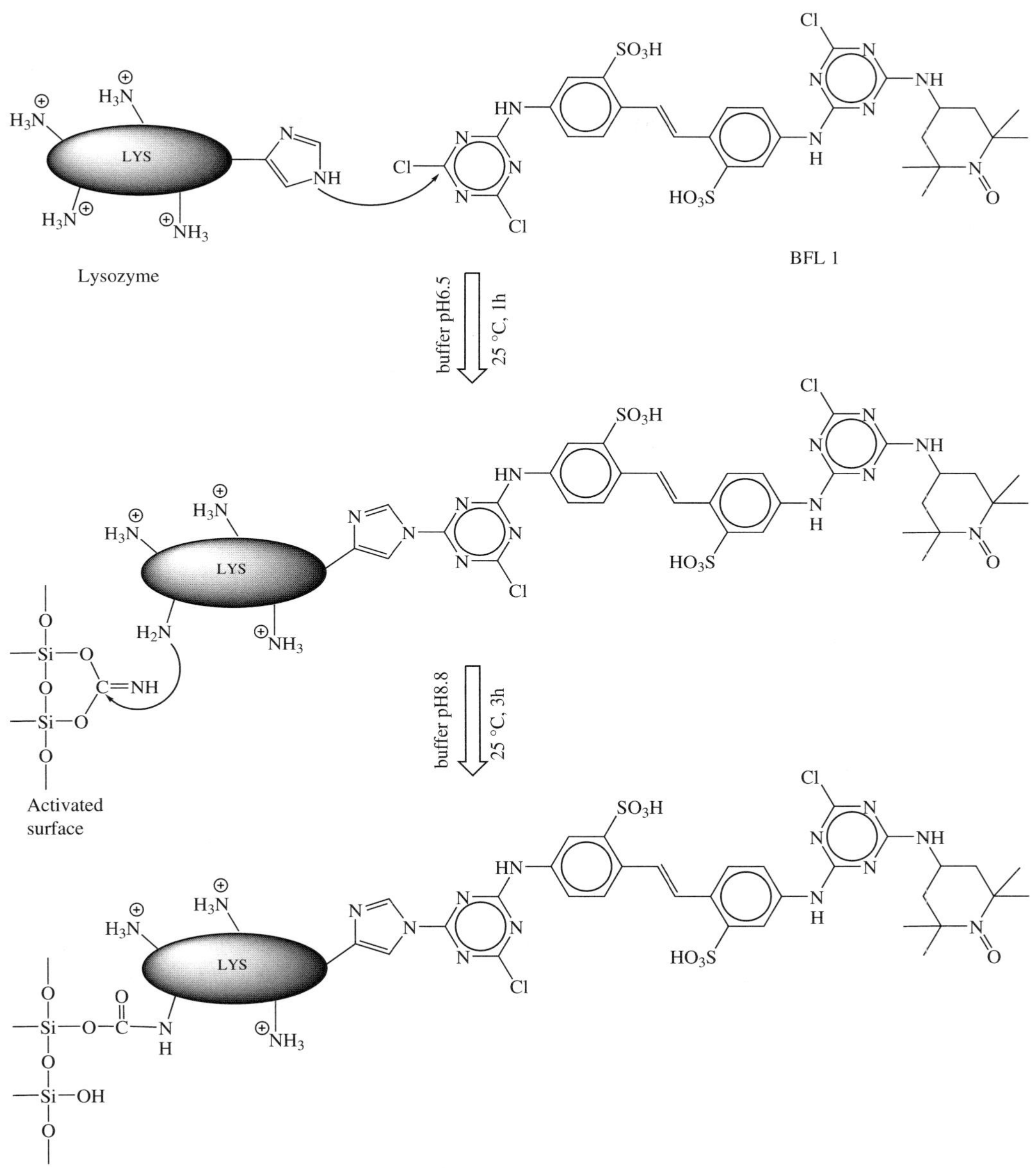

Figure 1.25 Scheme of immobilization of BFLT on quartz plate by BrCn with lysozyme as a tether [150]. (Reproduced with permission from Elsevier.)

analytes were detected by mass spectrometer with electron impact source in selected ion monitoring mode (EI/SIM) and quantified with an external standard calibration curve method. Linear calibration curves were obtained in the concentration ranges from 5 to 500 µg/l for HS and from 10 to 1000 µg/l for DES and DIS.

Figure 1.26 Preparation of virus–stilbene conjugate. (Reproduced with permission from Ref. [151].)

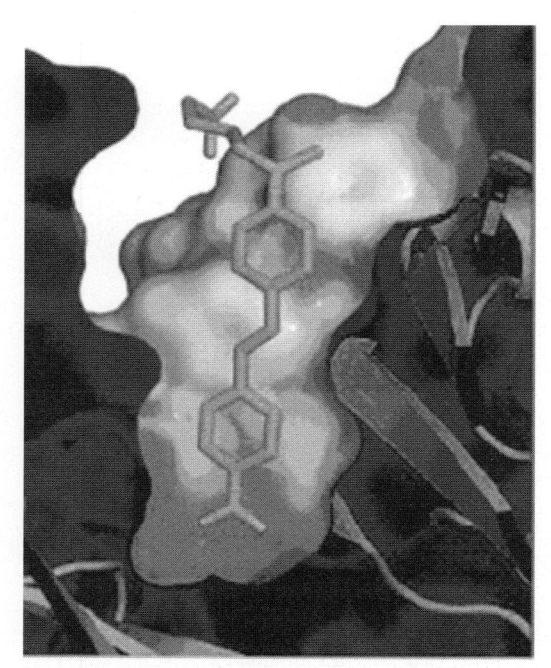

Figure 1.27 Antibody combining site of 11G10 in complex with hapten with electron density contoured at 1.5σ. Representation of the electrostatic surface [152].

A comprehensive method for the detection of four stilbene-type disulfonate agents and one distyrylbiphenyl-type fluorescent whitening agent (FWA) in paper materials (napkin and paper tissue) and infant clothes using the newly developed Oasis WAX (mixed mode of weak anion exchange and reversed-phase sorbent) solid-phase extraction cartridge was proposed [154]. The analytes were detected by ion-pair chromatography coupled with negative electrospray ionization-tandem mass spectrometry (HPLC–ESI-MS/MS), applying a di-n-hexyl-ammonium acetate (DHAA) as the ion-pairing reagent in mobile phase. The method was applied to commercial samples, showing that two stilbene-type disulfonates were predominant FWAs detected in napkin and infant cloth samples.

Three different sample preparation techniques, solid-phase extraction, reverse osmosis, and vacuum distillation, were studied and the recoveries were compared for detecting highly water-soluble stilbene sulfonic acids by liquid chromatography with photodiode array (PDA) and electrospray ionization-tandem mass spectrometry (LC–ESI-MS/MS) [155]. The detection limits were 1–28 µg/l with LC–ESI-MS. The sample collected from wastewater treatment plant contained 21.1, 13.3, 12.1, 41.8, and 9.9 µg/l of *cis*-4,4′-diaminostilbene-2,2′-disulfonic acid (*cis*-DASDA), *trans*-4,4′-diaminostilbene-2,2′-disulfonic acid (*trans*-DASDA), 3-amino acetanilide-4-sulfonic acid (3-AASA), 4-chloroaniline-2-sulfonic acid (4-CASA), and 2-chloroaniline-5-sulfonic acid (2-CASA), respectively. Reversed-phase high-performance liquid chromatography (RP-HPLC) with PDA and MS detection was employed to study the accumulation of stilbenes and other naturally occurring polyphenol intermediates of flavonoid pathway in tomato fruits of plants genetically modified to synthesize resveratrol [156]. The results of these analysis revealed that the genetic modification of the tomato plants originated from different levels of accumulation of *trans*- and *cis*-piceid and *trans*- and *cis*-resveratrol in their fruit depending on the stages of ripening. Determination of stilbenes in *Sicilian pistachio* by high-performance liquid chromatographic diode array (HPLC-DAD/FLD) was carried out [157]. The presence of several natural stilbenes in 12 samples of pistachios harvested from 10 different farms of Sicily (Bronte and Agrigento) was detected and two types of stilbenes in the samples of pistachios examined, *trans*-resveratrol and *trans*-resveratrol-3-O-β-glucoside (*trans*-piceid), were found. HPLC methods for the detection of 2,3,5,4′-teterahydroxystilbene-2-O-β-D-glucoside in Yangyan Pills were used [158]. The 2,3,5,4′-teterahydroxystilbene-2-O-β-D-glucoside sample showed a good linear relationship in the range of 0.05–0.40 mg/ml.

Determination of glyoxal (Go), and methylglyoxal (MGo), in the serum of diabetic patients by MEKC, using stilbenediamine as derivatizing reagent, was reported [159]. Uncoated fused silica capillary, effective length 50 cm × 75 µm i.d., applied voltage 20 kV, and photodiode array detection were used. Calibration was linear within 0.02–150 µg/ml with detection limits of 3.5–5.8 ng/ml. An HPLC method with photodiode array detection and ESI/MS detection was developed for the qualitative and quantitative analyses of stilbenes, stilbene glycosides, and other compounds in the dried rhizome of *Polygonum cuspidatum* [160]. Five samples of *Rhizoma polygoni cuspidati* from different regions were analyzed by this method. The major constituents piceid, resveratrol, emodin-8-O-β-D-glucoside, and emodin were selected to provide an index for the quality assessment of the herbal drug. A simple, sensitive, and specific HPLC method was developed and applied for simultaneous detection of the six major active constituents in *Smilax china*, namely, taxifolin-3-O-glycoside, piceid, oxyresveratrol, engeletin, resveratrol, and scirpusin A [161]. The samples were separated on an Aglient Zorbax XDB-C18 column with gradient elution of acetonitrile and 0.02% (v/v) phosphoric acid at a flow rate of 1.0 ml/min and detected at 300 nm.

A liquid chromatography–andem mass spectrometry method was proposed for simultaneous detection of stilbenes, diethylstilbestrol, hexestrol, and dienestrol in animal tissue [162]. Sample cleanup and analyte enrichment was performed by automated solid-phase extraction (ASPE) with a silica gel cartridge. The recovery level of the method was 84–108% for DES and DIS between 0.5 and 5 ng/g, and 59–87% for HS between 0.25 and 2.5 ng/g. A facile method based on liquid chromatography coupled with electrospray ionization-tandem mass spectrometry has been established for the analysis of bioactive phenolic compounds in rhubarbs [163]. From six rhubarb species (*Rheum officinale*, *R. palmatum*, and *R. tanguticum* and unofficial *R. franzenbachii*, *R. hotaoense*, and *R. emodi*), a total of 107 phenolic compounds were identified or tentatively characterized based on their mass spectra. Stilbenes, which are the major constituents of unofficial rhubarbs, were found to be different among the species. Seven prenylated stilbenes were identified by combined HPLC-PAD-APCI/MSn analysis of an extraction of mucilage isolated from peanut (*Arachis hypogaea* L.) root tips [164]. The principal constituent was assigned the structure 4-(3-methyl-but-1-enyl)-3,5-dimethoxy-4'-hydroxy-*trans*-stilbene (I). The common name mucilagin A was proposed for this novel compound, with its concentration in the mucilage established at 250 µg/g (wet weight basis). The authors suggest that compounds detected in peanut mucilage may play a role in regulating root–soil pathogen interactions.

A combination of reversed-phase HPLC with UV-diode array detection and electrospray ionization-tandem mass spectrometry ion-trap detection was used for characterization of a photochemical mixture of *trans*-resveratrol and its derivatives, including oligomers and glucosides [165]. As the polyphenol source, the stems of three frost-hardy grapevine varieties (*Hasaine (Hasansky) sladki*, *Zilga*, and *Yubilei Novgoroda*) were used. A quantitative determination of stilbene oligomers in *Jin Que-gen* collected from different regions was performed [166]. An HPLC method has been developed for efficiently quantifying two stilbene tetramers, carasinol B (**1**) and kobophenol A (**2**), and one stilbene trimer, (+)-α-viniferin (**3**), in the plant. A simultaneous determination of the contents of two stilbene tetramers, carasinol B and kobophenol A, and one stilbene trimer, (+)-α-viniferin, in roots, tubers, and leaves of *Caragana sinica* in various seasons was performed using an improved HPLC method [167]. The contents of stilbene tetramers were maximal in winter while the contents of the stilbene trimer were maximal in summer.

A rapid analysis of resveratrol, *trans*-ε-viniferin, and *trans*-δ-viniferin from downy mildew-infected grapevine leaves by liquid chromatography–atmospheric pressure photoionization mass spectrometry was performed [168]. The characterization of unknown stilbene derivatives such as six resveratrol dimers, two dimethylated resveratrol dimers, and a resveratrol trimer is reported. ^{13}C NMR spectroscopy in combination with HPLC and spectrophotometry was used to complement HPLC or spectrophotometry to analyze stilbene and anthocyanin metabolites in grape cell cultures [169]. The effect of various elicitors such as sucrose and methyl jasmonate and fungal elicitor on stilbene and anthocyanin biosynthesis was investigated. Methyl jasmonate and fungal elicitor strongly increased stilbene production through the activation of enzymes from phenylalanine ammonia lyase to stilbene synthase. A liquid chromatography–mass spectrometry method was employed to analyze total resveratrol (including free resveratrol and resveratrol from piceid) in fruit products and wine [170]. Samples were extracted using methanol, enzymatically hydrolyzed, and analyzed using reversed-phase HPLC with positive ion atmospheric pressure chemical ionization (APCI) mass spectrometric detection. Following APCI, the abundance of protonated molecules was recorded using selected ion monitoring of m/z 229. An external standard curve was used for quantitation, which showed a linear range of 0.52–2260 pmol of *trans*-resveratrol injected on-column. The extraction efficiency of the method was detected to be 92%. Resveratrol was detected in grape, cranberry, and wine samples. Concentration ranged from 1.56 to 1042 nmol/g in Concord grape products and from 8.63 to 24.84 µmol/l in Italian red wine. Concentrations of resveratrol were found to be similar in cranberry and grape juice at 1.07 and 1.56 nmol/g, respectively.

1.8.2
Miscellaneous Analytical Methods

X-ray structure of 1-(methylthio)-*cis*-stilbene-2-thiol, Ph(SCH$_3$)C:C(SH)Ph(HL), and 1-(benzylthio)-*cis*-stilbene-2-thiol, Ph(SCH$_2$Ph)C:C(SH)Ph, forming monomeric complexes with Sb^{3+} complex was established [172]. The structure of Moracin M, a stilbenoid extracted from the stem bark of *Milicia excelsa* (*Moraceae*), was obtained by a single-crystal X-ray analysis. A competitive ELISA method was developed for quantitative detection of hexestrol [173]. Polyclonal rabbit antisera, raised against protein conjugate hexestrol-mono-carboxyl-propyl-ethyl-bovine-serum-albumin (HS-MCPE-BSA), were used in immobilized antibody-based and competitive immunoassays. Assay conditions, including concentration of antisera and horseradish peroxidase (HRP)-HS, were optimized. The effects of incubation time, surfactant concentration, ionic strength, and pH of the medium were also investigated. The typical calibration curve gave an average IC$_{50}$ value of 2.4 ng/ml, calibration range from 0.2 to 30.5 ng/ml, and a detection limit of 0.07 ng/ml.

Stilbene-related heterocyclic compounds including benzalphthalide, phthalazinone, imidazoindole, and pyrimidoisoindole derivatives were tested for their anti-HIV activity [174]. Assays based on recombinant viruses were used to evaluate HIV replication inhibition, and stably transfected cell lines were used to evaluate inhibition of Tat and NF-κB proteins. Some of the stilbene-related heterocyclic compounds analyzed displayed anti-HIV activity through interference with NF-κB and Tat function. Near-IR spectroscopy (NIRS) and artificial neural networks were employed for quantitative detection of four active constituents in rhubarb:

anthraquinones, anthraquinone glucosides, stilbene glucosides, and tannins and related compounds [175]. The authors proposed that this method can be used for detecting the active constituents in Chinese herbal medicine. Simultaneous determination of stilbene and anthraquinone compound in *Polygonum cuspidatum* by ultraviolet spectrophotometry was performed [176]. A method based on *in vivo* fluorescence using commercial spectrofluorometers that allowed fast and local assessment of stilbene content in grapevine leaves was tested [177]. Synthesis of stilbenes in grapevine *Vitis vinifera* var. Muscat Ottonel leaves was induced by *Plasmopara viticola* inoculation or UV-C irradiation. Fluorescence was measured from both the abaxial and adaxial sides of leaves; then, stilbene content was analyzed by HPLC. The authors concluded that significant regressions were found between HPLC stilbene content and the corresponding leaf UV-induced blue fluorescence. The authors suggested that *in vivo* fluorescence is a good tool for the rapid study of stilbene synthesis in grapevine leaves that can potentially be extended to other fluorescent molecules.

An online Raman analyzer to quantitatively track the levels of *trans*-stilbene, benzaldehyde, and α-methoxybenzyl hydroperoxide in a continuous flow ozonolysis reactor was described [178]. The analysis was carried out using spectral stripping in order to overcome baseline artifacts inherent to simple peak area detections and to incorporate prior knowledge into the analytical model. The performance of spectral stripping was compared to partial least squares (PLS) analysis. Two new dihydrostilbenes, stilbostemins H (**1**) and I (**2**), were isolated and identified from the roots of *Stemona sessilifolia*, together with known stilbostemins B, D, and G, and stemanthrenes A and C (**4–8**) [179]. Structures of new stilbenoids were established by 1D and 2D ^{1}H NMR and ^{13}C NMR.

A rapid and sensitive capillary electrophoretic method for analysis of resveratrol in wine was proposed [180]. The protocol consists of sample preparation using a C-18 solid-phase extraction cartridge. The limits of detection for *trans*- and *cis*-resveratrol were 0.1 and 0.15 µmol/l, respectively. These procedures were used to analyze the *trans*- and *cis*-resveratrol levels in 26 wines. It was found that the concentration of *trans*-resveratrol ranged from 0.987 to 25.4 µmol/l, whereas the concentration of *cis*-resveratrol was much lower. The adsorptive voltammetric behavior of resveratrol was studied at a graphite electrode in B-R buffer (pH 6.0) solution using adsorptive cyclic voltammetric technique [181]. The oxidation of resveratrol was an irreversible adsorption-controlled process. It was found that in the range from 8.0×10^{-9} to 2.0×10^{-6} mol/l, the currents measured by differential pulse voltammetries presented a good linear property as a function of the resveratrol concentration. The proposed method was also applied for the determination of resveratrol in Chinese patent medicine with good results.

An amperometric biosensor for *trans*-resveratrol determination in aqueous solutions by means of carbon paste electrodes modified with peroxidase basic isoenzymes (PBIs) from *Brassica napus* was developed [182]. Catalytic properties of PBIs from *Brassica napus* toward *trans*-resveratrol oxidation were demonstrated by conventional UV–vis spectroscopic measurements. The enzymatic reaction rate was studied and kinetics parameters were detected. An amperometric biosensor based on *Brassica napus* PBIs to detect reservatrol was also proposed. The method employed a dialysis membrane covered, PBIs entrapped, and ferrocene (Fc)-embedded carbon paste electrode (PBIs-Fc-CP) and was based on the fact that the decreased amount of H_2O_2 produced by the action of PBIs was proportional to the oxidized amount of H_2O_2 in the solution. The lowest resveratrol value measured for a signal-to-noise ratio of 3:1 was 0.83 µM.

Supported liquid membranes (SLMs) consisting of 5% tri-*n*-octylphosphine oxide (TOPO) dissolved in di-*n*-hexylether/*n*-undecane (1:1) have been used in the simultaneous extraction of a mixture of three stilbene compounds (dienestrol, diethylstilbestrol, and hexestrol) in cow's milk, urine, bovine kidney, and liver tissue matrices [183]. The efficiencies obtained after the enrichment of 1 ng/l stilbenes in a variety of biological matrices of milk, urine, liver, kidney, and water were 60–70, 71–86, 69–80, 63–74, and 72–93%, respectively. A new method to contribute to the discrimination of polyphenols including resveratrol with synthetic pores was proposed [184]. The work [185] evaluated two types of commonly available chiral detectors for their possible use in chiral method development and screening: polarimeters and CD detectors. Linearity, precision, and the limit of detection (LOD) of six compounds (*trans*-stilbene oxide, ethyl chrysanthemate, propranolol, 1-methyl-2-tetralone, naproxen, and methyl methionine) on four common detectors (three polarimeters and one CD detector) were experimentally determined and the limit of quantitation calculated from the experimental LOD. *trans*-Stilbene oxide worked well across all the detectors, showing good linearity, precision, and low detection limits. However, the other five compounds proved to be more discriminating and showed that the CD detector performed better as a detector for chiral screens than the polarimeters.

As described in this chapter, stilbenes synthetic chemistry has been making gradual progress for the past two decades, paving the way for more fundamental uses and applications of stilbenes in research fields of both materials and life sciences. We could not give all details and references to cover this vast area in limited pages. Therefore, readers who want to study stilbenes chemistry further are recommended to consult relevant papers, reviews, and books cited in references.

References

1 Block, J. (2004) *Wilson & Gisvold's Textbook of Organic Medicinal and Pharmaceutical Chemistry*, Lippincott Williams & Wilkins, Hagerstown, MD.

2 Waldeck, D.H. (1991) *Chemical Reviews*, **91**, 415–436.

3 Whitten, D.G. (1993) *Accounts of Chemical Research*, **26**, 502–509.

4 Papper, V. and Likhtenshtein, G.I. (2001) *Journal of Photochemistry and Photobiology A: Chemistry*, **140**, 39–52.

5 Polo, A.S., Itokazu, M.K., Frin, K.M., Patrocinio, A.O.T., Murakami, I., and Neyde, Y. (2007) *Coordination Chemistry Reviews*, **251**, 255–281.

6 Polo, A.S., Itokazu, M.K., Frin, K.M., Patrocinio, A.T., Murakami, I., and Neyde, Y. (2006) *Coordination Chemistry Reviews*, **250**, 1669–1680.

7 Gutlich, P., Garcia, Y., and Woike, T. (2001) *Coordination Chemistry Reviews*, **219–221**, 839–879.

8 Grabowski, R., Rotkiewicz, W., and Rettig, W. (2003) *Chemical Reviews*, **103**, 3899–4032.
9 Meier, H. (1992) *Angewandte Chemie – International Edition in English*, **31**, 1399–1420.
10 Momotake, A. and Arai, T. (2004) *Journal of Photochemistry and Photobiology C: Photochemistry Reviews*, **5**, 1–25.
11 (a) Wadsworth, W.S. Jr. (1977) *Organic Reactions*, **25**, 73–253; (b) Kelly, S.E. (1991) *Comprehensive Organic Synthesis*, **1**, 729–817.
12 Ververidis, F., Trantas, E., Douglas, C., Vollmer, G., Kretzschmar, G., and Panopoulos, N. (2007) *Biotechnology Journal*, **2**, 1214–1234.
13 Cassidy, A., Hanley, B., and Lamuela-Raventos, R.M. (2000) *Journal of the Science of Food and Agriculture*, **80**, 1044–1062.
14 (a) Kimura, Y. (2005) *Journal of Traditional Medicines*, **22**, 154–161; (b) Kimura, Y. (2005) *In Vivo*, **9**, 37–60.
15 Cai, Y., Luo, Q., Sun, M., and Corke, H. (2004) *Life Sciences*, **74**, 2157–2184.
16 Seeram, N.P. (2008) *Journal of Agricultural and Food Chemistry*, **56**, 630–635.
17 Lin, M. and Yao, C.-S. (2006) *Studies in Natural Products Chemistry*, **33**, 601–644.
18 Saiko, P., Szakmary, A., Jaeger, W., and Szekeres, T. (2008) *Mutation Research: Reviews in Mutation Research*, **658**, 68–94.
19 Seeram, N.P. (2008) *Journal of Agricultural and Food Chemistry*, **56** (3), 630–635.
20 Brown, T., Holt, H. Jr., and Lee, M. (2006) *Heterocyclic Chemistry*, **2**, 1–51.
21 Iriti, M. and Faoro, F. (2006) *Medical Hypotheses*, **67**, 833–838.
22 Nichenametla, S.N., Taruscio, T.G., Barney, D.L., and Exon, J.H. (2006) *Critical Reviews in Food Science and Nutrition*, **46**, 161–183.
23 Hensley, K., Mou, S., Pye, Q.N., Dixon, R.A., Summner, L.W., and Floyd, R.A. (2004) *Current Topics in Nutraceutical Research*, **2**, 13–25.
24 Likhtenshtein, G.I., Papper, V., Pines, D., and Pines, E. (1997) Photochemical and photophysical characterization of 4,4′-substituted stilbenes: linear free energy, in *Recent Research Development in Photochemistry and Photobiology*, vol. 1 (ed. S.G. Pandalai), Transworld Research Network, Trivandrum, India, pp. 205–250.
25 Ferré-Filmon, K., Delaude, L., Demonceau, A., and Noels, A.F. (2004) *Coordination Chemistry Reviews*, **248**, 121–124.
26 Ketcham, R. and Martinelly, L. (1962) *The Journal of Organic Chemistry*, **27**, 466–472.
27 Siergrist, A.E. (1967) *Helvetica Chimica Acta*, **50**, 906–957.
28 Kretzcshmann, H. and Meier, H. (1991) *Tetrahedron Letters*, **32**, 5059–5063.
29 Wittig, G. and Schöllkopf, U. (1954) *Chemische Berichte*, **87**, 1318–1324.
30 Taber, D.F. and Nelson, C.G. (2006) *The Journal of Organic Chemistry*, **71**, 8973–8974.
31 Manecke, G. and Luttke, S. (1970) *Chemische Berichte*, **103**, 700–707.
32 Horner, L., Hoffmann, H.M.R., and Wippel, H.G. (1958) *Chemische Berichte*, **91**, 61–63.
33 Wadsworth, W.S. Jr. and Emmons, W.D. (1961) *Journal of the American Chemical Society*, **83**, 1733–1739.
34 Wadsworth, W.S. Jr. (1977) *Organic Reactions*, **25**, 73–253.
35 Kelly, S.E. (1991) *Comprehensive Organic Synthesis*, **1**, 729–817.
36 Ianni, A. and Waldvogel, S.R. (2006) *Synthesis*, 2103–2112.
37 Hilt, G. and Hengst, C. (2007) *The Journal of Organic Chemistry*, **72** (19), 7337–7342.
38 Simoni, D., Rossi, M., Rondanin, R., Mazzali, A., Baruchello, R., Malagutti, C., Roberti, M., and Invidiata, F.P. (2000) *Organic Letters*, **2**, 3765–3768.
39 Patois, C., Savignac, P., About-Jaudet, E., and Collignon, N. (1998) *Organic Syntheses*, **9**, 88–93.
40 Ando, K. (1997) *The Journal of Organic Chemistry*, **62**, 1934–1939.

41 Heck, R.F. and Nolley, J.P. Jr. (1972) *The Journal of Organic Chemistry*, **37**, 2320–2322.
42 Mizoroki, T., Mori, K., and Ozaki, A. (1971) *Bulletin of the Chemical Society of Japan*, **44**, 581–586.
43 Liu, S., Berry, N., Thomson, N., Pettman, A., Hyder, Z., Mo, J., and Xiao, J. (2006) *The Journal of Organic Chemistry*, **71**, 7467–7470.
44 Heck, R.F. (1982) *Organic Reactions*, **27**, 345–390.
45 de Meijere, G.A. and Meyer, F.E. Jr. (1994) *Angewandte Chemie – International Edition in English*, **33**, 2379–2411.
46 Belestskaya, I.P. and Cheprakov, A.V. (2000) *Chemical Reviews*, **100**, 3009–3066.
47 Choudary, B.M., Madhi, S., Kantam, M.L., Sreedhar, B., and Iwasawa, Y. (2004) *Journal of the American Chemical Society*, **126**, 2292–2293.
48 Ozawa, F., Kubo, A., and Hayashi, T. (1992) *Chemistry Letters*, 2177–2180.
49 Kiss, L., Kurtan, T., Antus, S., and Brunner, H. (2003) *ARKIVOC*, GB-653.
50 Hagiwara, H., Sugawara, Y., Hoshi, T., and Suzuki, T. (2005) *Chemical Communications*, (23), 2942–2944.
51 Mitsuru, K., Daisuke, K., and Koichi, N. (2005) *ARKIVOC*, JC-1563.
52 Casares, J.A., Espinet, P., Fuentes, B., and Salas, G. (2007) *Journal of the American Chemical Society*, **129**, 3508–3509.
53 Stille, J.K., Echavarren, A.M., Williams, R.M., and Hendrix, J.A. (1998) *Organic Syntheses*, **9**, 553–559.
54 Stille, J.K. (1986) *Angewandte Chemie – International Edition in English*, **25**, 508–524.
55 (a) Casado, A.L. and Espinet, P. (1998) *Organometallics*, **17**, 954–959; (b) Casado, A.L. and Espinet, P. (1998) *Journal of the American Chemical Society*, **120**, 8978–8985.
56 Gallagher, W.P. and Maleczka, R.E. Jr. (2005) *The Journal of Organic Chemistry*, **70**, 841–846.
57 Barton, D.H.R. and Willis, B.J. (1970) *Journal of the Chemical Society D*, 1225.
58 Kellogg, R.M. and Wassenaar, S. (1970) *Tetrahedron Letters*, **11**, 1987–1991.
59 ter Wiel, M.K.J., Vicario, J., Davey, S.G., Meetsma, A., and Feringa, B.L. (2005) *Organic & Biomolecular Chemistry*, 28–30.
60 Shimasaki, T., Kato, S.-I., and Shinmyozu, T. (2007) *Journal of Organic Chemistry*, **72**, 6251–6254.
61 McMurry, J.E. and Fleming, M.P. (1974) *Journal of the American Chemical Society*, **96**, 4708–4709.
62 Ephritikhine, M. (1998) *Chemical Communications*, 2549–2554.
63 Suzuki, A.J. (1999) *Organometallic Chemistry*, **576**, 147–168.
64 Rele, S.M., Nayak, S.K., and Chattopadhyay, S. (2008) *Tetrahedron*, **64**, 7225–7233.
65 Perkin, W.H. (1868) *Journal of the Chemical Society*, **21** (53), 181–186.
66 Rosen, T. (1991) *Comprehensive Organic Synthesis*, **2**, 395–408.
67 Sinha, A.K., Kumar, V., Sharma, A., Sharma, A., and Kumar, R. (2007) *Tetrahedron*, **63**, 11070–11077.
68 Lebel, H., Ladjel, C., and Brethous, L. (2007) *Journal of the American Chemical Society*, **129**, 13321–13326.
69 Saumitra, S. and Subir, K.S. (2004) *Organic Syntheses*, **10**, 263–268.
70 Zuev, V.V. (2006) *Russian Journal of General Chemistry*, **76**, 839–840.
71 Wang, J.-X., Wang, K., Zhao, L., Li, H., Fu, Y., and Hu, Y. (2006) *Advanced Synthesis & Catalysis*, **348**, 1262–1270.
72 Dong, C.-G., Yeung, P., and Hu, Q.-S. (2007) *Organic Letters*, **9**, 363–366.
73 Andrus, M.B. and Liu, J. (2006) *Tetrahedron Letters*, **47**, 5811–5814.
74 Sugiono, E. and Detert, H. (2006) *Silicon Chemistry*, **3**, 31–42.
75 Echavarren, A.M. and Porcel, S. (2006) *Science of Synthesis*, **28**, 507–560.
76 Bohm, V.P.W. and Herrmann, W.A. (2000) *Chemistry: A European Journal*, **6**, 1017–1025.
77 Shirakawa, E., Otsuka, H., and Hayashi, T. (2005) *Chemical Communications*, (47), 5885–5886.

78 Shimasaki, T., Kato, S.-I., Ideta, K., Goto, K., and Shinmyozu, T. (2007) *Journal of Organic Chemistry*, **72**, 1073–1087.

79 Tudose, A., Maj, A., Sauvage, X., Delaude, L., Demonceau, A., and Noels, A.F. (2006) *Journal of Molecular Catalysis A: Chemical*, **257**, 158–166.

80 Buchacher, P., Helgeson, R., and Wudl, F. (1998) *The Journal of Organic Chemistry*, **63**, 9698–9702.

81 McDonald, O., Lackey, K., Davis-Ward, R., Wood, E., Samano, V., Maloney, P., Deanda, F., and Hunter, R. (2006) *Bioorganic & Medicinal Chemistry Letters*, **16**, 5378–5383.

82 Prukala, W., Majchrzak, M., Pietraszuk, C., and Marciniec, B. (2006) *Journal of Molecular Catalysis A: Chemical*, **254**, 58–63.

83 Gester, S., Pietzsch, J., and Wuest, F.R. (2007) *Journal of Labelled Compounds and Radiopharmaceuticals*, **50**, 105–113.

84 Mariaca, R., Behrnd, N.-R., Eggli, P., Stoeckli-Evans, H., and Hulliger, J. (2006) *Crystal Engineering Communications*, **8**, 222–232.

85 Jin, F., Zhou, H.-P., Wang, X.-C., Hu, Z.-J., Wu, J.-Y., Tian, Y.-P., and Jiang, M.-H. (2007) *Polyhedron*, **26**, 1338–1346.

86 (a) Kott, L., Holzheuer, W.B., Wong, M.M., and Webster, G.K. (2007) *Journal of Pharmaceutical and Biomedical Analysis*, **43**, 57–65; (b) Ianni, A. and Waldvogel, S.R. (2006) *Synthesis*, 2103–2112.

87 Sun, X., Zhu, J., Zhong, C., Izumi, K.J., and Zhang, C. (2007) *Chinese Journal of Chemistry*, **25**, 1866–1871.

88 Lenardao, E.J., Cella, R., Jacob, R.G., da Silva, T.B., and Perin, G. (2006) *Journal of the Brazilian Chemical Society*, **17**, 1031–1038.

89 Tomasulo, M., Sortino, S., and Raymo, F.M. (2008) *Journal of Organic Chemistry*, **73**, 118–126.

90 Chidichimo, G., De Filpo, G., Salerno, G., Veltri, L., Gabriele, B., and Nicoletta, F.P. (2007) *Molecular Crystals and Liquid Crystals*, **465**, 165–174.

91 Breton, G.W., Oliver, L.H., and Nickerson, J.E. (2007) *Journal of Organic Chemistry*, **72**, 1412–1416.

92 Sokolov, M.N., Fyodorova, N.E., Paetow, R., Fenske, D., Ravelo, A.G., Naumov, D.Yu., and Fedorov, V.E. (2007) *Inorganica Chimica Acta*, **360**, 2192–2196.

93 Stuhr-Hansen, N., Christensen, J.B., Harrit, N., and Bjørnholm, T. (2003) *The Journal of Organic Chemistry*, **68**, 1275–1282.

94 Gao, M., Wang, M., Miller, K.D., Sledge, G.W., Hutchins, G.D., and Zheng, Q.-H. (2006) *Bioorganic & Medicinal Chemistry Letters*, **16**, 5767–5772.

95 Eisnor, C.R., Gossage, R.A., and Yadav, P. (2006) *Tetrahedron*, **62**, 3395–3401.

96 Neighbors, J.D., Salnikova, M.S., Beutler, J.A., and Wiemer, D.F. (2006) *Bioorganic & Medicinal Chemistry*, **14** (6), 1771–1784.

97 Oelgemöller, M., Brem, B., Frank, R., Schneider, S., Lenoir, D., Hertkorn, N., Origane, Y., Lemmen, P., Lex, J., and Inoue, Y. (2002) *Journal of the Chemical Society, Perkin Transactions 2*, 1760–1771.

98 Kabalka, G.W., Wu, Z., and Ju, Y. (2001) *Tetrahedron Letters*, **42** (29), 4759–4760.

99 Wang, Z.-X., Zhang, X.-J., Zhou, Y., and Zou, Y. (2005) *Journal of Chinese Pharmaceutical Sciences*, **14**, 204–208.

100 Pratap, R., Kumar, R., Maulik, P.R., and Ram, V.J. (2006) *Tetrahedron Letters*, **47**, 2949–2952.

101 Wyrzykiewicz, E., Wendzonka, M., and Kedzia, B. (2006) *European Journal of Medicinal Chemistry*, **41**, 519–525.

102 Kung, H.F., Kung, M.-P., and Zhuang, Z.-P. (2006) *PCT International Patent Application*, pp. 99

103 Li, Y.-Q., Li, Z.-L., Zhao, W.-J., Wen, R.-X., Meng, Q.-W., and Zeng, Y. (2006) *European Journal of Medicinal Chemistry*, **41**, 1084–1089.

104 Batema, G.D., van de Westelaken, K.T.L., Guerra, J., Lutz, M., Spek, A.L., van

Walree, C.A., de Mello Donega, C., Meijerink, A., van Klink, G.P.M., and van Koten, G. (2007) *European Journal of Inorganic Chemistry*, (10), 1422–1435.
105 Guang, S., Yin, S., Xu, H., Zhu, W., Gao, Y., and Song, Y. (2007) *Dyes and Pigments*, **73**, 285–291.
106 Murray, D.H., Fletcher, K.B., and Jiyani, R.W. (2007) Abstracts of Papers, 233rd ACS National Meeting, Chicago, IL, March 25–29, 2007.
107 Benhamou, L., Cesar, V., Lugan, N., and Lavigne, G. (2007) *Organometallics*, **26**, 4673–4676.
108 Cho, C.-H. and Park, K. (2007) *Bulletin of the Korean Chemical Society*, **28**, 1159–1166.
109 Nicolaou, K.C., Lister, T., Denton, R.M., and Gelin, C.F. (2007) *Angewandte Chemie – International Edition*, **46**, 7501–7505.
110 Sun, D., Hubig, S.M., and Kochi, J.K. (1999) *The Journal of Organic Chemistry*, **64**, 2250–2258.
111 Jaiyu, A., Rojanathanes, R., and Sukwattanasinitt, M. (2007) *Tetrahedron Letters*, **48**, 1817–1821.
112 Miljanic, S., Frkanec, L., Meic, Z., and Zinic, M. (2006) *European Journal of Organic Chemistry*, (5), 1323–1334.
113 Grad, M.E., Raditoiu, V., Wagner, L., Raditoiu, A., and Alfa, X. (2007) *Revista de Chimie*, **58**, 786–790.
114 Jin, L.-W., Kung, H.F., and Kung, M.-P. (2007) (USA). US Pat. Appl. Publ. pp. 60, cont. -in-part of US Ser. No. 218,587.
115 Du, L.-H. and Wang, Y.-G. (2007) *Synthetic Communications*, **37**, 217–222.
116 Kumar, V., Sharma, A., Sharma, A., and Sinha, A.K. (2007) *Tetrahedron*, **63**, 7640–7764.
117 Kureshy, R.I., Prathap, K.J., Singh, S., Agrawal, S., Khan, N.-U.H., Abdi, S.H.R., and Jasra, R.V. (2007) *Chirality*, **19**, 809–815.
118 Sinha, A.K., Kumar, V., and Sharma, A. (2007) PCT International Patent Application, WO2007110881(A1), pp. 37.
119 Cella, R. and Stefani, H.A. (2006) *Tetrahedron*, **62**, 5656–5661.
120 Hara, M., Samori, S., Cai, X., Tojo, S., Arai, T., Momotake, A., Hayakawa, J., Uda, M., Kawai, K., Endo, M., Fujitsuka, M., and Majima, T. (2004) *Journal of the American Chemical Society*, **126**, 14217–14223.
121 Lehmann, M., Koehn, C., Meier, H., Renker, S., and Oehlhof, A. (2006) *Journal of Materials Chemistry*, **16**, 441–451.
122 Alper, H., Arya, P., Bourque, C., Jefferson, G.R., and Manzer, L.E. (2000) *Canadian Journal of Chemistry*, **78**, 920–924.
123 Ogawa, M., Atsuya Momotake, A., and Arai, T. (2004) *Tetrahedron Letters*, **45**, 8515–8518.
124 Lijanova, I.V., Moggio, I., Arias, E., Vazquez-Garcia, R., and Martinez-Garcia, M. (2007) *Journal of Nanoscience and Nanotechnology*, **7**, 3607–3614.
125 Schulz, A. and Meier, H. (2007) *Tetrahedron*, **63**, 11429–11435.
126 Jiang, Y., Wang, J.-Y., Ma, Y., Cui, Y.-X., Zhou, Q.-F., Pei, J., and Key, L. (2006) *Organic Letters*, **8**, 4287–4290.
127 Watanabe, S., Ikegami, M., Nagahata, R., and Arai, T. (2007) *Bulletin of the Chemical Society of Japan*, **80**, 586–588.
128 Soomro, S.A., Schulz, A., and Meier, H. (2006) *Tetrahedron*, **62**, 8089–8094.
129 Lehmann, M., Fischbach, I., Spiess, H.W., and Meier, H. (2004) *Journal of the American Chemical Society*, **126**, 772–784.
130 Liu, J.-J. and Tian, H. (2007) *Yingyong Huaxue*, **24**, 863–867.
131 Klotz, E.J.F., Claridge, T.D.W., and Anderson, H.L. (2006) *Journal of the American Chemical Society*, **128**, 15374–15375.
132 Onagi, H., Carrozzini, B., Cascarano, G.L., Easton, C.J., Edwards, A.J., Lincoln, S.F., and Rae, A.D. (2003) *Chemistry: A European Journal*, **9**, 5971–5977.
133 Zhao, S.F., Zhou, R.X., and Zheng, X.M. (2006) *Indian Journal of Chemistry, Section A: Inorganic, Bio-inorganic, Physical,*

Theoretical & Analytical Chemistry, **45,** 2215–2217.
134 Song, D.H., Yoo, H.Y., and Kim, J.P. (2007) *Dyes and Pigments,* **75,** 727–731.
135 Rameshbabu, K. and Kannan, P. (2007) *Journal of Applied Polymer Science,* **104,** 2760–2768.
136 Kurata, T., Pu, Y.-J., and Nishide, H. (2007) *Polymer Journal,* **39,** 675–683.
137 Buruiana, E.C., Zamfir, M., and Buruiana, T. (2007) *European Polymer Journal,* **43** (10), 4316–4324.
138 Glusac, K.D. and Schanze, K.S. (2002) *Polymer Preprints (American Chemical Society, Division of Polymer Chemistry),* **43,** 87–88.
139 Jin, F., Zhou, H.-P., Wang, X.-C., Hu, Z.-J., Wu, J.-Y., Tian, Y.-P., and Jiang, M.-H. (2007) *Polyhedron,* **26,** 1338–1346.
140 Vaish, N., Hu, Y., Capaldo, K.P., Yeung, C.H., Garg, N., Montgomery, S.J., and Kannan, G. (2007) *US Patent Application Publication,* pp. 11, cont.-in-part of US Ser. No. 204, 277.
141 Kannan, G. and Montgomery, S.J. (2006) *US Patent Application Publication,* US7105627(B1), pp. 5.
142 Datar, Y. and Muralidharan, S. (2007) *Abstracts of Papers, 233rd ACS National Meeting, Chicago, IL, March 25–29, 2007,* COLL-420.
143 Wang, C. and Muralidharan, S. (2007) *Abstracts of Papers, 233rd ACS National Meeting, Chicago, IL, March 25–29, 2007,* COLL-426.
144 Tagaya, A., Ohkita, H., Harada, T., Ishibashi, K., and Koike, Y. (2006) *Macromolecules,* **39,** 3019–3023.
145 Morisaki, Y., Ishida, T., and Chujo, Y. (2006) *Organic Letters,* **8** (6), 1029–1032.
146 Ding, L. and Russell, T.P. (2006) *Macromolecule,* **39,** 6776–6780.
147 Martin, S., Cea, P., Pera, G., Haro, M., and Lopez, M.C. (2007) *Journal of Colloid and Interface Science,* **308,** 239–248.
148 Zhang, L., Zhu, H., Sajimon, M.C., Stutz, J.A.R., Siegmund, K., Richert, C., Shafirovich, V., and Lewis, F.D. (2006) *Journal of the Chinese Chemical Society,* **53,** 1501–1507.
149 Zhang, L., Long, H., Schatz, G.C., and Lewis, F.D. (2007) *Organic & Biomolecular Chemistry,* **5,** 450–456.
150 Parkhomyuk-Ben Arye, P., Strashnikova, N., and Likhtenshtein, G.I. (2002) *Journal of Biochemical and Biophysical Methods,* **51,** 1–15.
151 Wang, Q., Raja, K.S., Janda, K.D., Lin, T., and Finn, M.G. (2003) *Bioconjugate Chemistry,* **14,** 38–43.
152 Tian, F., Debler, E.W., Millar, D.P., Deniz, A.A., Wilson, I.A., and Schultz, P.G. (2006) *Angewandte Chemie – International Edition,* **45,** 7763–7765.
153 Wu, Y., Liu, S., Hou, D., Shen, J., Wang, H., and Shan, J. (2006) *Sepu,* **24,** 462–465.
154 Chen, H.-C. and Ding, W.-H. (2006) *Journal of Chromatography A,* **1108,** 202–207.
155 Rao, R.N., Venkateswarlu, N., Khalid, S., Narsimha, R., and Sridhar, S. (2006) *Journal of Chromatography A,* **1113,** 20–31.
156 Nicoletti, I., De Rossi, A., Giovinazzo, G., and Corradini, D. (2007) *Journal of Agricultural and Food Chemistry,* **55,** 3304–3311.
157 Grippi, F., Crosta, L., Aiello, G., Tolomeo, M., Oliveri, F., Gebbia, N., and Curione, A. (2007) *Food Chemistry,* **107,** 483–488.
158 Chen, X. and Luo, Y. (2007) *Zhongyao Xinyao Yu Linchuang Yaoli,* **18,** 482–484.
159 Mirza, M.A., Kandhro, A.J., Memon, S.Q., Khuhawar, M.Y., and Arain, R. (2007) *Electrophoresis,* **28,** 3940–3947.
160 Yi, T., Zhang, H., and Cai, Z. (2007) *Phytochemical Analysis,* **18,** 387–392.
161 Shao, B., Guo, H.-Z., Cui, Y.-J., Liu, A.-H., Yu, H.-L., Guo, H., Xu, M., and Guo, D.-N. (2007) *Journal of Pharmaceutical and Biomedical Analysis,* **44,** 737–742.
162 Xu, H., Gu, L., He, J., Lin, A., and Tang, D. (2007) *Journal of Chromatography B,* **852,** 529–533.
163 Ye, M., Han, J., Chen, H., Zheng, J., and Guo, D. (2007) *Journal of the American Society for Mass Spectrometry,* **18,** 82–91.

164 Sobolev, V.S., Potter, T.L., and Horn, B.W. (2006) *Phytochemical Analysis*, **17**, 312–322.
165 Puessa, T., Floren, J., Kuldkepp, P., and Raal, A. (2006) *Journal of Agricultural and Food Chemistry*, **54**, 7488–7494.
166 Shu, N., Zhou, H., Huang, H., and Hu, C. (2006) *Chemical & Pharmaceutical Bulletin*, **54**, 878–881.
167 Shu, N., Zhou S H., and Hu, C. (2006) *Biological & Pharmaceutical Bulletin*, **29**, 608–612.
168 Jean-Denis, J.B., Pezet, R., and Tabacchi, R. (2006) *Journal of Chromatography A*, **111**, 263–268.
169 Saigne-Soulard, C., Richard, T., Merillon, J.-M., and Monti, J.-P. (2006) *Analytica Chimica Acta*, **563**, 137–144.
170 Wang, Y., Catana, F., Yang, Y., Roderick, R., and van Breemen, B. (2002) *Journal of Agricultural and Food Chemistry*, **50**, 431–435.
171 Reddy, K.H. (2006) *Journal of the Indian Chemical Society*, **83**, 1031–1033.
172 Kapche, G.D., Waffo-Teguo, P., Massip, S., Guillon, J., Vitrac, C., Krisa, S., Ngadjui, B., and Merillon, J.-M. (2007) *Analytical Sciences: X-Ray Structure Analysis Online*, **23**, 59–60.
173 Xu, C.L., Peng, C.F., Liu, L., Wang, L., Jin, Z.Y., and Chu, X.G. (2006) *Journal of Pharmaceutical and Biomedical Analysis*, **41**, 1029–1036.
174 Bedoya, L.M., Del Olmo, E., Sancho, R., Barboza, B., Beltran, M., Garcia-Cadenas, A.E., Sanchez-Palomino, S., Lopez-Perez, J.L., Munoz, E., San Feliciano, A., and Alcami, J. (2006) *Bioorganic & Medicinal Chemistry Letters*, **16**, 4075–4079.
175 Yu, X.-H., Zhang, Z.-Y., Ma, Q., and Fan, G.-Q. (2007) *Guangpuxue Yu Guangpu Fenxi*, **27** (3), 481–485.
176 Ni, W.-D., Man, R.-L., Li, Z.-M., and Lu, H.-M. (2006) *Huaxue Gongchengshi*, **20**, 35–39.
177 Poutaraud, A., Latouche, G., Martins, S., Meyer, S., Merdinoglu, D., and Cerovic, Z.G. (2007) *Journal of Agricultural and Food Chemistry*, **55**, 4913–4920.
178 Pelletier, M.J., Fabilli, M.L., and Moon, B. (2007) *Applied Spectroscopy*, **61**, 1107–1115.
179 Yang, X.-Z., Tang, C.-P., Ke, C.-Q., and Ye, Y. (2007) *Journal of Asian Natural Products Research*, **9**, 261–266.
180 Gu, X., Creasy, L., Kester, A., and Zeece, M. (1999) *Journal of Agricultural and Food Chemistry*, **47**, 3223–3227.
181 Liu, J.-X., Wu, Y.-J., Wang, F., Gao, L., and Ye, B.-X. (2008) *Journal of the Chinese Chemical Society*, **55**, 264–270.
182 Granero, A.M., Fernandez, H., Agostini, E., and Zon, M.A. (2008) *Electroanalysis*, **20**, 858–864.
183 Msagati, T.A.M. and Nindi, M.M. (2006) *Annali di Chimica*, **96**, 635–646.
184 Hagihara, S., Tanaka, H., and Matile, S. (2008) *Organic & Biomolecular Chemistry*, **6**, 2259–2262.
185 Kott, L., Holzheuer, W.B., Wong, M.M., and Webster, G.K. (2007) *Journal of Pharmaceutical and Biomedical Analysis*, **43**, 57–65.

2
Stilbene Chemical Reactions

Phenyl segments of stilbene (1,2-diphenylethylene) are chemically unreactive, and only carbon–carbon double bond may be easily involved in several reactions. To activate stilbenes, it is generally necessary to introduce more reactive functional groups (Chapter 1).

2.1
Halogenation of Stilbenes

An example of halogenation is the bromination of (E)-stilbene (Figure 2.1).

Since elemental bromine (Br_2) is volatile and highly corrosive, special systems, for example, pyridinium tribromide and related compound, are commonly used to generate Br_2 *in situ* [2]. A method for stereoselective bromination of stilbene and chalcone in a water suspension medium using pyridinium tribromide has been reported [2]. Bromination reactions of (E)-stilbene and (E)-chalcone in a water suspension medium by salts

$$\text{Py}\overset{+}{N}HBr^- + Br_2 \quad \text{Ph}-\overset{+}{N}Me_3Br^- + Br_2 \quad (n\text{-Bu})_4\overset{+}{N}Br^- + Br_2$$

proceeded efficiently and stereoselectively, and the reaction products were easily collected by filtration.

For the generation of Br_2, hydrobromic acid, which is oxidized by hydrogen peroxide, may be used:

$$2HBr + H_2O_2 \rightarrow Br_2 + 2H_2O$$

In this reaction, both (E)-stilbene and (Z)-stilbene produce 1,2-dibromo-1,2-diphenylethane. However, bromination of the (Z) isomer results in a racemic mixture of DL-stilbene dibromide, while the bromination of an (E) isomer results in a majority of *meso*-stilbene dibromide along with small amount of DL-enantiomers.

A number of methods of bromine production *in situ* have been reported. A regioselective and stereoselective methoxy-bromination of olefins has been accomplished using LiBr and (diacetoxyiodo)benzene as oxidants [3]. Zinc bromide

Stilbenes. Applications in Chemistry, Life Sciences and Materials Science. Gertz Likhtenshtein
Copyright © 2010 WILEY-VCH Verlag GmbH & Co. KGaA, Weinheim
ISBN: 978-3-527-32388-3

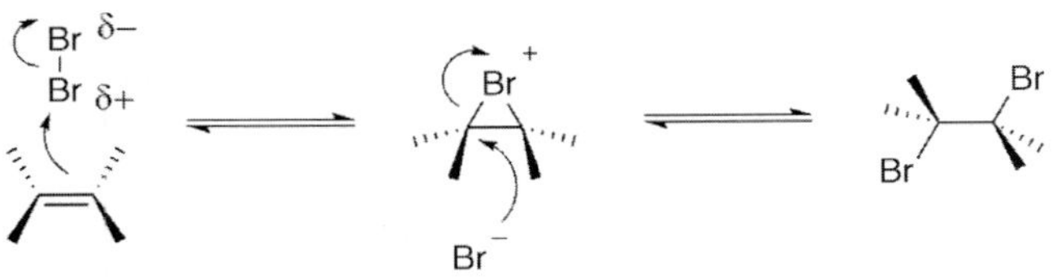

Figure 2.1 Bromination of (*E*)-stilbene [1].

and lead tetraacetate were used to brominate in chloroform a variety of alkenes including stilbenes to corresponding vicinal dibromoalkanes [4]. Olefins were subjected to oxidative bromination using Selectfluor/KBr [5]. Conversion of stilbene into trisubstituted olefins has been investigated in Ref. [6]. It was shown that the bromination–dehydrobromination of (*E*)-stilbene gave 64% (*E*)-BrCPh:CHPh (I), while the reaction of (*Z*)-stilbene gave 85% (*Z*)-BrCPh:CHPh (II). I and II were heated with arylzinc chlorides to give the corresponding 1-aryl-1,2-diphenylethene derivatives. Effects of media, reaction conditions, and mechanism of bromination have been a matter of interest to researchers. The possibility that intermediate molecular complexes were involved in the liquid-phase halogenation of unsaturated compounds including stilbenes was examined [7]. Application of Hammett equation in bromine addition to *para*-substituted stilbenes has been reported [8]. It was shown [9] that in the electrophilic addition of Br to *trans*-stilbene (*t*-St) in nonpolar solvents, stereoselectivity decreases in cyclodextrin (CD) cavities. In this reaction, a significant yield of DL-stilbene dibromide was obtained. The authors attributed the reversal of stereoselectivity to the polar environment provided by the secondary hydroxyl groups of cyclodextrins that stabilizes the acyclic α-halocarbonium ion intermediate and also hinders the approach of the incoming tribromide ion.

Ratios of *meso*- to DL-1,2-dibromo-1,2-diphenylethane obtained in the bromination of *cis*- and *trans*-stilbenes in 1,2-dichloroethane have been measured as a function of the reagent concentration [10]. With stoichiometric reagents, reactions were stereospecific and stereoconvergent to give the *meso*-dibromide at [Br$_2$] > 3×10^{-3} M. Kinetic measurements have shown that under all conditions the reaction was occurring through the same rate-determining step. Influence of the bromide–tribromide–pentabromide equilibrium on the counteranion of ionic intermediates in the reaction of electrophilic bromine addition to stilbenes in chloroform has been demonstrated [11]. Two equilibria were found in chloroform solutions of Bu$_4$N$^+$Br$^-$ and Br$_2$, leading to tribromide and pentabromide salts. The formation constants of both intermediates ($K_1 = 2.77 \times 10^4$ M^{-1} and $K_2 = 3.51 \times 10^5$ M^{-2} at 25 °C) were obtained. The stability of the Br$_3^-$ species in chloroform was at least three orders of magnitude lower than that in 1,2-dichloroethane. Formation of DL-1,2-dibromo-1,2-diphenylethane and *cis–trans* isomerization of the unreacted olefin were found to depend on the reagent concentration and nature of counteranions, either Br$_3^-$ or Br$^-$. The relative reversibility of bromonium and β-bromocarbonium ion formation and the product distribution of the reaction of Br$_2$ with *cis*- and *trans*-stilbenes and with their *p*-Me (I), *p*-(trifluoromethyl), and *p*,*p*′-bis(trifluoromethyl) derivatives have been investigated [12]. In the 10^{-1}–10^{-4} M concentration range, the rate constants for the bromination of these stilbenes at 25 °C spanned seven orders of magnitude. The observed dibromide ratios showed that an open β-bromocarbonium ion was the intermediate in the bromination of cis- and trans-stilbenes. It was also shown that the extent of bridging increased, and equilibration to the more stable trans-bromonium ion occurred at the lowest reagent concentration to give the erythro- (or meso-) dibromide as the main product. The formation of the β-bromocarbonium ion intermediate was a rate-determining stage.

A mechanism of stilbene bromination has been studied [13]. On the basis of kinetic data on bromination of disubstituted stilbenes, $XC_6H_4C_xH:C_yHC_6H_4Y$ ($X = OH$, OMe, Me, Cl, Br, H, $CHMe_2$; $Y = OH$, OMe, Me, H, Cl, NO_2, $CHMe_2$, Br, CF_3), in MeOH, a dual-path addition mechanism was suggested. According to the mechanism, two pathways leading to discrete carbonium ions are involved. Substituent effects were found to be only approximately additive. The linear free energy relationship analysis showed that in CCl_4 the structures of the transition states are symmetrical, whereas in MeOH, the transition states are carbonium ion-like. A catalytic oxidative bromination of arenes, alkenes, and alkynes in aqueous media was achieved by using NH_4VO_3 catalyst combined with H_2O_2, HBr, and KBr [14]. Dodecyltrimethylammonium bromide was found to serve as an efficient surfactant to facilitate the NH_4VO_3-catalyzed bromination in aqueous media. A scheme of the oxidative bromination of stilbenes in water in the presence of an additive is shown in Figure 2.2.

A comparative analysis of three bromination laboratory procedures was done [15]. These processes of alkene bromination were introduced to students through the use of a safe, effective, and modern procedure. Several general reactions of alkene and aryl alkenes bromination have been reported. The reaction of alkenes and aryl alkenes with Br_2 in damp MeCN occurred with solvent incorporation to give corresponding alkanes, loxazolines, alkylamine hydrobromides, and alcohols [16]. The bromination of styrene, using tetrabutylammonium tribromide, can be effected under mild conditions with ultrasonic irradiation to give the corresponding vicinal dibromide [17]. In the work [18], electrophilic brominations of organic compounds by employing tantalum(V)-catalyzed oxidation of bromonium ion (Br^+) by hydrogen peroxide were accomplished.

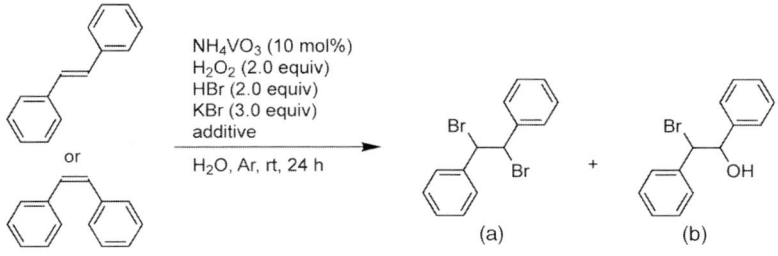

Figure 2.2 Oxidative bromination of stilbenes in water in the presence of an additive [14]. (Reproduced with permission from Elsevier.)

2.2
Oxidation of Stilbenes

2.2.1
Epoxidation

Epoxidation of stilbenes and other alkenes occupies a great deal of attention. Asymmetric epoxidation is an industrially important method for synthesizing epoxides from readily available olefins. In particular, the use of coordination complexes of transition metals as catalysts is of abiding importance, as it proffers an effective possibility for the synthesis of enantiomerically pure compounds ([19, 20] references therein). The manganese(III) complex with a diamide ligand was found to catalyze both the epoxidation of (Z)- and (E)-stilbenes with high conversion and the oxidation of benzyl alcohol to benzaldehyde (Figure 2.3) [21].

It was found that immobilized $CoCl_2$ within nanoreactors of Si-MCM-48 catalyzed the oxidation of *trans*-stilbene and series of olefins to corresponding epoxides with 35–95% conversion and 75–100% selectivity [22]. Selective epoxidation of olefins including catalysis by silver-doped manganese oxide catalysts (Ag-MnO$_x$) has been reported [23]. *tert*-Bu hydroperoxide (TBHP) was found to give the highest conversion and selectivity in the catalysis for the epoxidation of *trans*-stilbene and *tert*-Bu hydroperoxide. Epoxidation of *cis*- or *trans*-stilbenes with molecular oxygen and isobutyraldehyde as coreductant catalyzed by immobilized vitamin B12 within Al-MCM-41 has been carried out with good to fair selectivity [24]. Homogeneous salen Mn(III) complex was found to be a catalyst for the epoxidation of *trans*-stilbene and other alkenes with oxygen/sacrificial-isobutanal, $PhI(OAc)_2$, or H_2O_2 as oxidant [25]. Reactions of *cis*- and *trans*-stilbenes, with stereospecific epoxidation in acetonitrile catalyzed by $Fe(ClO_4)_3/H_2O_2/MeCN$ system, have been investigated [26]. The reaction was found to occur as a nonradical process and a nonstereospecific reaction. A six-coordinated Mn(II) species of a macrocyclic ligand N,N'-cyclohexanebis(3-formyl-5-methylsalicylaldimine) having two H_2O molecules in axial positions was used for the epoxidation of *E*-stilbene [27]. PhO and NaOCl as terminal oxidants and MeCN and CH_2Cl_2 as solvents were used. It was shown that the gold catalyst Au/TiO_2 exhibited high activity in the stereoselective epoxidation of *trans*-stilbene in methylcyclohexane (MCH) in the presence of 5 mol% *tert*-Bu hydroperoxide [28]. The reaction occurred via a mechanism of chain reaction involving the activation of molecular oxygen by a radical produced from methylcyclohexane (Figure 2.4).

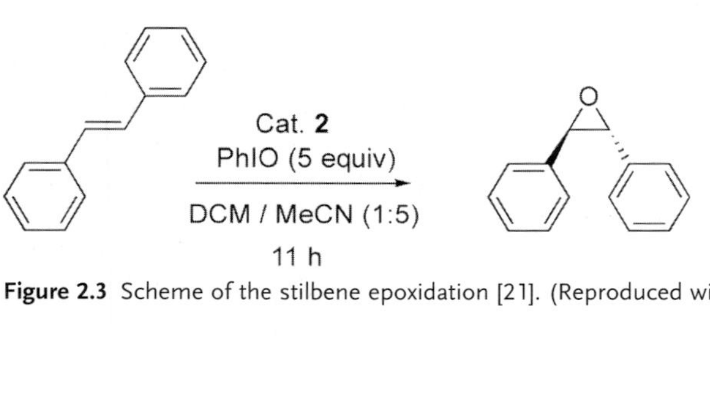

Figure 2.3 Scheme of the stilbene epoxidation [21]. (Reproduced with permission.)

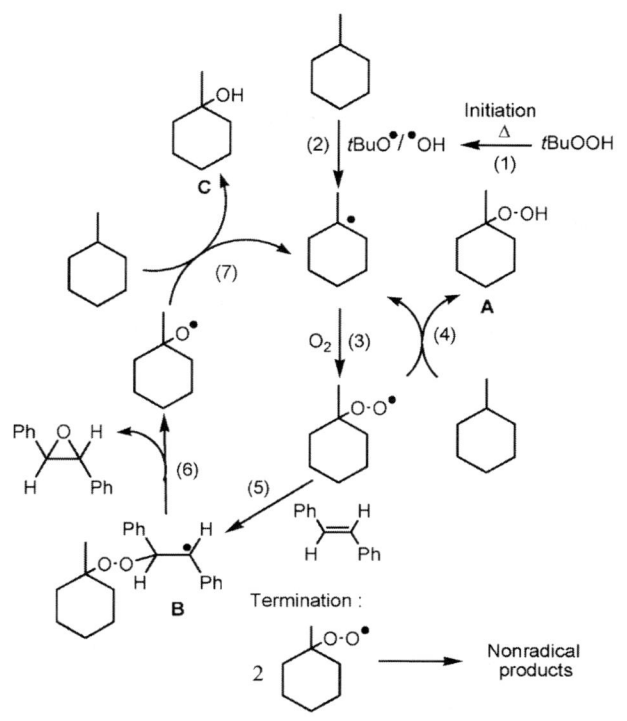

Figure 2.4 Scheme of the stilbene oxidation by dioxygen [28]. (Reproduced with permission.)

A free radical mechanism has been proposed in the stereoselective epoxidation of trans-stilbene by tert-Bu hydroperoxide in methylcyclohexane in the presence of gold catalysts [29]. trans-Stilbene oxide was the major reaction product observed, with selectivities up to 88% when using Au/TiO$_2$. The selectivity decreased when Au/C was used instead of oxide-supported gold catalysts or H$_2$O$_2$ instead of TBHP. The authors suggested that TBHP is the radical source while MCH propagates the active radical. Tetra-n-butylammonium salts of transition metal-substituted heteropolytungstates, PW$_{11}$MO$_{39}$$^{n-}$, catalyzed the epoxidation of stilbenes by t-butyl hydroperoxide (M = Co^{II}, Mn^{II}, Cu^{II}, Ti^{IV}, Ru^{IV}, V^{V}, Nb^{V}) and hydrogen peroxide (M = Zr^{IV}) in acetonitrile [30]. It was shown that the epoxidation of cis-stilbene is nonstereospecific. In the work [31], the synthesis of a family of new Ru complexes containing meridional or facial tridentate ligands [Ru^{II}(T)(D)(X)]$^{n+}$ (T = 2,2':6',2''-terpyridine or tripyrazolylmethane; D = 4,4'-dibenzyl-4,4',5,5'-tetrahydro-2,2'-bioxazole (S,S-box-C) or 2-[((1'S)-1'-(hydroxymethyl)-2'-phenyl)ethylcarboxamide]-(4S)-4-benzyl-4,5-dihydrooxazole (S, S-box-O); X = Cl, H$_2$O, MeCN, or pyridine) was described. The reactivity of the Ru (OH)$_2$ complexes was tested with regard to the epoxidation of trans-stilbene. The reactivity did not depend on the redox potentials of the catalyst, but it strongly depended on the geometry of the tridentate ligands.

[Cr(III)(α-TDL$_1^*$)(bipy)(Cl)] (I) and [Cr(III)(TDL$_2^*$)(bipy)(Cl)] (II) complexes (H$_2$TDL$_1^*$ = N-3, 5-di-(tert-butyl)salicylideneglucosamine, H$_2$TDL$_2^*$ = N-35-di-(tert-,

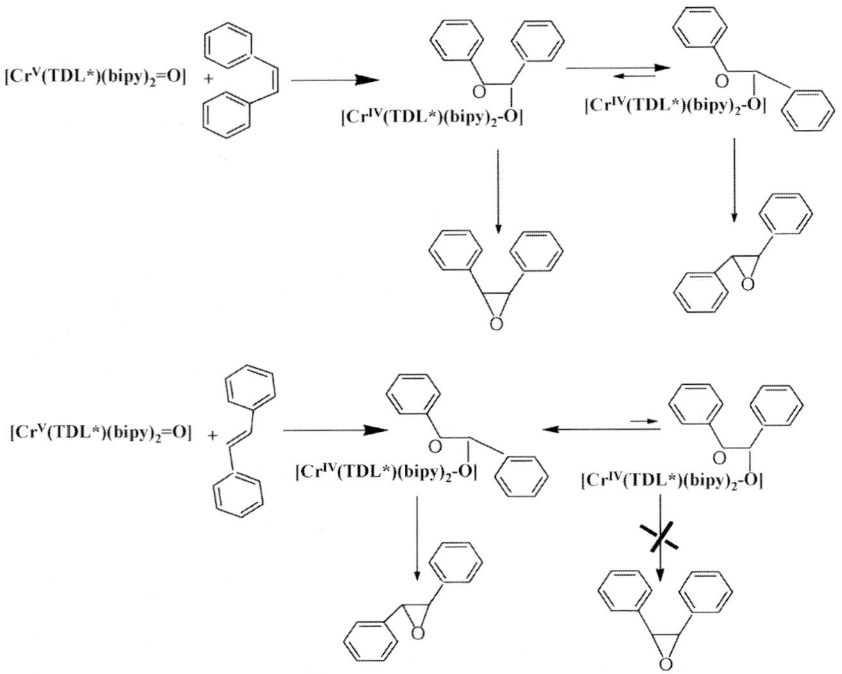

Figure 2.5 Proposed mechanism of epoxidation of E- and Z-stilbenes [32].

butyl)sali-cylidenealanine, bipy = 2,2′-bipyridyl) have been synthesized and characterized by physical methods [32]. It was shown that the complexes I and II catalyzed the epoxidation of stilbenes using aqueous Me_3COOH as terminal oxidant. Alkenes were converted to their epoxides exhibiting moderate enantioselectivity at ambient temperature. The proposed mechanism of epoxidation of E- and Z-stilbenes is shown in Figure 2.5.

Several general methods of olefin epoxidation have been developed in Ref. [33]. $FeCl_3 \cdot 6H_2O$ in combination with pyridine-2,6-dicarboxylic acid and different amines showed high reactivity and selectivity toward the epoxidation of aromatic olefins and moderate reactivity toward that of aliphatic olefins. Asymmetric epoxidation of olefins by manganese(III) complexes stabilized on nanocrystalline magnesium oxide has been accomplished [34]. In the presence of (1R,2R)-(−)-diaminocyclohexane as a chiral ligand, the reaction led to good yields and up to 91% enantiomeric excess.

2.2.2
Other Oxidation Reactions

Early data on the mechanism of the reaction of ozone with double bonds of organic compounds have been reviewed [35]. On the basis of kinetics analysis, it was suggested that this reaction does not occur by a synchronous addition mechanism but through a reversible stage involving the formation of an intermediate complex of ozone with the double bond. It was suggested that the subsequent stages of the reaction involve the formation of a primary ozonide, its decomposition into two fragments, their combination, and a number of other transformations. In a gas–solid organic state reaction, the ozonolysis of crystalline (*E*)-stilbene gave PhCHO·BzOH, diethylstilbestrol (DES) gave *p*-HOC$_6$H$_4$COEt, and Ph$_2$C:CPh$_2$ gave PhCO [36]. Online analysis of a continuous-flow ozonolysis of *trans*-stilbene using Raman spectroscopy was reported [37]. Raman analyzer quantitatively tracked the levels of *trans*-stilbene, benzaldehyde, and α-methoxybenzyl hydroperoxide in a continuous-flow ozonolysis reactor.

Regioselective oxidative coupling of 4-hydroxystilbenes has been investigated [38]. It was shown that treating 5-[2-(4-hydroxyphenyl)vinyl]benzene-1,3-diol (resveratrol) with an equimolar amount of silver(I) acetate in dry MeOH at 50 °C for 1 h, followed by chromatographic purification with a short silica gel column, allowed the isolation of its (*E*)-dehydrodimer, 5-[5-[2-(3,5-dihydroxyphenyl)vinyl]-2-(4-hydroxyphenyl)-2,3-dihydrobenzofuran-3-yl]benzene-1,3-diol, as a racemic mixture in high yield. The present method was applicable to the oxidative dimerization of 4-hydroxystilbenes such as *trans*-styrylphenol and 5-[6-hydroxy-2-(4-hydroxyphenyl)-4-[2-(4-hydroxyphenyl)-vinyl]-2,3-dihydrobenzofuran-3-yl]benzene-1,3-diol (epsilon-viniferin) leading to the corresponding 2-(4-hydroxyphenyl)-2,3-dihydrobenzofurans possessing various types of biological activities.

It was reported [39] that osmium tetroxide promoted catalytic oxidative cleavage of *cis*-stilbene and other olefins. The *cis*-stilbene catalytic oxidative cleavage gave benzoic acid in 95% yield. The process for the preparation of substituted aromatic and heteroaromatic aldehydes and carboxylic acids by the oxidation of substituted stilbenes has been patented [40]. Stilbenes of various substituents (R$_1$, R$_2$ = H, C1–4 alkyl, C1–4 alkoxy, OH, NO$_2$, CN, CO$_2$H, CONH$_2$, SO$_3$H, halogen; X = C, N; Z = CHO, CO$_2$H; R$_1 \neq$ R$_2$ = H) (e.g., disodium 4,4′-dinitrostilbene-2,2′-disulfonate) were prepared in high yield and selectivity by the oxidation of the corresponding stilbene derivatives. A high oxidation potential triphenylamine electrocatalyst was used for the electrocatalytic anodic oxidative cleavage of symmetrical and unsymmetrical stilbenes bearing ≥2 strong electron-withdrawing groups [41]. It was suggested that the oxidation in acetonitrile–water solution involves the corresponding 1,2-diols, which are converted to aldehydes in high yield. The oxidation of olefins using binaphthyl-ruthenium(III) complexes and iodobenzene diacetate was found to lead to the formation of α-ketoacetates that were formed together with corresponding epoxides [42]. The Ti(I)/Ti(II) mixed-valence toluene complex {2,5-[(C$_4$H$_3$N)CPh$_2$]$_2$[C$_4$H$_2$N(Me)]}Ti(μ,η^6-C$_7$H$_8$)Ti[{2,5-[(C$_4$H$_3$)CPh$_2$]$_2$[C$_4$HN(Me)]}][K(DME)$_2$]·toluene (**1**) with an inverse sandwich-type structure was used for oxidative additions to *trans*-stilbene [43]. The olefin adducts {2,5-[(C$_4$H$_3$N)CPh$_2$]$_2$[C$_4$H$_2$N]}Ti(η^2-*trans*-PhHC:CHPh)[K(DME)] and {2,5-[(C$_4$H$_3$N)CPh$_2$]$_2$[C$_4$H$_2$N(Me)]}Ti(η^2-*trans*-PhHC:CHPh) were obtained from reactions carried out in the presence of *trans*-stilbene. It was found that photo-oxygenation under visible light irradiation of stilbene derivatives catalyzed by 9-mesityl-10-methylacridinium ion (Acr$^+$-Mes) was accompanied by an efficient *cis–trans* isomerization [44]. The reaction afforded corresponding benzaldehydes via electron transfer from Acr$^+$-Mes to stilbene derivatives and oxygen.

Oxidation of 4-hydroxystilbenes in methanol using a hypervalent iodine-based oxidant led to the formal 1,2-addition of methoxy groups across the central stilbene

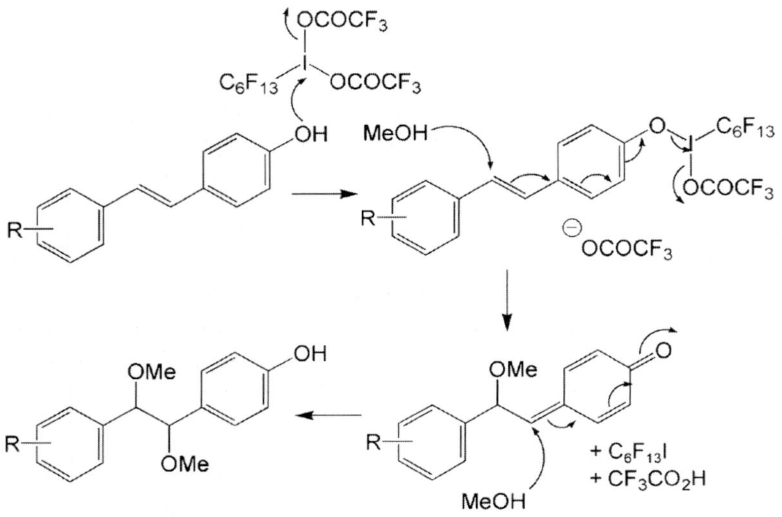

Figure 2.6 Oxidation of 4-hydroxystilbenes in methanol using a hypervalent iodine-based oxidant leading to the 1,2-addition of methoxy groups [45]. (Reproduced with permission.)

double bond [45]. Treating the structurally related 4-hydroxyisoflavone with di(trifluoroacetoxy)iodobenzene led to the formation of 2,4′-dihydroxybenzyl (Figure 2.6). Oxidation of the corresponding 3-hydroxystilbenes and 3-hydroxyisoflavone gave conventional dienone oxidation products.

Catalytic activities of 1,10-phenanthroline (Phen), oxalic acid (Oxa), and picolinic acid (PA) in the chromium(VI) oxidation of trans-stilbene to diketone have been investigated in acidic solution [46]. The authors suggested that the Cr(VI)–Phen/Oxa/PA complex is a reactive electrophile and that the oxidation of trans-stilbene involves the nucleophilic attack of the ethylenic bond on Cr(VI)–Phen/Oxa/PA complex with the formation of a ternary complex. Catalytic asymmetric dihydroxylation of olefins using polysulfone-based microencapsulated osmium tetroxide has been reported [47]. The catalyst was employed in the dihydroxylation of various olefins, using $(DHQD)_2PHAL$ as the chiral ligand and N-methyl-morpholine-N-oxide as the co-oxidant in H_2O–acetone–CH_3CN (1: 1: 1). This catalyst was recovered by simple filtration and was reused to obtain products with good enantioselectivity up to five times. Oxovanadium(IV) complex of β-alanine-derived ligand immobilized on polystyrene was found to be useful for the oxidation of trans-stilbene to various organic substrates in the presence of 30% H_2O_2 as an oxidant [48]. The electrochemical and chemical (with $NOBF_4$) oxidation of 3,4′-bis[bis(p-t-butylphenyl)amino]stilbene afforded the formation of the bis(cation radical) [49].

Magnetization, magnetic susceptibility, and the ESR $\Delta M_s = \pm 2$ signal of the biradicals indicated triplet ground states with a large triplet–singlet energy gap.

Results of theoretical calculation for model processes of epoxidation and hydroxylation of alkenes using the density function theory have been described [50].

2.3 Stilbene Reduction

The α,α-disubstituted stilbenes were reduced with lithium, sodium, or potassium in anhydrous liquid ammonia [51]. The reaction was carried out by two procedures: (a) the alkali metal was added to a solution of the stilbene in ammonia/tetrahydrofuran and (b) a solution of the stilbene in tetrahydrofuran was added to a preformed blue solution of the alkali metal in ammonia. The ratio of bibenzyl diastereoisomers formed depended on the alkali metal and the reduction procedure used. The results were exploited in the synthesis of 2-^{14}C-labeled *meso*-hexoestrol. In the work [52], it was shown that stilbene formed stilbene–chromium tricarbonyl complexes that underwent heterogeneous hydrogenation by H_2 on skeletal Ni and Pd on carbon more slowly than free stilbene. For the homogeneous hydrogenation of these complexes using a H_2PtCl_6–$SnCl_2$–LiBr system, styrene and η^6-styrene chromium tricarbonyl were reduced with a high rate, whereas stilbene and its chromium tricarbonyl complex were hydrogenated very slowly. A Pd/P(*t*-Bu)$_3$ catalyst for selective reduction of various alkenes under transfer hydrogen conditions has been developed leading to the corresponding saturated derivatives [53, 54]. Recently, the SmI_2/H_2O/amine and YbI_2/H_2O/amine systems for reduction of unsaturated hydrocarbons have been developed [55].

A high selectivity and activity for Z-stilbene was observed in the titanium nitride-catalyzed reduction in solution with hydride NaAlH$_4$ as a reducing agent (Figure 2.7) [56]. Filtration tests, recycling of the catalysts, and NMR data indicated that the reaction mechanism is heterogeneous, with hydroalumination intermediates formed in the course of the reduction.

Supported liquid-phase (water–ethylene glycol) catalyst containing Pd complexes was used for hydrogenation of stilbene in toluene [57]. It was found that zeolite CaY acted, in the presence of Bronsted acid sites, either as a reagent for reducing stilbenes to 1,2-diarylethanes or as a catalyst for isomerizing *cis*-stilbenes to the more stable *trans* form [58]. In contrast, the Lewis acid sites generated by the activation process yield radical cations from stilbenes, but these did not yield any stable products.

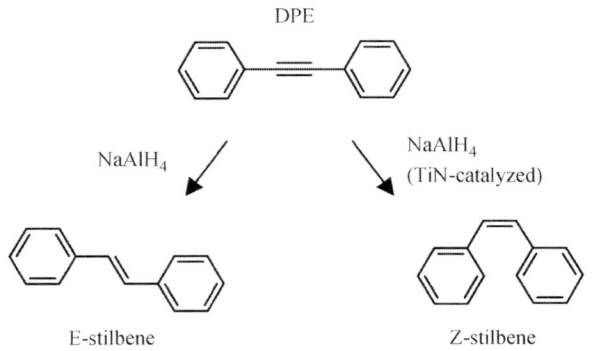

Figure 2.7 Titanium nitride-catalyzed reduction in solution with hydride NaAlH$_4$ [56]. (Reproduced with permission from Elsevier.)

2.4
Other Reactions

2.4.1
Vinyl Lithiation

A direct vinyl lithiation of *cis*-stilbene and a directed vinyl lithiation of an unsymmetrical *cis*-stilbene have been reported [59]. The reactions run by the scheme presented in Figure 2.8.

It was shown that vinyl deprotonation of unsymmetrical *cis*-stilbene 2-styryl-phenyl-carbamic acid *tert*-butyl ester can be achieved using *s*-BuLi in THF at −25 °C. The generated 1-lithio-1,2-diphenylethene undergoes an *in situ* Z-to-E isomerization, and subsequent reaction with electrophiles results in stereoselective synthesis of trisubstituted alkenes.

2.4.2
Carbolization

A regioselective carbolithiation of *o*-amino-(*E*)-stilbenes has been achieved with a series of alkyllithiums when THF was employed as reaction solvent (Figure 2.9) [60]. High levels of diastereoselectivity have been obtained following the reaction of the lithiated intermediate in THF with different electrophiles such as MeOD, CO_2, and Bu_3SnCl. It was shown that diastereoselectivity was influenced by the *ortho*-amino substituent and the alkyllithium used for carbolithiation.

Substituted 3,4-dihydro-1*H*-quinolin-2-ones, 1,2,3,4-tetrahydroquinolines, 1,4-dihydroquinolines, and quinolines were synthesized by the regioselective

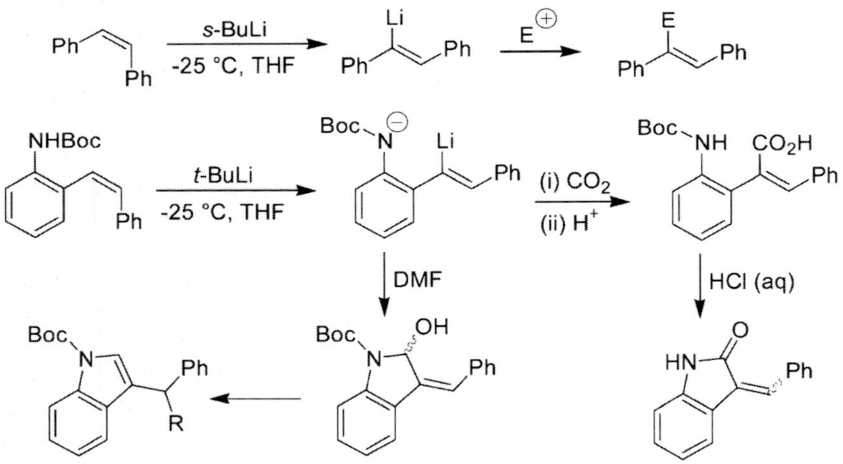

Figure 2.8 Direct vinyl lithiation of *cis*-stilbene and a directed vinyl lithiation [59]. Direct vinyl lithiation of *cis*-stilbene. (Reproduced with permission.)

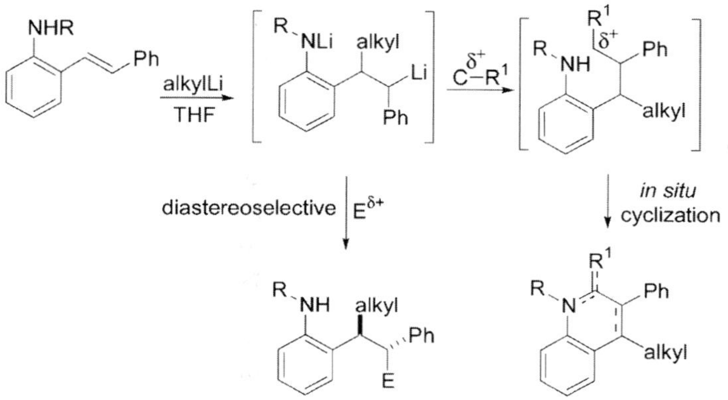

Figure 2.9 A regioselective carbolithiation of o-amino-(E)-stilbenes [60]. (Reproduced with permission.)

carbolithiation of substituted *ortho*-amino-(*E*)-stilbenes [61]. For example, the reaction of (*E*)-2-BnNHC$_6$H$_4$CH:CHPh with *t*-BuLi, followed by addition of DMF, led to 1,4-dihydroquinoline, while the reaction of (*E*)-2-BnNHC$_6$H$_4$CH:CHPh with *t*-BuLi gave 2-BnNHC$_6$H$_4$CH(CMe$_3$)CH$_2$Ph regioselectively.

2.4.3
Addition Reactions

The reversible addition of thiolate ions RS$^-$ (R = Et, HOCH$_2$CH$_2$, MeO$_2$CCH$_2$CH$_2$, MeO$_2$CCH$_2$) to α-nitrostilbene and of HOCH$_2$CH– to PhCH:C(NO)C$_6$H$_4$R$_1$ (R$_1$ = 4-Me, H, 4-Br, 3-NO$_2$, 4-NO$_2$) has been studied [62]. On the basis of the process kinetics measured in 50% aqueous Me$_2$SO, the Bronsted β values (β$_{nuc}$, β$_{lg}$, β$_{eq}$), Hammett ρ values [ρ(k$_1$), ρ(k$_{-1}$), ρ(K$_1$)], and the intrinsic rate constants (κ$_0$ = κ when K = 1) have been calculated. For a given pK$_a$ of the nucleophile, thiolate ion addition was thermodynamically and kinetically much more favorable than amine addition because, as the authors suggest, of soft–soft type interactions in the adduct and in the transition state. Rate constants for carbon protonation of the HOCH$_2$CHS– adducts of the substituted α-nitrostilbenes by AcOH were measured. The stereoselective/stereospecific Michael addition of *ortho*-lithiated stilbene oxides to α,β-unsaturated Fischer carbene complexes followed by an unusual cyclization of the corresponding intermediate in a 6-*endo*-tet mode was described [63]. Polysubstituted tetrahydronaphthols were found to be reaction products. Hydroboration of stilbenes and related disubstituted alkenes catalyzed by QUINAP complexes has been reported [64]. The reaction proceeded with high enantio- and regioselectivity. Rhodium and iridium catalysts gave the same product regioisomer but opposite enantiomers (Figure 2.10).

Reactions of additions of 3-(2-chloro-1-azaazulenyl)methylene with *cis*- and *trans*-stilbenes and *p*-substituted styrenes to form 3-cyclopropyl-1-azaazulenes have been accomplished [65]. The reaction proceeded in a stereospecific manner to afford the

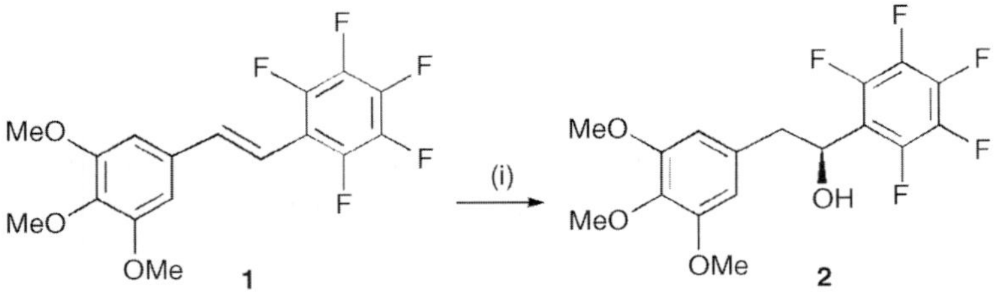

Figure 2.10 Hydroboration of stilbenes catalyzed by QUINAP complexes [64]. (Reproduced with permission.)

QuinapRh⁺	>98%; 88% e.e.; 77% yield
DppbRh⁺	45%; 55% yield
Thermal	95%; 60% yield

corresponding cyclopropane derivatives. Substitution effects on the relative rate ratio of addition reactions of carbene with styrenes having various substituents on p-positions were measured to investigate the electronic nature of carbene. It was found that cis- and trans-stilbenes reacted with fluoroxytrifluoromethane to afford, mainly by cis-addition, products of electrophilic fluorination [66]. The reaction has been shown to proceed through discrete carbocations analogous to the intermediates well known from studies of conventional halogenation. The degree of cis-addition in the electrophilic fluorination depended on tight ion pairs and neighboring group participation.

2.4.4
Substituted Groups Reactions

Antioxidative effects of resveratrol (3,5,4′-trans-trihydroxystilbene) and its analogues against free radical-induced peroxidation of human low-density lipoprotein (LDL) were studied [67]. These effects were apparently related to the stilbenes' reactivity toward the hydrogen abstraction from hydroxyl groups. The following analogues were synthesized: 3,4,3′,4′-tetrahydroxy-trans-stilbene (3,4,3′,4′-THS), 3,4,4′-trihydroxy-trans-stilbene (3,4,4′-THS), 2,4,4′-trihydroxy-trans-stilbene (2,4,4′-THS), 3,3′-dimethoxy-4,4′-dihydroxy-trans-stilbene (3,3′-DM-4,4′-DHS), 3,4-dihydroxy-trans-stilbene (3,4-DHS), 4,4′-dihydroxy-trans-stilbene (4,4′-DHS), 3,5-dihydroxy-trans-stilbene (3,5-DHS), and 2,4-dihydroxy-trans-stilbene (2,4-DHS). Kinetic analysis of the antioxidation process demonstrated that these trans-stilbene derivatives are effective antioxidants against both AAPH- and Cu^{2+}-induced LDL peroxidation with the activity sequence of 3,4,3′,4′-THS ∼ 3,3′-DM-4,4′-DHS > 3,4-DHS ∼ 3,4,4′-THS > 2,4,4′-THS > resveratrol ∼ 3,5-DHS > 4,4′-DHS ∼ 2,4-HS and 3,4,3′,4′-THS ∼ 3,4-DHS ∼ 3,4,4′-THS > 3,3′-DM-4,4′-DHS > 4,4′-DHS > resveratrol ∼ 2,4-HS > 2,4,4′-THS ∼ 3,5-DHS, respectively. Molecules bearing ortho-dihydroxyl or 4-hydroxy-3-methoxyl groups possess significantly higher antioxidant activity than those bearing no such functionalities.

Results of a study of the hydrolysis mechanism of substituted α-nitrostilbenes (NS-Z with Z = 4-Me, H, 4-Br, 3-NO_2, and 4-NO_2) in 50% Me_2SO–50% (v/v) water at

20 °C were reported [68]. On the basis of the kinetics data, it was suggested that the mechanism consists of four steps: (i) nucleophilic addition to NS-Z of water and OH$^-$ to form PhCH(OH)C(Ar)NO$_2^-$ (TOH$^-$); (ii) carbon protonation of TOH$^-$ by water and buffer acids to form PhCH(OH)CH(Ar)NO$_2$; (iii) rapid oxygen deprotonation of TOH^0 to form PhCH(O$^-$)CH(Ar)NO$_2$ (TO$^-$); and (iv) collapse of TO$^-$ into benzaldehyde and arylnitromethane anion. The acid form of TOH^0, PhCH(OH)CH(Ar)NO$_2$H (TOH, aci^0), can also be generated as a transient by reaction of TOH$^-$ with strong acid. The rate constants of most of the individual steps have been measured. The nucleophilic vinylic substitution reactions of amines with α-nitro-β-substituted stilbenes in 50% dimethyl sulfoxide–50% water have been investigated [69]. The kinetics of the reaction of Ph(OMe)C=CPh(NO$_2$) with piperidine, morpholine, pyrrolidine, and n-butylamine and that of the reactions of piperidine with Ph(Cl)C=CPh(NO$_2$), Ph(I)C=CPh(NO$_2$), and Ph(SEt)C=CPh(NO$_2$) were measured in 50% Me$_2$SO–50% water at 20 °C. It was shown that the nucleophilic attack is rate limiting in the reactions of piperidine with Ph(LG)C=CPh(NO$_2$), when LG = Cl, I, EtS, and in the reactions of Ph(OMe)C=CPh(NO$_2$) with all the amines at high pH. An intermediate in nucleophilic vinylic substitution in the reaction of methoxyamine with β-methoxy-α-nitrostilbene was observed [70].

The diazonium salt of 4,4′-diamino-stilbene-2,2′-disulfonic acid was found to be coupled with different acetoacetarylides to give a series of azo dyes [71]. (E)-4-Stilbenethiol reacted with dibromoalkanes to produce new stilbenes [72]. Conversion of stilbenols to stilbenethiols via N,N-dimethylthiocarbamates has been reported [73]. A series of O-stilbenyl N,N-dimethylthiocarbamates have been prepared by the reaction of N,N-dimethylthiocarbamoyl chloride with stilbenols in the presence of DABCO and DMF. O-Stilbenyl N,N-dimethylthiocarbamate was converted to S-stilbenyl N,N-dimethylthiocarbamate by thermal rearrangement. The latter compound gave the corresponding stilbenethiols.

Stilbene derivative 6b,10b-dihydrobenzo[j]cyclobut[a]acenaphthylene (DBCA) was

R$_1$ = CH$_3$O, R$_2$ = CH$_3$(CH$_2$)$_3$CH(C$_2$H$_5$)CH$_2$O

converted into biradical species (pleiadene) upon thermolysis or photolysis [74].

In the next step, pleiadene dimerizes. Synthesis of either (E)-3,5-dimethoxystilbenes or 3,5-dimethoxydibenzyls was carried out by selective removal of the 4-methoxy group of 3,4,5-trimethoxystilbenes I (R$_1$ = MeO; Ar = Ph, 2-MeOC$_6$H$_4$, 3-MeOC$_6$H$_4$, 4-MeOC$_6$H$_4$, 3,4,5-(MeO)3C$_6$H$_2$) [75]. The reaction was performed under electron transfer conditions with Na metal in THF. It was shown that 4′,4″-diethylstilbestrol quinone reacted with nucleosides, nucleotides, and amines

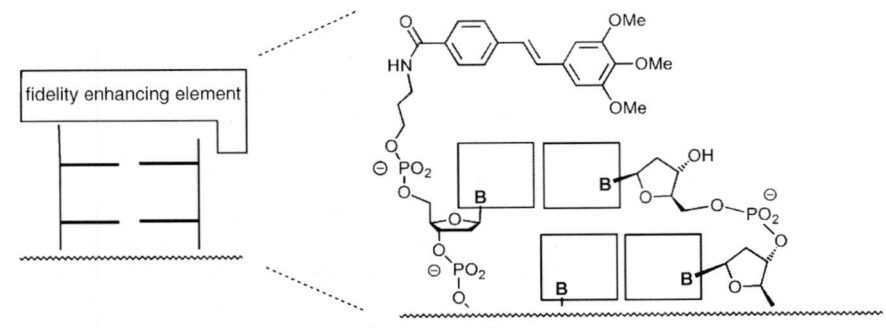

Figure 2.11 Proposed scheme of the fidelity- and affinity-enhancing modulators of DNA duplex stability by trimethoxystilbene[77]. (Reproduced with permission.)

n-pentyl amine giving Z,Z-dienestrol [76]. Rat liver cytochrome P-450 reductase reduced 4′,4″-diethylstilbestrol quinone to E- and Z-diethylstilbestrol.

Several works were concerned with the interaction of stilbene derivatives with DNA. A series of 5′-linked stilbene–DNA conjugates with different substituents in the distal aromatic ring of the stilbene was prepared [77]. Trimethoxystilbene and aminomethylstilbene were found to be the fidelity- and affinity-enhancing modulators of DNA duplex stability. The phosphoramidite of the trimethoxystilbene was employed in automatic DNA synthesis, facilitating the generation of DNA chips with improved fidelity. Figure 2.11 shows a scheme of the fidelity- and affinity-enhancing modulators of DNA duplex stability by trimethoxystilbene.

The structure and properties of 18 hairpin-forming bis(oligonucleotide) conjugates possessing stilbene diether linkers have been described [78]. It was shown that conjugates possessing bis(2-hydroxyethyl)stilbene 4,4′-diether linkers formed the most stable DNA hairpins. Hairpins with two T:A base pairs or four noncanonical G:G base pairs were stable at room temperature. The length of the hydroxyalkyl groups resulted in a decrease in hairpin thermal stability. Diethylstilbestrol has been shown by ^{32}P-postlabeling analysis to bind covalently to DNA *in vivo* and *in vitro* [79]. Influence of DES dose, age of animals, and organ specificity on adduct formation in hamsters has been examined. The covalent binding of DES to DNA catalyzed by hamster liver microsomes required cumene hydroperoxide as a cofactor. It was suggested that stilbene–DNA adduction may occur only under oxidative stress conditions.

2.5
Stilbenes in Polymerization

2.5.1
Radical Polymerization

It was shown that E-isomers of 2-, 3-, and 4-fluorostilbene, 4,4′-difluorostilbene, and Z-isomers of 4-fluorostilbene were involved as initiators in radical copolymerization with styrene at 60 °C [80]. Polymers of methyl methacrylate and styrene have been prepared in the presence of low concentrations of various *para*-derivatives of stilbene, using ^{13}C-benzoyl peroxide as the source of the initiating radicals [81]. Stilbenes were found to have high reactivity toward the benzoyloxy radical. Effects of *cis*- and *trans*-4-methoxy-4′-chlorostilbene on radical polymerization of methyl methacrylate and styrene, using Bz$_2$O$_2$ and 2-cyano-2-propylazoformamide, both enriched with ^{13}C as initiators, have been investigated [82]. The ^{13}C NMR spectra of the polymers showed that stilbene derivative units were present at many of the sites adjacent to benzoate end groups. The presence of the substituents in the stilbene molecules markedly enhanced its reactivity toward the BzO radical. Radical copolymerization of the *E*-isomers of 2-, 3-, and 4-fluorostilbene, 4,4′-difluorostilbene, and also the *Z*-isomers of 4-fluorostilbene with styrene at 60 °C has been accomplished [83]. Reactivities of various fluorostilbenes toward the polystyrene radical were discussed and compared with those toward the benzoyloxy radical. The reactivities of 4-fluoro- and 4,4′-difluorostilbene, 4-chloro- and 4,4′-dichlorostilbene, and 4-phenyl- and 4,4′-diphenylstilbene toward the benzoyloxy radical were found to be similar [84]. It was shown that the introduction of fluorine or chlorine at the 4-position in stilbene had little effect on the reactivity, but quite a large increase was caused by 4-Ph substitution. The modification of poly(*p*-bromostyrene), which was synthesized using benzoyl peroxide and TEMPO as initiators by free radical polymerization, with *p*-stilbenylboronic acid was reported [85].

Upon pulse radiolysis of *trans*-stilbene (*t*-St) solutions in THF, the radical anion of *trans*-stilbene was demonstrated to be formed by the reaction of electrons with *t*-St (reaction with the rate constant $k_5 = (1.16 \pm 0.03) \times 10^{11}$ dm^3/(mol s)) [86]. The transient absorption spectrum observed with λ_{max} at 500 and 720 nm was attributed to the unassociated radical anion St$^{\bullet-}$. This species reacted with the countercation of THF formed upon radiolysis and with radiolytically generated radicals. Addition of sodium tetrahydridoaluminate (NAH) resulted in the radical anion being associated with Na$^+$ as a contact ion pair. In the presence of the lithium salt, formation of solvent-separated ion pairs has been detected.

2.5.2
Anionic Polymerization

Nonempirical calculations of reaction complexes formed during anionic polymerization of butadiene, in the presence of stilbenes, have been made [87]. The mechanism of *cis–trans* isomerization in the terminal unit of the living polymer consisted in concerted rotation about the C$_\beta$—C$_\gamma$ bond, and the migration of Li between C$_\alpha$ and C$_\gamma$ atoms was proposed. *Ab initio* calculations of electron transfer in anionic polymerization were described [88].

Anionic copolymers, *trans*-stilbenebutadiene copolymer, *trans*-stilbene-isoprene copolymer, and *trans*-stilbene-2,3-dimethylbutadiene copolymer, copolymerized using BuLi initiator, were studied in THF at 0 °C and in benzene at 40 °C [89]. It was shown that the rate of monomer consumption (excluding stilbene) decreased as follows: butadiene > isoprene > 2,3-dimethylbutadiene. Anionic copolymerization

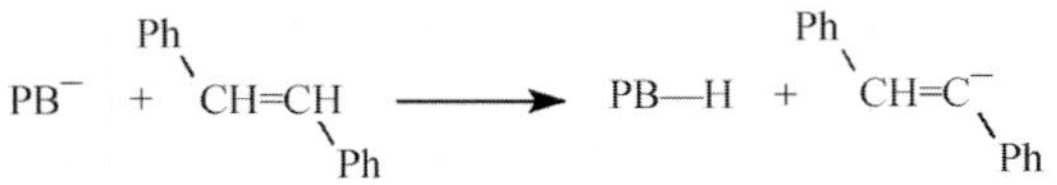

$$\text{PB}^- + \underset{\text{Ph}}{\overset{\text{Ph}}{\text{CH=CH}}} \longrightarrow \text{PB—H} + \underset{\text{Ph}}{\overset{\text{Ph}}{\text{CH=C}^-}}$$

Figure 2.12 Hydrogen abstraction from *trans*-stilbene by living anionic polybutadiene [93]. (Reproduced with permission.)

of styrene and *trans*-stilbene with *n*-butyllithium [90] and reaction of *trans*-stilbene with *n*-butyllithium in tetrahydrofuran have been investigated [91]. *trans*-Stilbene reacted with butyllithium to yield a mixture of 1,2-diphenylhexyllithium and 1,2,3,4-tetraphenyloctyllithium [92]. The reaction products 1,2-dilithio-1,2-diphenylhexane and 1,2-dilithio-1,2,3,4-tetraphenyloctane were efficient initiators for the anionic polymerization of styrene, yielding polymers with predictable molecular weights and narrow molecular weight distributions. The reaction of poly(butadienyl)lithium with 1,2-diphenylethylenes, particularly *trans*-stilbene, in the process of living polymerization resulted in monoaddition to the poly(butadienyl)lithium chain ends [93]. Nearly quantitative block copolymer formation was achieved, with an average styrene block size of four monomer units and a polydispersity index of 1.19 for the polystyrene block. A scheme of the hydrogen abstraction from *trans*-stilbene by living anionic polybutadiene (PB) is shown in Figure 2.12.

Several polymerization processes involving stilbene have been patented. A large series of stilbene compounds has been used in anionic polymerization as bifunctional initiators. Substituted or unsubstituted *trans*-stilbenes were contacted with nonpolymerizable olefins for living polymerization of olefins to produce copolymer [94]. In this kind of process, the initiators grew only one chain per initiator molecule and the polymerization continued until monomer was exhausted. The (un)substituted *trans*-stilbene was involved in the process of the capping of a living polymer with a number of nonpolymerizable monomers such as derivatives of diphenylalkylene, α-methoxystyrene, and phenylnaphthalene [95]. The charge-transporting polymers and molecular glasses were synthesized by anionic polymerization involving stilbenes [96]. In the work [97], block copolymers could be synthesized by anionic polymerization. Polymers obtained were used as layered photoreceptor structures with overcoatings.

2.6
Complexation

2.6.1
Complexation with Small Molecules

It was demonstrated that a complexation of *trans*-stilbene with cyclodextrin led to a decrease in stereoselectivity of additive bromination and to a significant yield of DL-stilbene dibromide in contrast to the formation of *meso*-stilbene dibromide in nonpolar solvents [98]. The authors suggested that this reversal of stereoselectivity was attributed to the polar environment provided by the secondary hydroxyl groups of cyclodextrins that stabilizes the acyclic α-halocarbonium ion intermediate and also hinders the approach of the incoming tribromide ion. Cyclodextrin-based supramolecular complexes have been prepared using a stilbene bis(β-CD) dimer guest [99]. When the host cyclodextrin dimer was in *trans* conformation, supramolecular dimers or small assemblies were formed in solution, whereas in its *cis* conformation supramolecular linear polymers with high molecular weight were observed.

It was shown that bis(18-crown-6)stilbene formed bimolecular (D–A) and sandwich-like (D–A–D) complexes with a di(quinolyl)ethylene derivative [100]. Unusual three-decker structures of a D–A–D complex were shown. Chiral stilbenophanes with small and large rigid cavities and bis-cyclophanes with a stilbene-bridging unit formed charge transfer complexes with either TCNQ or TCNE [101]. Stilbenophanes formed complexes with TCNQ [102].

Stilbene derivatives formed complexes with metal-containing compounds. For example, stilbene formed stilbene Cr tricarbonyl complexes that underwent heterogeneous hydrogenation by H_2 on skeletal Ni and Pd on carbon more slowly than free stilbene [103]. For the homogeneous hydrogenation of these complexes using a H_2PtCl_6–$SnCl_2$–LiBr system, styrene and η^6-styrene Cr tricarbonyl were reduced with a high rate, whereas stilbene and its Cr tricarbonyl complex were hydrogenated very slowly. Chiral Pd(0) *trans*-stilbene complexes Pd(diphos*)(*trans*-stilbene) (diphos* = (R,R)-Me-Duphos, (R,R)-Et-Duphos, (R,R)-i-Pr-Duphos, (R,R)-Me-BPE, (S,S)-Me-FerroLANE, (S,S)-Me-DuXantphos, (S,S)-Et-FerroTANE, (R,S)-CyPF-t-Bu, (R,S)-PPF-t-Bu, (R,S)-BoPhoz) and Ni(R,R)-Me-Duphos)(*trans*-stilbene) were prepared by $NaBH(OMe)_3$ reduction of the corresponding M(diphos*)Cl_2 compounds in the presence of *trans*-stilbene (Figure 2.13) [104]. These complexes underwent oxidative addition with PhI.

Dinuclear Ba(II) complexes of *cis*- and *trans*-stilbenobis(18-crown-6) have been prepared [105]. These complexes catalyzed size-selective ester and anilide cleavage. The *cis* form of the catalyst was more efficient than the *trans* form and much more sensitive to the size of the substrate. A series of the 4′-substituted stilbenoid-NCN-pincer platinum(II) complexes containing the donor PtCl grouping were synthesized by Horner–Wadsworth–Emmons reaction [106]. The series includes [PtCl(NCN-R-4)] (NCN-R-4 = [$C_6H_2(CH_2NMe_2)_2$-2,6-R-4]− in which R = $C_2H_2C_6H_4$-R′-4′ with R′ = NMe_2, OMe, $SiMe_3$, H, I, CN, NO_2). Synthesis of chromium tricarbonyl complexes with stilbene in solution was reported [107]. It was shown that 1-(methylthio)-*cis*-stilbene-2-thiol, $Ph(SCH_3)C:C(SH)Ph$(HL), and 1-(benzylthio)-*cis*-stilbene-2-thiol, $Ph(SCH_2Ph)C:C(SH)Ph$, formed monomeric complexes with Sb^{3+} [108]. Two supramolecular complexes, $Co(en)_3(dasb)(ClO_4) \cdot H_2O$ and $Ni(en)_3(H_2O)_2(dasb)$, constructed by 4,4′-diazido-2,2′-stilbenedisulfonate anion (dasb)$_2^-$ through weak

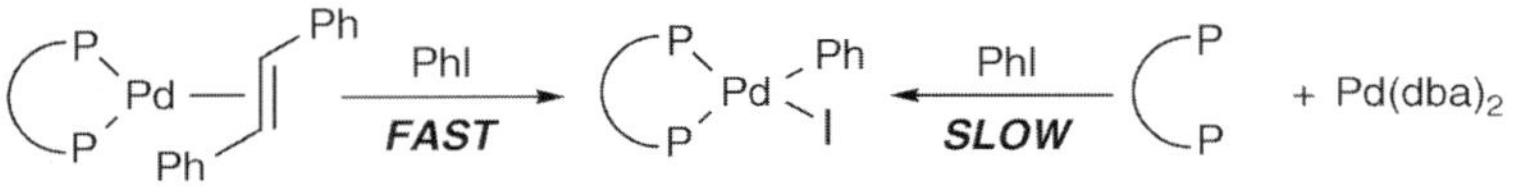

Figure 2.13 Formation of chiral Pd(0) *trans*-stilbene complexes [104]. (Reproduced with permission.)

interactions, were synthesized [109]. In these structures, sulfonate groups of the anion form strong H bonds with the coordinated H_2O molecules, as the H-bond acceptor and components were connected by H bonds to three-dimensional networks.

2.6.2
Complexation with Proteins

2.6.2.1 Tubulins

Tubulin is one of the several members of a small family of globular proteins. The most common members of the tubulin family are α-tubulin and β-tubulin. Each has a molecular weight of approximately 55 kDa. Proteins make up microtubules that are assembled from dimers of α- and β-tubulin bonded to guanosine-5′-triphosphate. Tubulins are targets for anticancer, antigout, and antifungal drugs inhibiting microtubule formation.

The interaction of a series of stilbenes, based on combretastatin A-4 with tubulin, has been investigated [110]. It was found that the substitution of small alkyl substituents for the 4′-MeO group of combretastatin A-4 and the loss of the 3′-OH group do not have a major effect on the interaction with tubulin. An idealized structure for a tubulin binding agent of this type was proposed. 4-Arylcoumarin analogues of combretastatin A-4 were shown to interact with purified tubulin [111]. Thermodynamic parameters of their interaction with purified tubulin were measured by fluorescence spectroscopy and isothermal microcalorimetry. The proposed three-dimensional structure model of the tubulin–colchicine complex allowed authors to identify the pharmacophore of the combretastatin A-4 analogues responsible for their biological activity. A new class of combretastatin A-4 analogues, which interact with tubulin, was used to inhibit tubulin polymerization [112]. Some of these compounds incorporated the benzo[b]thiophene ring system. Seventeen natural products and twenty-two synthetic agents, analogues of combretastatin, were examined for the effect of tubulin polymerization and colchicine binding [113]. The (cis)-stilbene derivative (cis)-1-(3,4,5-trimethoxyphenyl)-2-(3′-hydroxy-4′-methoxyphenyl)ethene (combretastatin A-4) was found to be the most promising compound.

2.6.2.2 Antibodies

A complex of a donor–acceptor-substituted stilbene with antibodies has been prepared [114]. Monoclonal antibodies (Mabs) were generated against the trans-4-N,N-dimethylamino-4′-nitrostilbene. To characterize the nature of an antibody–stilbene complex, the crystal structure of Fab 11G10 in complex with hapten at 2.75 Å resolution has been detected. Stilbene moiety was bound in a planar conformation. Rotation around the excited-state C=C bond and the styryl-anilino C−C bond was proposed to be restricted by interactions between the ligand and the antibody residues Phe^{L94}, Leu^{L89}, Leu^{L36}, Tyr^{H33}, Tyr^{H95}, and Val^{H93}. The binding pocket possesses a relatively high percentage of polar residues in immediate vicinity to the ligand. Complex formation between antibody EP2-19G2 and trans-stilbene was enabled by a deeply penetrating ligand binding pocket, which in turn results from a noncanonical interface between the two variable domains of the antibody [115]. It was shown that the prolonged luminescence is a result of charge recombination in a charge transfer excited complex of an anionic stilbene and a cationic, parallel π-stacked tryptophan.

A complex between *trans*-4-(N-2,4-dinitrophenylamino)-4'-(N,N-dimethylamino) stilbene (StDNP) and *anti*-2,4-dinitrophenyl antibody (*anti*-DNP) was prepared and investigated [116–118] (see also Section 10.3.6). Computer modeling suggested that the dinitrophenyl segment of StDNP was squeezed between two tryptophans, while there was still sufficient space for stilbene double-bond twisting during photoisomerization without steric hindrance in the excited singlet state of the *anti*-DNP-bound StDNP tracer (Figure 2.14).

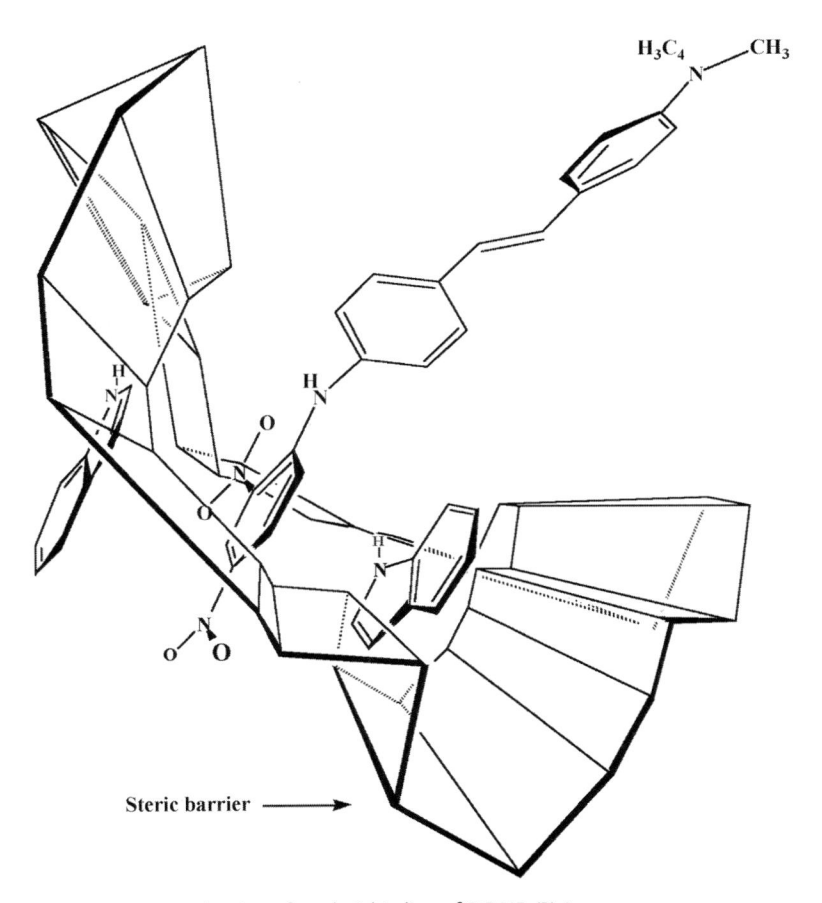

Figure 2.14 Iconic drawing of mode 2 binding of StDNP (**1**) in *anti*-DNP antibody binding site viewed down the cleft canyon [116]. (Reproduced from Ref. [118] with permission from Elsevier.)

The rate constants measured for *trans–cis* stilbene photoisomerization were found to be similar for both the free and the *anti*-DNP-bound StDNP tracer. Computer modeling suggested that the dinitrophenyl segment of StDNP is squeezed between two tryptophans.

In conclusion, stilbenes involve in miscellaneous chemical reactions. For non-substituted stilbenes, the most chemically reactive part is double bond, which relatively easily undergoes the halogenation, epoxidation, oxidation, reduction, and addition. The chemistry of substituted stilbenes is in principle as rich as organic chemistry. Including stilbenes in dendrides, dextrins, polymers, and surfaces led to a sufficient change in their chemical, photochemical, photophysical, and mechanical properties and, therefore, establishes the basis for design of new materials.

References

1 Vollhardt, P.C. and Schore, N.E. (2002) *Organic Chemistry: Structure and Function*, 4th edn, W. H. Freeman & Co.
2 Tanaka, K., Shiraishi, R., and Toda, F. (1999) *Journal of the Chemical Society, Perkin Transactions 1*, (21), 3069–3070.
3 Karade, N.N., Gampawar, S.V., and Tiwari, G.B. (2007) *Letters in Organic Chemistry*, **4**, 419–422.
4 Muathen, H.A. (2004) *Synthetic Communications*, **34**, 3545–3552.
5 Ye, C. and Shreeve, J.M. (2004) *Journal of Organic Chemistry*, **69**, 8561–8563.
6 Al-Hassan, M.I. (1989) *Synthetic Communications*, **19**, 463–472.
7 Sergeev, G.B., Serguchev, Y.A., and Smirnov, V.V. (1973) *Russian Chemical Reviews*, **42**, 697–712.
8 Heublein, G. and Shutz, E. (1969) *Zietshrift fur Chemie*, **9**, 147–152.
9 Manickam, M.C.D., Pitchumani, K., and Srinivasan, C. (2002) *Journal of Inclusion Phenomena and Macrocyclic Chemistry*, **43**, 207–211.
10 Bellucci, G., Bianchini, R., Chiappe, C., and Marioni, F. (1990) *Journal of Organic Chemistry*, **55**, 4094–4098.
11 Bianchini, R. and Chiappe, C. (1992) *Journal of Organic Chemistry*, **57**, 6474–6478.
12 Bellucci, G., Bianchini, R., Chiappe, C., Brown, R.S., and Slebocka-Tilk, H. (1991) *Journal of the American Chemical Society*, **113**, 8012–8016.
13 Dubois, J.E. and Ruasse, M.F. (1973) *Journal of Organic Chemistry*, **38**, 493–499.
14 Moriuchi, T., Yamaguchi, M., Kikushima, K., and Hirao, T. (2007) *Tetrahedron Letters*, **48**, 2667–2670.
15 McKenzie, L.C., Huffman, L.M., and Hutchison, J.E. (2005) *Journal of Chemical Education*, **82**, 306–310.
16 Bellucci, G., Bianchini, R., and Chiappe, C. (1991) *Journal of Organic Chemistry*, **56**, 3067–3073.
17 Berthelot, J., Benammar, Y., and Lange, C. (1991) *Tetrahedron Letters*, **32**, 4135–4136.
18 Kirihara, M., Okubo, K., Koshiyama, T., Kato, Y., and Hatano, A. (2004) *ITE Letters on Batteries, New Technologies & Medicine*, **5**, 279–281.
19 Xia, Q.H., Ge, H.Q., Ye, C.P., Lui, Z.M., and Su, K.X. (2005) *Chemical Reviews*, **105**, 1603–1662.
20 McGarrigle, E.M. and Gilheany, D.G. (2005) *Chemical Reviews*, **105**, 1563–1620, and references therein.
21 Tran, L.-H., Eriksson, L., Sun, L., and Aakermark, B. (2008) *Journal of Organometallic Chemistry*, **693**, 1150–1153.
22 Nouroozi, F., Farzaneh, F., and Khosroshahi, M. (2006) *Reaction Kinetics and Catalysis Letters*, **89**, 139–147.

23 Sithambaram, S., Calvert, C., Opembe, N., and Suib, S.L. (2007) *Preprints American Chemical Society, Division of Petroleum Chemistry*, **52**, 243–245.
24 Farzaneh, F., Tayebi, L., and Ghandi, M. (2007) *Reaction Kinetics and Catalysis Letters*, **91**, 333–340.
25 Xiong, D., Fu, Z., Zhong, S., and Jiang, X. (2007) *Catalysis Letters*, **113**, 155–159.
26 Muto, T., Urano, C., Hayashi, T., Miura, T., and Kimura, M. (1983) *Chemical & Pharmaceutical Bulletin*, **31**, 1166–1171.
27 Chattopadhyay, T., Islam, S., Nethaji, M., Majee, A., and Das, D. (2007) *Journal of Molecular Catalysis A: Chemical*, **267**, 255–264.
28 Lignier, P., Morfin, F., Mangematin, S., Massin, L., Rousset, J., and Caps, V. (2007) *Chemical Communications*, (2), 186–188.
29 Lignier, P., Morfin, F., Piccolo, L., Rousset, J.-L., and Caps, V. (2007) *Catalysis Today*, **122**, 284–291.
30 Kholdeeva, O.A., Maksimov, G.M., Fedotov, M.A., and Grigoriev, V.A. (1994) *Kinetics and Catalysis Letters*, **53**, 331–337.
31 Serrano, I., Sala, X., Plantalech, E., Rodriguez, M., Romero, I., Jansat, S., Gomez, M., Parella, T., Stoeckli-Evans, H., Solans, X., Font-Bardia, M., Vidjayacoumar, B., and Llobet, A. (2007) *Inorganic Chemistry*, **46**, 5381–5389.
32 Chatterjee, D., Basak, S., and Muzart, J. (2007) *Journal of Molecular Catalysis A: Chemical*, **271**, 270–276.
33 Bitterlich, B., Anilkumar, G., Gelalcha, F.G., Spilker, B., Grotevendt, A., Jackstell, R., Tse, M.K., and Beller, M. (2007) *Chemistry – An Asian Journal*, **2**, 521–529.
34 Choudary, B.M., Pal, U., Kantam, M.L., Ranganath, K.V.S., and Sreedhar, B. (2006) *Advanced Synthesis and Catalysis*, **348**, 1038–1042.
35 Razumovskii, S.D. and Zaikov, G.E. (1980) *Russian Chemical Reviews*, **49**, 1163–1180.
36 Bouas-Laurent, H., Desvergne, J.P., Lapouyade, R., and Thomas, J.M. (1976) *Molecular Crystals and Liquid Crystals*, **32**, 143–146.
37 Pelletier, M.J., Fabilli, M.L., and Moon, B. (2007) *Applied Spectroscopy*, **61**, 1107–1115.
38 Sako, M., Hosokawa, H., Ito, T., and Iinuma, M. (2004) *Journal of Organic Chemistry*, **69**, 2598–2600.
39 Travis, B.R., Narayan, R.S., and Borhan, B. (2002) *Journal of the American Chemical Society*, **124**, 3824–3825.
40 Steinbauer, G., Giselbrecht, K., Schoftner, M., and Reiter, K. (2000) (DSM Fine Chemicals Austria GmbH, Austria). Eur. Patent Appl., EP1016654(A2), pp. 6.
41 Wu, X., Davis, A.P., and Fry, J. (2007) *Organic Letters*, **9**, 5633–5636.
42 Provins, L., and Murahashi, S.-I. (2007) *ARKIVOC*, (10), 107–120.
43 Nikiforov, G.B., Crewdson, P., Gambarotta, S., Korobkov, I., and Budzelaar, P.H.M. (2007) *Organometallics*, **26**, 48–55.
44 Ohkubo, K., Nanjo, T., and Fukuzumi, S. (2006) *Catalysis Today*, **117**, 356–361.
45 Lion, C.J., Vasselin, D.A., Schwalbe, C.H., Matthews, C.S., and Stevens, M.F.G. (2005) *Organic & Biomolecular Chemistry*, **3**, 3996–4001.
46 Meenakshisundaram, S.P., Gopalakrishnan, M., Nagarajan, S., and Sarathi, N. (2006) *Catalysis Communications*, **7**, 502–507.
47 Reddy, S.M., Srinivasulu, M., Reddy, Y.V., Narasimhulu, M., and Venkateswarlu, Y. (2006) *Tetrahedron Letters*, **47**, 5285–5288.
48 Maurya, M.R., Sikarwar, S., and Kumar, M. (2007) *Catalysis Communications*, **8**, 2017–2024.
49 Michinobu, T., Tsuchida, E., and Nishide, H. (2000) *Bulletin of the Chemical Society of Japan*, **73**, 1021–1027.
50 Khenkin, A.M., Kumar, D., Shaik, S., and Neumann, R. (2006) *Journal of the American Chemical Society*, **128**, 15451–15460.
51 Collins, D.J. and Hobbs, J.J. (1983) *Australian Journal of Chemistry*, **36**, 619–625.

52 Artemov, A.N., Sazonova, E.V., and Lomakin, D.S. (2007) *Russian Chemical Bulletin*, **56**, 45–48.

53 Brunel, J.M. (2007) *Tetrahedron*, **63**, 3899–3906.

54 Brunel, J.M. (2007) *Synlett*, (2), 330–332.

55 Dahlen, A., Nilsson, A., and Hilmersson, G. (2006) *Journal of Organic Chemistry*, **71**, 1576–1580.

56 Kaskel, S., Schlichte, K., and Kratzke, T. (2004) *Journal of Molecular Catalysis A: Chemical*, **208**, 291–298.

57 Fujita, S.-I., Sano, Y., Bhanage, B.M., and Arai, M. (2006) *Applied Catalysis A: General*, **314**, 89–93.

58 Pitchumani, K., Joy, A., Prevost, N., and Ramamurthy, V. (1997) *Chemical Communications*, (1), 127–128.

59 Cotter, J., Hogan, A.-M.L., and O'Shea, D.F. (2007) *Organic Letters*, **9**, 1493–1496.

60 Hogan, A.-M.L., and O'Shea, D.F. (2007) *The Journal of Organic Chemistry*, **72**, 9557–9571.

61 Hogan, A.-M.L., and O'Shea, D.F. (2006) *Organic Letters*, **8**, 3769–3773.

62 Bernasconi, C.F. and Killion, R.B. (1988) *Journal of the American Chemical Society*, **110**, 7506–7512.

63 Capriati, V., Florio, S., Luisi, R., Perma, F.M., Salomone, A., and Gasparrini, F. (2005) *Organic Letters*, **7**, 4895–4898.

64 Black, A., Brown, J.M., and Pichon, C. (2005) *Chemical Communications*, (42), 5284–5286.

65 Saito, K., Fushihara, H., and Abe, N. (1993) *Chemical & Pharmaceutical Bulletin*, **41**, 752–754.

66 Barton, D.H.R., Hesse, R.H., Jackman, G.P., Ogunkoya, L., and Pechet, M.M. (1974) *Journal of the Chemical Society, Perkin Transactions 1*, 739–742.

67 Cheng, J.C., Fang, J.G., Chen, W.F., Zhou, B., Yang, L., and Liu, Z.L. (2006) *Bioorganic Chemistry*, **34**, 142–157.

68 Bernasconi, C.F. and Fassberg, J. (1994) *Journal of the American Chemical Society*, **116**, 514–522.

69 Bernasconi, C.F., Fassberg, J., Killion, R.B., Jr., and Rappoport, Z. (1990) *Journal of Organic Chemistry*, **55**, 4568–4575.

70 Bernasconi, C.F., Leyes, A.E., Rappoport, Z., and Eventova, I. (1993) *Journal of the American Chemical Society*, **115**, 7513–7514.

71 Grad, M.E., Raditoiu, V., Wagner, L., Raditoiu, A., and Lupea, A.X. (2007) *Revista de Chimie*, **58** (8), 786–790.

72 Nowakowska, Z. (2006) *Phosphorus, Sulfur, and Silicon and the Related Elements*, **181**, 1789–1799.

73 Nowakowska, Z. (2006) *Phosphorus, Sulfur, and Silicon and the Related Elements*, **181**, 707–715.

74 Buchacher, P., Helgeson, R., and Wudl, F. (1998) *The Journal of Organic Chemistry*, **63**, 9698–9702.

75 Azzena, U., Dettori, G., Idini, M.V., Pisano, L., and Sechi, G. (2003) *Tetrahedron*, **59**, 7961–7966.

76 Liehr, J.G., DaGue, B.B., and Ballatore, A.M. (1985) *Carcinogenesis*, **6**, 829–836.

77 Dogan, Z., Paulini, R., Stuetz, J.A.R., Narayanan, S., and Richert, C. (2004) *Journal of the American Chemical Society*, **126**, 4762–4763.

78 Lewis, F.D., Wu, Y., and Liu, X. (2002) *Journal of the American Chemical Society*, **124**, 12165–12173.

79 Bhat, H.K., Han, X., Gladek, A., and Liehr, J.G. (1994) *Carcinogenesis*, **15**, 2137–2142.

80 Barson, C.A., Bevington, J.C., and Breuer S.W. (1994) *Polymer Bulletin*, **32**, 625–628.

81 Bevington, J.C., Breuer, S.W., and Huckerby, T.N. (1989) *Macromolecules*, **22**, 55–61.

82 Bevington, J.C., Breuer, S.W., and Huckerby, T.N. (1984) *Polymer Bulletin*, **12**, 531–534.

83 Bevington, J.C., Breuer, S.W., Huckerby, T.N., and Jones, R. (1995) *Polymer Bulletin*, **34**, 37–42.

84 Barson, C.A., Bevington, J.C., and Breuer, S.W. (1996) *Polymer Bulletin*, **36**, 423–426.

85 Possidonio, S., Peres, L.O., and Hui, W.S. (2005) Synthesis and characterization of

modified polymer of p-bromostyrene by "living" free radical and Suzuki reaction, Proceedings of the International Symposium on Electrets (ISE 12), 12th, Salvador, Brazil, September 11–14, 2005 (eds R.M. Faria, J.A. Giacometti, and O.N. Oliveira Jr.), pp. 515–517.
86 Langan, J.R. and Salmon, G.A. (1982) *Journal of Chemical Society, Faraday Transactions 1*, **78**, 3645–3657.
87 Kalnin'sh, K.K. and Podolsky, A.F. (2002) *International Journal of Quantum Chemistry*, **88**, 624–633.
88 Kalnin'sh, K.K. and Panarin, E.F. (2001) *Doklady Physical Chemistry*, **377**, 786–787.
89 Yuki, H., Okamoto, Y., Tsubota, K., and Kosai, K. (1970) *Polymer Journal*, **1**, 145–147.
90 Yuki, H., Kato, M., Okamoto, Y. (1968) *Bulletin of the Chemical Society of Japan*, **41**, 1940–1944.
91 Okamoto, Y., Kato, M., Yuki, H. (1969) *Bulletin of the Chemical Society of Japan*, **42**, 760–765.
92 Wyman, D.P. and Altares, T., Jr. (2003) *Die Makromolekulare Chemie*, **72**, 68–75.
93 Donkers, E.H.D., Willemse, R.X.E., and Klumperman, B. (2005) *Journal of Polymer Science Part A: Polymer Chemistry*, **43**, 2536–2545.
94 Faust, R. and Zsolt, F. (1995) US Patent, PCT International Patent Application, WO9510554(A1), 1995.
95 Faust, R. (1997) US Patent, Cont.-in-part of U.S. Ser. No. 137, 684, pp. 4.
96 Bronstert, K., Knol, K., and Haedicke, E. (1991) US Patent No. 764,870, filed on September 24, 1991
97 Damodar, M., Pai, J.F., Yanus, P.J., and DeFeo, D.S. (1996) US Patent No. 720,121 filed on September 27, 1996.
98 Manickam, M.C.D., Pitchumani, K., and Srinivasan, C. (2002) *Journal of Inclusion Phenomena and Macrocyclic Chemistry*, **43**, 207–211.
99 Kuad, P., Miyawaki, A., Takashima, Y., Yamaguchi, H., and Harada, A. (2007) *Journal of the American Chemical Society*, **129**, 12630–12631.
100 Vedernikov, A.I., Kuz'mina, L.G., Lobova, N.A., Ushakov, E.N., Howard, J.A.K., Alfimov, M.V., and Gromov, S.P. (2007) *Mendeleev Communications*, **17**, 151–153.
101 Rajakumar, P. and Selvam, S. (2007) *Tetrahedron*, **63**, 8891–8901.
102 Rajakumar, P., Swaroop, M.G., Jayavelu, S., and Murugesan, K. (2006) *Tetrahedron*, **62**, 12041–12050.
103 Artemov, A.N., Sazonova, E.V., and Lomakin, D.S. (2007) *Russian Chemical Bulletin*, **56**, 45–48.
104 Brunker, T.J., Blank, N.F., Moncarz, J.R., Scriban, C., Anderson, B.J., Glueck, D.S., Zakharov, L.N., Golen, J.A., Sommer, R.D., Incarvito, C.D., and Rheingold, A.L. (2005) *Organometallics*, **24**, 2730–2746.
105 Cacciapaglia, R., Di Stefano, S., and Mandolini, L. (2002) *Journal of Organic Chemistry*, **67**, 521–525.
106 Batema, G.D., van de Westelaken, K.T.L., Guerra, J., Lutz, M., Spek, A.L., van Walree, C.A., de Mello Donega, C., Meijerink, A., van Klink, G.P.M., and van Koten, G. (2007) *European Journal of Inorganic Chemistry*, (10), 1422–1435.
107 Mishchenko, O.G., Klement'eva, S.V., Maslennikov, S.V., Artemov, A.N., and Spirina, I.V. (2006) *Russian Journal of General Chemistry*, **76**, 1907–1910.
108 Reddy, K.H. (2006) *Journal of the Indian Chemical Society*, **83**, 1031–1033.
109 Wang, Y., Cao, R., Bi, W., Li, X., Li, X., and Wang, Y. (2005) *Zeitschrift fuer Anorganische und Allgemeine Chemie*, **631**, 2309–2311.
110 Woods, J.A., Hadfield, J.A., Pettit, G.R., Fox, B.W., and McGown, A.T. (1995) *British Journal of Cancer*, **71**, 705–711.
111 Rappl, C., Barbier, P., Bourgarel-Rey, V., Gregoire, C., Gilli, R., Carre, M., Combes, S., Finet, J.-P., and Peyrot, V. (2006) *Biochemistry*, **45**, 9210–9218.
112 Mocharla, V.P. and Pinney, K.G. (1998) Book of Abstracts, 215th ACS National Meeting, Dallas, TX, March 29–April 2, 1998, ORGN-041.

113 Lin, C.M., Singh, S.B., Chu, P.S., Dempcy, R.O., Schmidt, J.M., Pettit, G.R., and Hamel, E. (1988) *Molecular Pharmacology*, **34**, 200–208.

114 Tian, F., Debler, E.W., Millar, D.P., Deniz, A.A., Wilson, I.A., and Schultz, P.G. (2006) *Angewandte Chemie – International Edition*, **45**, 7763–7765.

115 Debler, E.W., Kaufmann, G.F., Meijler, M.M., Heine, A., Mee, J.M., Pljevaljcic, G., Di Bilio, A.J., Schultz, P.G., Millar, D.P., Janda, K.D., Wilson, I.A., Gray, H.B., and Lerner, R.A. (2008) *Science*, **319** (5867), 1232–1235.

116 Ahluwalia, A., Papper, V., Chen, O., Likhtenshtein, G.I., and De Rossi, D. (2002) *Analytical Biochemistry*, **305**, 121–134.

117 Chen, O., Glaser, R., and Likhtenshtein, G.I. (2003) *Biophysical Chemistry*, **103**, 139–156.

118 Chen, O., Glaser, R., and Likhtenshtein, G.I. (2008) *Journal of Biochemical and Biophysical Methods*, **70**, 1073–1079.

3
Stilbene Photophysics

3.1
General

Stilbenes possess a number of photophysical properties, including fluorescence and phosphorescence behavior (Figure 3.1). Because of their high sensitivity, fluorescence and phosphorescence techniques are especially useful for solving many problems of the structure and dynamics of the chemical, physical, and biological molecular system.

The main luminescence parameters traditionally measured are the frequency of maximal intensity v_{max}, intensity I, the quantum yield ϕ, the lifetime of the exited state τ, polarization, parameters of Raman spectroscopy, and excited-state energy migration. The usefulness of the fluorescence methods has been greatly enhanced with the development of new experimental techniques such as nano-, pico-, and femtosecond time-resolved spectroscopy, single-molecule detection, confocal microscopy, and two-photon correlation spectroscopy.

Owning these properties, stilbenes exhibit a considerable potential for miscellaneous traditional and new applications, such as systems involving photoredox processes usable for solar energy conversion, information storage systems, chemical sensors and biosensors, photovoltaic devices, and so on. In this chapter, we focus on recent developments in stilbene photophysics.

3.2
Stilbene Excited States

3.2.1
Excited Singlet State

The excited-state optimized structures and the computed absorption and emission frequencies of stilbene were calculated [2]. The stilbene orbitals and the S_1 and S_2 energy profiles for both isomers of stilbene are shown in Figure 3.2.

Stilbenes. Applications in Chemistry, Life Sciences and Materials Science. Gertz Likhtenshtein
Copyright © 2010 WILEY-VCH Verlag GmbH & Co. KGaA, Weinheim
ISBN: 978-3-527-32388-3

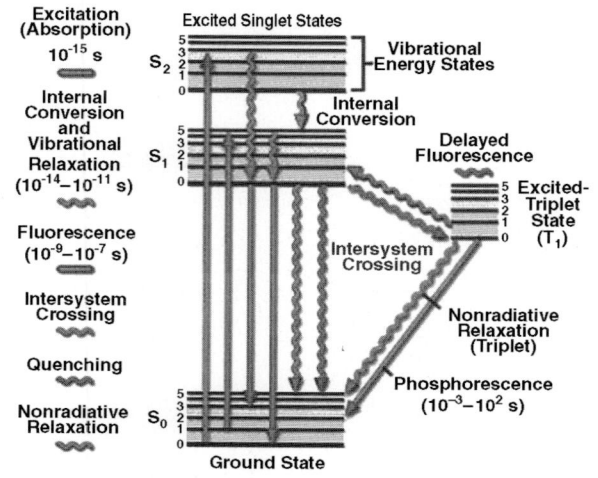

Figure 3.1 The Jablonski energy diagram [1].

Time-dependent density functional theory calculation, together with simulations of the electron energy distribution, allowed to estimate selective photoelectron energies of the S_0, S_1, S_2, and D_0 electronic states in *trans*-stilbene (Figure 3.3) [3]. The theory calculations of S_0, S_1, S_2, and D_0, together with simulations of the electron energy distribution, supported the experimental findings for selective photoelectron energies of the S_0, S_1, ... electronic states.

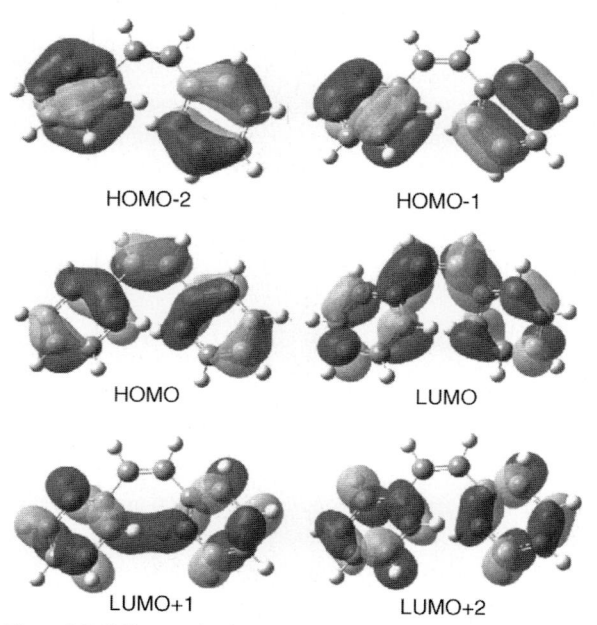

Figure 3.2 Stilbene orbitals [2]. (Reproduced with permissions.)

Excited-state θ-constrained optimizations

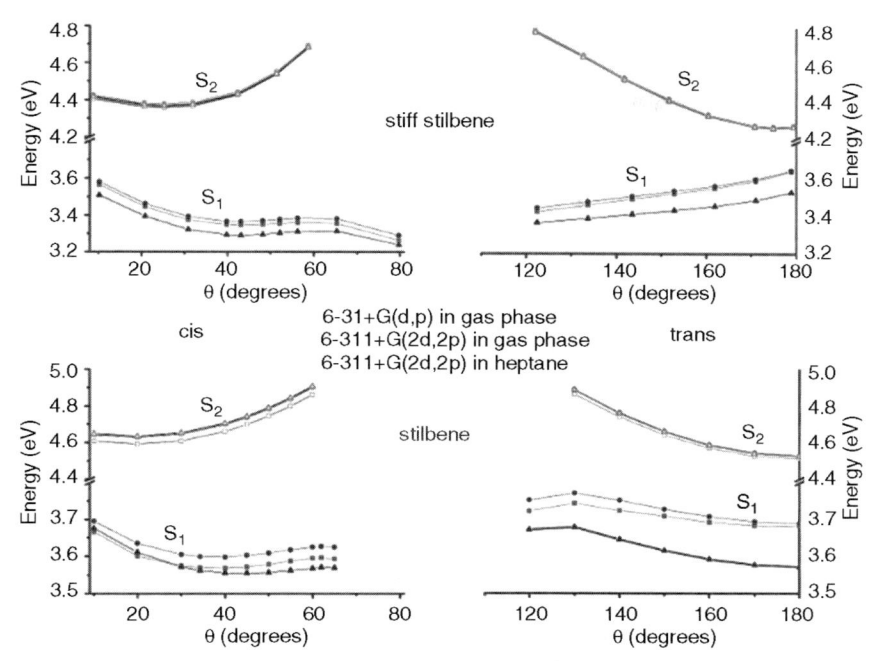

Figure 3.3 TD-DFT transition energies at different geometries. Geometries (geo) are optimized with the 6-31G(d) basis set, and transition energies are computed with the 6-31 + G(d,p) one. Vertical excitation energies are read as subtraction of the values reported on the same column (fixed geometry), while AED are read as subtraction of the energies of two states, each at its equilibrium position. The energies of S_3 and D_2, not explicitly reported, are always within 0.03 and 001 eV with respect to S_2 and D_1 ones, respectively [3]. (Reproduced with permission from Elsevier.)

A map of the singlet–singlet excitation and photoisomerization potential energy surface for tetraphenylethylene in alkane solvents were prepared using fluorescence and picosecond optical calorimetry (Figure 3.4) [4]. The line shapes of the vertical and relaxed excited-state emissions at 294 K in methylcyclohexane were obtained from the steady-state emission spectrum, the wavelength dependence of the time-resolved fluorescence decays, and the temperature dependence of the vertical and relaxed state emission quantum yields and of the time-resolved fluorescence decays.

3.2.2
Excited Triplet State

A schematic picture of the *cis*-stilbene spectroscopic triplet energy, ($E_T^{spectro}$), is given in Figure 3.5. The scheme has been supported by quantum chemical density function theory (DFT) calculations [5].

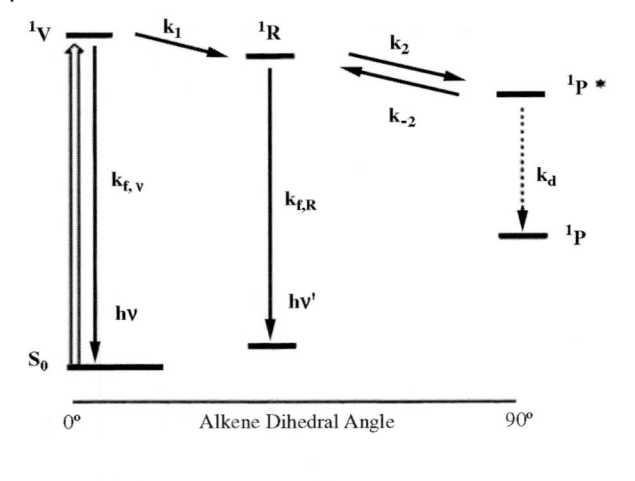

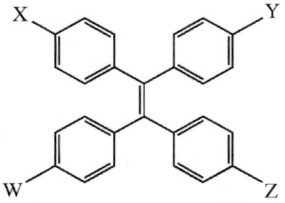

Figure 3.4 A map of the photoisomerization potential energy surface for tetraphenylethylene in alkane solvents [4]. (Reproduced with permission.)

1: W=X=Z=H
2: W=Z=H X=Y=CH$_3$
3: W=H Z=CH$_3$ X=Y=(CH$_3$, H)
4: W=X=Y=Z=CH$_3$

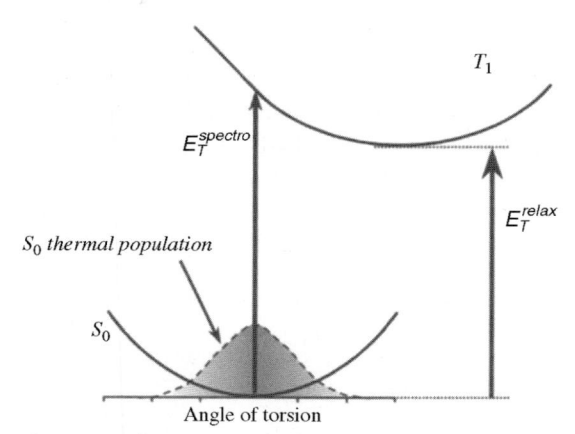

Figure 3.5 Schematic picture of the spectroscopic triplet energy (E_T^{spectro}), the relaxed triplet energy (E_T^{relax}), and the thermal population of ground-state potential energy surface [5]. (Reproduced with permission from Elsevier.)

Results of theoretical calculation at the B3LYP-DFT level using the 6-31 + G(d,p) basis set in ground and triplet states of stilbenes (c-1, t-1) and of 2,3-diphenylnorbornene (**2**) were described [6]. Pronounced pyramidalization at the olefinic C atoms giving a PhCCPh dihedral angle of 51.0° in 3$2^*$ was shown.

3.3
Absorption Spectra

3.3.1
Singlet–Singlet Absorption Spectra

The work in Ref. [7] focused on the computation of UV–vis electronic absorption spectra for different methoxylated stilbenes. The calculations were performed using the time-dependent formalism of density functional theory (TD-DFT) and the B3LYP hybrid functional. For the strongly absorbing first excited singlet state (HOMO → LUMO excitation) of methoxylated stilbenes, calculated transition energies were in agreement with experimental data. It was shown [8] that the Pt-stilbene compounds feature an array of absorptions in the UV–vis region that arise from intraligand (IL) π,π^* transitions. Absorption of these compounds was redshifted compared to that of 4-ethynylstilbene, which implies that there is conjugation between ligands that is conducted by the Pt center. Calculation of absorption spectra of *cis* and *trans* forms of stilbene by the quantum chemical method of INDO with spectroscopy parameterization was carried out [9]. The electron structure of a stilbene molecule and energy-level diagrams were drawn and analyzed. On the basis of the results obtained, the energy-level diagrams, the most probable configurations of photoisomer molecules in ground-excited states, were considered.

According to [10], the electronic spectra of liquid crystalline films of polymers with photoreactive stilbene 4,4′-dicarboxylate containing rod-like mesogenic groups were highly perturbed by the formation of chromophore H-aggregates. Olefins with a phenanthroline ring were prepared and exhibited photochromic behavior with its absorption maximal changing between 380 and 440 nm (Figure 3.6) [11].

3.3.2
Triplet–Triplet Absorption Spectra

The absorption and emission spectra and transient triplet–triplet (T–T) absorption spectra of five fluorostilbenes were studied over a wide temperature range, down to 90 K, and compared with stilbene [12]. The phosphorescence spectra of 5,10-dihydro-5,10-dimethylindolo[3,2-b]indole, its 2,7-dicarbethoxy derivatives, and various S and Se analogues of the latter were measured in liquids, glasses, and low-temperature inert gas or N_2 matrices [13]. Properties of the triplet state (T–T) absorption spectra (Figure 3.7) and triplet lifetimes of *trans*-1-(R)-2-(2,4-dinitrophenyl)ethylenes (R: 1-naphthyl: **IIa**, 9-anthryl: **IIb**, styryl: **IIc** and 4-(2′,4′-dinitro)-*trans*-stilbenyl: **IId**) at 25 and −196 °C were investigated [14]. Experiments showed that the decay of the triplet state follows a first-order law, and the triplet lifetime ($\tau_T = 1/k_{obs}$) in argon-saturated toluene at room temperature was equal to 4 µs for **IId**.

3.3.3
Two-Photon Absorption and Fluorescence

Nonlinear optics (NLO) is the branch of optics that describes the behavior of light in *nonlinear media*, in which the dielectric polarization **P** responds nonlinearly to the electric field **E** of the light. This nonlinearity is typically observed only at very high light intensities such as those provided by pulsed lasers.

Two-photon absorption (TPA) is the simultaneous absorption of two photons of identical or different frequencies to excite a molecule from one state (usually the ground state) to a higher energy state. Two-photon absorption is many orders of magnitude weaker than linear absorption. It differs from linear absorption in that the strength of absorption depends on the square of the light intensity, thus it is a nonlinear optical process. The selection rules for TPA are different from those for one-photon absorption (OPA). For example, in a centrosymmetric molecule, one- and two-photon allowed transitions are mutually exclusive. Since photons have a spin of ±1, one-photon absorption requires excitation to involve an electron changing its molecular orbital to one with a spin different from ±1. Two-photon absorption requires a change of +2, 0, or −2. Two-photon absorption can be measured by photon-excited fluorescence (TPEF) and nonlinear transmission (NLT). If there were an intermediate state in the gap, this could happen via two separate one-photon transitions in a process described as resonant TPA. In nonresonant TPA, the transition occurs without the presence of the intermediate state.

Time-dependent density functional theory studies of the photoswitching of the two-photon absorption spectra in stilbenes were reported [15]. The one- and two-photon absorption characteristics of the open- and closed-ring isomers of **1–3** have been

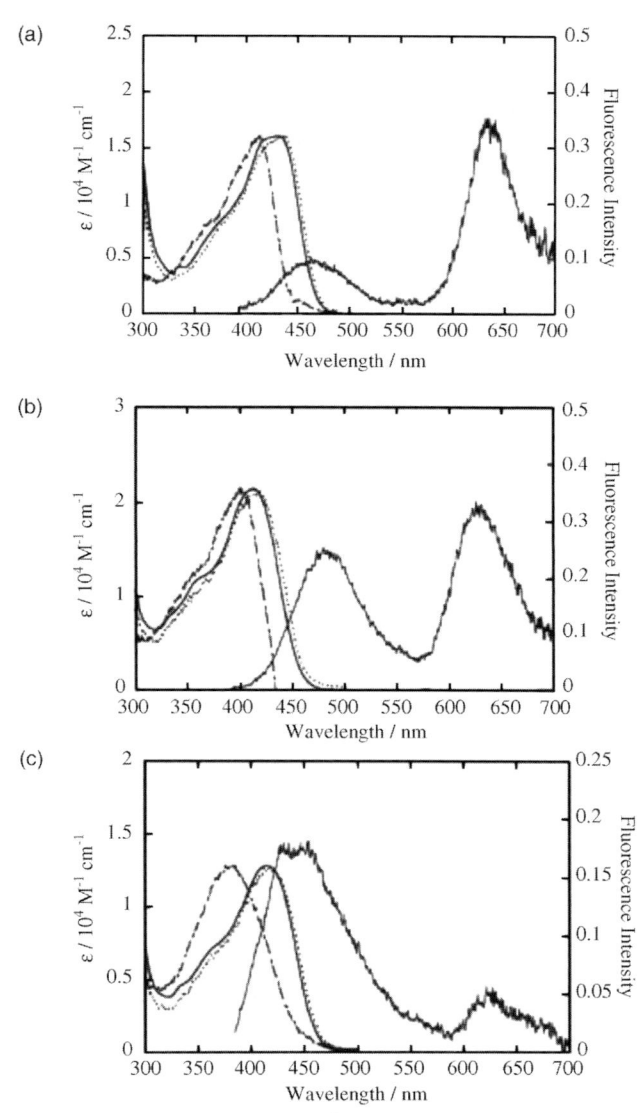

Figure 3.6 Absorption (thick line), fluorescence at 340 nm (thin line), and fluorescence excitation spectra at 510 nm (dot-dashed line). (a and b) 640 nm and (c) 638 nm (dotted line) of phenathroline derivative in benzene (a), in acetonitrile (b), and in methanol (c) [11]. (Reproduced with permission.)

investigated. It was found that the excited states populated by two-photon absorption were nearly 1 eV higher in energy than the lowest energy excited state populated by one-photon absorption and the states excited by OPA had $\pi\pi^*$ character about the C–C framework associated with the bond formation/scission of the central C–C bond. In

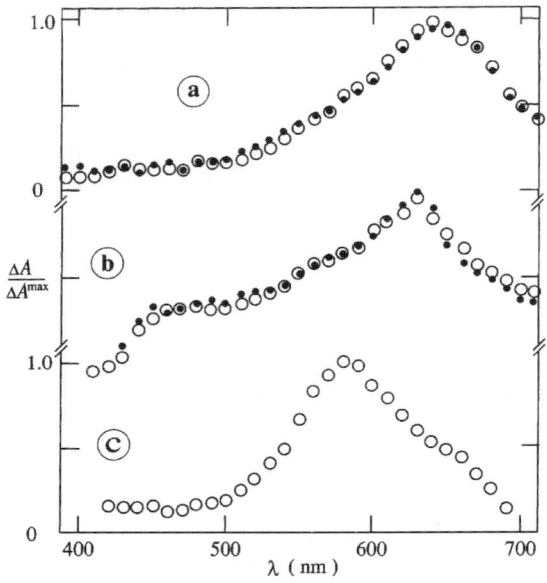

Figure 3.7 T–T-absorption spectra in argon-saturated MTHF at 25 and −196 °C (open and filled symbols, respectively). (a) **IId**, (b) **IIa**, and (c) **IIb**; $\lambda_{exc} = 354$ nm [14]. (Reproduced with permission from Elsevier.)

contrast, states populated by TPA have $\pi\pi^*$ character along the C—C skeletal periphery, including Ph excitations. *trans*-4-(Dimethylamino)-4′-[N-ethyl-N-(2-hydroxyethyl)amino]stilbene (DMAHAS) has been synthesized and characterized by ^{1}H NMR and IR spectra and elemental analysis [16]. Linear absorption, single-photon-induced fluorescence, and two-photon-induced fluorescence in the *trans*-4-(dimethylamino)-4′-[N-ethyl-N-(2-hydroxyethyl)amino]stilbene were detected. The dye has a moderate two-photon absorption cross section of $\sigma^2 = 0.91 \times 10^{-46}$ cm^4 s/photon at 532 nm and shows a strong two-photon-induced blue fluorescence of 432 nm when pumped with 800 nm laser irradiation.

The one- and two-photon excited fluorescence property and crystal structure of a substituted stilbene-type compound *trans*-4-diethylamino-4′-bromostilbene (DEABS) has been reported [17]. Results indicate that this compound has a strong two-photon-excited blue fluorescence at 440 nm when the 700 nm laser is used as the pump source. The one- and two-photon absorption and fluorescence properties of a free radical photopolymerization initiator, (E,E)-4-{2-[p′-(N,N-di-n-butylamino)stilben-p-yl]vinyl}pyridine (Figure 3.8), in various solvents have been investigated [18]. The dye has a moderate two-photon absorption cross section of $\sigma^2 = 0.91 \times 10^{-46}$ cm^4 s/photon at 532 nm. This compound showed a strong two-photon-induced blue fluorescence of 432 nm when pumped with 800 nm laser irradiation. Quantum chemical calculation indicated that the new initiator possesses a large delocalized π-electron system, a large change in dipole moment upon transition to the excited state, and a large transition moment. The two-photon absorption cross section was calculated as high as 881.34×10^{-50} cm^4 s/photon.

Both the electronic and the vibronic contributions to one- and two-photon absorption of a D-π-D charge transfer molecule (4-dimethylamino-4'-Me-*trans*-stilbene) were studied by density functional response theory combined with a linear coupling mode (Figure 3.8) [18].

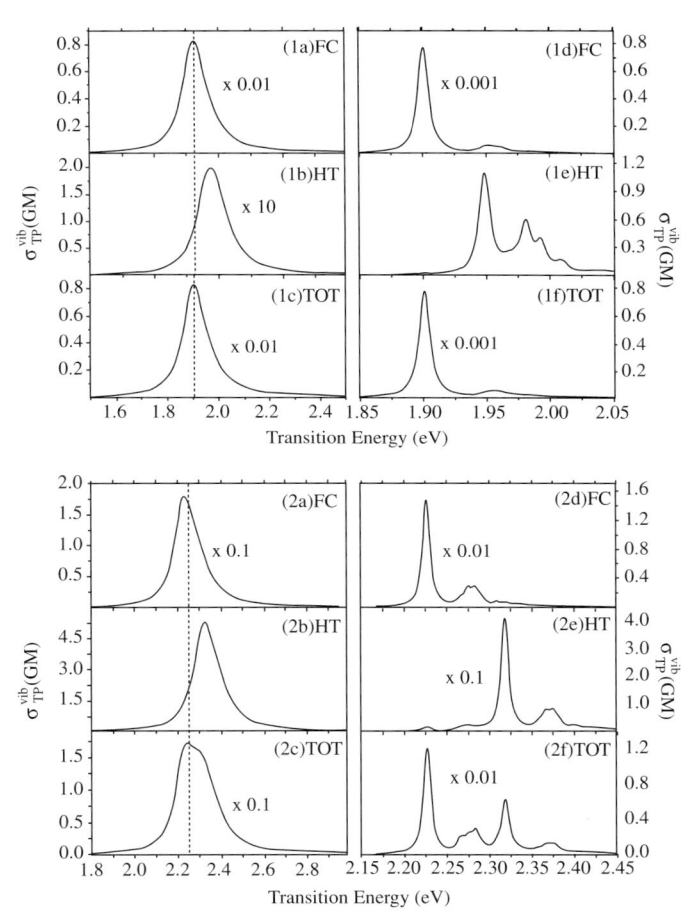

Figure 3.8 Vibronic two-photon absorption spectra of S_1 **1a–1f** and S_2 **2a–2f** of 4-dimethylamino-4'-Me-*trans*-stilbene calculated with B3LYP at the HF geometry. Franck–Condon contribution, Herzberg–Teller contribution, and total spectra are shown for lifetime broadening of 0.10 eV (left panels) and 0.01 eV (right panels), respectively. The dashed line corresponds to the vertical transition [18]. (Reproduced with permission.)

3.4
Fluorescence from Excited Singlet States

3.4.1
Fluorescence Behavior and Hammett Relationships

The Hammett concept of the linear free energy relationships (LFERs) or the Hammond plot is widely used in physical organic chemistry [19–22]. There are several reasons to apply LFERs or the Hammond plot to stilbene photophysics and photochemistry. First, the correlation approach to stilbene photochemistry makes it possible to elucidate the contribution of substituent or solvent effects to various steps of the processes, which take place in the elementary act of a chemical reaction but are undetectable by direct experimental measurements. Second, these quantitative relationships based on the Hammett-like structure–reactivity correlation with σ values of substituents can indicate mechanisms for various substituted stilbenes. It is reasonable to assume that molecules lying on the same Hammett plot belong to the same reaction series and mechanism. Thus, it is possible to quantitatively predict the photophysical parameters of trans–cis photoisomerization and assume its mechanism for a substituted stilbene in arbitrarily chosen media by considering the donor–acceptor properties of its substituents. Additional effects that generate an additional series of reactions may explain deviations from such linear relationships. If all such observed effects are classified and quantified, retrospective rationalization and prediction of photochemical processes is possible. Experimental LFERs data can serve as a base for a profound theoretical investigation. To establish the reaction series, small changes can be introduced in two ways:

1. **Modification of the stilbene molecule by introducing different 4,4'-substituents**. This leads to a Hammett-like relationship. Although LFERs usually deal with reaction rate and equilibrium data of chemical reactions, this approach can be extended to various photophysical parameters of the excited molecules.

2. **Solvent effects**. Thermodynamically, solvation may be viewed along the same lines as substituent effects, the solvating molecules being equivalent to loosely attached substituents.

3.4.1.1 Excitation to the Franck–Condon State
Excitation of the investigated stilbene molecule from its ground state ^1t to the Franck–Condon state $^1t^*_{FC}$ occurs in a few femtoseconds. As a result, only fast electronic polarization techniques can follow a drastic change of the charge distribution around the zwitterionic exited FC state. The latter has been proved particularly by the excitation energy dependence on the solvent refractive index [23]. The first step after excitation to the Franck–Condon state of the trans-stilbene configuration is vibrational relaxation followed by solvent–solute relaxation that leads to a rapid population of the $^1t^*$ state from which fluorescence occurs. These relaxation processes result in a Stokes shift (ΔE).

Absorption and emission spectra of 20 trans-4,4′-disubstituted stilbenes have been measured in four solvents: cyclohexane (CH), chlorobenzene (CB), 2-butanone (methylethylketone, MEK), and dimethylsulfoxide (DMSO) at room temperature [24]. Fluorescence quantum yields (Φ_f) and fluorescence lifetimes (τ_f) have been measured for these stilbenes. Substituent effects on the Stokes shift were described by a spectroscopic Hammett equation

$$\frac{\Delta E}{2.3kT} = \sigma\rho, \tag{3.1}$$

where σ is the Hammett substituent constant and ρ the slope, which can indicate the general mechanism or some general effects of the investigated reaction.

Figure 3.9 shows the plots of Stokes shift versus Hammett σ-constant difference ($\sigma_X - \sigma_Y$) of the 4,4′-substituents for a series of *trans*-4,4′-disubstituted stilbenes

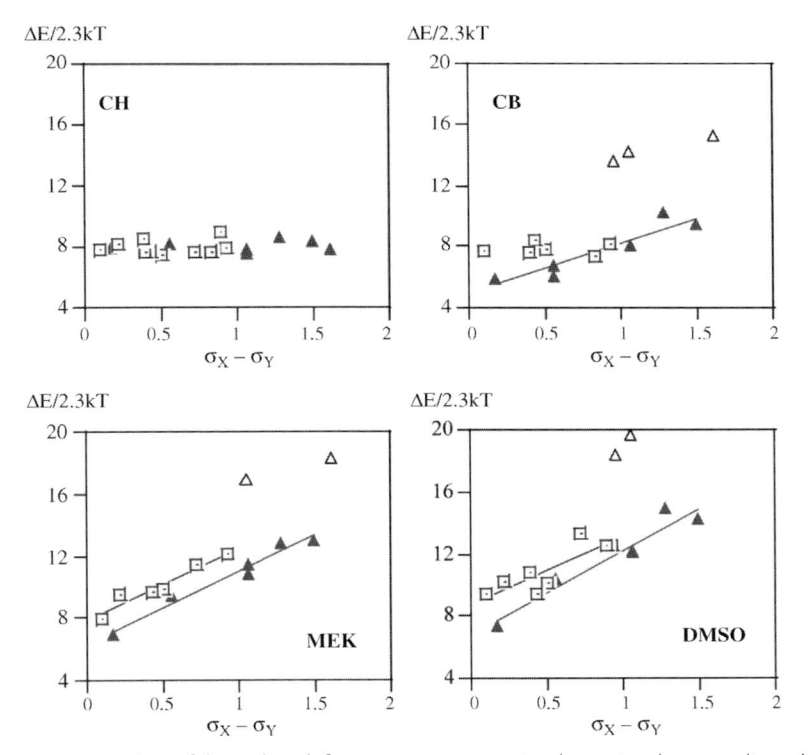

Figure 3.9 Plots of the Stokes shift versus Hammett σ-constants difference of the two 4,4′-positioned stilbene substituents (X and Y) (taking into account their relative sign) in cyclohexane, chlorobenzene, methylethylketone, and DMSO. The first group is assigned by open squares and the second by filled triangles. The open triangles assign the anomalies, which persist only in polar solvents, and occasionally may form a third group of stilbenes. The uncertainty in estimation of $\Delta E/2.3kT$ was found to be ±0.58 [24]. (Reproduced with permission from Elsevier.)

Table 3.1 Sensitivity (ρ) of the Stokes shift, fluorescence decay rate constant, fluorescence quantum yield, and radiative deactivation rate constant to intramolecular substituent effects for two different groups of *trans*-4,4'-disubstituted stilbenes [87].

	Group I				Group II			
	CH	CB	MEK	DMSO	CH	CB	MEK	DMSO
$\Delta E/2.3kT$	0	0	4.67	6.21	0	2.65	4.64	5.45
$\log 1/\tau_{fl}$	1.23	0.88	0.92	0.50	0.86	0.48	0.49	0.33
$\log \Phi_{fl}$	—	−0.79	−0.78	−0.86	—	−1.17	−1.21	−1.43
$\log k_r$	—	0.09	0.14	−0.36	—	−0.68	−0.72	−1.1

Reproduced with permission from Elsevier.

that exhibit a linear behavior with scatter in polar solvents, meaning that the solvent–solute relaxation is sensitive to the substituent effects [24, 25].

These relationships have an unusually high slope (ρ-constant) in polar solvents (Table 3.1, MEK, 4.64; DMSO, 6.21). Such high values are typical for chemical reactions running through charged intermediates. In nonpolar solvent, cyclohexane, $\rho \approx 0$.

These data can be explained in terms of the high stabilization energy resulting from solvation of the excited $^1t^*$ state. High ρ values in these cases indicate that the polar solvent–solute intermolecular stabilization of the zwitterionic excited $^1t^*$ state is very sensitive to intramolecular substituent effects. In contrast, there is no dependence of Stokes shifts on σ-constants in cyclohexane, which is nonpolar aprotic solvent, where the vibrational relaxation of the Franck–Condon state plays a primary role in stabilizing the excited state. This implies that the vibrational relaxation is not sensitive to intramolecular donor–acceptor interactions. The observations showed that the ultrafast intra- and intermolecular electronic polarization plays a major role in determining the position of the Franck–Condon zwitterionic state and its sensitivity to the relaxation of polar-substituted stilbenes.

Stilbene molecules substituted with the 4-NO_2 group II exhibit unusually large Stokes shifts and large deviations from the plot of ΔE on the Hammett σ-constants. These deviations can be attributed to polar and specific intramolecular effects of the 4-NO_2 substituent, which is able to quench the charge transfer state emission [26].

3.4.1.2 Radiative Deactivation

The fluorescence lifetime (τ_{fl}) and fluorescence quantum yield Φ_{fl} [24, 25] may be expressed as follows:

$$\tau_{fl} = (k_r + k_{nr} + k_{t \to c})^{-1} \equiv k_d^{-1}, \tag{3.2}$$

$$\Phi_{fl} = \frac{k_r}{k_r + k_{nr} + k_{t \to c}}, \tag{3.3}$$

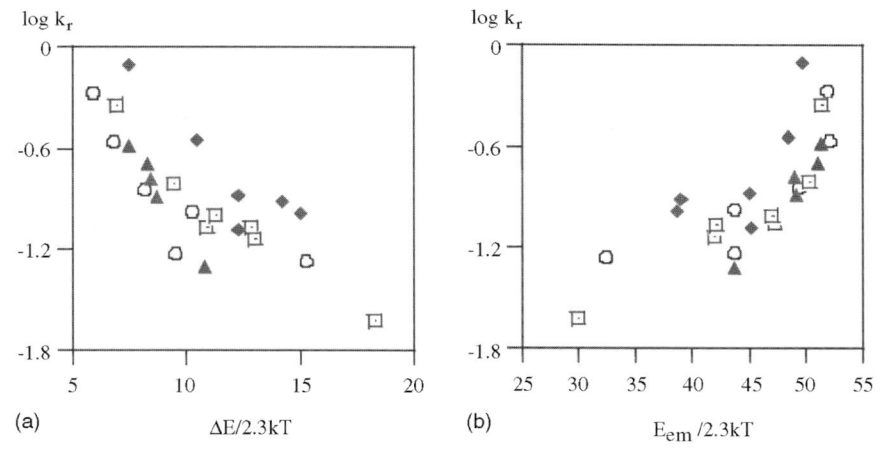

Figure 3.10 Plots of the Stokes shift (a) and emission energy (b) versus logarithmic value of the radiative rate constant (k_r in ns^{-1}) for the strong donor–acceptor disubstituted stilbenes from the second and third groups in CB (open circles), MEK (squares), DMSO (rhombus), and poly(vinyl alcohol) films (triangles) [24]. (Reproduced with permission from Elsevier.)

where k_d is the excited-state decay rate constant, k_r and k_{nr} are the radiative and nonradiative decay rate constants, respectively, and $k_{t \to c}$ is the rate constant of the $^1t^* \to {}^1p^*$ transition.

In most cases studied in low viscous solutions and in organized media of low viscosity, $k_{t \to c}$ was found much higher than the corresponding k_r and k_{nr} [27–29]. Therefore, in a good approximation, k_d is close to the *trans–cis* photoisomerization rate constant. The radiative rate constant k_r may be calculated from experimental values of τ_{fl} and Φ_{fl} according to the following equation:

$$k_r \approx \Phi_{fl} k_d. \tag{3.4}$$

Figure 3.10 shows the logarithmic dependence of k_r on the Stokes shift and emission energy in CB, MEK, DMSO, and poly(vinyl alcohol) films [24]. These plots indicate the essential intermolecular polar effects on the radiative deactivation rate of stilbene molecules. The higher the Stokes shifts were, the slower the radiative deactivation of the excited $^1t^*$ state was. On the other hand, radiative deactivation rate constants increased with an increase in emission energy. The k_r values are found sensitive to polar substituents of stilbenes. In this case, the substituent effects are characterized by the negative ρ-constants (Table 3.1). These experimental data may be explained as follows. The substituents affect the Franck–Condon factor at the radiative deactivation transition $^1t^* \to {}^1t$, which appears to be vertical. This alteration of the Franck–Condon factor shifts the vibrational energy levels of the ground state and varies the probability of this transition.

3.4.2
Miscellaneous Data on Stilbenes Fluorescence

3.4.2.1 Stilbenes in Solution

A fluorescence technique for separating the radiative from the nonradiative energy transfer using essentially front-surface emission from a round cell containing high donor and acceptor concentration (0.1×10^{-2} M) was developed and applied to *trans*-stilbene–azulene systems [30]. The use of the Stern–Volmer plots corrected by the Beer–Lambert relationship allowed to calculate the quenching constants $kq\pi$ at 30 °C for *trans*-stilbene, *trans*-*p*-bromo-, and *trans*-*m,m'*-dibromostilbene in *n*-pentane that were found to be 20.7, 24.8, and 16.4 mol, respectively. Data on the electronic absorption and fluorescence spectra, quantum yields for fluorescence (Φ_f) and *trans*–*cis* photoisomerization (Φ_{tc}), and fluorescence lifetimes of a series of *trans*-stilbenes in various solvents were reported [31]. *trans*-4-(*N*-Arylamino)-4'-cyanostilbenes (2H, 2Me, 2OM, 2CN, and 2Xy with aryl = Ph, 4-methylphenyl, 4-methoxyphenyl, 4-cyanophenyl, and 2,5-dimethylphenyl, respectively), *trans*-4-(*N*-methyl-*N*-phenylamino)-4'-cyanostilbene (2MP), *trans*-4-(*N,N*-diphenylamino)-4'-cyanostilbene (2PP), *trans*-4-(*N*-methyl-*N*-phenylamino)-4'-nitrostilbene (3MP), and three ring-bridged analogues 2OMB, 2MPB, and 3MPB were investigated. Connection of the stilbenes' spectral properties and the isomerization-free twisted intramolecular charge transfer (TICT) state was discussed.

The absorption, fluorescence, and fluorescence excitation spectra of concentrated toluene solution of selected *para*-substituted *trans*-stilbene derivatives detected in [32] provided evidence for aggregation. A redshifted fluorescence spectrum peaking at 420 nm gained in intensity as the stilbene concentration was increased. The presence of polar substituents was found to be crucial to the formation of a fluorescent ground-state dimer (or higher aggregate). The time-resolved fluorescence behavior of two derivatives of 4-(dimethylamino)-4'-cyanostilbene (DCS) bearing a more voluminous (JCS) and a less voluminous anilino group (ACS) was studied in ethanol using picosecond time-resolved single-photon-counting technique [33]. For JCS, reconstructing emission spectra exhibited an isosbestic point that indicated level dynamics between two emitting excited singlet states (LE and CT). Kinetic evaluation yielded a precursor–successor relationship between LE and CT and CT formation time constants of 4 ps for ACS and 8 ps for JCS. *trans*-Stilbene and several cyclic derivatives with hindered free rotation around the C(vinyl)–C(phenyl) single bond were studied by various spectroscopic techniques [34]. Fluorescence properties of compounds under investigation were found to strongly depend on their chemical structure. Derivatives containing six- or seven-membered aliphatic rings did not exhibit any measurable $S_1 \rightarrow S_0$ fluorescence. The introduction of two methoxy groups into the six-membered aliphatic ring derivatives accelerated its photoreactivity to such an extent that fluorescence became too low to be measured. Only *trans*-stilbene and its 4-membered cyclic analogues were amenable to fluorescence lifetime measurements.

Fluorescence properties of two series of trinuclear diimine Re(I) tricarbonyl complexes [Re(CO)$_3$(NN)]$_3$(μ^3-L1,2) (PF$_6$)$_3$ (NN = ethylenediamine or substituted bipyridine (A) and phenanthroline; 1,3,5-tris(4-ethenylpyridyl)benzene = L1, 1,3,5-tris(4-ethynylpyridyl)benzene = L2) (B) linked by an isomerizable stilbene-like ligand were described [35]. It was detected that the L2-bridged complexes (group B) exhibit strong luminescence and long emission lifetimes at room temperature in solution that are typical of decay from 3MLCT excited states. The L1-bridged complexes (group A) showed only very weak luminescence and short lifetimes under the same conditions. The authors attributed the low emission quantum yields and short lifetimes in group A complexes to intramolecular sensitization of the $3\pi\pi^*$ excited state localized on the olefin link of the bridging ligand accompanied by a subsequent *trans–cis* isomerization process.

According to [36], the emission maxima of fac-[Re(CO)$_3$(NN)(L)]$^+$ complexes, NN = polypyridyl ligands and L = stilbene-like ligands, in acetonitrile solution and in poly(methyl methacrylate) (PMMA) polymer film exhibited hypsochromic shifts as the medium rigidity increases due to the luminescence rigidochromic effect. Time-resolved IR (TRIR) spectroscopy, in combination with other techniques, characterized the excited-state electronic properties of the fac-[Re(CO)$_3$(phen)(bpe)]PF$_6$ complex, where bpe is 1,2-bis(4-pyridyl)ethylene.

The fluorescence of *trans*-stilbene and four methoxy-substituted stilbene derivatives has been detected in a variety of solvents [37]. Compared to other stilbene derivatives, *trans*-3,5-dimethoxystilbene displayed a large quantum yield of fluorescence and a low quantum yield of *trans–cis* isomerization in polar organic solvents. The unique fluorescence properties of *trans*-3,5-dimethoxystilbene were attributed to the formation of a highly polarized charge transfer excited state ($\mu_e = 13.2$ D). The fluorescence of all five *trans*-isomers was quenched by 2,2,2-trifluoroethanol.

Fluorescence quantum yields of *cis*-stilbene-d$_0$ and -d$_2$ were measured as a function of temperature in *n*-hexane and *n*-tetradecane [38]. Emission contributions from ^{1}c*, φ_{fcc}, and adiabatically formed ^{1}t*, φ_{ftc}, to the *cis*-stilbene fluorescence quantum yields were revealed by the application of principal component analysis. The φ_{ftc} values were found to be temperature independent in both solvents. Medium and deuterium isotope effects on fluorescence quantum yields of *trans*-stilbene-d$_0$, *t*-d$_0$, and α,α'-d$_2$, *t*-d$_2$, in *n*-hexane and *n*-tetradecane and decay processes were investigated [39]. Radiative, k_f, and nonradiative, k_{tp}, decay rate constants of *trans*-stilbene were defined. The index of refraction dependence of k_f was given by $k_f = k_{fo}nx$, where $x = 1.65$ and $k_{fo} = 3.75 \times 10^8 \text{ s}^{-1}$, the known radiative rate constant for jet-cooled isolated *trans*-stilbene in the gas phase. Substitution of vinyl hydrogens with deuterium led to a 50–60% increase in fluorescence quantum yields. A matrix of fluorescence spectra at 30 °C from a *cis*-stilbene solution in *n*-hexane was obtained and a series of *cis*-stilbene solutions with added *trans*-stilbene concentration were investigated [40]. The analysis of experimental data indicated that the transmission from the pure *cis*-stilbene solution is composed of *trans*-stilbene fluorescence (51–54% of total area) and structureless *cis*-stilbene fluorescence, $\lambda_{max} = 408$ nm. The major source of the *trans* portion of the spectrum from the pure *cis*-stilbene solution (72%) was shown to be adiabatic $^1c^* \rightarrow {}^1t^*$ conversion by 0.16% of initially excited *cis* molecules, $^1c^*$. The index of refraction dependence of the radiative rate constant, k_f, fluorescence quantum yields of *trans*-4,4'-di-*tert*-butylstilbene in *n*-hexane and *n*-tetradecane, measured as a function of temperature in combination with experimental fluorescence lifetimes, was defined. The k_f values for *trans*-4,4'-di-*tert*-butylstilbene and *trans*-4,4'-dimethoxystilbene were measured to be 0.8×10^8 and $3.5 \times 10^8 \text{ s}^{-1}$, respectively.

The temperature dependence of fluorescence quantum yields was used for the evaluation of the participation of equilibration of planar, $^1t^*$, with twisted (phantom), $^1p^*$, conformations of the lowest excited singlet of *trans*-stilbene [41]. The majority of the fluorescence investigated in this work occurred with very short fluorescent lifetime at room temperature. The spectral luminescent characteristics of stilbene-substituted (1,4-bis(styryl)benzene, 4-phenylstilbene) were calculated [42]. Spectral properties of two conformers of pseudo-*p*-distyryl[2.2]paracyclophane, a model of a stilbene dimer in the ground state in THF, were studied [43]. It was shown that the conformer A with the smallest optical gap of 3 eV has a short radiative lifetime of 3.3 ns, while the conformer B with an optical gap 0.3 eV larger is much more abundant in solution and its radiative lifetime is 10 times longer. On the basis of *ab initio* calculations, A and B were assigned to flat and twisted conformations, respectively. Conformer B showed a partial decay of excited-state absorption and fluorescence with a time constant ranging from 5 to 30 ps, depending on excitation photon energy. Photophysical properties of 4-(2',4'-dinitro)-*trans*-stilbene were studied in solution [44]. The quantum yield of fluorescence was found to be small in 2-methyltetrahydrofuran at room temperature but enhanced upon approaching −196 °C.

A series of aminostilbenes (1A–C, 2A–C) have been synthesized to test the effect of substitution of the amino group upon the photophysics and photochemistry of stilbenes [45]. This study indicated that the photophysics properties of *trans*-2-aminostilbene, 1A, and *trans*-3-aminostilbene, 1B, were similar. *cis*-2-Aminostilbene, 2A, and *cis*-3-aminostilbene, 2B, showed similar fluorescent lifetimes. Anomalous behavior of emission anisotropy for short-living derivatives of stilbenes was described [46]. The absorption, fluorescence and polarization excitation, and emission spectra

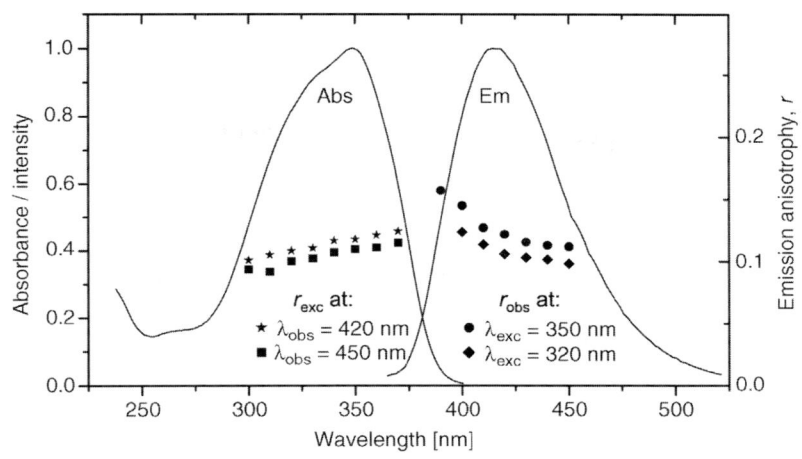

Figure 3.11 Absorption, emission, and polarization spectra as well as excitation and emission polarization spectra (points) of DMS in propanol [47]. (Reproduced with permission from Elsevier.)

for stilbenes in solvents of different viscosity and polarity were presented. Higher emission anisotropy values in the absorption band were observed that stay at the same level upon excitations different from those corresponding to the 0–0 transition wavelength. The authors explained this effect by the presence of two luminescent centers contributing to emitted light or by incomplete relaxation of the excited states before emission. Steady-state absorption, emission, polarization spectra, and time-resolved emission spectra (Figure 3.11) were used to study the spectroscopic properties of 4-dimethylamino-4′-methoxy-stilbene (DMS) in various solvents at 23 °C [47]. The short time components of fluorescence (τ less than 100 ps) were found for DMS in polar solvents. The long-time component shortens with decreasing viscosity of solvents. Emission anisotropy r values increase in the vicinity of the 0–0 transition wavelength. Figure 3.11 shows absorption, emission, and polarization spectra as well as excitation and emission polarization spectra (points) of DMS in propanol. Changes in emission of DMS in ethylene glycol spectra in time were detected.

Imide model compounds synthesized from 2,5-dimethoxy-4′-aminostilbene and different anhydride were photophysically characterized by UV absorption spectroscopy and stationary and time-resolved fluorescence spectroscopy [48]. Fluorescence measurements in solution showed the dominance of nonradiative processes at room temperature. The steady-state fluorescence spectra obtained from stilbenes incorporated in various polymer matrices indicated differences in the number of bands. The semiempirical quantum mechanical calculations demonstrated the dominance of a charge transfer for the HOMO–LUMO configuration from the stilbene skeleton to the imide subunit.

Luminescent stilbenoid chromophores with diethoxysilane end groups were prepared via Heck reactions and characterized [49]. Diethoxysilane-substituted styrenes were used as vinylic components, thus allowing the combined connection of the chromophore to the silane moiety with an extension of the π-system. Monodisperse oligo(phenylenevinylene)s of different conjugation lengths and bromine or iodine as reactive sites were used as coupling partners. Electric and optical properties were tuned via the length of the conjugated system, electron-withdrawing cyanide and electron-donating alkoxy side chains. Sixteen new fluorescent N4-(E)-stilbenyloxyalkylcarbonyl-cytosines and N4-(E)-stilbenyloxyalkylcarbonyl-1-methylcytosines were synthesized [50]. Assignment of signals in the spectra of compounds in NMR in DMSO-d_6 solution was made the basis of the homonuclear (COSY) and heteronuclear (HETCOR) spectra. The effect of the substituent (Cl, Br, and NO_2) on the stilbene moiety on the fluorescence spectrum of each compound was discussed.

3.4.2.2 Stilbene Dendrimer Fluorescence

Fluorescence properties of dendrimers (Gn is the dendrimer generation number) containing four different luminophores, namely, terphenyl (T), dansyl (D), stilbenyl (S), and eosin (E) (Figures 3.12 and 3.13), have been studied [51]. Depending on photophysical properties of fluorescent units and structures of dendrimers, different mechanisms of fluorescence depolarization were observed: (i) global rotation for GnT dendrimers; (ii) global rotation and local motions of the dansyl units at the periphery of GnD dendrimers; (iii) energy migration among stilbenyl units in G2S; and (iv) restricted motion when E is encapsulated in a dendrimer, coupled to energy migration if the dendrimer hosts more than one isolated eosin molecule. Emission spectra in acetonitrile solution are also reported, $\lambda_{ex} = 290$ nm.

Fluorescence emission of photoresponsive polyphenylene dendrimers of different generation with considerably high quantum efficiency was observed [52]. Experiments showed the fluorescence and fluorescence excitation spectra of polyphenylene-based stilbene dendrimers and G1–G3 in chloroform. G1–G3 exhibited fluorescence emission at longer wavelengths compared to stilbene. Excitation energies can be estimated to be 86, 82, and 81 kcal/mol for stilbenes, G1, G2, and G3, respectively. Fluorescence lifetimes of dendrimers were obtained considerably longer than that of stilbenes. The singlet lifetimes of G1, G2, and G3 were obtained at 2.6, 1.6, and 1.7 ns, respectively, in benzene, while the singlet lifetime of stilbene was reported to be 70 ps in pentane at room temperature, and the fluorescence quantum yields of the dendrimers (0.6) were much higher than that of stilbene. The fluorescence lifetime of G2 was not affected by temperature and was found to be 1.6 ns at 150 K. In the review [53], the syntheses and reactions of photoresponsive dendrimers were described. Dendrimers with photoreversible stilbene cores undergo mutual *cis–trans* isomerization in organic solvents within the lifetime of the excited singlet state to give photostationary state mixtures of *cis*- and *trans*-isomers. The large dendron group surrounding the photoreactive core may affect the excited-state properties of the core to increase the photoisomerization efficiency and/or reduce the fluorescence efficiency. The photochemistry of stilbene dendrimers, with various types of dendron groups, azobenzene dendrimers, and other photoresponsive dendrimers was discussed. It was also shown [53] that three polyphenylene-based stilbene dendrimers, G1, G2, and G3, exhibited fluorescence emission of considerably higher quantum efficiency and longer fluorescence lifetime compared to the parent stilbene.

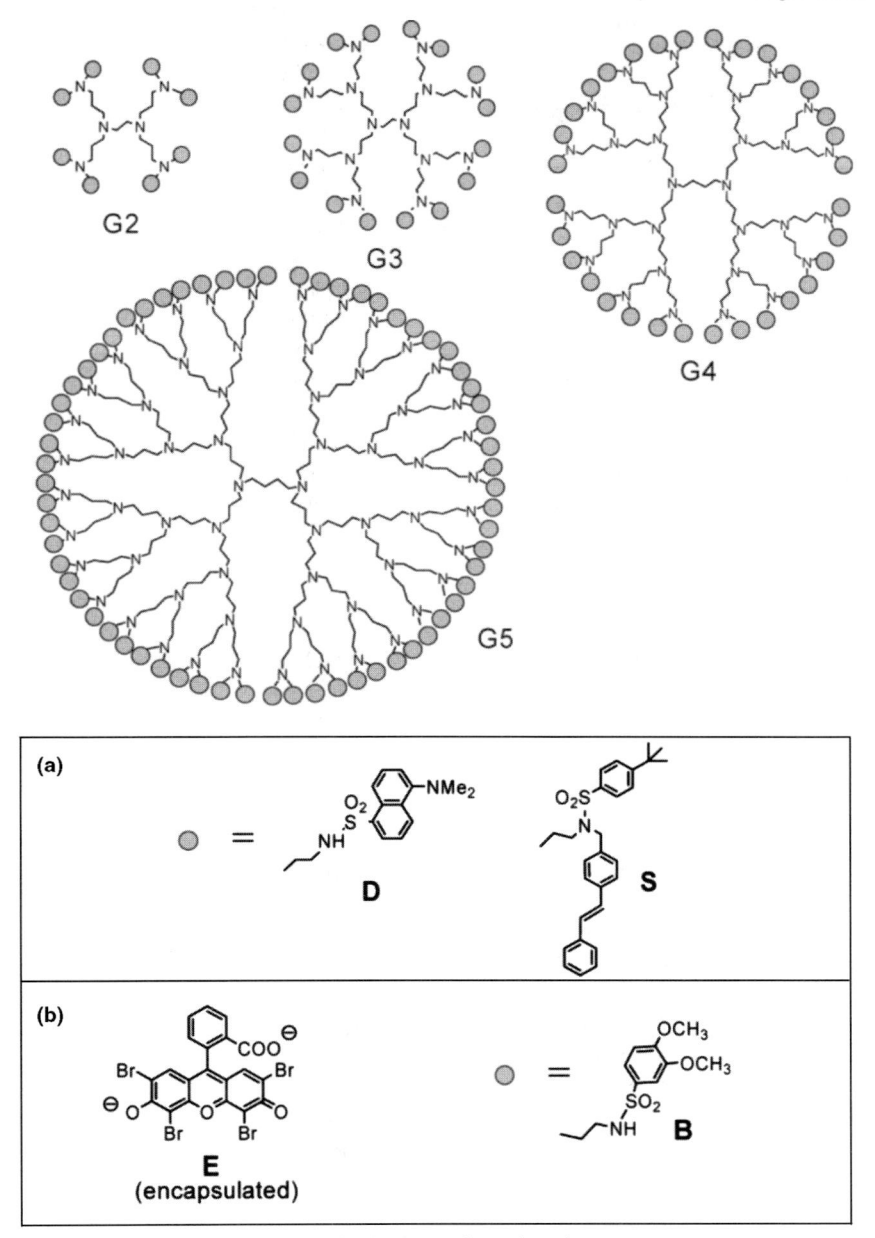

Figure 3.12 Poly(propylene amine) dendrimers with (a) dansyl (D) and stilbenyl (S) peripheral units; (b) eosin (E) encapsulated in G4 carrying 1,2-dimethoxybenezene (B) peripheral units [51]. (Reproduced with permission.)

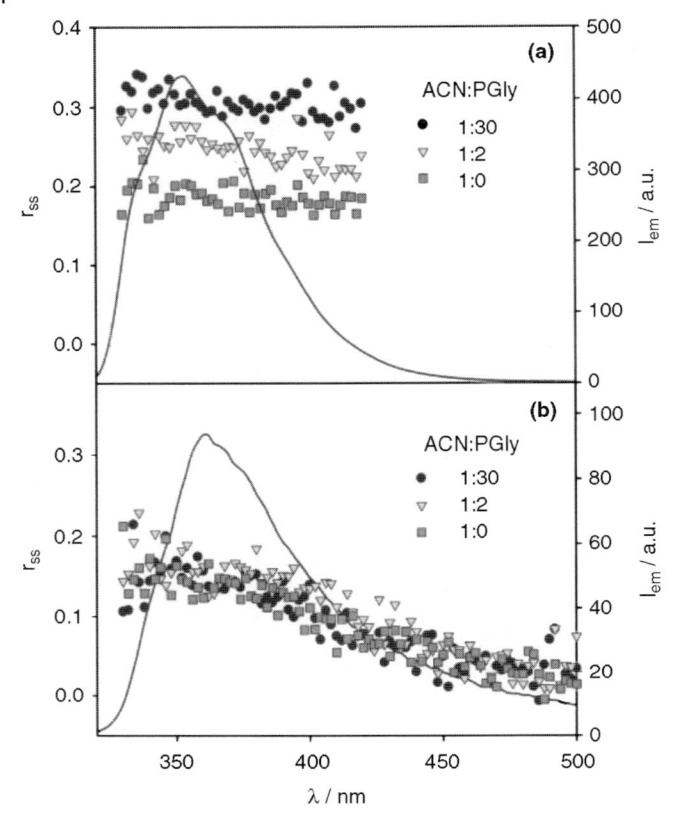

Figure 3.13 Steady-state anisotropy, r_{ss}, in acetonitrile, ACN (squares), ACN/PGly (propylene glycol) 1:2 (v/v) (triangles), and ACN/PGly 1:30 (v/v) (circles) solutions at 298 K of compounds S (a) and G2S (b) [51]. (Reproduced with permission.)

3.4.2.3 Stilbenes on Templates and in Proteins

The local media effect on stilbenes covalently immobilized on silica plate was investigated [54, 55] (see also Section 10.3.7.1). Several substituted *trans*-stilbene derivatives have been prepared and immobilized on a quartz surface. A number of immobilization methods have been tried including the silanization technique, cross-linking with cyanuric chloride, surface activation with cyanogen bromide, and surface smoothing with coating proteins (Figure 3.14). Studies of solvent polar effects on the fluorescence spectrum of the immobilized stilbenes indicated that the maximum wavelength of the fluorescence emission was not very sensitive to solvent polarity. The apparent local polarity of the medium in the vicinity of the stilbene label was estimated and parameter polarity (E30T value) was found to be close to 50 kcal/mol.

The effect of SiO_2 and SiO_2–TiO_2 surfaces (in the powder form) on the fluorescence and lifetimes of the adsorbed two push–pull stilbenes, *E,E*-1-(4-cyanophenyl)-4-(4-*N,N*-dimethylaminophenyl)-1,3-butadiene and *E*-9-(4-cyanostyryl)-2,3,6,7-tetrahydro-1*H*,5*H*-pyrido[3,2,1-*i,j*]-quinoline was observed [56]. The stilbene fluorescence was quenched effectively by mechanism of electron injection from excited stilbene

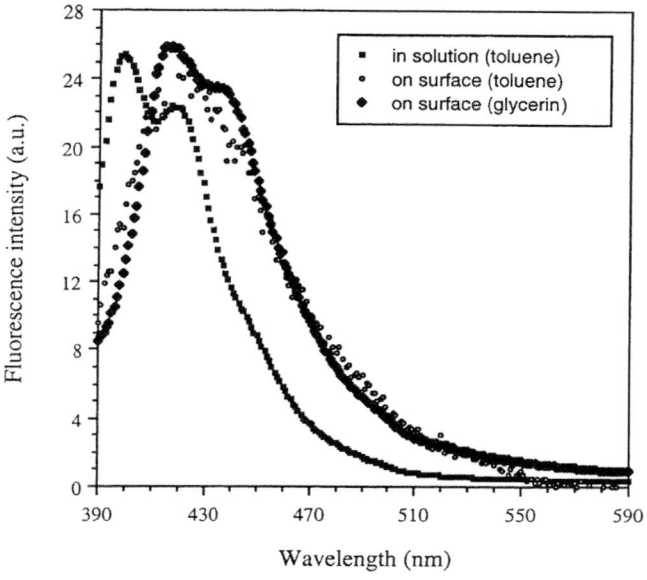

Figure 3.14 Fluorescence emission of *trans*-4-dimethylamino-4′-amino-stilbene in a free state in 8×10^{-9} M toluene solution (closed squares) and an immobilized state cross-linked with cyanuric chloride to the coating BSA protein in toluene (open squares) and in glycerin (rhombus) [55]. (Reproduced with permission from Elsevier.)

to TiO_2. The change in the fluorescence maxima of the stilbenes on the SiO_2 surface was suggested to be caused by a change in local polarity around the adsorbed species due to the competitive adsorption of water molecules from air and removal of adsorbed stilbenes to places with lower polarity. The fluorescent emission of the stilbene chromophore in polyurethane cationomers was studied comparatively with the urethane-stilbene [57]. It was shown that stilbene polycations absorbed at $\lambda_{max}A = 316$ nm and emitted violet-blue light with emission maxima at $\lambda_{max}F = 444$ nm (DMF solution) and $\lambda_{max}F = 465$ nm (solid state). The redshift of the fluorescent band of the polyurethane stilbene was attributed to the formation of small aggregates of stilbene molecules in thin films.

Several works were related to antibody–stilbene complexes. The process of preparation and the use of monoclonal anti-stilbene antibodies have been patented [58]. The invention provided hybridomas that produce and secrete antistilbene antibodies. An antibody of the present invention has particular utility in identifying and/or locating target moieties appended to or incorporating antigenic stilbene, in one embodiment, and therefore provides a method for detecting antigenic stilbene. The method includes the steps of exposing antigenic stilbene to an antistilbene antibody and detecting an antistilbene antibody–stilbene immunoconjugate. Such immunoconjugates were detected using fluoroscopic procedures. Examples of stilbene-tethered hydrophobic C-nucleosides were described [59]. Compounds of this type were targeted for use with "blue fluorescent antibodies" with the aim of probing

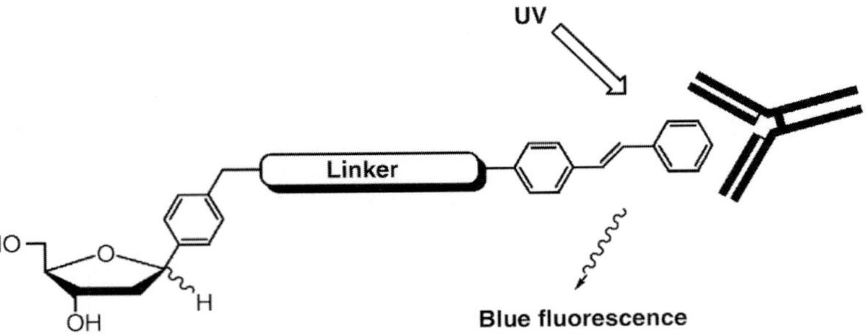

Figure 3.15 Scheme of stilbene-tethered hydrophobic C-nucleosides [59]. (Reproduced with permission.)

native and nonnatural DNA (Figure 3.15). The authors suggested that hydrophobic nucleosides will be useful in current native and nonnatural DNA studies and invaluable for investigating novel, nonnatural genomes in the future.

Fluorescence and fluorescence quenching properties of stilbene derivatives were used for the characterization of solid-phase antibodies modified by a stilbene-labeled hapten [60]. Measurements of quenching by kJ, rates of *cis–trans* photoisomerization, and photodestruction of a stilbene-labeled hapten by a quencher from solution were carried out. These experimental parameters enabled a quantitative description of the order of binding sites of antibodies immobilized on a surface and can be used to characterize the microviscosity and steric hindrance in the vicinity of the binding site. In this work, anti-dinitrophenyl antibodies and stilbene-labeled dinitrophenyl (DNP) were used to investigate three different protein immobilization methods: physical adsorption, covalent binding, and the Langmuir–Blodgett technique. Blue fluorescent antibodies as reporters of steric accessibility in virus conjugates were described [61]. The attachment of organic compounds to either the inside or the outside surface of the cowpea mosaic virus (CPMV) coat protein was verified with an indicative antibody–antigen interaction. Antibody binding was subsequently blocked by installing poly(ethylene glycol) chains. These results typify the type of site-specific control that is available with CPMV and related virus building blocks.

The deoxynucleotide analogue–stilbene conjugate was incorporated into nascent DNA by DNA polymerase activity and its fluorescence properties were investigated [62]. It was shown that the blue fluorescence of stilbene was detected only upon binding of an antibody that specifically recognizes the stilbene-modified nucleotide (Figure 3.16).

A complex between *trans*-stilbene and antibody (EP2-19G2-1) was prepared and its fluorescence properties were investigated [63]. These complexes strongly fluoresced with colors ranging from blue to green.

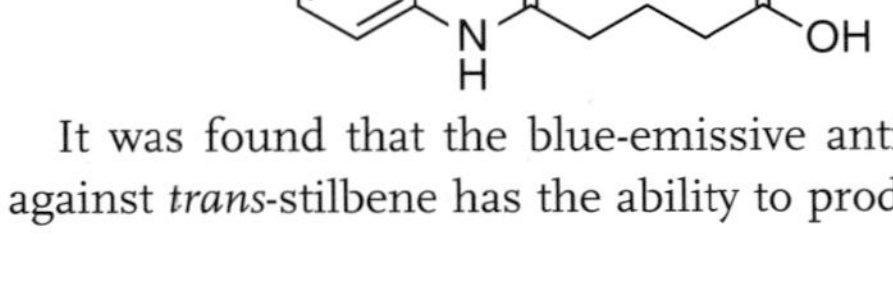

It was found that the blue-emissive antibody EP2-19G2 that has been elicited against *trans*-stilbene has the ability to produce bright luminescence and has been

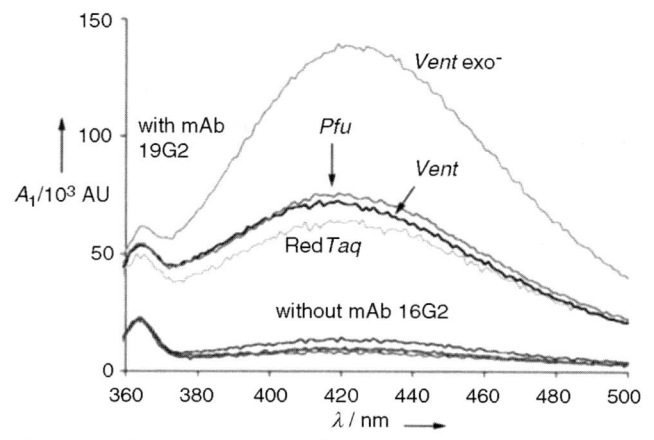

Figure 3.16 Fluorescence spectra of DNA-containing deaza-ATP-PEG-stilbene (sDNA). sDNA was obtained by PCR with four different DNA polymerases and a mixture of dATP and **9**, with the template plasmid, pCGMT-92H2 [62].

used as a biosensor in various applications. The authors showed that the prolonged luminescence was not stilbene fluorescence. Instead, the emissive species is a charge transfer excited complex of an anionic stilbene and a cationic, parallel π-stacked tryptophan. Upon charge recombination, this complex generates exceptionally bright blue light. Structural analysis indicated that the complex formation was enabled by a deeply penetrating ligand binding pocket, which in turn results from a noncanonical interface between the two variable domains of the antibody. Detailed structural and photophysical studies revealed that the antibody stabilizes the planar excited-state configuration through the formation of an exciplex between Trp^{H103} of the antibody and stilbene. The authors concluded that the very bright blue luminescence of EP2-19G2-1 is attributable to electron–hole recombination of the Trp:stilbene charge transfer excited state held in the rigid EP2-19G2 matrix that disfavors nonradiative decay and the loss of the stilbene fluorescence is due to very rapid charge recombination.

Antibodies to the donor–acceptor-substituted stilbene *trans*-4-*N*,*N*-dimethylamino-4′-cyanostilbene (DCS) and other donor–acceptor-substituted stilbenes were

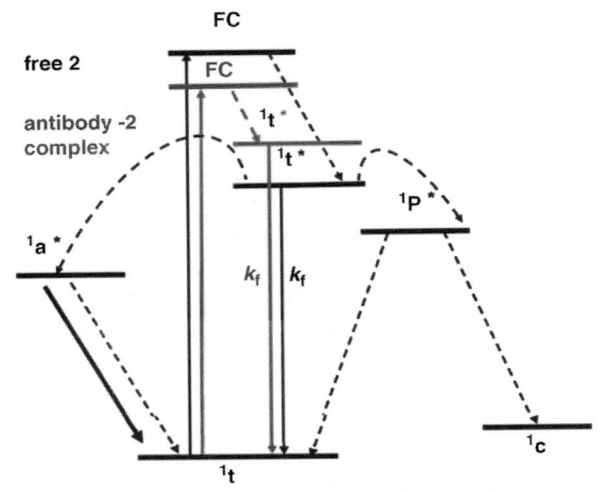

Figure 3.17 Excited-state photochemistry of free **2** and hypothesized primary pathway of the antibody–2 complexes (gray). Dashed lines: nonradiative transition; solid lines: radiative transitions. FC: Franck–Condon excited state [64].

generated [64]. Figure 3.17 shows a scheme of the excited-state photochemistry of free **2** and hypothesized primary pathway of the antibody–2 complexes.

On the basis of spectroscopic and structural data, a model in which the relatively rigid protein environment sterically increases the activation barriers connecting the planar $^1t^*$ to the twisted excited states $^1p^*$ (and possibly $^1a^*$) and modulates the energetics of the fluorescent $^1t^*$ state was suggested. The antibody was suggested to be more effective in stabilizing the polar FC excited state of **2** compared to the ground state, but less effective than bulk water in selectively stabilizing the light-emitting state. As the authors suggest, excitation and emission wavelengths of the antibody–2 complexes may make them useful as fluorescent biosensors for *in vitro* and *in vivo* applications.

3.5
Interactions Involving Triplet State and Phosphorescence

The role of the triplet state in the *cis–trans* isomerization of stilbenes effected by photosensitizers, such as acetophenone, benzophenone, or anthraquinone, which have large $S_0 \rightarrow T_1$ excitation energies, was first revealed in Ref. [65]. Theoretical considerations and experimental data on intermolecular triplet–triplet energy transfer leading to the sensitized stilbene photoisomerization are described in Section 4.2.2. It was shown that data on positional dependence of the heavy-atom effect on the *cis–trans* photoisomerization of bromostilbenes were consistent with the fact that, in contrast to the *para* position, the *meta* position is near a node in the highest occupied and the lowest unoccupied MO of stilbene [66]. According to [67], internal and external heavy-atom effects induce phosphorescence in *trans*-stilbene and in indeno[2,1-a]indene, a rigid *trans*-stilbene analogue, in glass media at 77 K. The origin of these emissions was found in the narrow range of 580–600 nm.

It was found [68] that dioxygen strongly attenuated the slopes of lines obtained by plotting the *trans/cis* stilbenes photostationary ratios versus azulene concentration, indicating dioxygen interactions with twisted triplets. A stilbene triplet lifetime in benzene was estimated at 120 ns at 77 K. Comparison of the interactions of dioxygen and β-carotene with stilbene triplets indicated that electronic excitation was not transferred to dioxygen [69]. The authors suggested that effective spin-exchange interactions may proceed via a triplet encounter complex that gives O(3Σ) and twisted stilbene ground-state molecules. Spin-exchange quenching of alpha-methylstilbene triplets by molecular oxygen and by the free radical di-*tert*-butyl nitroxide was reported [70]. The effect of the two quenching events was found to be identical. The data obtained indicated that in the case of stilbene spin-exchange quenching by O_2 at the twisted geometry favored the *cis*-isomer but occurred in competition with excitation transfer from transoid triplets that led to the *trans*-isomer and to singlet oxygen. Quenching rate constants were close to diffusion-controlled and predicted singlet O_2 quantum yields of 0.08 and 0.13 in the presence of air and under an O_2 atmosphere, in good agreement with experimental measurements. A scheme of the dioxygen effect of the stilbene photoisomerization is shown in Figure 3.18.

The electronic spectroscopy of *trans*-isomers of 3-(*N*-phenylamino)stilbene (m1c), 3-(*N*-methyl-*N*-phenylamino)stilbene (m1d), 3-(*N*,*N*-diphenylamino)stilbene (m1e), and 3-(*N*-(2,6-dimethylphenyl)amino)stilbene (m1f) and their double-bond constrained analogues, m2a–m2c and m2e, were studied [71]. When compared with *trans*-3-aminostilbene (m1a), m1c–m1e displayed a redshift of the $S_0 \rightarrow S_1$ absorption and fluorescence spectra, lower oscillator strength and fluorescence rate constants, and more efficient $S_1 \rightarrow T_1$ intersystem crossing. The N-Ph derivatives m1c–m1e had lower fluorescence quantum yields and higher photoisomerization quantum yields. The role of $S_1 \rightarrow T_1$ transition in the amino-substituted stilbenes as the predominant nonradiative decay pathway was discussed. The excited triplet (T_1) state formation of stilbene dendrimers (tetramethoxystilbene (generation G) G0, G1,

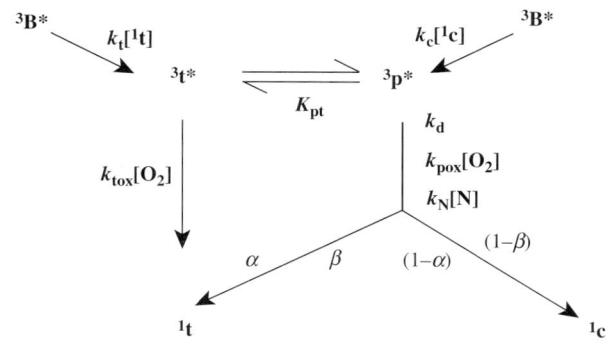

Figure 3.18 Scheme of the dioxygen effect of the stilbene photo isomerization [70]. (Reproduced with permission.)

G2, and G4) was observed [72]. From time-resolved studies on energy and molecular volume changes, it was found that the conformational change completes with the decay of the T_1 state for G0–G2. The dynamics was slightly slower for G4, which is attributed to the conformational change of the dendron part.

3.6
Fluorescence of Excimers and Exciplexes

High chemical reactivity of molecules in the singlet-excited states causes the formation of complexes with other molecules in the ground state. According to existing terminology, complexes between identical molecules (monomer) are named excimer, whereas those between different molecules are referred to as exciplexes. The characteristic feature of these complexes is the large value of the dipole moments and corresponding relaxation shift. Thus, exciplexes can serve as sensitive indicators of micropolaritiy and the relaxation dynamics of the environment. The difference in fluorescence parameters of monomers, excimers, and exciplexes allows to detect the fluorescence spectra of these compounds in a single experiment.

Analysis of fluorescence of urethane acrylic monomer copolymerized with *trans*-4-(2-methacryloyloxyethylcarbamoyloxymethyl)stilbene (SUM) was performed (Figure 3.19) [73].

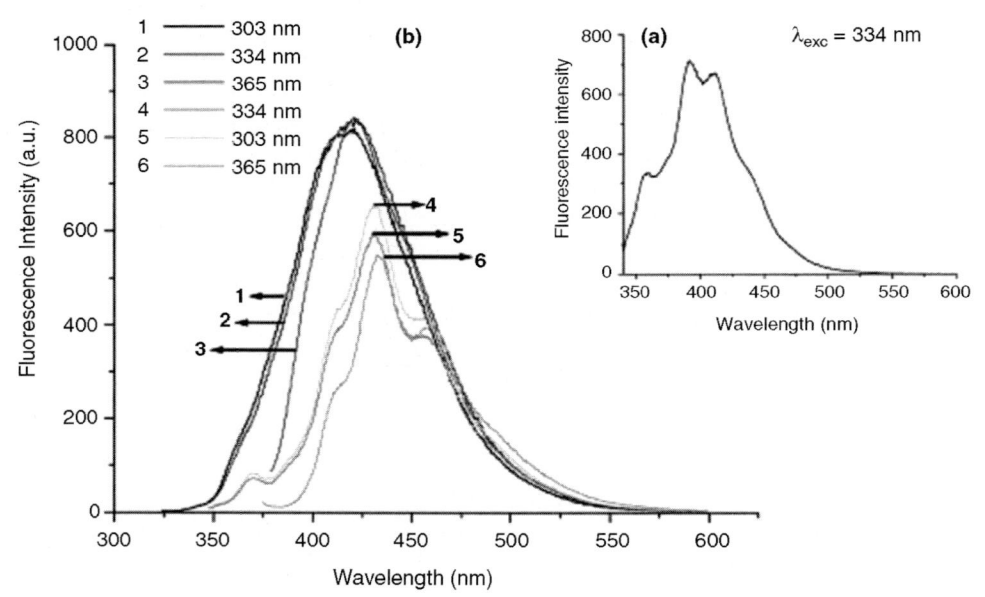

Figure 3.19 Fluorescence spectrum for SUM in DMF solution (a) and (b), stilbene copolymer (SUMMA) film (**1–3**) and monomer (SUM) in solid state (**4–6**) at various excitation wavelengths [73]. (Reproduced with permission from Elsevier.)

A well-defined excimer green band was observed in polypropylene (PP) films containing different concentrations of 4,4'-bis(2-benzoxazolyl)stilbene (BBS) [74]. During drawing (130 °C), the PP reorganization broke the BBS excimer-type arrangement, leading to the prevalence of the blue emission of the single molecules. The photophysics of this stilbene derivative was efficiently applied for the detection of tensile deformation of PP films. In perdeuterated *trans*-stilbene grafted polystyrene, the chromophore concentration led to aggregation and, consequently, to excimer formation [75]. In macrocyclic and medium-size stilbenophanes tethered by silyl chains, excimer emission was observed when the distances between two stilbene units in the stilbenophanes were sufficiently small [76]. It was shown [53] that upon addition of N,N-dimethylaniline in benzene, the fluorescence intensity of the G3 stilbene core decreases, with a concomitant increase in the fluorescence emission in the longer wavelength region. This region was assigned to the emission of the exciplex; the quenching profile and the wavelength of the exciplex emission are similar for G1–G3.

3.7
Energy Transfer

A new fluorescence method based on the singlet–singlet energy transfer between stilbene label and heme group in myoglobin was developed (Section 10.3.7) [77]. The authors of the work [78] reported that both singlet and triplet energy transfers in stilbene-cored benzophenone dendrimers (*trans*-BPST) took place efficiently (Figure 3.20). Upon excitation (290 nm) of stilbene group, the intramolecular singlet energy transfer from the excited core stilbene to the benzophenone part (99.7%) was confirmed by quenching of the fluorescence from the core stilbene. The very weak phosphorescence from benzophenone part in *trans*-BPST was observed even at 77 K.

The intramolecular energy transfer from the dendron subunit to the stilbene core in water-soluble stilbene dendrimer was described [79]. Addition of KCl to the solution of water-soluble stilbene dendrimers resulted in fluorescence and fluorescence excitation spectral changes due to diminishing interaction between the core of the dendrimers and the water. By salt addition, the energy transfer efficiency increased from 49 to 100%. It was shown that the second-generation dendrimer could transfer excited energy from the dendron to the core most efficiently because the dendron subunit was folded up by the effect of added salt. Thus, the energy transfer efficiency from the dendron to the core depended on the generation or distance between the core and the outermost aryl groups. Steady-state photophysics studies on stilbene–anthracene dendrimer dyad showed energy transfer from the peripheral stilbene units to the anthracene core [80]. The competitive intramolecular energy transfer was investigated by analyzing photophysical and photochemical properties of binuclear complexes containing the (phen)ReI(CO)$_3$ subunit bridged by *trans*-1,2-bis(4-pyridyl)ethylene [81]. The fluorescence properties of dendrimers (Gn) of generation number n, containing luminophore stilbenyl (S), were investigated.

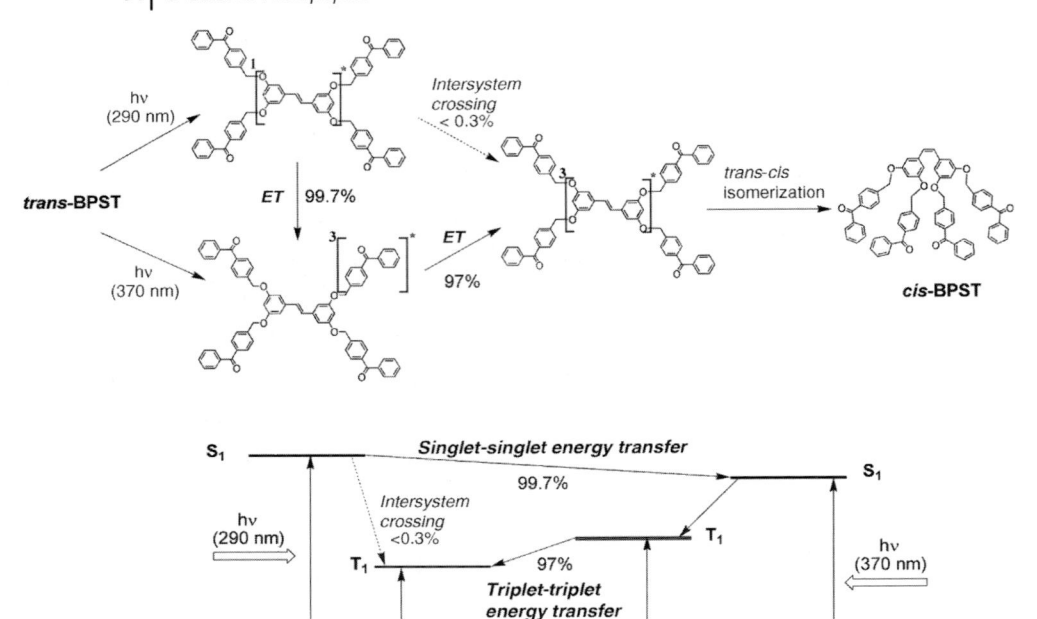

Figure 3.20 Mechanism of photochemical isomerization and energy diagram for *trans*-BPST [78]. (Reproduced with permission from Elsevier.)

In the study [82], the photophysics of a series of transition metal complexes containing a (diimine)ReI(CO)$_3$(py) chromophore covalently linked to *trans*-stilbene via a semirigid amide spacer was examined. In this series of complexes, moderately exothermic triplet–triplet energy transfer from the 3dπ^* charge transfer excited state of the diimine-Re chromophore to *trans*-stilbene was competitive with normal radiative and nonradiative decay of the 3dπ^* state. The electron transfer driving force (ΔEEnT) varied from -29 to -38 kJ/mol, while the activation energy ($E_a \leq 2$ kJ/mol) and the frequency factor ($A \approx 10^6$ s^{-1}) were found to be low. The authors suggested a mechanism in which energy transfer from the 3dπ^* state to *trans*-stilbene occurred at nearly the optimal driving force (i.e., ΔEEnT $\approx \lambda$, where λ is the reorganization energy) and the exchange coupling matrix element (V_{TT}) is small, leading to low A values.

The role of the central bond torsion and of the double bond and phenyl–vinyl torsions in nonvertical triplet excitation transfer to stilbenes was stressed [83]. The pivotal role of triplet–triplet energy transfer in photochemical and photophysical processes is discussed in Section 4.2.2.

3.8
Intramolecular Charge Transfer

It was reported [84] that compared to other stilbene derivatives, *trans*-3,5-dimethoxystilbene displayed a large quantum yield of fluorescence and a low quantum yield of *trans–cis* isomerization in polar organic solvents. According to the authors, the unique photophysical properties of *trans*-3,5-dimethoxystilbene were attributed to the formation of a highly polarized charge transfer excited state ($\mu_e = 13.2$ D). The charge transfer transitions relevant to single- and double-bond photochemistry twisting have been studied in the framework of the biradicaloid state theory using the AM1 method for a family of donor–acceptor-substituted stilbenoids and a series of sparkle-simulated model stilbenes [85–87]. Particular attention was given to the occurrence of S_0–S_1 state conical intersections. The difference in critical points at which the conical intersections occur for double-bond and single-bond twisted stilbenoids was shown to be related to the splitting of the cyanine limit of their planar counterparts.

This chapter briefly reviews recent progress in the investigation of fluorescent and phosphorescent properties of stilbenes as well as such phenomena as triplet–triplet and singlet–singlet energy transfer and Raman scattering. The trends in this area include the use of a wide arsenal of stilbenes, employment of elaborated experimental methods such as nano and picosecond time-resolved absorption and fluorescent spectroscopy, and the use of modern theoretical calculations, for example, density function theory. The importance of these research endeavors for further basic and applied applications of stilbenes cannot be overestimated.

References

1 Likhtenshtein, G.I. (1988) *Chemical Physics of Redox Metalloenzymes*, Springer-Verlag, Heidelberg.
2 Improta, R. and Santoro, F. (2005) *Journal of Physical Chemistry A*, **109**, 10058–10067.
3 Dietl, C., Papastathopoulos, E., Niklaus, P., Improta, R., Santoro, F., and Gerber, G. (2005) *Chemical Physics*, **310**, 201–211.
4 Ma, J., Dutt, G.B., Waldeck, D.H., and Zimmt, M.B. (1994) *Journal of the American Chemical Society*, **116**, 10619–10629.
5 Lalevee, J., Allonas, X., and Fouassier, J.P. (2005) *Chemical Physics Letters*, **401**, 483–486.
6 Catalan, J. and Saltiel, J. (2001) *Journal of Physical Chemistry A*, **105**, 6273–6276.
7 Durbeej, B. and Eriksson, L.A. (2005) *Journal of Physical Chemistry A*, **109** (25), 5677–5682.
8 Glusac, K.D. and Schanze, K.S. (2002) *Polymer Preprints (American Chemical Society, Division of Polymer Chemistry)*, **43**, 87–88.
9 Dolgova, O.V., Vasilèva, N.Yu., and Sokolova, I.V. (2002) *Optics and Spectroscopy (Translation of Optika i Spektroskopiya)*, **93**, 19–26.
10 Creed, D., Somlai, A.M., Hoyle, C.E., and Page, K.A. (2003) *Polymer Preprints*

(*American Chemical Society Division of Polymer Chemistry*), **44**, 84–85.
11 Azechi, Y., Takemura, K., Shinohara, Y., Nishimura, Y., and Arai, T. (2007) *Journal of Physical Organic Chemistry*, **20**, 864–871.
12 Muszkat, K.A., Castel, N., Jakob, A., Fischer, E., Luettke, W., and Rauch, K. (1991) *Journal of Photochemistry and Photobiology A: Chemistry*, **56**, 219–226.
13 Dobrin, S., Starukhin, A., Kaszynski, P., and Waluk, J. (1997) *Optika i Spektroskopiya*, **83**, 669–673.
14 Gorner, H. (1999) *Journal of Photochemistry and Photobiology A: Chemistry*, **126**, 15–21.
15 Clark, A.E. (2006) *Journal of Physical Chemistry A*, **110**, 3790–3796.
16 Yan, Y.-X., Wang, D., Zhao, X., Tao, X.-T., and Jiang, M.-H. (2003) *Chinese Journal of Chemistry*, **21**, 626–629.
17 Liu, Z.Q., Fang, Qi., Yu, W.T., Xue, G., Cao, D.X., Wang, D., Xia, G.M., and Jiang, M.H. (2002) *Chinese Chemical Letters*, **13**, 997–1000.
18 Lin, N., Zhao, X., Rizzo, A., and Luo, Y. (2007) *Journal of Chemical Physics*, **126**, 244509/1–244509/8.
19 Kosower, E.M., Hofmann, D., and Wallenfels, K. (1962) *Journal of the American Chemical Society*, **84**, 2755–2760.
20 Leffler, E. and Grunwald, E. (1963) *Rates and Equilibria of Organic Reactions*, John Wiley & Sons, Inc., New York.
21 Pross, A. (1995) *Theoretical and Physical Principles of Organic Reactivity*, Wiley–Interscience, New York, pp. 159–182.
22 Stephenson, L. and Hammond, G.S. (1969) *Angewandte Chemie – International Edition in English*, **8**, 261–270.
23 Anderton, R. and Kaufman, J. (1995) *Chemical Physics Letters*, **237**, 145–151.
24 Papper, V., Pines, D., Likhtenshtein, G.I., and Pines, E. (1997) *Journal of Photochemistry and Photobiology A: Chemistry*, **111**, 87–96.
25 Papper, V., Pines, D., Likhtenshtein, G.I., and Pines, E. (1997) *Recent Development in Photochemistry and Photobiology*, vol. 1, Transworld Research Network, pp. 205–250.
26 Görner, H. (1980) *Journal of Photochemistry*, **13**, 269–294.
27 (a) Saltiel, J. and Sun, Y.-P. (1990) *Photochromism, Molecules and Systems* (eds H. Dürr and H. Bouas-Laurent), Elsevier, Amsterdam, p. 64; (b) Saltiel, J. and Charlton, J.L. (1980) *Rearrangements in Ground and Excited States*, vol. 3 (ed. P. DeMayo), Academic Press, New York, p. 25.
28 Saltiel, J. and Sun, Y.-P. (1989) *The Journal of Physical Chemistry*, **93**, 6246–6250.
29 Saltiel, J., Waller, A.S., Sears, D.F., and Garrett, C.Z. (1993) *Journal of Chemical Physics*, **97**, 2516–2522.
30 Marinari, A. and Saltiel, J. (1976) *Molecular Photochemistry*, **7**, 225–249.
31 Yang, J.-S., Liau, K.-L., Hwang, C.-Y., and Wang, C.-M. (2006) *Journal of Physical Chemistry A*, **110**, 8003–8010.
32 Aguiar, M., Akcelrud, L., Pinto, M.R., Atvars, T.D.Z., Kadrasz, F.E., and Saltiel, J. (2003) *Journal of Photoscience*, **10**, 149–155.
33 Pines, D., Pines, E., and Rettig, W. (2003) *Journal of Physical Chemistry A*, **107**, 236–242.
34 Oelgemoeller, M., Brem, B., Frank, R., Schneider, S., Lenoir, D., Hertkorn, N., Origane, Y., Lemmen, P., Lex, J., and Inoue, Y. (2002) *Journal of the Chemical Society, Perkin Transactions 2*, 1760–1770.
35 Sun, S.-S. and Lees, A.J. (2002) *Organometallics*, **21**, 39–49.
36 Polo, A.S., Itokazu, M.K., Frin, K.M., Patrocinio, A.O.T., and Murakami, I. (2006) *Coordination Chemistry Reviews*, **250**, 1669–1680.
37 Roberts, J.C. and Pincock, J.A. (2006) *Journal of Organic Chemistry*, **71** (4), 1480–1492.
38 Saltiel, J., Waller, A.S., and Sears, D.F. Jr. (1993) *Journal of the American Chemical Society*, **115**, 2453–2456.
39 Saltiel, J., Waller, A.S., Sears, D.F. Jr., and Garrett, C.Z. (1993) *Journal of Physical Chemistry*, **97**, 2516–2522.

40 Saltiel, J., Waller, A., Sun, Y.P., and Sears, D.F. Jr. (1990) *Journal of the American Chemical Society*, **112**, 4580–4581.

41 Charlton, J.L. and Saltiel, J. (1977) *Journal of Physical Chemistry*, **81**, 1940–1944.

42 Dolgova, O.V., Sokolova, I.V., and Vasilèva, N.Yu. (2007) *Atmospheric and Oceanic Optics*, **20**, 345–347.

43 Ruseckas, A., Namdas, E.B., Lee, J.Y., Mukamel, S., Wang, S., Bazan, G.C., and Sundstroem, V. (2003) *Journal of Physical Chemistry A*, **107**, 8029–8034.

44 Gorner, H. (1999) *Journal of Photochemistry and Photobiology A: Chemistry*, **126**, 15–21.

45 Kalgutkar, R.S., Yang, J.-S., and Lewis, F.D. (1998) Book of Abstracts, 216th ACS National Meeting, Boston, MA, August 23–27, 1998, PHYS-290.

46 Kubicki, A.A. (2003) *Chemical Physics Letters*, **373**, 471–474.

47 Kubicki, A.A. (2008) *Chemical Physics Letters*, **457**, 246–249.

48 Strehmel, V., Frank, C.W., and Strehmel, B. (1997) *Journal of Photochemistry and Photobiology A: Chemistry*, **105**, 353–364.

49 Sugiono, E. and Detert, H. (2006) *Silicon Chemistry*, **3**, 31–42.

50 Prukala, D. (2006) *Journal of Heterocyclic Chemistry*, **43**, 337–344.

51 Vicinelli, V., Bergamini, G., Ceroni, P., Balzani, V., Voegtle, F., and Lukin, O. (2007) *Journal of Physical Chemistry B*, **111**, 6620–6627.

52 Imai, M. and Arai, T. (2002) *Tetrahedron Letters*, **43**, 5265–5268.

53 Momotake, A. and Arai, T. (2004) *Journal of Photochemistry and Photobiology C: Photochemistry Reviews*, **5**, 1–25.

54 Parkhomyuk-Ben Arye, P., Strashnikova, N., and Likhtenshtein, G.I. (2002) *Journal of Biochemical and Biophysical Methods*, **51**, 1–15.

55 Strashnikova, N.V., Papper, V., Parhomyuk, P., Ratner, V., Likhtenshtein, G.I., and Marks, R. (1999) *Journal of Photochemistry and Photobiology A: Chemistry*, **122**, 133–142.

56 Eremenko, A., Smirnova, N., Rusina, O., Rechthaler, K., Koehler, G., Ogenko, V., and Chuiko, A. (1999) Special Publication – *Royal Society of Chemistry (Fundamental and Applied Aspects of Chemically Modified Surfaces)*, **235**, 333–340.

57 Strat, G., Buruiana, E., Buruiana, T., Pohoata, V., and Strat F M. (2005) *Journal of Optoelectronics and Advanced Materials*, **7**, 925–928.

58 Wirsching, P., Janda, K.D., and Lerner, R.A. (2002) Anti-stilbene Antibodies. WO/2002/022629, filed Sept. 13, 2001 and issued March 21, 2002.

59 Chen, D.-W., Beuscher, I.V., Stevens, R.C., Wirsching, P., Lerner, R.A., and Janda, K.D. (2001) *The Journal of Organic Chemistry*, **66**, 1725–1732.

60 Ahluwalia, A., De Rossi, D., Giusto, G., Chen, O., Papper, V., and Likhtenshtein, G.I. (2002) *Analytical Biochemistry*, **305**, 121–134.

61 Wang, Q., Raja, K.S., Janda, K.D., Lin, T., and Finn, M.G. (2003) *Bioconjugate Chemistry*, **14**, 38–43.

62 Kaufmann, G.F., Meijler, M.M., Sun, C., Chen, D.-W., Kujawa, D.P., Mee, J.M., Hoffman, T.Z., Wirsching, P., Lerner, R.A., and Janda, K. (2005) *Angewandte Chemie – International Edition*, **44**, 2144–2148.

63 Debler, E.W., Kaufmann, G.F., Meijler, M.M., Heine, A., Mee, J.M., Pljevaljcic, G., Di Bilio, A.J., Schultz, P.G., Millar, D.P., Janda, K.D., Wilson, I.A., Gray, H.B., and Lerner, R.A. (2008) *Science*, **319** (5867), 1232–1235.

64 Debler, E.W., Millar, D.P., Deniz, A.A., Wilson, I.A., and Schultz, P.G. (2006) *Angewandte Chemie – International Edition*, **45**, 7763–7765.

65 Hammond, S. and Saltiel, J. (1962) *Journal of the American Chemical Society*, **84**, 4983–4984.

66 Saltiel, J., Chang, D.W.L., and Megarity, E.D. (1974) *Journal of the American Chemical Society*, **96**, 6521–6522.

67 Saltiel, J., Khalil, G.E., and Schanze, K. (1980) *Chemical Physics Letters*, **70**, 233–235.

68 Saltiel, J. and Thomas, B. (1974) *Journal of the American Chemical Society*, **96**, 5660–5661.
69 Saltiel, J. and Thomas, B. (1976) *Chemical Physics Letters*, **37**, 147–149.
70 Saltiel, J. and Klima, R.F. (2006) *Photochemistry and Photobiology*, **82**, 38–42.
71 Yang, J.-S., Liau, K.-L., Tu, C.-W., and Hwang, C.-Yu. (2005) *Journal of Physical Chemistry A*, **109**, 6450–6456.
72 Tatewaki, H., Mizutani, T., Hayakawa, J., Arai, T., and Terazima, M. (2003) *Journal of Physical Chemistry A*, **107**, 6515–6521.
73 Buruiana, E.C., Zamfir, M., and Buruiana, T. (2007) *European Polymer Journal*, **43**, 4316–4324.
74 Pucci, A., Bertoldo, M., and Bronco, S. (2005) *Macromolecular Rapid Communications*, **26**, 1043–1048.
75 Ding, L. and Russell, T.P. (2006) *Macromolecules*, **39**, 6776–6780.
76 Maeda, H., Nishimura, K., Mizuno, K., Yamaji, M., Oshima, J., and Tobita, S. (2005) *Journal of Organic Chemistry*, **70**, 9693–9700.
77 Chen, O., Urlander, N., Likhtenshtein, G.I., and Priel, Z. (2008) *Journal of Biochemical and Biophysical Methods*, **70**, 1006–10013.
78 Miura, Y., Momotake, A., Shinohara, Y., Wahadoszamen, M., Nishimura, Y., and Arai, T. (2007) *Tetrahedron Letters*, **48**, 639–641.
79 Hayakawa, J., Momotake, A., Nagahata, R., and Arai, T. (2003) *Chemistry Letters*, **32**, 1008–1009.
80 Sengupta, S. and Pal, N. (2002) *Tetrahedron Letters*, **43**, 3517–3520.
81 Polo, A.S., Itokazu, M.K., Frin, K.M., Patrocinio, A.O.T., and Murakami, I. (2006) *Coordination Chemistry Reviews*, **250**, 1669–1680.
82 Schanze, K.S., Lucia, L.A., Cooper, M., Walters, K.A., Ji, H.-F., and Sabina, O. (1998) *Journal of Physical Chemistry A*, **102**, 5577–5584.
83 Saltiel, J., Mace, J.E., Watkins, L.P., Gormin, D.A., Clark, R.J., and Dmitrenko, O. (2003) *Journal of the American Chemical Society*, **125**, 16158–16159.
84 Roberts, J.C. and Pincock, J.A. (2006) *Journal of Organic Chemistry*, **71**, 1480–1492.
85 Tajima, Y., Ishikawa, H., Miyazawa, T., Kira, M., and Mikami, N. (1997) *Journal of the American Chemical Society*, **119**, 7400–7401.
86 Dekhtyar, M. and Rettig, W. (2007) *Journal of Physical Chemistry A*, **111**, 2035–2039.
87 Papper, V. and Likhtenshtein, G.I. (2001) *Journal of Photochemistry and Photobiology A: Chemistry*, **140**, 39–52.

4
Stilbene Photoisomerization

4.1
General

Stilbenes exhibit a diverse photochemical behavior in solution such as reversible *cis/trans* isomerization, cyclization of *cis*-stilbene **1** to dihydrophenanthrene (DHP) and further oxidation to phenanthrene **3**, and dimerization of *trans*-stilbene to yield tetraphenylcyclobutane products **4** and **5** [1] (Figure 4.1).

An important class of photoinduced chemistry of organic molecules involves rearrangement along a double bond, usually referred to as *cis–trans* isomerization (Figure 4.2) [2–12].

Optically induced *cis–trans* isomerization is a key structural dynamic element for many types of photochromic switches as stilbenes and azobenzene derivatives and for photosensor proteins as bacteriorhodopsin, rhodopsin, and photoactive yellow protein.

The photoisomerization of stilbenes is found to be a simple and convenient model for a detailed study of factors affecting the unimolecular photoreaction dynamics. The steady-state equilibrium constant of the photoisomerization process resulting from monochromatic excitation is given as

$$K_{eq} = \frac{[cis]}{[trans]} = \frac{Q_{t \to c} \varepsilon_t}{Q_{c \to t} \varepsilon_c}, \qquad (4.1)$$

where Q is quantum yield and ε is absorption coefficient. The quantum yields of both *cis–trans* and *trans–cis* photoisomerizations are similar. That is why one can obtain one diastereometric form in preference to other by choosing an appropriate wavelength.

From Figure 4.3, one can see that the fluorescence radiation and the photoisomerization are competitive processes.

That is why slowing down the photoisomerization process will cause an increase in fluorescence. The observed rate constant of the steady-state photoisomerization is given as

$$k_{iso}^{app} = \sigma \phi I_{in} \frac{k_{iso} k_r}{\left(\frac{1}{\tau_d} + k_{iso}\right)\left(\frac{1}{\tau_d} + k_r\right)} \qquad (4.2)$$

Stilbenes. Applications in Chemistry, Life Sciences and Materials Science. Gertz Likhtenshtein
Copyright © 2010 WILEY-VCH Verlag GmbH & Co. KGaA, Weinheim
ISBN: 978-3-527-32388-3

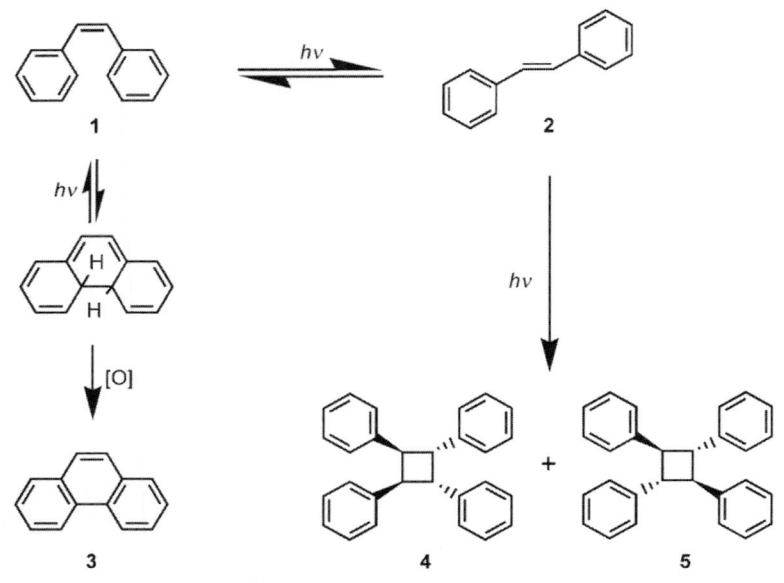

Figure 4.1 Photochemical transformation of stilbenes [1].

or

$$k_{iso}^{app} = \sigma\phi I_{in} \frac{k_r}{\tau_C} \quad \text{when } k_{iso} \gg \frac{1}{\tau_d}, \tag{4.3}$$

where $k_{iso}, k_r \ll 1/\tau_d$, and σ is the square section of the absorption, ϕ is the quantum yield of the exited state, I_{in} is the incident light power, k_{iso} is the isomerization rate of T^{**}, k_r is the solvent reorganization rate, and τ_d is the lifetime of T^*. One can see that the rate of isomerization depends upon k_{iso} and k_r. Since k_r is the characteristic of the media, the isomerization rate in certain conditions strongly depends on the properties of the medium when $k_r \gg 1/\tau_d$, k_{iso} does not depend on media dynamics and, consequently, on temperature.

Saltiel and coworkers proposed and developed first classical mechanisms for the light-induced *trans–cis* photoisomerization of *trans*-stilbene that have formed the basis for subsequent works [2–10]. According to pioneering works of the Saltiel group [2, 3], the *trans–cis* photoisomerization of stilbenes can proceed by two in-principle mechanisms: (i) the direct process involving exited single state

Figure 4.2 Scheme of p-orbitals in *trans*-stilbene (a), intermediate excited state (b), and *cis*-stilbene (c) [12]. (Reproduced from Ref. [116].)

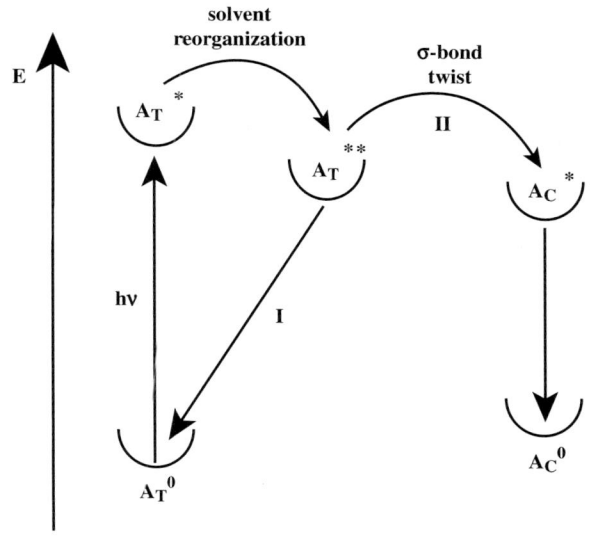

Figure 4.3 Illustration of the photoisomerization process of a chromophore (A) in the context of the properties of the medium, where T = *trans*- and C = *cis*-olefin diastereomers. (Reproduced from Ref. [116].)

(Figure 4.3) and (ii) the sensitized mechanism via triplet exited triplet state. Their first models assumed one-dimensional reaction coordinates. Reducing the reaction coordinate to a single molecular parameter, the torsion angle about the olefinic double bond, was a simplification. Nevertheless, the sections through the potential energy surfaces (PESs) correctly describe the basic features of the photoisomerization. A detailed picture of the photoisomerization that is complicated by the presence of multiple electronic states and multiple degrees of freedom has long been a matter of interest to researchers.

Recent data on stilbene photoisomerization dynamics have been reviewed in Refs [12–20].

4.2
Mechanisms of Photoisomerization

4.2.1
Ideas, Concepts, and Theoretical Calculations

In the first step of the direct stilbene photoisomerization, the stilbene molecule is excited to the singlet or triplet state when it absorbs a quantum of energy. After that, solvent molecules undergo reorganization, thus reducing the energy of the system. Next step is the excited-state decay by two competitive processes: radiative (fluorescence or phosphorescene emission) or nonradiative (photochemical) deactivation.

The following possible photochemical deactivation channels have been discussed:

1. Through double-bond twisting leading to *trans–cis* that consists of an initial emissive $^1t^*$ state and subsequently a populated nonemissive $^1p^*$ state (which is twisted by 90° about the C=C bond and undergoes internal conversion to give a mixture of *trans-* and *cis-*stilbene.

2. Nonradiative decay through single-bond twisting. This mechanism suggested an additional excited state, $^1a^*$, which corresponds to a relatively low-lying, twisted intramolecular charge transfer (TICT) state in which the styryl-anilino C—C bond is rotated. This process does not lead to a distinguishable photoisomer.

3. Planar TICT (PICT) mechanism.

4. Linear mechanism that suggests formation of a linear conjugated structure between amino substituent and phenyl group keeping the double bond free for twisting.

5. A volume-conserving mechanism such as hula-twist (H-T) that requires a concomitant twisting of the double bond and the adjacent single bond to accomplish the double-bond *cis–trans* isomerization, different from the usual one-bond rotation mechanism.

6. A "media-melting" mechanism that involves a nonradiative energy transfer from the vibrational modes of the excited stilbene to rotational or/and translational modes of surrounding molecules to give a space for inevitable move of the phenyl segments.

7. Synthesized nonvertical triplet excitation transfer (NVET), attributed to large structural changes between ground and triplet states.

8. Both the single- and the double-bond torsions in NVET.

4.2.1.1 Through Double-Bond Twisting (Saltiel) Mechanism

According to pioneering works of the Saltiel group [2, 3], the direct process occurs in three steps (Figure 4.4): (i) the stilbene is converted to the excited singlet or triplet state upon absorbance of light energy; (ii) the absorbance of light energy by the π-system causes the π-bond to break and thus results in partial rotational twisting about the σ-bond; and (iii) as a result of the 180° rotation about the remaining σ-bond, renewed π-bond formation is again symmetrically allowed, thus converting the *trans-*isomer into the *cis-*diastereomer, and vice versa.

According to this model, the reversible *trans–cis* photoisomerization proceeds from the lowest exited singlet $^1t^*$ configuration to the twisted singlet zwitterionic intermediate $^1p^*$ (phantom) where there is an avoided crossing with the ground state:

$$^1t^* \to {}^1p^* \to {}^1p \to (1-\beta)^1c + \beta^1t$$

or, alternatively, by the intersystem crossing (ISC) pathway to the biradical twisted triplet $^3p^*$ state (perpendicular with respect to the C=C double bond), which either

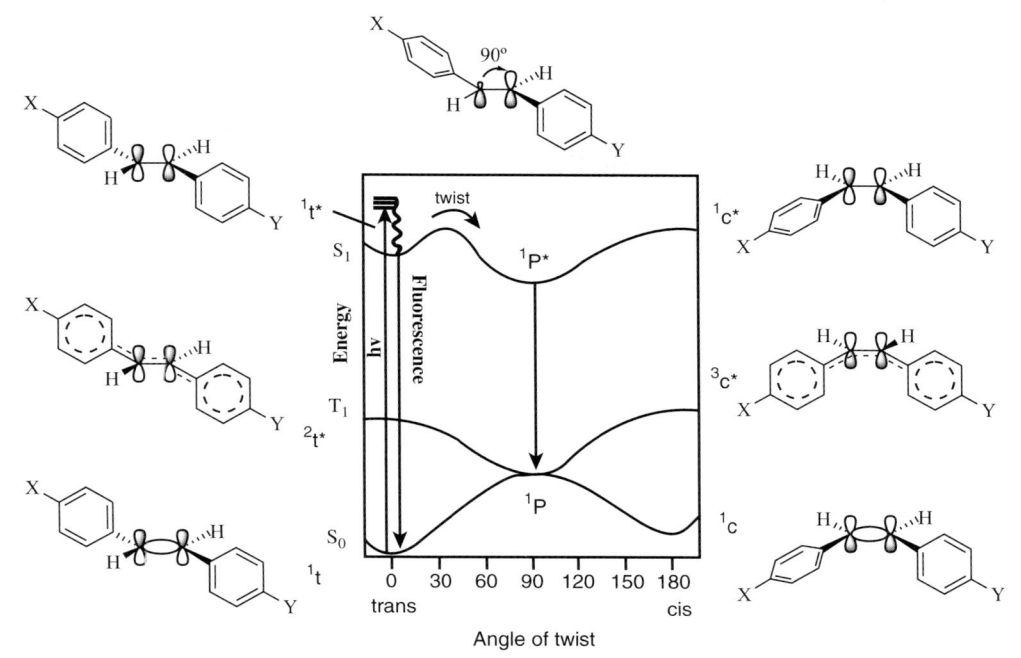

Figure 4.4 Graphical representation of the *trans-* and *cis-*stilbene orbitals and the electronic states participating in the *trans–cis* photoisomerization processes. S_0, S_1, and T_1 descriptors designate the potential energy surfaces of the ground-state singlet, first excited singlet, and triplet states, respectively [12]. (Reproduced from Ref. [31].)

crosses or nearly crosses the ground singlet surface:

$$^1t^* \to {}^3t^* \to {}^3p^* \to {}^1p \to (1-*)^1c + *\,^1t$$

where $^3t^*$ is the *trans*-configuration of the lowest triplet, ^1p is the twisted ground state $(1 - *)$ is the fraction of triplet decay into the *cis* form, and $(1 - \beta)$ is the fraction of perpendicular singlet configuration decaying into the *cis* form.

A map of the photoisomerization potential energy surface for tetraphenylethylene in alkane solvents was prepared using a fluorescence and picosecond optical calorimetry (Figure 3.4) [21]. Line shapes of the vertical and relaxed excited-state emissions at 294 K in methylcyclohexane were obtained from the steady-state emission spectrum, the wavelength dependence of the time-resolved fluorescence decays, the temperature dependences of the vertical and relaxed state emission quantum yields, and of the time-resolved fluorescence decays. Analysis of these data in conjunction with values of the twisted excited-state energy provided values for the energies of the vertical, conformationally relaxed, and twisted excited states on the photoisomerization surface, as well as the barriers to their interconversion. The energy difference between the last two states is found to be 1.76 ± 0.15 kcal/mol in methylcyclohexane.

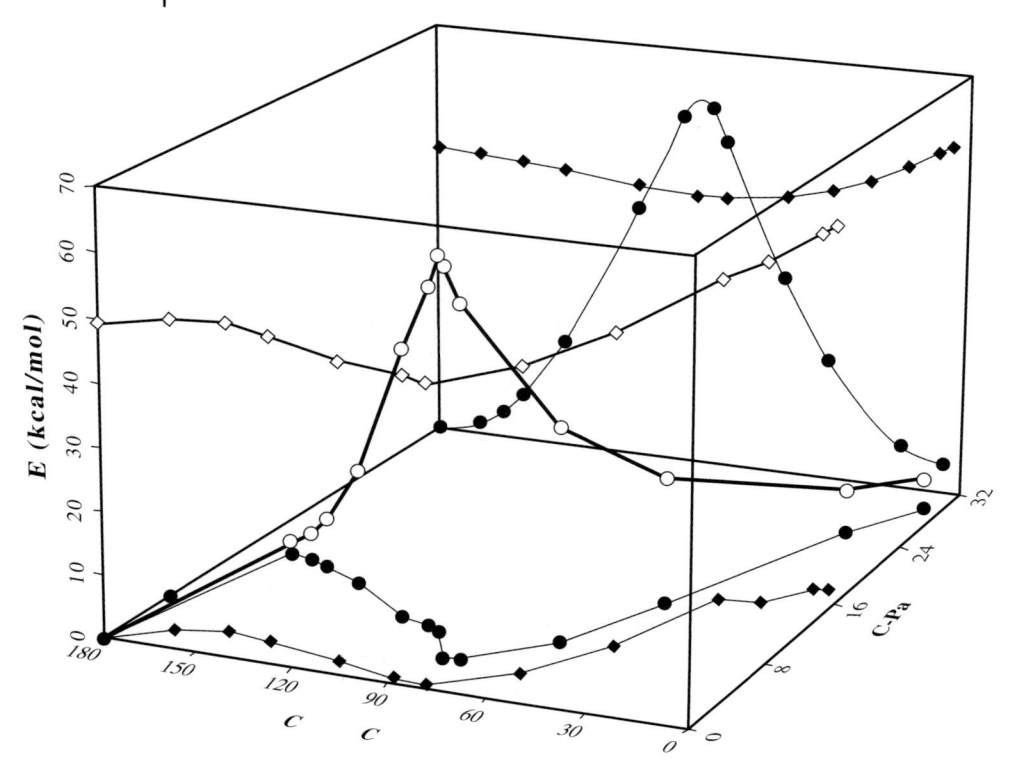

Figure 4.5 Calculated potential energy curves for twisting about the central bond in S_0 (○) and T_1 (◇) of stilbene. Except for the global minima at c-1 and t-1 in S_0 and at ^3p in T_1 for which geometries are fully optimized, $\theta_1 = \theta_2$ is fixed. Open and closed symbols distinguish points plotted in three dimensions from those that are projected on coordinate planes. Projections (solid symbols) on the $E = 0$ kcal/mol plane show the variation in $(\theta_3 + \theta_4)/2$ as a function of θ_1. Projections on a $(\theta_3 + \theta_4)/2 =$ constant plane give curves in good agreement with those previously proposed. [22]. (Reproduced with permission.)

Results of calculation of potential energy curves for twisting about the central bond in S_0 and T_1 of stilbene were reported (Figure 4.5) [22]. The origin of nonvertical triplet excitation transfer and the relative role of double-bond and phenyl–vinyl torsions in stilbenes were described.

Detailed simulations for the dynamics of electrons and nuclei during the *cis–trans* photoisomerization of stilbene employing a semiclassical description were reported [23]. These calculations indicated that after excitation of electrons from the HOMO to the LUMO by a femtosecond-scale laser pulse, two principal avoided crossings take place between the HOMO and the LUMO levels, each of which leads to substantial depopulation of the LUMO. The authors proposed that the first such HOMO–LUMO coupling can lead to the formation of 4a,4b-dihydrophenanthrene, while the second coupling leads to the formation of *trans*-stilbenes. These calculations also indicated that the pyramidalization of the two carbon atoms of the vinyl group is involved in both couplings and that the rotation of the two phenyl rings, together with their interaction, plays an important role in the first coupling.

Time-dependent density functional theory (TD-DFT) calculations, together with simulations of the electron energy distribution, allowed to estimate selective photoelectron energies of the S_0, S_1, S_2, and D_0 electronic states (Figure 3.3) [24].

4.2.1.2 Single-Bond Twisting Mechanism

It was suggested [18] and supported in Refs [25–30] that for stilbenes possessing high dipole moment in the process of isomerization, there are two possible photochemical deactivation channels: nonradiative decay through single-bond twisting that does not lead to a distinguishable photoisomer and through double-bond twisting leading to *trans–cis* photoisomerization. According to the authors, torsional motion from the exited single state resulted in a low-lying twisted intramolecular charge transfer. An example of the twisting structures is presented in Figure 4.6.

The double-bond twisting was proposed to be more than one order of magnitude slower than single-bond twisting for these compounds. The comparison of steady-state and time-resolved fluorescence studies of selectively bridged stilbazolium dyes as a function of temperature allowed authors to develop a kinetic model. Proposed reaction scheme leading from the Franck–Condon state E to an in-plane-relaxed state R, with subsequent nonradiative relaxation to the funnel states A* and P* reached by single- or double-bond (P*) twisting, is shown in Figure 4.7. Recent high-level quantum chemical calculations could even identify the structure of such a conical intersection (COI). They pointed to the importance of more than one bond being involved in the twisting process and highlighted the close relationship between COI and twisted (TICT) structure.

4.2.1.3 Planar Intramolecular Charge Transfer Precursor Mechanism

An important role of a planar intermolecular charge transfer was suggested (Figure 4.6) [18]. The authors took in consideration the following facts. The electronic properties such as photoluminescence, energy migration, electron transfer, conductivity, and nonlinear optics for a π-conjugated system strongly depend on

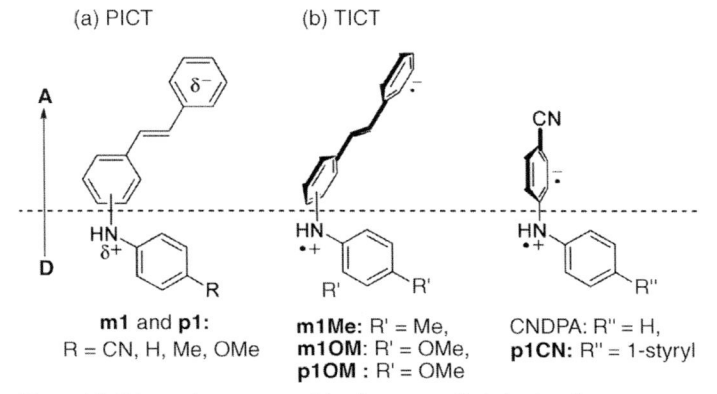

Figure 4.6 Schematic summary of the charge transfer behavior of the PICT and TICT states of m1 and p1 and the TICT states of CNDPA in acetonitrile [30]. (Reproduced with permission.)

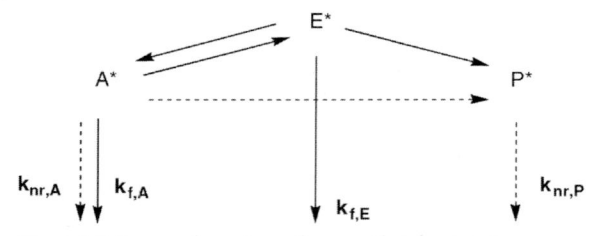

Figure 4.7 Proposed reaction scheme with subsequent nonradiative relaxation to the funnel states A* and P* reached by single- or double-bond (P*) twisting [18]. (Reproduced with permission from Elsevier.)

the degree of electronic coupling (delocalization) between subunits. Another crucial factor that determines the degree of electronic coupling in π-conjugated systems is the conformation, namely, the more planar the molecular structure, the stronger the coupling. The photochemical behavior of a series of *trans*-3-(*N*-arylamino)stilbenes (m1, aryl = 4-substituted phenyl with a substituent of cyano (CN), hydrogen (H), methyl (Me), or methoxy (OM)) in both nonpolar and polar solvents was reported and compared with that of the corresponding *para* isomers (p1CN, p1H, p1Me, and p1OM). The DFT (B3LYP level of theory 6-31G(d,p) basis set)-derived frontier molecular orbitals (FMOs) of m1CN, p1CN, m1Me, and p1Me are shown in Figure 4.8.

A schematic of the charge transfer behavior of the PICT and TICT states of m1 and p1 and the TICT states of CNDPA in acetonitrile is presented in Figure 4.6.

According to the authors, the poor charge redistribution (delocalization) ability is responsible for the unfavorable TICT-forming process and facilitates the competition of the single-bond torsional reaction. In contrast, the quinoidal character of a molecule in the PICT state kinetically favors both fluorescence and photoisomerization but disfavors the single-bond torsion. The concept of thermodynamically allowed, but kinetically inhibited, TICT formation was formulated.

4.2.1.4 Double-Bond Twisting Mechanism in Linear Quinoid Structure

All 4,4′-substituted stilbenes investigated in [12, 31] were divided into three groups according to the intramolecular stabilization of the excited $^1t^*$ state (Section 4.2.1.1)

Group I: Stilbene molecules have the weak donor and acceptor substituents. Their excited singlet $^1t^*$ state is relatively weakly stabilized by solvents.

Group II: Stilbene molecules have the strong donor substituent $(CH_3)_2N$ in the 4-position of the aromatic ring. They are characterized by a large redshift in both the absorption and the fluorescence spectra of *trans*-stilbenes. The $(CH_3)_2N$ group considerably stabilizes the $^1t^*$ state compared to that of the first group.

Group III: Stilbene molecules have the strong donor–acceptor pairs of 4,4′-substituents and exhibit very large redshifts, compared to the first and second

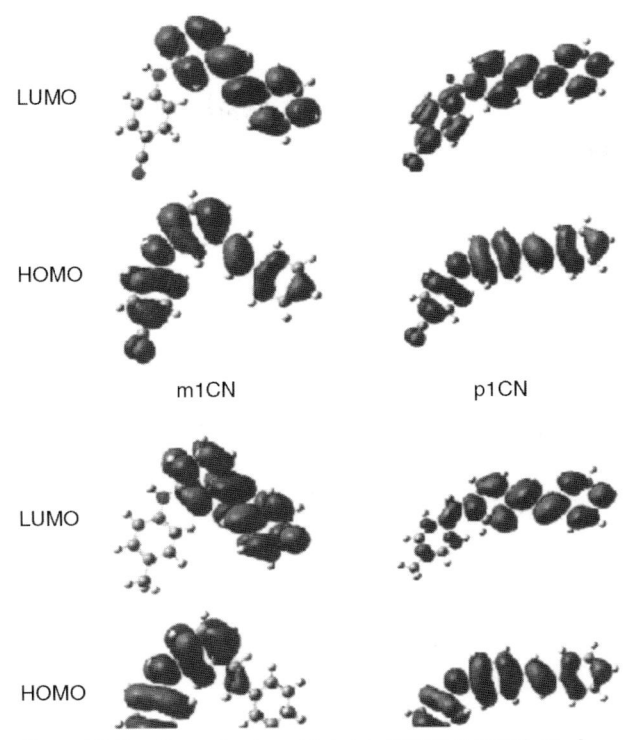

Figure 4.8 Frontier molecular orbital plots (B3LYP/6-31G(d,p)) of optimized structures of m1CN, p1CN, m1Me, and p1Me [30]. (Reproduced with permission.)

groups. The highly polarized excited state, which creates a huge dipole moment, is extensively stabilized by polar interactions.

As one can see in Figure 4.9, the high electric dipole moment for the excited $^1t^*$ state of the stilbene groups 2 and 3 may be attributed to the intramolecular charge transfer accompanying the formation of double (quinoid) bond between substituents and carbon atom and simultaneous breaking of double bonds. However, this synchronous process requires a significant reorganization of the stilbene molecule that can result in slowing down the reaction.

The rate of the concerted chemical reactions depends on the number of degree of freedom of nuclei involved in the transition [32–34]. According to a model developed in [34], a concerted reaction occurs as a result of the simultaneous transition (taking approximately 10^{-13} s) of a system of independent oscillators, with the mean displacement of nuclei φ_0, from the ground state to the activated state in which this displacement exceeds for each nucleus a certain critical value (φ_{cr}). If $\varphi_{cr} > \varphi_0$ and the activation energy of the concerted process $E_{syn} > nRT$, the theory gives the following expression for the synchronization factor that is the ratio of the pre-exponential factors of synchronous and simple processes:

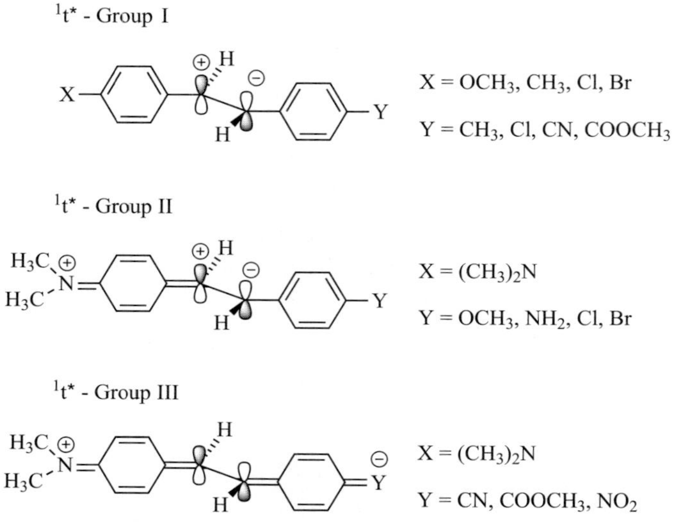

Figure 4.9 Classification of *trans*-4,4′-disubstituted stilbenes according to the donor–acceptor abilities of their substituents to stabilize the excited $^1t^*$ state [12]. (Reproduced with permission from [31].)

$^1t^*$ - Group I: X = OCH$_3$, CH$_3$, Cl, Br; Y = CH$_3$, Cl, CN, COOCH$_3$

$^1t^*$ - Group II: X = (CH$_3$)$_2$N; Y = OCH$_3$, NH$_2$, Cl, Br

$^1t^*$ - Group III: X = (CH$_3$)$_2$N; Y = CN, COOCH$_3$, NO$_2$

$$\alpha_{syn} = \frac{n}{2^{n-1}} \left(\frac{nRT}{\pi E_{syn}} \right)^{(n-1)/2}, \quad (4.4)$$

where n is the number of vibrational degrees of freedom of the nuclei participating in the concerted transition. An analysis of Equation (4.4) provides a clear idea of the scale of the synchronization factor and the dependence of this factor on the number of n and therefore on the number of broken bonds and the energy activation.

For stilbenes of groups 2 and 3, the electron transfer from the amino group to an acceptor segment is more thermodynamically preferable than that for the localized zwitter ion formation. Nevertheless, the light-induced one-step synchronous formation of structures with breaking double bond and formation of several new double bonds required the reorganization of 12–18 nuclei and, therefore, appears to be forbidden. Thus, the more realistic is a multistep (at least two steps) process involving electron transfer from amino group to acceptor appearing with the TICT structure a probable first intermediate. In each step, only optimum reorganization takes place.

4.2.1.5 A Volume-Conserving Mechanism

For *cis–trans* isomerization in highly condensed media at very low temperature ≤ 77 K, Liu and Hammond postulated mechanisms called one-bond flip and hula-twist mechanism [35–38]. According to the H-T mechanism, isomerization takes place not by the one-bond rotation around the double bond but by the concomitant twist of the double bond and the adjacent single bond to accomplish the double-bond isomerization (Figure 4.10). These mechanisms were assumed to reduce free volume

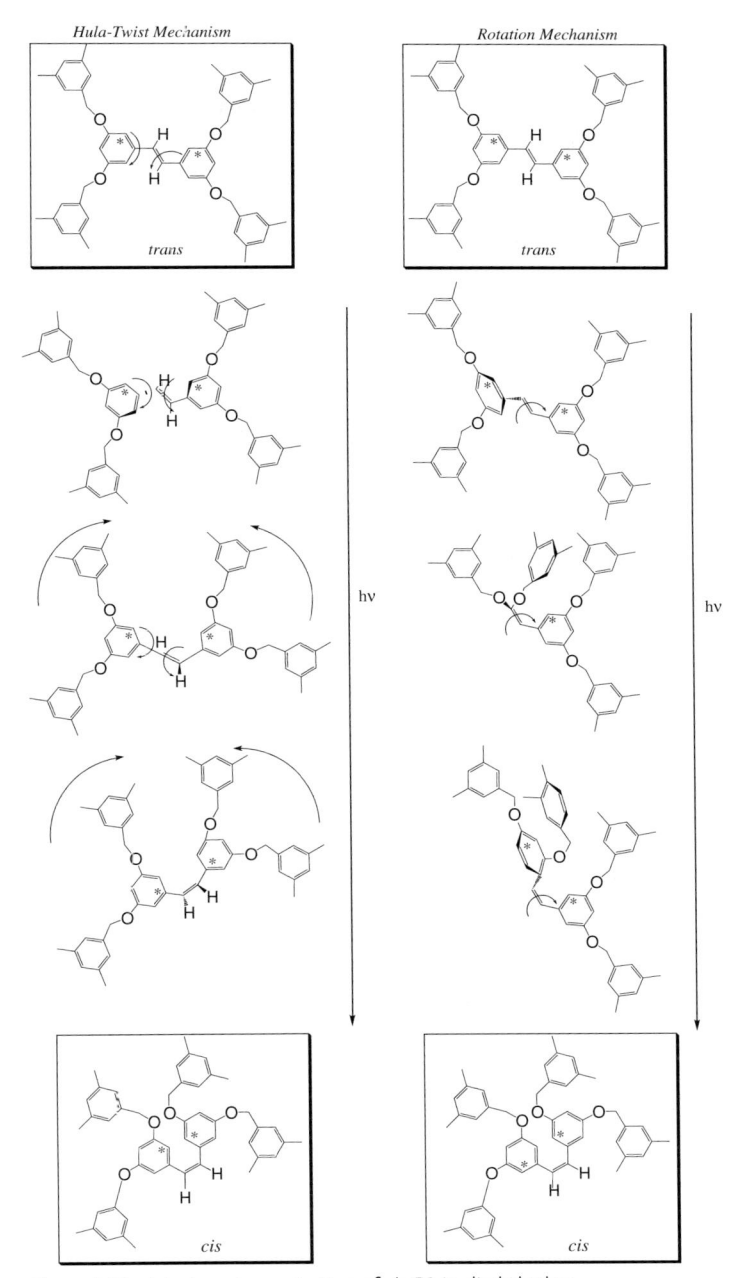

Figure 4.10 cis-to-trans isomerization of cis-G2 in diethyl ether-isopentane (1:1) at 77 K by a volume-conserving mechanism [38]. (Reproduced with permission.)

requirements by confining motion to the vicinity of the isomerizing double bonds while minimizing the motion of bulky substituents.

4.2.1.6 Media "Melting" Mechanism: Photoisomerization in Rigid Surroundings

Twisting stilbene phenyl segments in excited molecules at very low temperatures remains a challenging problem. In any way, the isomerization requires the bulky segment motion and, accordingly, a free space. Multiple experiments ([39–41] and references therein) unambiguously indicated that even such a small molecule as nitroxide is immobilized in rigid solvents and proteins at low temperatures $T < 200$ K. An analysis of ESR spectra of nitroxides in this condition revealed only a small libration with the amplitude $A < 2$ Å. To reconcile these two apparently conflicting facts, we suggest the following mechanism for stilbene photoisomerization in rigid media. After excitation, a part of the vibrational energy of several kcal/mol, which corresponds to the 40–60 nm Stokes shift, may be transferred to rotational and twisting modes of surrounding molecules to animate its dynamics. In other words, the stilbene twisting may take place in "a melting drop" of media.

4.2.1.7 Nonvertical Energy Transfer

The nonvertical energy transfer is attributed to large structural changes between ground and triplet states, to triplet–triplet energy transfer (TTET), and to spin-exchange (SE) [42–45]. This mechanism was first proposed to account for larger than expected triplet–triplet energy transfer rate constants from energy-deficient donors to cis-stilbene (c-St). According to the Fermi golden rule, the rate constant of triplet–triplet energy transfer

$$k_{TT} = \frac{2\pi}{h} J_{TT} FC, \tag{4.5}$$

where J_{TT} is the TT exchange integral and FC is the Franck–Condon factor. It means that at the moment of energy transition, energies of donor and acceptor should be equal. It was shown that NVET is characterized by a decrease in the triplet–triplet energy transfer rate constants, which remains slower than expected in the apparent endothermic region, as calculated from the Sandros plot of the triplet–triplet energy constant versus the gap between donor and acceptor energies (Figure 4.11). It was proposed that in the sensitized stilbene photoisomerization, energy requirements for the transfer are reduced by the population of ground-state vibrational modes of the acceptor that provide access to geometries that destabilize the ground state (S_0) while stabilizing the triplet state (T_1). The proposition that double-bond torsion is involved in nonvertical triplet excitation transfer to cis-stilbene was based on expected destabilization along $^1c \to {}^1p$ (cis to perpendicular in S_0) and stabilization along $^3c^* \to {}^3p^*$ (in T_1) coordinates. A single potential energy minimum at the perpendicular geometry, more shallow on the trans side, was also postulated on the T_1 surface. In other words, stilbenes probably function as nonvertical triplet excitation (NVET) acceptors for energy-deficient donors because rotation about the

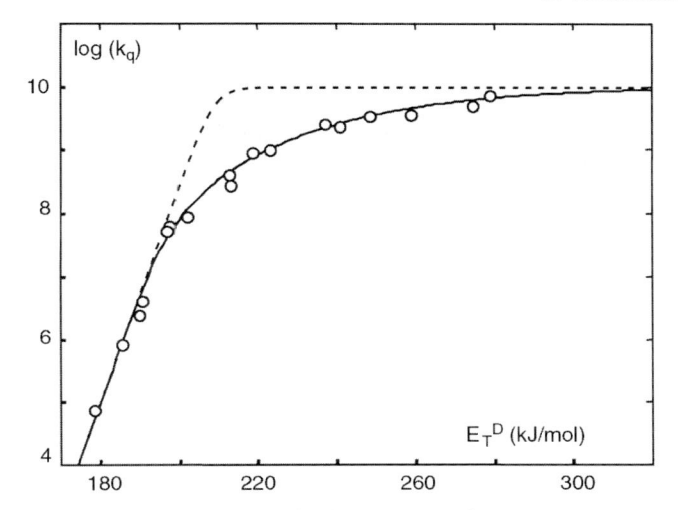

Figure 4.11 Experimental data for the energy transfer rate constants from donors to cis-stilbene (from dotted line: Sandros plot; plain line: thermal bond activation model fitted to experimental data [46]. (Reproduced with permission from Elsevier.)

central bond diminishes the energy gap between ground and triplet energy surfaces [45j].

Figure 4.11 demonstrates a difference between experimental data for the energy transfer rate constants from donors to cis-stilbene and theoretically predicted vertical excitation of the stilbene acceptor. This difference appears to be a clear indication in favor of the nonvertical triplet–triplet energy transfer mechanism [46].

Results of theoretical calculation at the B3LYP-DFT level using the 6-31 + G(d,p) basis set in the ground and triplet states of the (c-1, t-1) and of 2,3-stilbenes were

described in Ref. [22]. These results indicated that a pronounced pyramidalization at the olefinic C atoms gives a PhCCPh dihedral angle of 51.0° in 32* and the dihedral

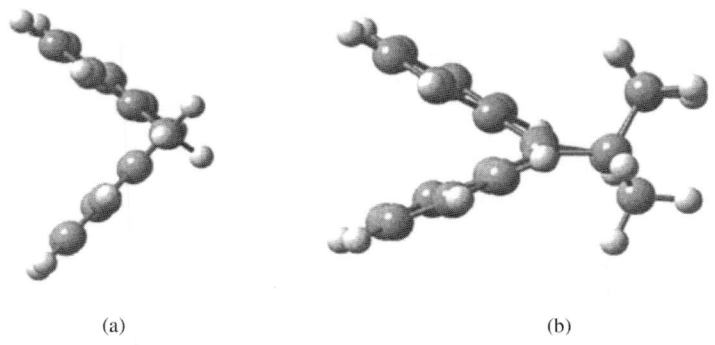

Figure 4.12 Calculated structures for the global energy minima of the triplet states of (a) **1** and (b) **2** [22]. (Reproduced with permission.)

angles of 32* that involve phenyl–vinyl torsions, 22.3 and 26.5° (Figure 4.12). The authors suggested that **2** was improperly used as a rigid model for c-**1**. The calculation predicted that in **2** and in **1** torsional motion about the CC double bond affords the most stabilization in the triplet state and that motion and not phenyl–vinyl torsion that most facilitates nonvertical energy transfer to stilbenes. A single energy minimum was found on the stilbene triplet energy surface, close to the postulated geometry of the phantom (perpendicular, 3p-1*) triplet.

4.2.1.8 A Dual Thermal Bond Activation Mechanism

Results obtained in [46] allowed the authors to suggest the large effects of both the single- and the double-bond torsions in the NVET behavior of cis-stilbenes. This conclusion was based on the calculation of the potential energy surfaces for cis-stilbene, using GAUSSIAN 98 suite of programs on the IBM supercomputer of IDRIS (CNRS, France) and the hybrid B3LYP functional with the 6-31 + G(d) basis. Spectroscopic triplet energies corresponding to these structures were then computed with time-dependent DFT method (TDB3LYP/6-31 + G*). It was assumed that the ground-state PESs are thermally populated according to a Boltzmann distribution: this leads to different molecular conformations exhibiting a wide change in spectroscopic triplet energies, as depicted in Figure 4.13. Potential energy surfaces associated with the double- and single-bond torsions (involved in the structural changes between relaxed ground and excited triplet states) are also shown in this figure. Assuming 5 kJ/mol of thermal activation at room temperature, it was observed that the ground-state molecules' distribution led to variations of the spectroscopic triplet energy within 29 kJ/mol along the single-bond torsion coordinate (β). The parameters deduced from the experimental data can be compared with those extracted from the calculated PESs. The authors focused on the contribution of the two major molecular coordinates involved in the singlet–triplet electronic transition of cis-stilbene by using the thermal activation model to fit the experimental energy transfer rate constants.

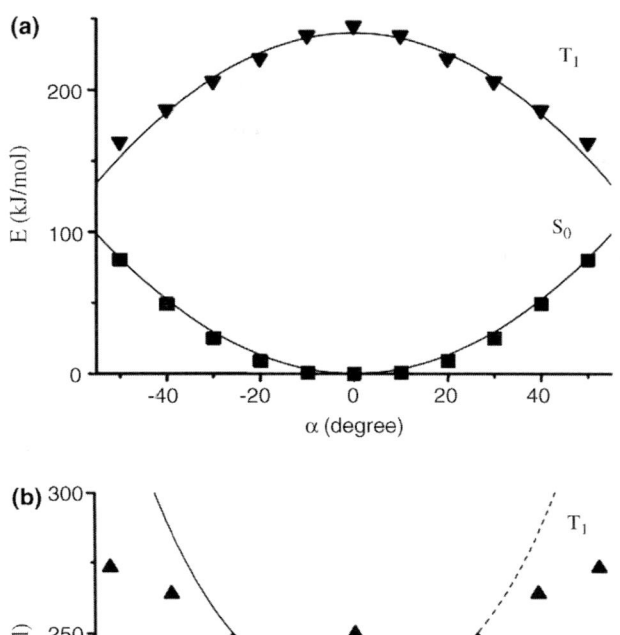

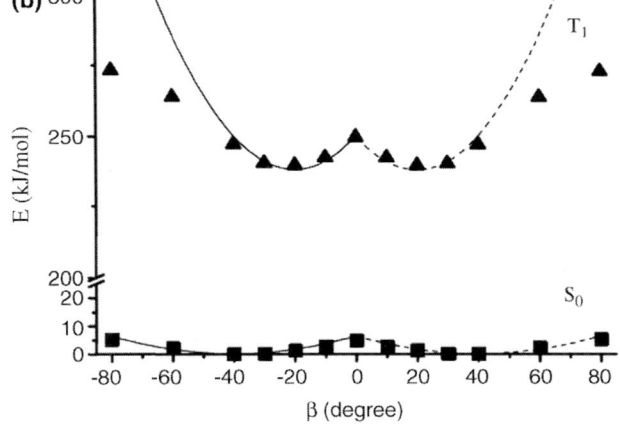

Figure 4.13 Potential energy surfaces for *cis*-stilbene along (a) the double-bond torsion coordinate (∗) and (b) the single-bond torsion coordinate (β) [46]. (Reproduced with permission from Elsevier.)

4.2.2
Experimental and Theoretical Studies of the Photoisomerization Mechanisms

It was reported that the picosecond formation and nanosecond decay of a transient absorption were produced by irradiation of *trans*-stilbene in solvent [48, 49]. This transient was referred to as the $^1p^*$ state. The $^1p^*$ state lifetime rapidly decreased in solvents of increasing polarity [50]. It was suggested to be the zwitterionic character of the $^1p^*$ state, that is, the formation of a diphenylmethyl anion tethered to diphenylmethyl cation.

The role of central bond torsion in nonvertical triplet excitation transfer to stilbenes, an example of stilbene analogues, biindanylidenes, was recently discussed in detail [45]. Comparison of the activation parameters for the two rigid stilbene analogues, *cis*- and *trans*-1,1′-biindanylidene (*c*-Bi and *t*-Bi)

with those for the stilbenes showed that the excitation transfer processes remain nonvertical despite the strong structural inhibition of phenyl–vinyl torsion. The relatively small pre-exponential factors of the respective isomers were found to be almost identical. Their magnitude was taken as a measure of the attenuation introduced by Franck–Condon overlap factors that decrease as the torsional state quantum number corresponding to the transition state increases. These results and results from theoretical calculations were consistent with central bond torsion as the key reaction coordinate in NVET to biindanylidenes and stilbenes.

Stationary point geometries on Bi S_0 and T_1 surfaces are presented in Figure 4.14. The crystal structure of *t*-Bi shows it to be strictly planar, eliminating phenyl–vinyl torsion toward planarity as a crucial NVET reaction coordinate. The authors emphasized that while experiment and theory support the initial proposition that double-bond torsion is the key reaction coordinate enabling NVET to stilbenes, as multidimensional surfaces are involved, other vibrations, including bond stretching, will also contribute.

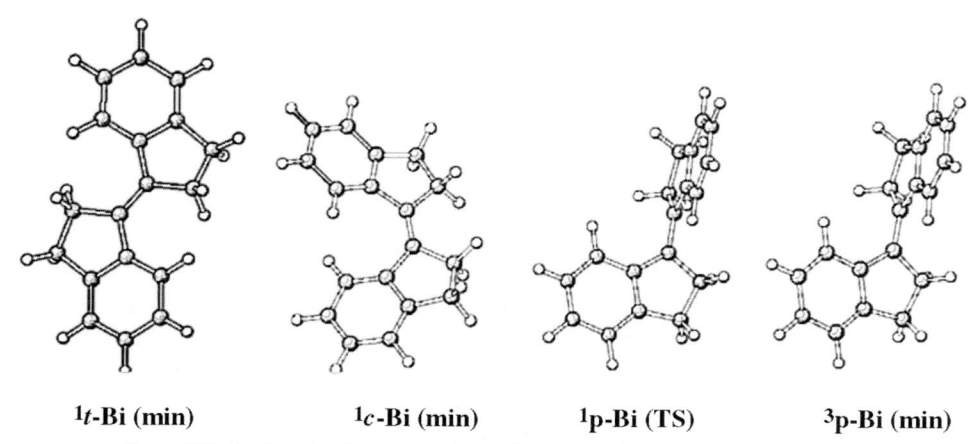

^1t-Bi (min) $\quad$ ^1c-Bi (min) $\quad$ ^1p-Bi (TS) $\quad$ ^3p-Bi (min)

Figure 4.14 Stationary point geometries on the Bi S_0 and T_1 surfaces [45]. (Reproduced with permission.)

Pyridinium derivatives of stilbenes have been shown to emit from several excited species multiple fluorescence [18, 51]. These systems possess two single bonds and one double bond, which can, in principle, twist in the excited state, and in addition the bond linking the dimethyl- (or dialkyl-)amino group. Selectively bridged derivatives such as **223** versus **222** allowed to suggest that the main nonradiative channel is connected to single-bond twisting and that double-bond twisting is only a minor nonradiative pathway in these compounds. Time-resolved studies not only established the large contribution of solvation dynamics to the Stokes shift but also indicated some temperature dependence of the fluorescence rate constant, consistent with the involvement of several emitting species. A reduced fluorescence quantum yield was also detected for donor–acceptor stilbene. In the flexible dyes related to **220**, the nonradiative channel from the single-bond-twisted species A* was suggested to be dominating. Fixing the conformation around the single bonds (**223**) led to a fluorescence enhancement (loose-bolt effect) and paved the way for photochemical *trans–cis* isomerization.

218 219 220 221

222 223

The photochemical behavior of a series of *trans*-3-(N-arylamino)stilbenes (m1, aryl = 4-substituted phenyl with a substituent of cyano (CN), hydrogen (H), methyl (Me), or methoxy (OM)) in both nonpolar and polar solvents was reported and compared with that of the corresponding *para* isomers (p1CN, p1H, p1Me, and p1OM) [52]. Electronic properties such as photoluminescence and energy migration strongly depended on the degree of electronic coupling (delocalization) between the subunits. Another crucial factor that determines the degree of electronic coupling in π-conjugated systems was the conformation, namely, the more planar the molecular structure, the stronger the coupling. The authors proposed a guideline based on the values of the fluorescence quantum yields Φ_f and *trans* → *cis* photoisomerization quantum yield (Φ_{tc}) Φ_{tc} for judging whether a TICT state is actively invoked in the excited decay of aminostilbenes. According to this proposal, to claim an important TICT state formation, two phenomena should be observed: (i) the value of Φ_f should be significantly smaller in polar versus nonpolar solvents, since the propensity of TICT state formation is larger in more polar solvents and the fluorescence quantum yield for TICT states is much lower than the precursor PICT state; (ii) fluorescence and *trans–cis* photoisomerization cannot account for all the decay processes (i.e., $\Phi_f + 2\Phi_{tc} \ll 1.0$) because the process of photoisomerization is essentially decoupled with the TICT state deactivating processes. More evidence for TICT state formation of p1CN included (a) the observation of PICT–TICT dual fluorescence in acetonitrile, (b) a significant reduction in the fluorescence rate constant (k_f) upon going from hexane ($5.8 \times 10^8\,s^{-1}$) to acetonitrile ($2.0 \times 10^7\,s^{-1}$), (c) the dual fluorescence disappears and normal deactivating mode for *trans*-stilbenes (i.e., $\Phi_f + 2\Phi_{tc} \approx 1.0$) is recovered when the torsional motion is inhibited through constrained ring bridging between D and A groups (i.e., p3CN), and (d) the fluorescence intensity increases upon increasing the temperature. The distinct TICT-forming propensity between the *meta* and the *para* isomers of *trans*-aminostilbenes 1CN and 1Me has been revealed. According to their energy diagrams, the PICT → TICT process was shown to be exothermic for all four cases, and the thermodynamic driving force is larger for the *para* isomers (i.e., p1CN > m1CN and p1Me > m1Me). The strong resonance-induced quinoidal character in the PICT state of p1Me disfavors the torsion of the stilbenyl-anilino C—N bond toward TICT but favors the torsion of the C=C bond for photoisomerization.

The first part of the isomerization path on the two lowest excited states of *trans* and *cis*-isomers of stilbene and stiff stilbene was studied by TD-PBE0 calculation in the gas phase and in heptane solution [53]. The authors performed a density functional theory (DFT) and time-dependent DFT study of S_0, S_1, and S_2 in the gas phase and in apolar solution. Solvent effects were taken into account by the PCM model. The excited-state optimized structures and the computed absorption and emission frequencies of stilbene were calculated. The stilbene orbitals and the S_1 and S_2 energy profiles for both isomers of stilbene are presented in Figures 3.2 and 3.3. On the basis of the present computational results and the available experiments, several arguments suggested that the dynamics after the optical excitation of *cis*-stilbene until the barrier crossing involves only the $S_1 B_{HL}$. Computations concerning *trans*-stilbene and stiff stilbene were in agreement with the available experimental results, providing a unifying picture of the photoisomerization path of *trans*- and *cis*-isomers of stilbene and stiff stilbene, consistent with the experimental determination of the energy barrier and of the excited-state lifetimes. In all of the examined compounds, the first part of the isomerization path (i.e., for twisting of θ up to ~50°) occurs in the B_{HL} bright state. That is, the isomerization proceeded before barrier crossing occurs on the HOMO → LUMO bright state, whereas the role played by other single-excitation states was suggested to be negligible.

4.3
Effect of Substituents and Polarity

The photochemistry of the substituted stilbenes opens up a unique possibility to follow the different timescale processes (in the femto-, pico-, and nanosecond regions) occurring in the molecules after irradiation. The investigated processes are the electronic polarization, vibrational and polar relaxation, radiative and nonradiative decay of the excited state, and twisting transition in the excited state. All these processes take place in the elementary act of a chemical reaction, but these are "overlapped" by each other and thus are undetectable by direct experimental measurements. It means that it is practically impossible to elucidate and differentiate the contribution of such factors as substituent or solvent effects to the above-mentioned processes. However, a Hammett-like correlation approach in photochemistry allowed one to elucidate and differentiate the contribution of these factors.

The various processes occurring in the stilbene molecule after its excitation exhibit different sensitivity to intramolecular donor–acceptor effects of substituents. This sensitivity was quantitatively characterized by ρ-constant of the linear Hammett-like relationships [54]. It has been shown that the Stokes shift in nonpolar cyclohexane was not dependent on the structure of the stilbene molecule (Section 3.4.1) [12, 31]. Therefore, the substituent effects on vibrational relaxation in the nonpolar solvent can be neglected. Nevertheless, these intramolecular electronic effects on the excitation energy of the substituted stilbenes were found to be very essential even in the nonpolar media (the excitation energy difference between stilbenes substituted with weak and strong donor–acceptor groups can reach 20 kcal/mol).

The nature of the substituent can have a profound effect on the excited state isomerization mechanism. While in the case of 4-nitrostilbene, the triplet pathway dominates the singlet route, and in the case of alkyl/alkoxy substituents the singlet mechanism is prevalent, and the isomerization of 4-bromostilbene tends to proceed by a combination of these two processes [55–59]. Asymmetrically substituted stilbenes can display a behavior that is even more varied. "Push–pull" stilbenes, which possess the strong donor–acceptor pairs of 4,4'-substituents on their aromatic rings, can form charge transfer states as a result of a certain geometrical distortion of the excited stilbene molecule. Formation of the charge transfer state strongly depends on the nature of substituents and may lead to the "dual fluorescence" phenomenon [60–65].

One of the features of stilbene photochemistry is its essentially strong dependence on medium polarity and temperature; the competition between fluorescence and *trans–cis* isomerization has been shown to be extremely sensitive to medium viscosity. Solvent polarity can affect both the dynamics and the pathway of the reaction. The dipolar character of asymmetrically substituted stilbenes and polarizability of the *trans*-stilbene transition state can explain the sensitivity of the photoisomerization rate to medium polarity [5, 6, 12, 31, 66–69].

The polarity effect strongly depends on the nature of substituents in a stilbene molecule. The data obtained in Refs [12, 31, 69] may be served as an example of such relationships. The data on the Hammett plot for a series of substituted stilbenes are presented in Figure 4.16. Donor–acceptor pairs of the 4,4′-substituents in the group I (Figure 4.9) usually increase the rate of the $^1t^* \rightarrow {}^1p^*$ transition owing to a higher stabilization of the more polar $^1p^*$ state, which appears to be zwitterionic, and consequent reduction of the intrinsic barrier to this reaction. Solvent polarity affects this transition in a similar way, so the rate of the *trans–cis* isomerization is increased by both polar solvents and polar substituents. The strong donor $(CH_3)_2N$ substituent tends to efficiently participate in a charge delocalization of the $^1t^*$ state compared to the $^1p^*$ state (group II). This electronic resonance interaction in the $^1t^*$ state destabilizes the activated transition $^1t^* \rightarrow {}^1p^*$ increasing the activation barrier and retards the photoisomerization rate. In fact, the quantum chemical calculations for the strong donor–acceptor-substituted stilbenes predicted the low polarity of the $^1p^*$ state, whereas for the nonpolar stilbenes a very highly polar $^1p^*$ state is expected [70, 71]. This switching from high to low polarity of the $^1p^*$ state is probably attributed to the phenomenon of "sudden polarization" [72]. Dependence of the $^1t^*$ decay rate constant (k_d in ns^{-1}) on the substituents σ-constants in different solvents is presented in Figure 4.15.

The photoisomerization rate of the 4-$(CH_3)_2N$-substituted stilbenes from the second stilbene group decreases considerably compared to the weak donor–acceptor-substituted stilbenes from the first group, but usually follows the same trends of

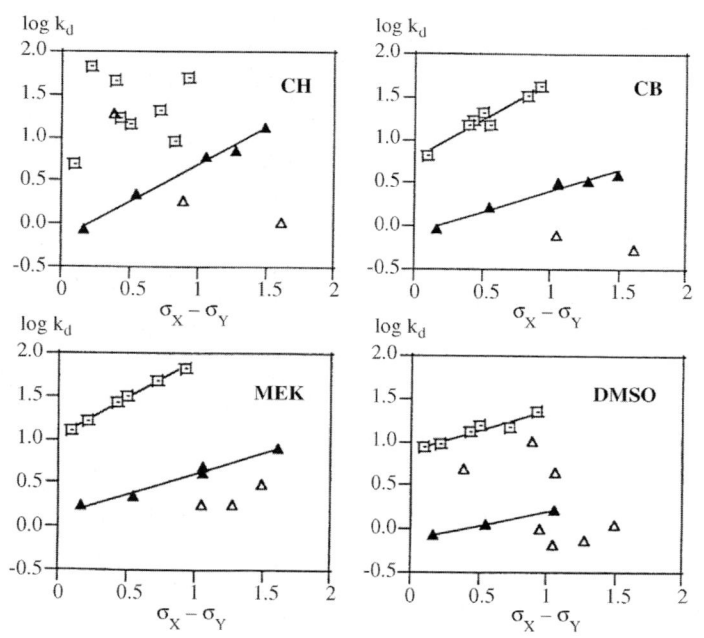

Figure 4.15 Dependence of the $^1t^*$ decay rate constant (k_d, in ns^{-1}) on the substituents σ-constants in different solvents. The stilbene groups I–III are designated by squares, filled triangle, and open triangles, respectively [12]. (Reproduced from [31].)

reactivity with a smaller ρ value (Table 3.1). This value is markedly higher for the first stilbene group (0.50–1.23) compared to the second group (0.33–0.86) and increases as the solvent polarity decreases. A similar behavior of the ρ-constant, being higher in solvents of lower polarity, was observed for dissociation of the substituted benzoic acids that is a classic example of the Hammett-like correlations [54]. In fact, this relatively low ρ value characterizes the low sensitivity of the twisting transition to the intra- and intermolecular electronic effects, probably because of the similar substituent effects on the energy of both $^1t^*$ and $^1p^*$ states.

The effect of thienyl groups on the photoisomerization and rotamerism of symmetric and asymmetric stilbenes has been studied [73]. Stationary and pulsed fluorimetric techniques, laser flash photolysis, and conventional photochemical methods and theoretical calculations were used for investigating photochemistry of five symmetric (bis-substituted) and asymmetric (mono-substituted) analogues of E-stilbene, where one or both side aryls are 2′-thienyl or 3′-thienyl groups. It was shown that the presence of one or two thienyl groups and their positional isomerism affect the spectral behavior, the relaxation properties (radiative/reactive competition), the photoisomerization mechanism (singlet/triplet), and the ground-state rotamerism. The photochemical behavior of trans-isomers of the four α,α′-di-X-stilbenes (X = F, Cl, Br, or I in olefinic positions) was studied in solution at room temperature [73]. For stilbenes with X = Br or I, the quantum yields of Br_2 and I_2 elimination were typically 0.2 and 0.6, respectively, and were essentially independent of the kind of solvent and the irradiation wavelength. The same transient ($\lambda_{max} \approx 410$ nm), observed by laser flash photolysis ($\lambda_{exc} = 248$ nm) of either Br_2- or I_2-stilbene, was assigned to the lowest triplet state of diphenylacetylene (DPA). This state is formed by consecutive absorption of two photons: the first photon generates DPA via the excited singlet state of trans-X2-stilbene by homolytic cleavage of the two CX bonds and the second photon within the 20 ns laser pulse yields $^3DPA^*$ via intersystem crossing.

The effect of solvent polarity on photoisomerization of trans-4,4′-bis(benzoxazolyl) stilbene has been reported [74]. On the basis of fluorescence quantum yield and lifetime measurements, the rate constants of radiative and nonradiative decays were calculated. In a polar solvent, a high fluorescence quantum yield was observed whereas a high rate of photoisomerization occurred in a nonpolar solvent.

4.4
Viscosity Effect

The trans–cis photoisomerization process following the solvent–solute relaxation competes with the radiative decay of the stilbene molecule. The $^1t^* \rightarrow {}^1p^*$ transition for trans-stilbene in the gas phase is very fast with a rate constant $k_{t \rightarrow c}$ of about $1.4 \times 10^{10}\,s^{-1}$ [66]. In viscous condensed phases, the transition is mostly governed by the media relaxation rate. The $k_{t \rightarrow c}$ rate constant strongly depends on solvent viscosity and temperature [75].

The trans–cis photoisomerization rate of the 4-dimethylamino-4′-aminostilbene strongly depends on temperature and, hence, on viscosity up to a temperature about

50 °C and above this temperature tends to be independent. A correlation between the rotational frequency of a nitroxide spin label (v_c) and the rate constant k_{iso} of 4-dimethyl-aminostilbene *trans–cis* photoisomerization has been found [76]. The label volume is similar to the stilbene fragments volume in the twisted state. The calibration allowed using the correlation for the estimation of the rotational correlation time of the stilbene fragments in the excited state of PSS for the fixed angle 180° (Figure 4.2).

It was shown [77] that the azulene and Oil Yellow effect on *cis–trans* stationary states for direct stilbene photoisomerization was independent of solvent viscosity, while the azulene effect on the sensitized stilbene photoisomerization was viscosity dependent. The observations supported the authors' proposal that the azulene effect on the direct photoisomerization was due to long-range (approximately 15 Å) nonradiation excitation transfer from *trans*-stilbene singlets, whereas the azulene effect on the sensitized photoisomerization was due to diffusion-controlled excitation transfer from *trans*-stilbene triplets. The unexpectedly large effect of azulene on the sensitized photoisomerization in a viscous solvent (Me_3COH) was proposed to be attributed to a ninefold increase of the effective lifetime of stilbene triplets in this solvent. A method for separation of viscosity and temperature effects on the singlet pathway to stilbene photoisomerization has been developed [75]. Contributions from the thermal and the viscosity-dependent processes of geometric twist were detected by employing a model that assumes that the overall activation energy observed in a viscous solvent is the sum of an inherent thermal barrier (E_t) and a solvent-dependent viscosity barrier (E_v). Arrhenius and Andrade equations were used to describe the variations of rate constant and viscosity with temperature. The method was applied to processes of fluorescence and *trans–cis* isomerization of *trans*-stilbene (I) and *trans*-1,1'-biindanylidene (II). The following values of the energy activation were obtained in glycerol: $E_v = 6.2$ kcal/mol and $E_t = 3.5$ kcal/mol for I, while $E_v = 9.9$ kcal/mol and $E_t = 2.4$ kcal/mol for II.

Temperature dependence of the fluorescence quantum yields and fluorescence lifetimes of *trans*-4,4'-di-*tert*-butylstilbene in *n*-hexane and *n*-tetradecane allowed to define the index of refraction dependence of the radiative rate constants, $k_f = (3.9 - 1.8) \times 10^8 \, s^{-1}$, and fluorescence lifetime [78]. This relationship was used to calculate torsional relaxation rate constants k_{tp}, for *trans*-4,4'-dimethyl- and *trans*-4,4'-di-*tert*-butylstilbene in the *n*-alkane solvent series. It was found that activation parameters for k_{tp}, based on Eyring's transition state theory, adhered to the medium-enhanced thermodynamic barrier model relationship, $\Delta H_{tp} = \Delta H_t + aE\eta$, and to the isokinetic relationship. The isokinetic relationship between the activation parameters for the parent *trans*-stilbene led to an isokinetic temperature of $\beta = 600$ K and brings it into agreement with the isokinetic temperature for activation parameters based on estimated microviscosities, $\eta\mu$, experienced by stilbene in its torsional motion. The authors concluded that only microviscosities rather than shear viscosities, η, can be employed in the expression $k_{tp} = k_t B\eta - b$, when $a = b$. These data clearly indicated the important role of the media dynamics in the stilbene *cis–trans* photoisomerization.

The S_1-photoisomerization of *cis*-stilbene was investigated by femtosecond pump-probe absorption spectroscopy in compressed solvents [79]. The viscosity dependence confirmed the existence of two pathways of the reaction. One showed an inverse viscosity dependence and led to *trans*-stilbene, the other one indicated no viscosity dependence and led to dihydrophenathrene.

4.5
Miscellaneous Experimental Data on Photoisomerization

4.5.1
Photoisomerization in Solutions

4.5.1.1 Direct Photoisomerization

Data on the temperature dependence of fluorescence quantum yields for the stilbenes and the *p*-, *m*-, and *m,m′*-Br derivatives in pentane obtained in Ref. [80] indicated that (i) intersystem crossing yields based on photostationary states can be unreliable, (ii) the singlet mechanism ($^1t^* \to {}^1p^*$) was confirmed as the major isomerization pathway of *trans*-stilbene singlets, (iii) significant fractions of bromostilbene $^1t^*$ molecules underwent intersystem crossing, and (iv) almost all bromostilbene singlets intersystem cross following twisting to $^1p^*$. Determination of the enthalpy and reaction volume changes in organic photoreactions using photoacoustic calorimetry was reported [81]. The photoreactions in *trans*-stilbene were investigated by photoacoustic calorimetry (PAS). This technique was used to measure both the thermal and the reaction volume changes for photoinitiated reactions. The following experimental values were found: $\Delta H = 7.5$ kcal/mol and $\Delta V = 5.6$ ml/mol.

Femtosecond photoelectron spectroscopy was employed to study the excitation of *trans*-stilbene above the isomerization reaction barrier [82]. Apart from the S_1 contribution, evidence of a second electronic state was found on the basis of two different transients measured across the photoelectron spectrum. Time-dependent density functional theory calculations on S_0, S_1, S_2, and D_0, together with simulations of the electron energy distribution, supported the experimental findings for selective photoelectron energies of the S_0, S_1, ... electronic states. The photoelectron spectra of *trans*-stilbene following the excitation with 266 nm laser pulses consisting of a pronounced three-peak structure were subjected to a substantial broadening, due to the large number of closely spaced vibrational states involved in the excitation scheme.

A series of four platinum acetylide complexes that contain 4-ethynylstilbene (4-ES) ligands have been subjected to a detailed photochemical and photophysical investigation [83]. Using absorption, variable temperature photoluminescence, and transient absorption spectroscopy, UV–vis absorption, and NMR spectroscopy, it was shown that these compounds undergo *trans*–*cis* photoisomerization from the triplet excited state. The obtained experimental data indicated that in all of the complexes, excitation led to a high yield of a $3\pi,\pi^*$ excited state that is localized on one of the 4-ES ligands. At low temperature, the $3\pi,\pi^*$ state exhibits strong phosphorescence that was very similar to the phosphorescence of *trans*-stilbene. At temperatures above the glass-to-fluid temperature of the solvent medium, the $3\pi,\pi^*$ state decays rapidly (40 ns). The decay pathway was suggested to involve rotation around the C: C bond of one of the 4-ES moieties. The steady-state photolysis led to *trans–cis* isomerization of one of the 4-ES ligands with a quantum efficiency of 0.4. The Pt-stilbene compound in which a conjugation exists between *trans*-stilbene as a ligand and Pt center in platinum–acetylide complexes underwent *trans–cis* photoisomerization of the stilbene ligand(s) [84].

It was found that the fluorescence lifetime of *trans*-3,3′,5,5′-tetramethoxystilbene (*trans*-TMST) experienced a large solvent effect changing from 2.3 ns in cyclohexane to 16.6 ns in acetonitrile [85]. This data indicated that the excited singlet state of *trans*-TMST has a charge transfer (CT) character. On the basis of the results obtained, the interior polar environment of a water-soluble TMST dendrimer was discussed. Effect of solvent polarity on photoisomerization of *trans*-4,4′-bis(benzoxazolyl) stilbene has been reported [86]. On the basis of fluorescence quantum yield and lifetime measurements, the rate constants of radiative and nonradiative decays were calculated. In a polar solvent, a high fluorescence quantum yield was observed whereas a high rate of photoisomerization occurred in a nonpolar solvent. Fluorescence lifetimes of *trans*-TMST were measured [87]. A large solvent effect ranging from 2.3 ns in cyclohexane to 16.6 ns in acetonitrile indicating the excited singlet state of *trans*-TMST has a charge transfer character.

Ultrafast torsional isomerization and ring closure reactions of photoexcited *cis*-stilbene have been examined in hexane, cyclohexane, methanol, and acetonitrile solvents [88]. The process was monitored by a combination of ground-state resonance Raman intensities, two-color UV picosecond anti-Stokes resonance Raman scattering, and product quantum yield measurements. Quantum yields for ring closure to ground-state dihydrophenanthrene were found to be two to three times higher in the nonpolar solvents than in the polar one. Vibrationally hot ground-state *trans*-stilbene was formed within the 10 ps time resolution of the time-resolved Raman experiments; cooling was about a factor of two faster in the former solvent. According to the authors, data on the time-resolved Raman experiments, the femtosecond transient absorption, and fluorescence properties indicated that increasing solvent polarity causes vertically excited *cis*-stilbene to distort more rapidly along the torsional isomerization coordinate, shortening the excited-state lifetime and decreasing the quantum yield for the competing ring closure reaction. It was shown in [89] that the excited states of stilbenylpyrroles I were deactivated by two photochemical processes: *cis–trans* isomerization and hydrogen transfer of NH to the stilbene double bond. NH-transfer results in the formation of two quinone dimethane intermediates and biradicals. Stilbenes also undergo intramolecular cyclization, giving rise to polycyclic compounds spiro-2*H*-pyrroles, pyrroloisoindoles, and pyrroloisoquinolines. Spiro-2*H*-pyrroles rearrange on silica gel, giving dihydroindoles.

Water-soluble *p*-sulfonato calix[*n*]arenes ($n = 8$, **1a–1b** and $n = 6$, **2a–2b**) was employed as host to control the outcome of the photodimerization and photoisomerization of 4-stilbazoles (**6a–6d**) [90]. A series of stilbenes with an alkyl group on the double bond of *cis*- or *trans*-stilbenes, with two or three alkyl group substituents on the stilbene A ring, and with an alkoxy group other than methoxy at position 3, 4, and/or 5 of the stilbene A ring have been synthesized [91]. It was found that prodrugs in which amino acid esters derivatives were formed with the phenolic hydroxyl at position 3 of the B ring. The photochemical release of an active form of the compound from a prodrug conjugate and the photochemical isomerization of the compounds, especially from a *trans* to *cis* form of compounds, have been demonstrated. The reactions can be used alone or in combination to convert inactive or comparatively less active forms of the compounds to more active forms.

The fluorescence of *trans*-stilbene and four methoxy-substituted stilbene derivatives

has been detected in a variety of solvents [92]. Compared to other stilbene derivatives, *trans*-3,5-dimethoxystilbene displayed a large quantum yield of fluorescence and a low quantum yield of *trans–cis* isomerization in polar organic solvents. The unique fluorescence properties of *trans*-3,5-dimethoxystilbene were attributed to the formation of a highly polarized charge transfer ($\mu_e = 13.2$ D). The fluorescence of all five *trans*-isomers was quenched by 2,2,2-trifluoroethanol.

The time-resolved fluorescence behavior of two derivatives of 4-(dimethylamino)-4′-cyanostilbene (DCS) bearing a more voluminous (JCS) and less voluminous anilino group (ACS) has been investigated.

For JCS, reconstructing emission (Figure 4.16) spectra exhibited an isosbestic point that indicated level dynamics between two emitting excited singlet states (LE and CT). Kinetic evaluation yielded a precursor–successor relationship between LE and CT and CT formation time constants of 4 ps for ACS and 8 ps for JCS. The authors suggested a twisting mechanism to be a major component of the reaction coordinate. An additional transient redshift of the CT band was attributed to the relatively slow solvation dynamics (ethanol).

4.5.1.2 Sensitized Photoisomerization

Data on the temperature dependence of the azulene (I) effect on the Ph_2CO-sensitized photoisomerization of stilbenes in PhMe, C_6H_6, Me_3COH, and MeCN were shown to be consistent with the formation of identical stilbene triplets from the two isomers [93]. Two mechanisms for the quenching interaction were discussed. The first mechanism assumed that the population of transoid triplet geometries, $3T^*$, is negligible and that quenching follows encounters of twisted stilbene triplets, $3P^*$, with I when they achieve $3T^*$ geometries in the encounter cage. According the second mechanism, the quenching was assumed to occur only by direct interaction between $3T^*$ and I. A lower enthalpy for the twisted triplets was calculated, $\Delta H_{tp} = -2.1$ and -1.6 kcal/mol in PhMe and Me_3COH, respectively, for the first mechanism and $\Delta H_{tp} = 0.5$ and 2.9 kcal/mol for PhMe and Me_3COH, respectively, for the second mechanism in the same solvents. Reasons for favoring the $3P^*$ quenching mechanism were presented. Experiments on the azulene effect on *p*-bromostilbene photoisomerization showed that singlet–triplet intersystem crossing was more efficient than in stilbene, but *meta*-substitution had no effect [94].

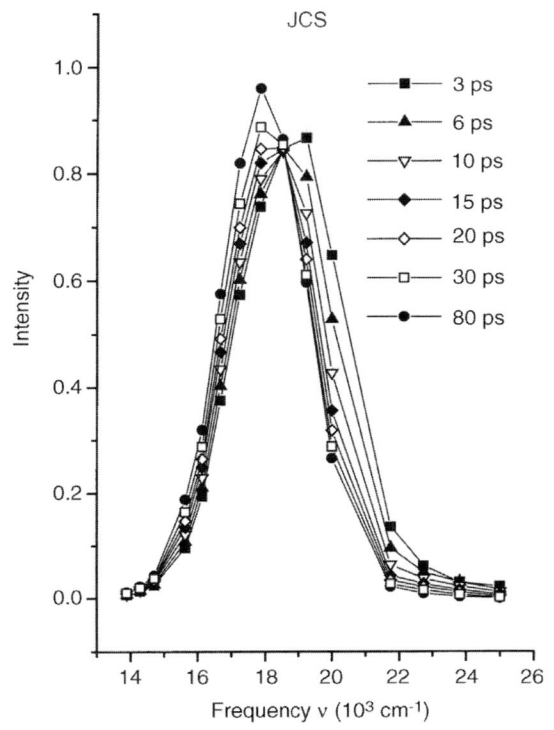

Figure 4.16 Reconstructed time-resolved emission spectra of JSC in ethanol at 298 K [28]. Reproduced wit permission.)

This indicated that the *meta*-position was near a node in the highest occupied and lowest unoccupied MOs of stilbene. Quenching experiments with dioxygen suggested a twisted geometry for the triplets in C_6H_6 with a lifetime of approximately 120 ns at 30 °C. It was shown that the anthracene ground-state triplet was responsible for sensitized *cis*-stilbene photoisomerization in C_6H_6 [95]. Singlet excited anthracene intersystem crosses to give, sequentially, excited anthracene triplet (with a lifetime approximately 30 ps) and ground-state triplet of anthracene. The effect of O_2 indicated that O_2 quenching of excited singlet led in part to adiabatic formation of excited triplet. Enthalpies (ΔH) for thermal *trans/cis*-stilbene isomerization attained, with iodine as catalyst in benzene or *tert*-butylbenzene solution, were measured in the 303.2–461.2 K range [96]. The temperature dependence of the equilibrium constant for the *trans–cis* reaction gave $\Delta H = 4.59 \pm 0.09$ kcal/mol and $\Delta S = 1.05 \pm 0.24$ eu. The ΔH value was sufficiently large to require intersection of the S_0 and T_1 potential energy curves close to the perpendicular geometry. Data on an intramolecular energy transfer within a triple chromophore from an optically excited fluorescein to an extremely low-lying *trans*-stilbene T_1 state were presented in Ref. [97]. Semiempirical calculations and sensitizing experiments were performed to obtain a good estimation of the $S_0 - T_1$ energy difference, which was found to be about 142 kJ/mol.

4.5.1.3 Photoisomerization of Stilbenophanes

The triplet-sensitized photoreaction of stilbenophanes that caused *trans–cis* photoisomerization of stilbenophanes (*trans,trans*-2, $Z = SiMe_2CH_2$-1,3-$C_6H_4CH_2SiMe_2$; *trans,trans*-1, $Z = CH_2SiMe_2CH_2$-1,4-CH_2; and *cis,cis*-3, $Z = CH_2$) was reported [98]. A series of macrocyclic and medium-size stilbenophanes tethered by silyl chains were synthesized and their photochemical and photophysical properties were examined. The triplet-sensitized photoreaction of stilbenophanes led to *cis–trans* photoisomerization. Photochemical properties of three isomers of [1.1]*meta*-stilbenophane, calixarene analogues, were investigated [99]. It was found that the conformational varieties of calixarene analogues having stilbene units could be controlled by photoirradiation using weak UV light. A ^{1}H NMR study indicated that a *trans–trans* isomer could be completely transformed into a 35 : 65 mixture of *cis–trans* and *cis–cis* isomers by photoirradiation at 254 nm. Conformational properties of [1.1]*meta*-stilbenophanes **2a**, **2b**, and **2c** in a solution and the effect of their functional groups on *cis–trans* photoisomerization have been discussed. The authors found this system of interest for application in molecular devices.

4.5.1.4 Stilbene Photoisomerization in Dendrimers

The review [100] described dendrimers with photoreversible stilbene cores that undergo mutual *cis–trans* isomerization in organic solvents to give photostationary state mixtures of *cis-* and *trans-*isomers. Stilbene dendrimers with molecular weights as high as 6500 underwent mutual *cis–trans* isomerization within the lifetime of the excited singlet state. The large dendron group surrounding the photoreactive core may affect the excited-state properties of the core to induce the efficiency of photoisomerization and/or reduce the fluorescence efficiency. The photochemistry of stilbene dendrimers, with various types of dendron groups, azobenzene dendrimers, and other photoresponsive dendrimers was discussed. It was shown that photoresponsive polyphenylene dendrimers *trans*-1 and 2 underwent photochemical isomerization to the *cis*-isomers [101]. For example, the polyphenylene dendrimer *trans*-2 (Figure 4.17) with the molecular weight as high as 4700 and with weak

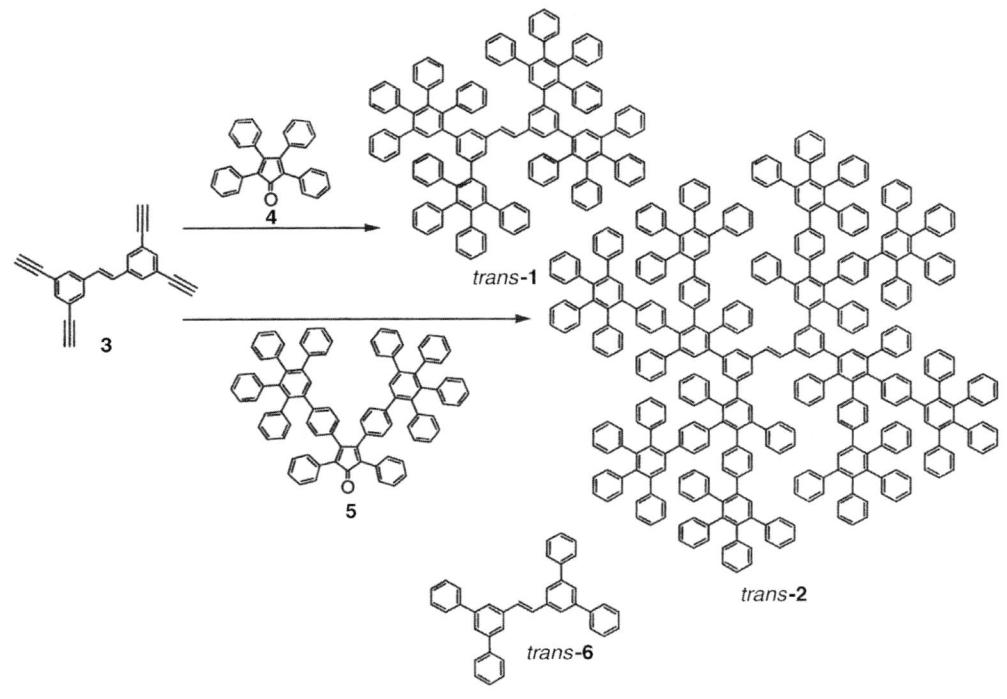

Figure 4.17 Synthesis and chemical structure of dendrimers investigated in Ref. [101]. (Reproduced with permission.)

conjugation throughout the molecule was found to undergo *trans–cis* photoisomerization within its excited-state lifetime.

Photochemical and photophysical properties of a poly(propylene amine) dendrimer (**2**) functionalized with *E*-stilbene units have been studied [102]. *Z*-photoisomerization and photocyclization of the *Z*-isomer of the stilbene units were investigated in air-equilibrated acetonitrile solutions. The quantum yields of the $E \rightarrow Z$ photoisomerization reaction and the fluorescence quantum yield of the *E* were found to be equal to 0.30 and 0.014, respectively. Stilbene dendrimers prepared by coupling 4,4′-dihydroxystilbene with first-, second-, third-, or fourth-generation benzyl ether-type dendrons underwent photoisomerization with the same efficiency as that of 4,4′-dimethoxystilbene [103]. The lifetime of the core structure was found to be shorter then 1 ns. According to [104], polyphenylene-based stilbene dendrimers, G1, G2, and G3, underwent mutual *cis–trans* isomerization upon direct irradiation with 310 nm light at room temperature. In a solvent glass at 77 K, one-way *cis–trans* isomerization was observed for G2.

Clear evidence for the large conformational change upon isomerization of stilbene (tetramethoxystilbene) dendrimers of generation G0, G1, G2, and G4 was presented [105]. From the time-resolved studies on energy and molecular volumes changes, it was found that the conformational change upon the *trans*- to *cis*-isomerization completed with the decay of the T_1 state for G0–G2. The diffusion coefficient (*D*) of the *trans* and *cis* forms of G0 was measured to be almost the same, whereas the difference becomes larger with increasing generation. As shown in Ref. [106], in the third generation of novel photoresponsive water-soluble stilbene dendrimers (*trans*- and *cis*-G3 WSD), unusual one-way *trans–cis* isomerization to give 100% of *cis*-isomer at the photostationary state upon UV irradiation in water took place. Photoisomerization in a series of stilbene dendrimers exhibiting fluorescence emission with considerably high quantum efficiencies was studied [107]. It was found that even the G4 stilbene dendrimer with the molecular weight as high as 6548 can undergo isomerization around the double bond within the lifetime of its excited singlet state. The photochemical isomerization of the C=C double bond in molecular weight of over 6000 took place efficiently within 10 ns timescale.

4.5.1.5 Stilbene Photoswitching Processes

Photoinduced gelation by stilbene oxalyl amide compounds was described [108]. Oxalyl amide derivatives bearing 4-dodecyloxy-stilbene as a *cis–trans* photoisomerizing unit were synthesized and their photochemical properties were investigated. The *trans* derivatives were found to act as versatile gelators of various organic solvents, whereas the corresponding *cis* derivatives showed a poor gelation ability. The FT-Raman, FT-IR, and ^{1}H NMR spectra demonstrated that the gelation process occurred because of a rapid *cis* → *trans* photoisomerization followed by a self-assembly of *trans* molecules. Apart from the formation of hydrogen bonding between the oxalyl amide parts of the molecules, confirmed by FT-IR spectroscopy, it was assumed that the π–π stacking between *trans*-stilbene units of the molecule and a lipophilic interaction between long alkyl chains were the interactions responsible for gelation.

trans-4-Me(CH$_2$)$_{11}$OC$_6$H$_4$CH:CHC$_6$H$_4$NHCOCOR [I R = OEt, NH-L-Leu-OMe] was proved to be efficient gelator of various organic solvents [109]. Considering the difference in gelation abilities of *trans*- and *cis*-I [R = OEt] and the photoresponsive conformational changes of the stilbene part of the molecule, a controlled gelation by light was achieved. FT-IR and ^{1}H NMR spectroscopic measurements supported the view that hydrogen bonding between oxamide fragments plays an important role in gel formation. A scheme of photoreversible and thermoreversible gelation of methanol by oxamide-based stilbene compounds is presented below.

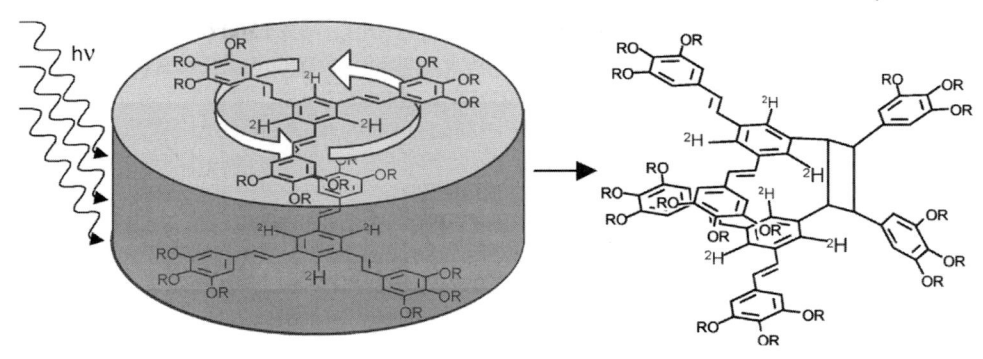

Figure 4.18 Schematic illustration of the dendrimer molecular motion that is necessary for the photochemical reaction [112]. (Reproduced with permission.)

The gelation ability of stilbene–cholesterol derivatives in different organic solvents was reported [110]. The photochemistry and the fluorescence of selectively deuterated at the α-position of the alkoxy chains of dodecyloxy-substituted stilbenoid dendrimers of the first and second generation were investigated in different crystal and liquid–crystal phases [111]. It was found that the photodegradation of double bonds begins when the material was heated to its liquid–crystal phase at irradiation λ_{350} nm for 2 h. No photoreactions occurred in the crystal state. Figure 4.18 illustrates the dendrimer motion that is necessary for the photochemical reaction.

Dynamics of the conformational change of water-soluble stilbene dendrimers upon photoexcitation monitored by the time-resolved transient grating method was investigated [112]. While the energy relaxation and the conformational change were completed within 30 ns for the first generation (W-G1), slower dynamics was observed for the second (W-G2) and third (W-G3) generations. The enthalpy change by the photoisomerization increases and the dynamics becomes slower with an increase in the generation. A photoswitchable stilbene-type β-hairpin mimetics, as a new tool in peptide engineering, was reported [113]. This system showed photoisomerization of the stilbene chromophore, resulting in a change in solution conformation between an unfolded structure and a folded β-hairpin.

A review of the recent progress in transition metal compounds involving photoinduced changes in the magnetic and/or optical properties to long-lived metastable states was presented in Ref. [114]. Photoswitchable compounds including stilbenoid complexes represented an attractive class of materials in coordination chemistry. The basic photophysical and photochemical phenomena, together with their representatives, were discussed. Some possible applications for energy and information storage were suggested.

4.5.1.6 Stilbene Photoisomerization on Templates

The synthesized dual stilbene-nitroxide probe (BFL1, Section 1.5) was covalently immobilized onto the surface of a quartz plate [115]. A few water/glycerol mixtures of different viscosities were prepared and the kinetics of photoisomerization for immobilized BFL1 was measured in these mixtures. Earlier, a similar experiment was done with immobilized 4,4′-bromomethylstilbene [67]. Initial fluorescence intensity was found to rise and the photoisomerization rate constant was found to fall with the increase in viscosity. This was expected because of the competition of the fluorescence emission and the *trans–cis* isomerization (Figure 3.2). The quantitative expression of this dependence was demonstrated by plotting the logarithm of photoisomerization rate constant $\{\log(k_{iso})\}$ versus initial intensity of fluorescence $\{I_0\}$.

A perdeuterated *trans*-stilbene grafted polystyrene has been synthesized and its photochemical and photophysical properties have been investigated [116]. The effects of chromophore concentration, solvent polarity, excitation energy, chromophore aggregation, and UV irradiation on photophysical properties of this compound have been studied. The photoexcitation of the stilbene chromophore caused the photoisomerization process, which involved low-bandgap *trans* form and high-bandgap *cis* form. Chromophore aggregation led to excimer formation and redshifts of the spectra, while the *trans*-stilbene grafted polystyrene emitted under UV-irradiation blue-green light in the solid state. The authors of the work [117] described a spectrophotometric investigation of the UV-light-induced *trans–cis* isomerization in poly(methyl methacrylate) (PMMA) films containing stilbene or stilbenecarboxaldehyde. The authors also reported, for the first time to their knowledge, halogenic storage in these materials that was stable for months. It was shown [118] that in the photosensitive polymer-dispersed liquid crystals (PDLCs) containing the nematic mixture E-5 doped with 2 wt% photochromic stilbene dye in a gelatin binder, the stilbene underwent *E–Z* photoisomerization upon irradiation. Irradiation of the stilbene-doped gelatin PDLC resulted in destroying the preferred director orientation of dopants. The optical properties and the Volta potential ΔV of stilbene-doped PDLC were changed. Differences were observed between hot- and cold-dried PDLC films. Polystyrenes naphthoylated with α- and β-naphthoyl chloride were used to sensitize the photoisomerization of *cis*- and *trans*-stilbene [119]. The resultant polymers had the same quantum efficiencies as their corresponding model compounds except for the isomerization of *trans*-stilbene by the α-naphthoylated polymer. The *trans–cis* photoisomerization of E-4-[2-(4-*tert*-butylphenyl)ethen-1-yl]benzoate photoisomerized to the Z-isomer and vice versa in the free state and in the binary complexes with N-(6A-deoxy-α-cyclodextrin-6A-yl)-N′-(6A-deoxy-β-cyclodextrin-6A-yl)urea and N,N-bis(6A-deoxy-β-cyclodextrin-6A-yl)urea has been described [120]. These systems functioned as molecular devices. The photochemical response was studied by UV–vis and fluorescence spectroscopy for two series that were liquid crystals of polyphosphates bearing dual photoreactive mesogenic units (stilbene and azobenzene/α-methylstilbene and azobenzene) [121]. The rate of the switching time for the conversion of *trans* to *cis* form of azobenzene unit was measured. Photochemical reactivities of stilbenes in organic glasses were compared with those in solution and in organized media [122]. The conclusion was that the preference for the most volume-conserving hula-twist mechanism isomerization in organic glasses appeared because of the close interaction between the guest and the host molecules. The authors suggested that in zeolites, crystals, and protein binding cavities, the residual empty space coupled with any specific guest–host interaction could lead to an involvement of the more volume-demanding torsional relaxation or bicycle-pedal or an extended process in photoisomerization.

Photochemical reactivities of *trans*-4,4′-dimethyl stilbene in organic glasses were first examined and compared with those in solution [123]. It was shown that the geometric isomerization of this stilbene was restricted. The excited-state chemistry of the system was different in this medium from that in organic solvents. This was attributed to the supramolecular effects of the host cavity. *trans*-4-Methyl-4′-(-S-$(CH_2)_n$-O-)stilbenes (2) ($n = 6$–9) were used to cap colloidal gold clusters, yielding composite shell-core nanostructures (3) and (4) [124]. Upon irradiation at 350 nm, photoisomerization of the appended *trans*-isomer to the corresponding *cis*-isomer took place both in solution and in the composite cluster. Quantum yields for photoisomerization of the composite clusters 3 and 4 were affected by the length of the linker because of the distance-dependent through-bond quenching by the metal core.

To evaluate the possible use of such molecules as molecular switches on semiconductor surfaces, the adsorption of *cis*- and *trans*-stilbene on Si(1 0 0) has been investigated [125]. For both isomers, bonding takes place via the C=C double bond to the Si dimer atoms allowing free movement of aromatic rings, a prerequisite for photoinduced isomerization on the surface. Phototransformation of stilbene in van der Waals nanocapsules was investigated [1]. *para*-Hexanoylcalix[4]arene nanocapsules were employed as hosts to carry out phototransformations of *cis*- and *trans*-stilbenes. Single-crystal X-ray diffraction studies were performed to define the location of encapsulated stilbenes inside the capsule and to analyze possible pathways of phototransformation. *cis*-Stilbene stacks as π–π dimers were found to be located at the center of the capsule, whereas *trans*-stilbene did not form such a dimer. Irradiation of the crystal inclusion complexes of each isomer of stilbene in the solid state led to the appearance of the second isomer, and after prolonged photolysis, photodimerization also occurred. The photochemistry and photophysics of the fac-[Re(CO)$_3$(NN)(L)]$^+$ complexes, NN = polypyridyl ligands and L = stilbene-like ligands, in acetonitrile solution and in PMMA polymer film have been reviewed [126, 127]. Under irradiation, the complexes exhibit *trans*–*cis* photoassisted isomerization of the coordinated stilbene-like ligand.

To summarize, in the present section, we have demonstrated that the stilbene photoisomerization is a "training area," a relatively simple and convenient model reaction for a thorough investigation of detailed mechanisms of photochemical reactions and factors affecting the photochemical conversion rate. Theoretical and experimental data in this area will pave the way for practical application of stilbenes as switching materials and biophysical probes (Chapter 10).

References

1 Ananchenko, G.S., Udachin, K.A., Ripmeester, J.A., Perrier, T., and Coleman, A.W. (2006) *Chemistry – A European Journal*, **12**, 2441–2447.

2 Hammond, G.S. and Saltiel, J. (1962) *Journal of the American Chemical Society*, **84**, 4983–4984.
3 Saltiel, J. (1968) *Journal of the American Chemical Society*, **90**, 6394–6400.
4 Saltiel, J. and D'Agostino, J. (1972) *Journal of the American Chemical Society*, **94**, 6445–6456.
5 Saltiel, J. and Sun, Y.-P. (1990) *Photochromism, Molecules and Systems* (eds H. Dürr and H. Bouas-Laurent), Elsevier, Amsterdam, p. 64.
6 Saltiel, J. and Charlton, L. (1980) *Rearrangements in Ground and Excited States*, vol. 3 (ed. P. DeMayo), Academic Press, New York, p. 25.
7 Saltiel, J. and Sun, Y.-P. (1989) *The Journal of Physical Chemistry*, **93**, 6246–6250.
8 Saltiel, J., Waller, A.S., Sears, D.F., and Garrett, C.Z. (1993) *Journal of Chemical Physics*, **97**, 2516–2522.
9 Sun, Y.-P., Saltiel, J., Park, N.S., Hoburg, E.A., and Waldeck, D.H. (1991) *The Journal of Physical Chemistry*, **95**, 10336–10344.
10 Saltiel, J., Waller, A.S., Sears, D.F., Hoburg, E.A., Zeglinski, D.M., and Waldeck, D.H. (1994) *The Journal of Physical Chemistry*, **98**, 10689–10698.
11 (a) Nibbering, E.T.J., Fidder, H., and Pines, E. (2005) *Annual Review of Physical Chemistry*, **56**, 337–367; (b) Dugave, C. and Demange, L. (2003) *Chemical Reviews*, **103**, 2475–2532, and references therein.
12 Papper, V. and Likhtenshtein, G.I. (2001) *Journal of Photochemistry and Photobiology A: Chemistry*, **140**, 39–52.
13 Waldeck, D.H. (1991) *Chemical Reviews*, **91**, 415–436.
14 Whitten, D.G. (1993) *Accounts of Chemical Research*, **26**, 502–509.
15 Polo, A.S., Itokazu, M.K., Frin, K.M., Patrocinio, A.O.T., Murakami, I., and Neyde, Y. (2007) *Coordination Chemistry Reviews*, **250 (13+14)**, 1669–1680.
16 Polo, A.S., Itokazu, M.K., Frin, K.M., Patrocinio, A.O.T., Murakami, I., and Neyde, Y. (2006) *Coordination Chemistry Reviews*, **250**, 1669–1680.

17 Gutlich, P., Garcia, Y., and Woike, T. (2001) *Coordination Chemistry Reviews*, **219–221**, 839–879.
18 Grabowski, R., Rotkiewicz, K., and Rettig, W. (2003) *Chemical Reviews*, **103**, 3899–4032.
19 Meier, H. (1992) *Angewandte Chemie – International Edition in English*, **31**, 1399–1420.
20 Momotake, A. and Arai, T. (2004) *Journal of Photochemistry and Photobiology C: Photochemistry Reviews*, **5**, 1–25.
21 Ma, J., Dutt, G.B., Waldeck, D.H., and Zimmt, M.B. (1994) *Journal of the American Chemical Society*, **116**, 10619–10629.
22 Catalan, J. and Saltiel, J. (2001) *Journal of Physical Chemistry A*, **105**, 6273–6276.
23 Dou, Y. and Allen, R.E. (2003) *Journal of Chemical Physics*, **119**, 10658–10666.
24 Dietl, C., Papastathopoulos, E., Niklaus, P., Improta, R., Santoro, F., and Gerber, G. (2005) *Chemical Physics*, **310**, 201–211.
25 Sczepan, M., Rettig, W., Tolmachev, A.I., and Kurdyukov, V.V. (2001) *Physical Chemistry Chemical Physics*, **3**, 3555–3561.
26 Lapouyade, L., Czeschka, K., Majenz, W., Rettig, W., Gilabert, E., and Rulliere, C. (1992) *The Journal of Physical Chemistry*, **96**, 9643–9650.
27 Rettig, W. (1994) *Topics in Current Chemistry*, **169**, 253–299.
28 Pines, D., Pines, E., and Rettig, W. (2003) *Journal of Physical Chemistry A*, **107**, 236–242.
29 Simeonov, A., Matsushita, M., Juban, E.A., Thompson, E., Hoffman, T.Z., Beuscher, A.E. IV, Taylor, M.J., Wirsching, P., Rettig, W., McCusker, J.K., Stevens, R.C., Millar, D.P., Schultz, P.G., Lerner, R.A., and Janda, K.D. (2000) *Science*, **290**, 307–313.
30 Yang, J.-S., Liau, K.-L., Li, C.-Y., and Chen, M.-Y. (2007) *Journal of the American Chemical Society*, **129 (43)**, 13183–13192, and references therein.
31 Papper, V., Pines, D., Likhtenshtein, G.I., and Pines, E. (1997) *Journal of Photochemistry and Photobiology A: Chemistry*, **111**, 87–96.

32 Likhtenshtein, G.I. (2003) *New Trends In Enzyme Catalysis and Mimicking Chemical Reactions*, Kluwer Academic/Plenum Publishers, New York.

33 Denisov, E.T., Sarkisov, O.M., and Likhtenshtein, G.I. (2003) *Chemical Kinetics. Fundamentals and Recent Developments*, Elsevier Science, Amsterdam.

34 Alexandrov, I.V. (1976) *Teoreticheskaya i Experimental naya Khimiay (Theoretical and Experimental Chemistry)*, **12**, 299–306.

35 (a) Liu, R.S.H. and Hammond, G.S. (2001) *Chemistry – A European Journal*, **7**, 4536–4544; (b) Liu, R.S.H. (2001) *Accounts of Chemical Research*, **34**, 555–562.

36 Liu, R.S.H. and Hammond, G.S. (2000) *Proceedings of the National Academy of Sciences of the United States of America*, **97**, 11153–11158.

37 Imai, M., Ikegami, M., Momotake, A., Nagahata, R., and Arai, T. (2003) *Photochemical & Photobiological Sciences*, **2**, 1181–1186.

38 Uda, M., Mizutani, T., Hayakawa, J., Ikegami, M., Momotake, A., Nagahata, R., and Arai, T. (2002) *Photochemistry and Photobiology*, **76**, 596–605.

39 Likhtenshtein, G.I. (1976) *Spin Labeling Method in Molecular Biology*, Wiley–Interscience, New York.

40 Likhtenshtein, G.I. (1993) *Biophysical Labeling Methods in Molecular Biology*, Cambridge University Press, Cambridge, NY.

41 Likhtenshtein, G.I., Yamauchi, J., Nakatuji, S., Smirnov, A., and Tamura, R. (2008) *Nitroxides: Application in Chemistry, Biomedicine, and Materials Science*, Wiley-VCH Verlag GmbH, Weinheim, and references therein.

42 Hammond, G.S. and Saltiel, J. (1963) *Journal of the American Chemical Society*, **85**, 2516–2517.

43 Hammond, G.S., Saltiel, J., Lamola, A.A., Turro, N.J., Bradshaw, J.S., Cowan, D.O., Counsell, R.C., Vogt, V., and Dalton, J.C. (1964) *Journal of the American Chemical Society*, **86**, 3197–3217.

44 Herkstroeter, W.G. and Hammond, G.S. (1966) *Journal of the American Chemical Society*, **88**, 4769–4777.

45 Saltiel, J., Mace, J.E., Watkins, L.P., Gormin, D.A., Clark, R.J., and Dmitrenko, O. (2003) *Journal of the American Chemical Society*, **125**, 16158–16159.

46 Lalevee, J., Allonas, X., and Fouassier, J.P. (2005) *Chemical Physics Letters*, **401**, 483–486.

47 Papper, V. and Likhtenshtein, G.I. (2001) *Photochemistry and Photobiology A: Chemistry*, **140**, 39–52.

48 Barbara, P.F., Rand, S.D., and Rentzepis, P. (1981) *Journal of the American Chemical Society*, **103**, 2156–2164.

49 Green, B.I. (1981) *Chemical Physics Letters*, **79**, 51–55.

50 Shilling, C.I. and Hiinski, E.F. (1988) *Journal of the American Chemical Society*, **110**, 2296–2298.

51 Sczepan, M., Rettig, W., Tolmachev, A.I., and Kurdyukov, V.V. (2001) *Physical Chemistry Chemical Physics*, **3**, 3555–3561.

52 Yang, J.-S., Liau, K.-L., Li, C.-Y., and Chen, M.-Y. (2005) *Chemical Physics Letters*, **401**, 483–486.

53 Improta, R. and Santoro, F. (2005) *Journal of Physical Chemistry A*, **109**, 10058–10067.

54 Pross, A. (1995) *Theoretical and Physical Principles of Organic Reactivity*, Wiley–Interscience, New York, pp. 159–182.

55 Malkin, S. and Fisher, E. (1964) *The Journal of Physical Chemistry*, **68**, 1153–1163.

56 Kim, S.K., Courtney, S., and Fleming, G.R. (1989) *Chemical Physics Letters*, **159**, 543–548.

57 Görner, H. (1987) *Journal of Photochemistry and Photobiology A: Chemistry*, **40**, 325–329.

58 Gurzadyan, G. and Görner, H. (2000) *Chemical Physics Letters*, **319**, 164–172.

59 Brown, P.E. and Whitten, D.G. (1985) *The Journal of Physical Chemistry*, **89**, 1217–1220.

60 Rettig, W. and Majenz, W. (1988) *Chemical Physics Letters*, **154**, 335–341.
61 Il'ichev, Y.V., Kühnle, W., and Zachariasse, K.A. (1996) *Chemical Physics*, **211**, 441–453.
62 Le Breton, H., Bennetau, B., Letard, J.-F., Lapouyade, R., and Rettig, W. (1996) *Journal of Photochemistry and Photobiology A: Chemistry*, **95**, 7–20.
63 Rechthaler, K. and Köhler, G. (1996) *Chemical Physics Letters*, **250**, 152–158.
64 Rettig, W., Majenz, W., Lapouyade, R., and Haucke, G. (1992) *Journal of Photochemistry and Photobiology A: Chemistry*, **62**, 415–421.
65 Eilers-König, N., Kühne, T., Schwarzer, D., Vöhringer, P., and Schroeder, J. (1996) *Chemical Physics Letters*, **253**, 69–76.
66 Schroeder, J., Schwarzer, D., Troe, J., and Voß, F. (1990) *The Journal of Physical Chemistry*, **93**, 2393–2404.
67 Strashnikova, N., Papper, V., Parkhomyuk, P., Likhtenshtein, G.I., Ratner, V., and Marks, R. (1999) *Journal of Photochemistry and Photobiology A: Chemistry*, **122**, 133–142.
68 Ahluwalia, A., Papper, V., Chen, O., Likhtenshtein, G.I., and De Rossi, D. (2002) *Analytical Biochemistry*, **305**, 121–134.
69 Papper, V., Pines, D., Likhtenshtein, G.I., and Pines, E. (1997) *Recent Development in Photochemistry and Photobiology*, vol. 1, Transworld Research Network, pp. 205–250.
70 Lapouyade, R., Czeschka, K., Majenz, W., Rettig, W., Gilabert, E., and Rulliere, C. (1992) *The Journal of Physical Chemistry*, **96**, 9643–9650.
71 Lapouyade, R., Kuhn, A., Letard, J.-F., and Rettig, W. (1993) *Chemical Physics Letters*, **208**, 48–64.
72 Bonacić-Koutecký, V., Bruckmann, P., Hiberty, P., Koutecký, J., Leforestier, C., and Salem, L. (1975) *Angewandte Chemie – International Edition in English*, **14**, 575–587.
73 Görner, H. (1995) *Journal of Photochemistry and Photobiology A: Chemistry*, **90**, 57–63.
74 Jiang, Y., and Wu, S. (2000) *Ganguang Kexue Yu Guang Huaxue*, **18**, 36–41.
75 Saltiel, J. and D'Agostino, J. (1972) *Journal of the American Chemical Society*, **94**, 6445–6456.
76 Likhtenshtein, G.I., Bishara, R., Papper, V., Uzan, B., Fishov, I., Gill, D., and Parola, A.H. (1996) *Journal of Biochemical and Biophysical Methods*, **33** (2), 117–133.
77 Saltiel, J. and Megarity, E.D. (1972) *Journal of the American Chemical Society*, **94**, 2742–2749.
78 Saltiel, J., Waller, A.S., Sears, D.F., Jr., Hoburg, E.A., Zeglinski, D.M., and Waldeck, D.H. (1994) *Journal of Physical Chemistry*, **98**, 10689–10698.
79 Nikowa, L., Schwarzer, D., Troe, J., and Schroeder, J. (1993) *Ultrafast Phenomena VIII, Springer Series in Chemical Physics*, Springer Verlag, vol. 55, pp. 603–605.
80 Saltiel, J., Marinari, A., Chang, D.W.L., Mitchener, J.C., and Megarity, E.D. (1979) *Journal of the American Chemical Society*, **101**, 2982–2996.
81 Herman, M.S. and Goodman, J.L. (1989) *Journal of the American Chemical Society*, **111**, 1849–1854.
82 Dietl, C., Papastathopoulos, E., Niklaus, P., Improta, R., Santoro, F., and Gerber, G. (2005) *Chemical Physics*, **310**, 201–211.
83 Haskins-Glusac, K., Ghiviriga, I., Abboud, K.A., and Schanze, K.S. (2004) *Journal of Physical Chemistry B*, **108**, 4969–4978.
84 Glusac, K.D. and Schanze, K.S. (2002) *Polymer Preprints (American Chemical Society, Division of Polymer Chemistry)*, **43**, 87–88.
85 Wahadoszamen, M., Momotake, A., Nishimura, Y., and Arai, T. (2006) *Journal of Physical Chemistry A*, **110**, 12566–12571.
86 Jiang, Y. and Wu, S. (2000) *Ganguang Kexue Yu Guang Huaxue*, **18**, 36–41.
87 Hayakawa, J., Ikegami, M., Mizutani, T., Wahadoszamen, M., Momotake, A.,

Nishimura, Y., and Arai, T. (2006) *Journal of Physical Chemistry A*, **110**, 12566–12571.

88 Myers, A.B., Rodier, J.-M., and Phillips, D.L. (1994) Reaction dynamics in clusters and condensed phases. Jerusalem Symposia on Quantum Chemistry and Biochemistry, 26, pp. 261–278.

89 Basaric, N., Marinic, E., and Sindler-Kulyk, M. (2006) *Journal of Organic Chemistry*, **71**, 9382–9392.

90 Kaliappan, R. and Ramamurthy, V. (2006) Abstracts of Papers, 231st ACS.

91 Hadfield, J.A., McGown, A.T., Mayalarp, S.P., Hamblett, I., Gaukroger, K., Lawrence, N.J., Hepworth, L.A., and Butler, J. (2007) US Patent, issued on May 22, 2007.

92 Roberts, J.C. and Pincock, J.A. (2006) *Journal of Organic Chemistry*, **71**, 1480–1492.

93 Saltiel, J. Rousseau, A.D., and Thomas, B. (1983) *Journal of the American Chemical Society*, **105**, 7631–7637.

94 Saltiel, J., Chang, D.W.L., Megarity, E.D., Rousseau, A., Shannon, P.T., Thomas, B., and Uriarte, A.K. (1975) *Pure and Applied Chemistry*, **41**, 559–579.

95 Saltiel, J., Townsend, D.E., and Sykes, A. (1983) *Journal of the American Chemical Society*, **105**, 2530–2538.

96 Saltiel, J., Ganapathy, S., and Werking, C. (1987) *Journal of Physical Chemistry*, **91**, 2755–2758.

97 Seydack, M. and Bendig, J. (2000) *Journal of Fluorescence*, **10**, 291–294.

98 Maeda, H., Nishimura, K., Mizuno, K., Yamaji, M., Oshima, J., and Tobita, S. (2005) *Journal of Organic Chemistry*, **70**, 9693–9701.

99 Sawada, T., Morita, M., Chifuku, K., Kuwahara, Y., Shosenji, H., Takafuji, M., and Ihara, H. (2007) *Tetrahedron Letters*, **48**, 9051–9055.

100 Momotake, A. and Arai, T. (2004) *Journal of Photochemistry and Photobiology C: Photochemistry Reviews*, **5**, 1–25.

101 Imai, M. and Arai, T. (2002) *Tetrahedron Letters*, **43**, 5265–5268.

102 Vicinelli, V., Ceroni, P., Maestri, M., Lazzari, M., Balzani, V., Lee, S.-K., van Heyst, J., and Voegtle, F. (2004) *Organic & Biomolecular Chemistry*, **2**, 2207–2213.

103 Watanabe, S., Ikegami, M., Nagahata, R., and Arai, T. (2007) *Bulletin of the Chemical Society of Japan*, **80**, 586–588.

104 Imai, M., Ikegami, M., Momotake, A., Nagahata, R., and Arai, T. (2003) *Photochemical & Photobiological Sciences*, **2**, 1181–1186.

105 Tatewaki, H., Mizutani, T., Hayakawa, J., Arai, T., and Terazima, M. (2003) *Journal of Physical Chemistry A*, **107**, 6515–6521.

106 Hayakawa, J., Momotake, A., and Arai, T. (2003) *Chemical Communications*, (1), 94–95.

107 Mizutani, T., Ikegami, M., Nagahata, R., and Arai, T. (2001) *Chemistry Letters*, 1014–1015.

108 Vicinelli, V., Ceroni, P., Maestri, M., Lazzari, M., Balzani, V., Lee, S.-K., van Heyst, J., and Voegtle, F. (2004) *Organic & Biomolecular Chemistry*, **2**, 2207–2213.

109 Miljanic, S., Frkanec, L., Meic, Z., and Zinic, M. (2005) *Langmuir*, **21**, 2754–2760.

110 Miljanic, S., Frkanec, L., Meic, Z., and Zinic, M. (2006) *European Journal of Organic Chemistry*, (5), 1323–1334.

111 Geiger, H.C., Geiger, D.K., and Baldwin, C. (2005) Abstracts of Papers, 229th ACS National Meeting, San Diego, CA, March 13–17, 2005, CHED-071.

112 Lehmann, M., Fischbach, I., and Spiess, H.W. (2004) *Journal of the American Chemical Society*, **126**, 772–784.

113 Tatewaki, H., Baden, N., Momotake, A., Arai, T., and Terazima, M. (2004) *Journal of Physical Chemistry B*, **108**, 12783–12789.

114 Erdelyi, M., Karlen, A., and Gogoll, A. (2006) *Chemistry – A European Journal*, **12**, 403–412.

115 Gutlich, P., Garcia, Y., and Woike, T. (2001) *Coordination Chemistry Reviews*, **219–221**, 839–879.

116 Parkhomyuk-Ben Arye, P., Strashnikova, N., and Likhtenshtein, G.I. (2002) *Journal of Biochemical and Biophysical Methods*, **51**, 1–15.

117 Ding, L. and Russell, T.P. (2006) *Macromolecules*, **39**, 6776–6780.
118 Ilieva, D., Nedelchev, L., Petrova, T., Dragostinova, V., Todorov, T., Nikolova, L., and Ramanujam, P.S. (2006) *Journal of Optics A: Pure and Applied Optics*, **8**, 221–224.
119 Zaplo, O., Stumpe, J., Seeboth, A., and Hermel, H. (1992) *Molecular Crystals and Liquid Crystals Science and Technology, Section A: Molecular Crystals and Liquid Crystals*, **213**, 153–161.
120 Hammond, H.A., Doty, J.C., Laakso, T.M., and Williams, J.L.R. (1970) *Macromolecules*, **3**, 711–715.
121 Lock, J., May, B., Clements, P., Lincoln, S., and Easton, C. (2004) *Journal of Inclusion Phenomena and Macrocyclic Chemistry*, **50**, 13–18.
122 Rameshbabu, K. and Kannan, P. (2007) *Journal of Applied Polymer Science*, **104**, 2760–2768.
123 Liu, R.S.H., Yang, L.-Y., and Liu, J. (2007) *Photochemistry and Photobiology*, **83**, 2–10.
124 Parthasarathy, A., Kaanumalle, L.S., and Ramamurthy, V. (2007) *Organic Letters*, **9**, 5059–5062.
125 Zhang, J., Whitesell, J.K., and Fox, M.A. (2001) *Chemistry of Materials*, **13**, 2323–2331.
126 Schmidt, P.M., Horn, K., Dil, J.H., and Kampen, T.U. (2007) *Surface Science*, **601**, 1775–1780.
127 Polo, A.S., Itokazu, M., Frin, K.M., Patrocinio, A.O.T., Murakami, I., and Neyde, Y. (2006) *Coordination Chemistry Reviews*, **250**, 1669–1680.

5
Miscellaneous Stilbene Photochemical Reactions

Besides photoisomerization, other rich arrays of photochemical and photophysical phenomena associated with stilbene and its substituents take place (Figure 4.1).
  Early data on photochemistry and photophysics of *trans*-stilbene and related alkenes in surfactant assemblies were summarized [1]. Photochemistry of stilbenoid compounds and their role in materials science was reviewed [2]. Due to their photophysics and photochemical properties stilbenoid compounds can be used for many applications in materials science. In particular, a nonlinear optics (NLO) system, a photoresist and photoconductive material, imaging and switching techniques, and stilbenoid liquid crystals (LCs) were discussed in this review. A review of the progress in transition metal compounds including stilbenoid complexes involving photoinduced changes in the magnetic and/or optical properties to long-lived metastable states was presented in Ref. [3]. Photoswitchable properties of compounds representing an attractive class of materials in coordination chemistry were considered.

5.1
Photocyclization

According to semiclassical simulations for the dynamics of the photocyclization of *cis*-stilbene, leading to the formation of 4a,4b-dihydrophenanthrene [4] photoexcited *cis*-stilbene simultaneously rotates about its vinyl and vinyl-Ph bonds. A series of strong couplings between the HOMO and the LUMO caused the formation of a new chimerical bond between the two Ph rings of stilbene. Calculated variations with time of torsional angles of stilbene and of HOMO-1, LUMO, and LUMO-1 energy levels are shown in Figure 5.1. The length changes in different C—C bonds, corresponding to the formation of the new molecules, were also presented.
  Experimental data on photocyclization of stilbenes and related indolic compounds such as bisphenylazostilbene were reported in Ref. [5]. The stereoselectivity of the formation of macrocyclic stilbenes as well as the regioselectivity of their photocyclization are strongly influenced by the length of the connecting alkanediyl chain [6]. Photochemical cyclization reaction in *cis*-3,3′,5,5′-tetramethoxystilbene was investigated [7]. The reaction occurred to give dihydrophenanthrene-type

Stilbenes. Applications in Chemistry, Life Sciences and Materials Science. Gertz Likhtenshtein
Copyright © 2010 WILEY-VCH Verlag GmbH & Co. KGaA, Weinheim
ISBN: 978-3-527-32388-3

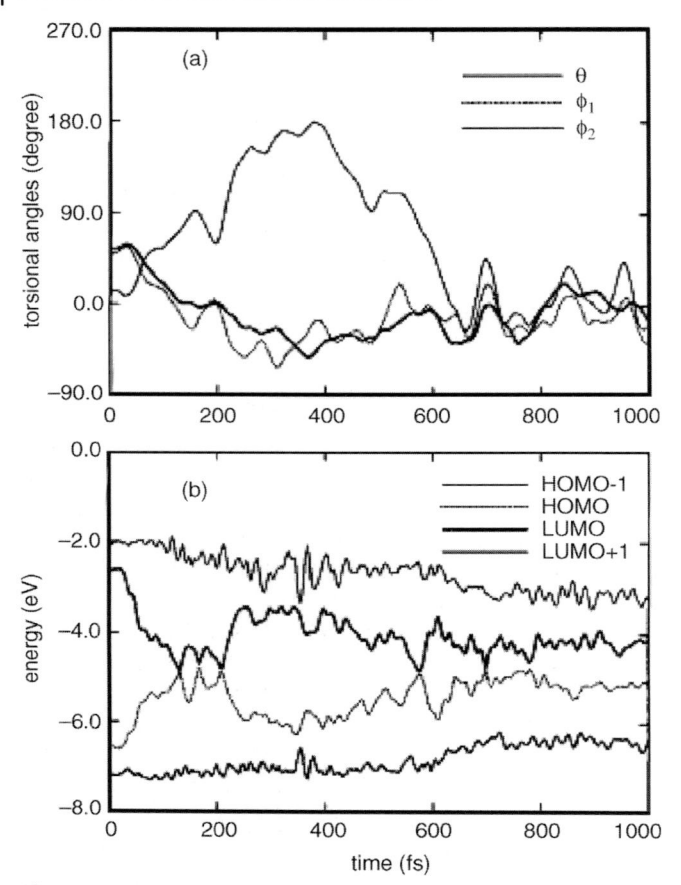

Figure 5.1 Variation with time of torsional angles of stilbene (a) and of HOMO−1, LUMO and LUMO + 1 energy levels (b) [4]. (Reproduced with permission.)

compound followed by oxidation to give 2,4,5,7-tetramethoxyphenanthrene. Photochemistry of o-pyrrolylstilbenes and formation of spiro-2H-pyrroles and their rearrangement to form dihydroindoles have been investigated [8]. It was shown that excited states of stilbenylpyrroles were deactivated by two photochemical processes: cis– trans-isomerization and hydrogen transfer of NH to the stilbene double bond. The NH-transfer results in the formation of two quinone dimethane intermediates and biradicals. Intramolecular cyclization of intermediates gave rise to polycyclic compounds such as spiro-2H-pyrroles **7**, pyrroloisoindoles **3**, dihydroindoles **2**, and pyrroloisoquinolines **8** (Figure 5.2).

Photophysics and photochemistry of ring-fluorinated stilbenes have been investigated.

The absorption and emission spectra, E–Z photointerconversion, photocyclization of the Z-isomers, and transient triplet–triplet (T–T) absorption spectra of five

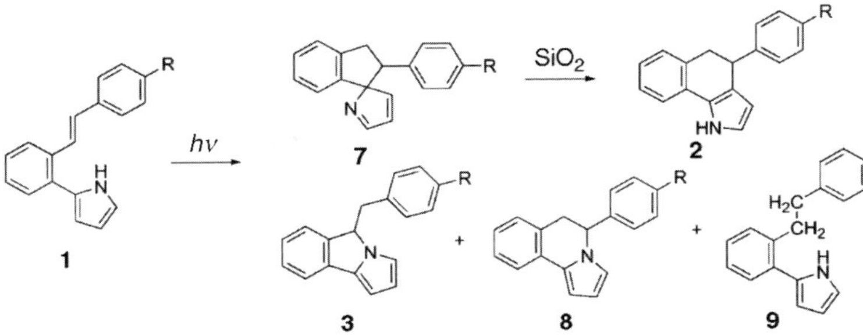

Figure 5.2 A scheme of intramolecular cyclization of intermediates gave rise to polycyclic compounds spiro-2H-pyrroles **7**, pyrroloisoindoles **3**, dihydroindoles **2**, and pyrroloisoquinolines **8** [8]. (Reproduced with permission.)

fluorostilbenes were studied over a wide temperature range, down to 90 K, and compared with stilbene [9]. All compounds under investigation form stable Z (*cis*) isomers upon irradiation, but photocyclization takes place only in the absence of fluorine substitution at the 2,6,2′,6′ positions or the 2,6 positions. T–T transients detected at −170 °C were found to be similar to those of *trans*-stilbene, whereas their lifetimes were lower than those of stilbene. The photochemical cyclization of several o-(aminoalkyl)stilbenes and their N-Me derivatives under conditions of direct and electron transfer-sensitized irradiation has been accomplished [10]. The cyclization of 2-PhCH:CHC$_6$H$_4$CH$_2$CH$_2$NHMe in deoxygenated MeCN gave 65% benzazepine, whereas similar photocyclization of 2-PhCH:CHC$_6$H$_4$CH$_2$CH$_2$NH$_2$ in the presence of *m*-dicyanobenzene as a photosensitizer led to 76% isoquinoline. The photocyclization of the stilbene-phenanthrene to [2,2]paracyclophane has been investigated [11]. For the model system 4-styryl[2,2]paracyclophane to [2,2]phenanthrenoparacyclophane, the reaction resulted in the introduction of alkyl substituents in the 6-, 7-, 8-, and 9-positions of the phenanthrene moiety. In addition, the side products of the process of ring cleavage of the cyclophane core have been characterized. It was shown [12] that the photoreactions of 2-substituted-1,4-naphthoquinones with *trans*-stilbene gave spiro-oxetanes as the main product. 2-Acetoxy-1,4-naphthoquinone reacted with *trans*-stilbene yielding spiro-oxetane II regiospecifically (2π + 2π).

Photocycloaddition, *cis–trans* photoisomerization, and photocyclization upon irradiation of stilbenophanes tethered by silyl chains have been studied [13]. The irradiation of macrocyclic *trans,trans*-I stilbenophanes *trans,trans*-I stereoselectively gave the corresponding intramolecular cyclobutane photocycloadducts, and the efficiency increased with decreasing distance between the two stilbene units, whereas photoreactions of *cis*-fixed stilbenophanes under an oxygen atmosphere selectively gave phenanthrenophanes as the result of *ortho–ortho* oxidative coupling. It was reported [14] that intramolecular (2 + 2) photocycloaddition of β-stilbazoles tethered by silyl chains took place with high efficiency. Complexation with dicarboxylic acid or catechol enhanced both the efficiency. The ultrafast torsional isomerization and ring closure reactions of photoexcited *cis*-stilbene have been examined in hexane, cyclohexane, methanol, and acetonitrile solvents [15]. The process was monitored by a combination of ground-state resonance Raman intensities, two-color UV picosecond anti-Stokes resonance Raman scattering, and product quantum yield measurements. Quantum yields for ring closure to ground-state dihydrophenanthrene were found to be two to three times higher in the nonpolar solvents than in the polar one. Vibrationally hot ground-state *trans*-stilbene was formed within the 10 ps time resolution of the time-resolved Raman experiments; cooling was about a factor of two faster in the former solvent. According to the authors, data of the time-resolved Raman experiments, the femtosecond transient absorption, and fluorescence properties indicated that increasing solvent polarity causes vertically excited *cis*-stilbene to distort more rapidly along the torsional isomerization coordinate, shortening the excited state lifetime, and decreasing the quantum yield for the competing ring closure reaction.

5.2
Bimolecular Reactions

5.2.1
Photodimerization

Photodimerization of stilbene in van der Waals nanocapsules has been studied [16]. *para*-Hexanoylcalix[4]arene nanocapsules (Figure 5.3) were used as hosts to carry out photodimerization of *cis*- and *trans*-stilbene to *syn*-tetraphenylcyclobutane. Single-crystal X-ray diffraction studies were performed to define precisely the location of encapsulated stilbenes inside the capsule. It was shown that *cis*-stilbene stacks as π–π dimers located at the center of the capsule, whereas *trans*-stilbene does not form such a dimer. A possible configuration of two stilbene molecules in hydrophobic nanocapsules based on amphiphilic *para*-hexanoylcalix[4]arene is presented in Figure 5.4. Because the molecules are shifted with respect to each other, they cannot yield a good π–π stack, although local stabilization is possible. The authors suggested that the entire structure is a compromise between π-electron interaction of a *trans*-stilbene molecule with the host and π–π interactions of the two *trans*-stilbene molecules.

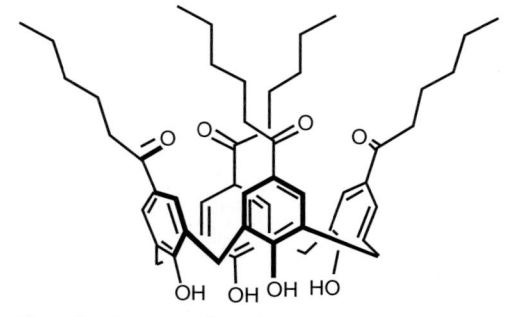

Figure 5.3 Structure of *para*-hexanoylcalix[4]arene nanocapsules [16]. (Reproduced with permission.)

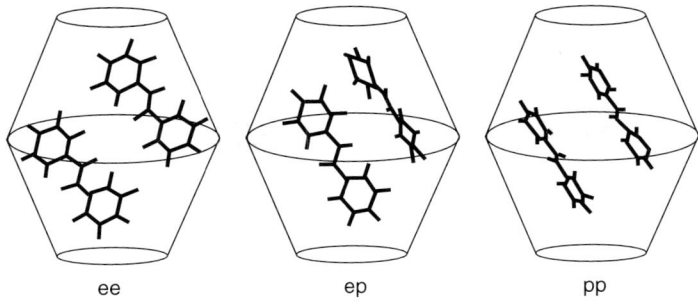

Figure 5.4 Possible configuration of two stilbene molecules in hydrophobic nanocapsules based on amphiphilic *para*-hexanoylcalix[4]arene [16]. (Reproduced with permission.)

Irradiation of the inclusion complexes of each isomer of stilbene in the solid state leads after prolonged photolysis to photodimerization.

Water soluble *p*-sulfonato calix[*n*]arenes ($n = 8$, **1a–1b** and $n = 6$, **2a–2b**) was employed as host to control the outcome of photodimerization and photoisomerization of 4-stilbazoles [17]. Novel macrocyclic and medium-size stilbenophanes tethered by silyl chains were synthesized, and their photochemical and photophysical properties were examined (Figure 5.5) [18]. Direct irradiation of macrocyclic stilbenophanes stereoselectively gave intramolecular photocycloadducts, and the efficiency increased with decreasing distance between the two stilbene units. The triplet-sensitized photoreaction of stilbenophanes caused *cis–trans* photoisomerization. Photoreactions of *cis*-fixed stilbenophanes under an oxygen atmosphere selectively led to phenanthrenophanes. Fluorescence quantum yields increased with the introduction of silyl substituents, and hence those of silyl-tethered stilbenophanes were larger than that of unsubstituted *trans*-stilbene. Intramolecular excimer emis-

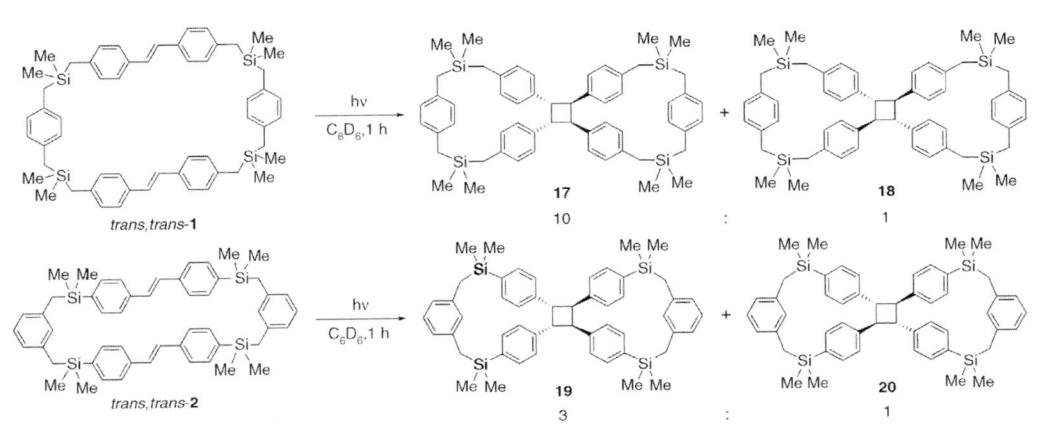

Figure 5.5 Photochemical reactions of stilbenophanes synthesized in Ref. [18]. (Reproduced with permission.)

sion was observed when the distance between two stilbene units in the stilbenophanes was sufficiently small.

Photochemical reactions of *trans*-stilbene have been studied at a silica gel–air interface [19]. It was shown that irradiation of *trans*-stilbene led to the formation of two dimers, along with the formation of *cis*-stilbene, phenanthrene, and a small amount of benzaldehyde. Isomerization of *trans*-stilbene competed efficiently with oxidation. Singlet molecular oxygen was found to be quenched in solution and at the silica gel–air interface at a faster rate by *cis*-stilbene than by *trans*-stilbene or 1,1-diphenylethylene. It has been demonstrated in Ref. [20] that the bis(dialkylammonium ion)-containing thread-like dication **1**-H_2·$2PF_6$ and the crown ether bis-*p*-phenylene[34]crown-10 (BPP34C10) formed a doubly encircled and doubly threaded 2:2 complex [(BPP34C10)$_2$·(**1**-H_2)$_2$][PF_6]$_4$ upon cocrystallization in the solid state (Figure 5.6). This result suggested a method for aligning stilbene derivatives in the solid state. By replacing the *p*-phenylene unit of 1^{2+} with a *trans*-stilbenoid unit (namely, producing *trans*-**2**-H_2·$2PF_6$), the authors anticipated that a 2:2 complex [(BPP34C10)$_2$·(**2**-H_2)$_2$][PF_6]$_4$ would form upon cocrystallization in which adjacent *trans*-stilbene olefinic bonds would be aligned in a manner suitable for a solid-state [2 + 2] cycloaddition to occur. Depending upon the alignment of the two *trans*-

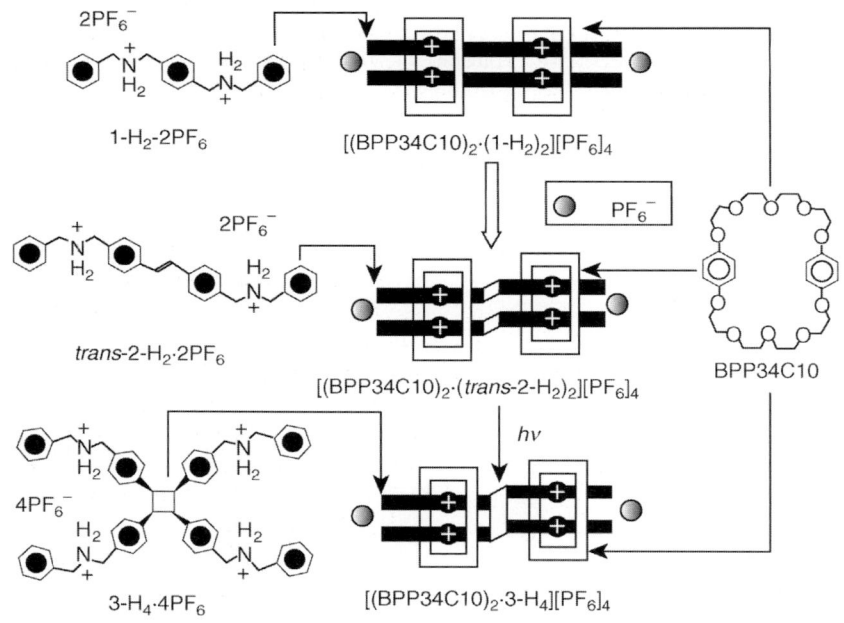

Figure 5.6 The solid-state supramolecular structure of the doubly threaded, doubly encircled [4] pseudorotaxane [(BPP34C10)$_2$(*trans*-**2**-H_2)$_2$]$^{4+}$. Hydrogen-bonding N···O, H···O distances (Å) and N—H···O angles (°) are (a) 2.96, 2.27, 133; (b) 3.02, 2.18, 155; (c) 2.90, 2.04, 161; (d) 2.94, 2.26, 131; and (e) 3.17, 2.34, 154 [20]. (Reproduced with permission.)

stilbenoid units in the complex, a single diastereoisomer of a tetrasubstituted cyclobutane derivative should be isolable after photochemical irradiation.

5.2.2
Reactions with Alkenes and Dienes

The kinetics of photochemical addition of *trans*-stilbene to olefins was studied [21]. The apparent rate constants (k_{app}) for the photochemical addition of *trans*-stilbene to tetramethylethylene (TME) and 1-methylcyclohexene ($+55$ to $-22\,°C$) were measured in the temperature range ($+55$ to $-22\,°C$). The k_{app} values were found to be close to diffusion controlled but differ in both magnitude and temperature dependence from the olefin used. The authors interpreted the observed dependence of k_{app} in the negative temperatures for the addition processes in terms of reversible exciplex formation. *trans*-Stilbene was involved in photochemical reaction with a oligomeric diene [22]. The reaction resulted in the formation of phenyl-substituted 2,3-dihydro-1,3-dithiolo[4,5-*e*][1,4]dithiin-6-thione. It was found that in the photoreactions of *trans*-stilbene and *cis*-stilbene, with 1-phenyl-1,2-propanedione in air, both stilbenes turned into *trans*-epoxide with a moderate yield and *cis*-epoxide with a very low yield, together with *cis*-stilbene and *trans*-stilbene, respectively. Photochemical reaction of *N*-methylnaphthalene-1,8-dicarboximide (I) with stilbenes was investigated [23]. Stereospecificity of cyclobutane formation was observed in the reaction with *cis*- and *trans*-but-2-ene. Irradiation of benzene solution of I in the presence of *trans*- and *cis*-stilbenes gave fragmentation products arising from precursor oxetanes.

5.2.3
Reactions with Amines, Imines, Nitroso Oxide, and Protic Solvents

Photochemical addition reactions of amines with alkenes, including stilbenes, were recently reviewed in Ref. [24]. The photostimulated electron transfer from amines to *trans*-stilbene has been investigated [25]. Time-resolved resonance Raman spectroscopy was used to monitor the decay of the *trans*-stilbene anion radical upon pulsed laser photolysis of MeCN solution of *trans*-stilbene and the tertiary amines ($Me_2CH)_2$-Net. The spectroscopic evidence for the formation of the stilbene anion radical via photostimulated electron transfer from the tertiary amines electron donor to stilbene was presented. Behavior of the exciplex, solvent-separated radical ion pair, and free radical ions during photochemical addition reaction of *trans*-stilbene with acyclic trialkylamines has been described [26]. It was shown that a solvent-separated radical ion pair, formed either directly from an encounter complex or via the exciplex, decayed by intersystem crossing giving the stilbene-amine adduct. The process occurred by dissociation to give free radical ions or by quenching by the ground state of acyclic trialkylamines. Intersystem crossing was found to be more rapid for diamines versus acyclic trialkylamines, resulting in a lower yield of free radical ions. Free *trans*-stilbene radical anion decayed by recombination with an amine radical cation; the decay was independent of amine structure or concentration. The same research group has reported the stereoelectronic control of amine dimer cation radical formation [27]. The fluorescence quenching rate constants of singlet *trans*-stilbenes (I) with trialkylamines, $NH_2(CH_2)_nNH_2$ (DA) in C_6H_6, was obtained to be equal to or exceed the diffusion rate. Intersystem crossing, to the stilbene triplet, was the predominant exciplex decay path at low amine concentration. At high amine concentration, exciplex quenching by ground-state amines occurred and was subjected to a pronounced steric effect with trialkylamines and depended on n of DA. It was suggested that the exciplex quenching occurred by the interaction of an amine cation radical with neutral amines to give a triplex of the stilbene anion radical and a 3-electron σ-bonded amine dimer cation radical.

It was shown [28] that the photoreaction of stilbene with caffeine, benzothiazole, 1-methylimidazole, and 2(methylthio)benzothiazole took place. The reactions gave [2 + 2] and [2 + 4]-cycloaddition, [2 + 2]-addition, noncatalyzed substitutive [2 + 2]-addition, with substituent migration and double-bond cleavage, substitutive [2 + 2]-addition with double bond cleavage, and intermolecular exchange of vinyl substituents. Results of the spin-trapping study of photochemical reactions of substituted *trans*-stilbenes with tertiary amines have been reported [29]. Seven 1,2-(p,p'-disubstituted phenyl)ethyl radicals were trapped by 2-methyl-2-nitrosopropane and isolated and identified by HPLC-EPR. The nitrogen hyperfine splitting constants (hfsc) of these radicals were found to be linearly correlated with the Hammett substituent constants $\sigma P'$, σR, and σI. The correlation between proton hfsc and a single parameter was not linear, while the correlation between the β-proton hfsc and the dual parameters $\sigma R'$ and σI was linear. The authors concluded that the inductive and resonance effects are of equal importance.

A series of aminostilbenes (1A–C, 2A–C) has been synthesized to test the effect of substitution of the amino group upon the photochemistry of stilbenes [30]. This study indicated that the photophysical properties of *trans*-2-aminostilbene, 1A, and *trans*-3-aminostilbene, 1B, were similar, and the regioselectivity of addition across the ethenyl unit was the same for 1A and *trans*-4-aminostilbene 1C. Photocyclization products formed via intermediate *cis*-aminostilbenes have been observed for 1A and 1C while 1B forms a stable dihydrophenanthrene in the absence of oxygen and light. While *cis*-2-aminostilbene, 2A, and *cis*-3-aminostilbene, 2B, showed similar fluorescent lifetimes, 2A and *cis*-4-aminostilbene, 2C, showed similar photochemistry. Photoadditions of alcohols to methoxy-substituted stilbenes have been described [31]. Upon irradiation of five substrates in 2,2,2-trifluoroethanol (TFE), products derived from photoaddition of the solvent were detected (Figure 5.7). The photoaddition of TFE proceeded with the general order of reactivity: styrenes > *trans*-1-arylpropenes *trans*-stilbenes. NMR spectroscopy of the products formed by irradiation in TFE indicated that the proton and nucleophile are attached to two adjacent atoms of the original alkene double bond. Transient carbocation intermediates were observed following laser flash photolysis of the stilbenes in 1,1,1,3,3,3-hexafluoro-2-propanol (HFIP). A mechanism that involves photoprotonation of the substrates by TFE or HFIP, followed by nucleophilic trapping of short-lived carbocation intermediates, has been proposed. Compared to other stilbene derivatives, *trans*-3,5-dimethoxystilbene displayed a large quantum yield of fluorescence and a low quantum yield of *trans–cis* isomerization in polar organic solvents. The unique photophysical

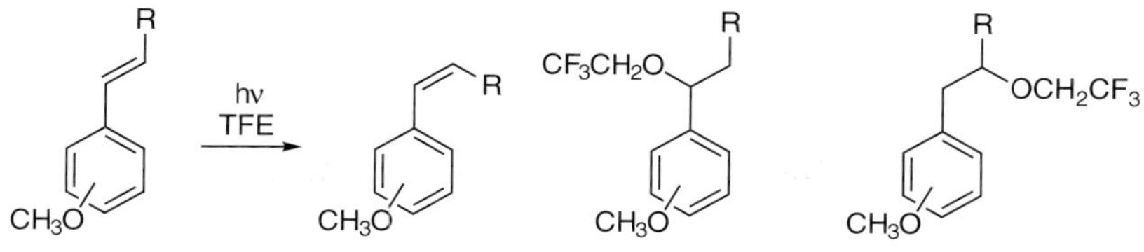

Figure 5.7 Products detected following irradiation of trans-1a–e in TFE [31]; ACS [31]. (Reproduced with permission.)

properties of trans-3,5-dimethoxystilbene were attributed to the formation of a highly polarized charge transfer excited state ($\mu_e = 13.2$ D).

Photolysis of trans-stilbene and triethylamine led to the formation of 1,2-diphenylethyl radical that was trapped by 2-methyl-2-nitrosopropane [32]. The process kinetics was followed by resolution-enhanced ESR spectroscopy. The spectral assignment was confirmed by preparation of the radical using the Grignard reaction of benzylmagnesium bromide and α-phenyl-tert-butylnitrone. As it was shown in Ref. [33], the photolysis of trans-stilbene and its derivatives I (R = MeO, Me, Me$_2$CH, Cl, Br, CN, R$_1$ = H; R = R$_1$ = MeO, Me$_2$CH, Me, H, Cl, Br, CN) with secondary amines as quenchers resulted in the formation of corresponding derivatives of 1,2-diphenylethyl radicals. The radicals were trapped by 2-methyl-2-nitrosopropane and were detected by the HPLC–EPR method. It was shown that the yield of the spin adduct was greater for tertiary amines. The smaller yields of the spin adduct formed from secondary amines were ascribed to greater reactivities of the 1,2-diphenylethyl and dialkylamino radicals within the solvent cage.

Data on photophysics and photochemistry of intramolecular stilbene–amine exciplexes and reactions have been reported [34]. The obtained data indicated that the photophysical and photochemical behavior of a series of trans-(aminoalkyl) stilbenes in which a primary, secondary, or tertiary amine is appended to the stilbene ortho position with a Me, Et, or Pr linker. The tertiary (aminoalkyl)stilbenes formed fluorescent exciplexes and underwent trans–cis isomerization but failed to undergo intramolecular N–H addition. The secondary (aminoalkyl)stilbenes did not form fluorescent exciplexes but underwent the addition to the stilbene double bond. Intramolecular reactions were highly selective, providing an efficient method for the synthesis of tetrahydrobenzazepines. Direct irradiation of the primary (aminoalkyl) stilbenes resulted only in trans–cis isomerization, while irradiation in the presence of the electron acceptor p-dicyanobenzene resulted in regioselective intramolecular N–H addition to the stilbene double bond.

Photochemical reactions of nitroso oxides at low temperatures with stilbenes have been reported [35]. Several singlet nitroso oxides were generated by the thermal reaction of triplet nitrenes with triplet oxygen at 95 K in 2-methyltetrahydrofuran. After photolysis of the nitroso oxides at 77 K, the formation of intermediates was observed. From spectroscopic and kinetics data, the authors postulated the formation

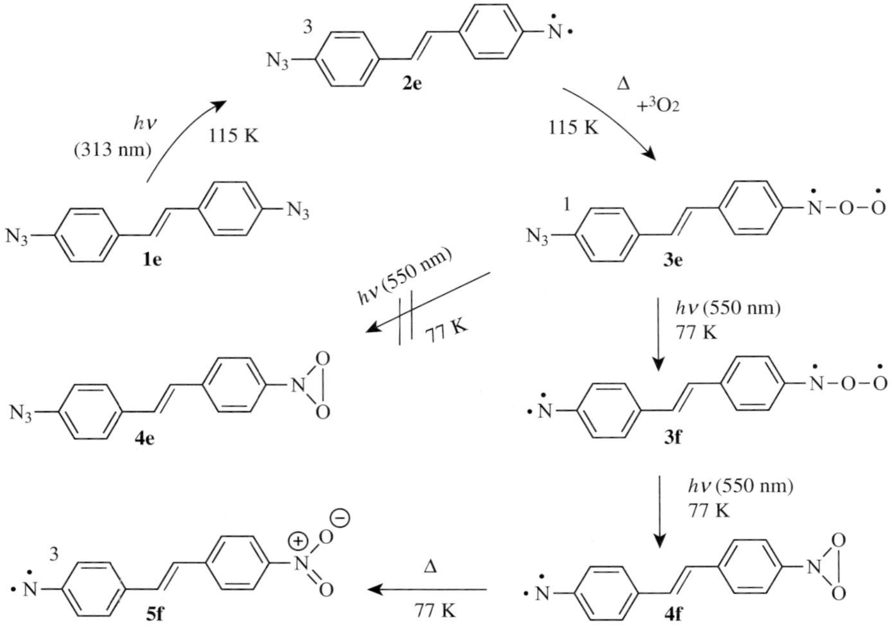

Figure 5.8 Proposed scheme of photogeneration of the nitroso oxides [35]. (Reproduced with permission.)

of the following dioxaziridines: 4-(dioxaziridine-yl)stilbene, 4-(dioxaziridine-yl)-4′-nitrostilbene, 4′-(dioxaziridine-yl)-4-(dimethylamino)stilbene, 4′-(dioxaziridine-yl)-4-aminobiphenyl, and 4-(dioxaziridine-yl)-4′-(nitrene-substituted)stilbene. All dioxaziridines reacted at 77 K thermally to form the corresponding nitro compounds. The rate constants of the ring opening reaction of the dioxaziridines were equal to 0.0030 ± 0.0005 s^{-1}. From *ab initio* calculation of the thermal reaction of the non-substituted dioxaziridine and *N*-phenyldioxaziridine, the authors concluded that dioxaziridines are separated from the corresponding nitro products by an orbital symmetry-forbidden barrier. The proposed scheme of photogeneration of the nitroso oxides is shown in Figure 5.8.

5.3
Photoreactions in Stilbene Dendrimers

Photochemistry and mobility of stilbenoid dendrimers in their neat phases have been studied [36]. The photochemistry and the fluorescence in different crystal and liquid crystal phases selectively containing deuterated, dodecyloxy-substituted stilbenoid dendrimers of the first and second generations were investigated. Molecules deuterated at the α-position of the alkoxy chains were involved in reactions of double bonds in the liquid crystal phases whereas no photoreactions occurred in the crystal

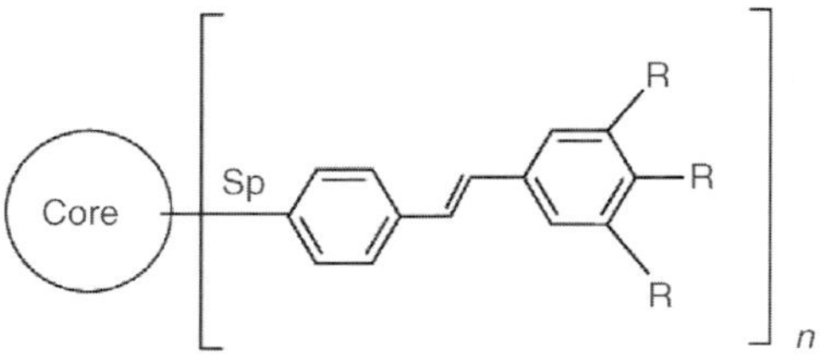

1 (R = H, alkyl, alkoxy)

Figure 5.9 Star-shaped compounds or dendrimers having (E)-stilbene chromophores linked by (branched) spacers Sp to the core [38]. (Reproduced with permissions from Elsevier.)

state. Photochemical conversion and fluorescence quenching for first- and second-generation dendrimers [all-(E)-1,3,5-tris[2-(3,4,5-tridodecyloxyphenyl)ethenyl]benzene] and [all-(E)-1,3,5-tris(2-{3,5-bis[2-(3,4,5-tridodecyloxyphenyl)ethenyl]phenyl} ethenyl)benzene] increased with increasing molecular motion and reach a maximum in the isotropic phase for fluid solution in benzene and toluene. Preparation and photochemistry of dendrimers with isolated stilbene chromophores have been reported [37]. Two dendrimers with *trans*-stilbene chromophores in the core and on the periphery of the dendrons and the model compounds, 4[$R_3C_6H_2CH_2OCH_2$] $C_6H_4CH:CHC_6H_4[CH_2OCH_2C_6H_2R_{3-3,4,5}]$-4 [R = H, OMe], were prepared and their photochemistry was studied in solution and in neat films. Due to the flexibility of the arms, intramolecular and intermolecular C—C bonds were formed upon irradiation. The same group [38] showed that dendrimers with terminal (E)-stilbene moieties based either on a hexamine core or on a benzenetricarboxylic acid core dendrimer (Figure 5.9) underwent photochemical reactions to yield cross-linked products lacking styryl moieties.

Dendrimers with terminal (E)-stilbene moieties based either on a hexamine core or on a benzenetricarboxylic acid core were prepared under irradiation [38]. Both types of dendrimers were found to undergo photochemical reactions to yield cross-linked products lacking styryl moieties.

5.4
Reactions in Polymers and Other Matrices

Photochemical and photophysical properties of a poly(propylene amine) dendrimer (**2**) functionalized with E-stilbene units have been studied [39]. The $E \rightarrow Z$ photoisomerization and photocyclization of the Z-isomer of the stilbene units were investigated in air equilibrated acetonitrile solutions. The quantum yields of the $E \rightarrow Z$ photoisomerization reaction and the fluorescence quantum yield of the E were found to be equal to 0.30 and 0.014, respectively. The stilbene moiety underwent the photocyclization to phenanthrene with quantum yield 0.015. Photochemical processes of the main-chain nematic LC polymer containing *trans*-4,4′-stilbene-bis-carboxylate chromophore/mesogen was studied in solvents and in spin cast thin films [40]. The film irradiation at 313 nm led to the loss of the stilbene absorption that was attributed to photodimers. The irradiation of the polymer in solution at 313 and 366 nm in the presence or absence of air indicated the formation of the 2 + 2 photocycloadduct. The photophysical and photochemical behavior of calamitic liquid crystal polymers with photoreactive stilbene 4,4′-dicarboxylate mesogen has been investigated at room temperature [41]. UV–vis spectra of the polymer in chloroform were found to be typical of simple 4,4′-stilbene dicarboxylate esters. Irradiation of pure films above 300 nm led to initial consumption of these aggregates and cross-linking most likely via 2 + 2 photocycloaddition with quantum yields less than 0.5. Emission spectra of polymer films were "excimer-like." Photochemical and photophysical effects in photoreactive liquid crystals in the series of dendritic and cyclic stilbenoid compounds have been investigated [42]. The dendrimers with $n = 1–5$) and [18]annulenes underwent dimerization, oligomerization, or the formation of charge carriers.

It was reported [43] that in the photosensitive polymer-dispersed liquid crystals (PDLC) containing the nematic mixture E-5 doped with 2 wt% photochromic stilbene dye in a gelatin binder, the stilbene segment underwent $E–Z$ photoisomerization upon irradiation. Irradiation of the stilbene-doped gelatin PDLC destroyed the preferred director orientation of dopants. Optical properties as well as the Volta potential ΔV of stilbene-doped PDLC were changed. Differences were observed between hot- and cold-dried PDLC films. The photocross-linking reaction of the two series of combined liquid crystal polyphosphates bearing dual photoreactive mesogenic units (stilbene and azobenzene/α-methylstilbene and azobenzene) was ascertained by spectroscopic and photolysis studies [44]. The photochemical properties of liquid crystal poly(bis-4,4′-oxy-α-methylstilbene-4-substituted (X) phenylazo-4′-phenyloxydecyl phosphate ester)s bearing photoreactive mesogenic units were studied by UV–vis and fluorescence spectroscopy [45]. The influence of the photoinduced $E–Z$ isomerization of various terminal substituents of the side-chain azobenzenes was investigated. The kinetics of the photoisomerization process reveals the switching times for the conversion between the *trans*- and *cis*-forms of the azobenzene units. The photo-optical properties of these polymers exhibited layered smectic phases and showed good photoinduced properties in their mesomorphic states.

5.5
Reaction Using Two-Photon Excitation

Nonresonant two-photon (NRTP) reaction of intramolecular charge transfer-type 4-disilanyl-4′-trifluoromethylstilbene (DTS) has been reported [46]. One-photon irradiation of a methanol solution of *trans*-DTS with 266-nm laser pulses induced solvolysis to give 4-dimethylsilyl-4′-trifluoromethylstilbenes (HTS). In contrast, NRTP excitation of *trans*-DTS in methanol with 532-nm laser pulses induced predominantly *cis–trans* isomerization and solvolytic reactions were suppressed. The photochemical decomposition of *trans*-stilbene and its derivatives in ethanol exposed to radiation of the second harmonic of a Nd:YAG laser (532 nm) of

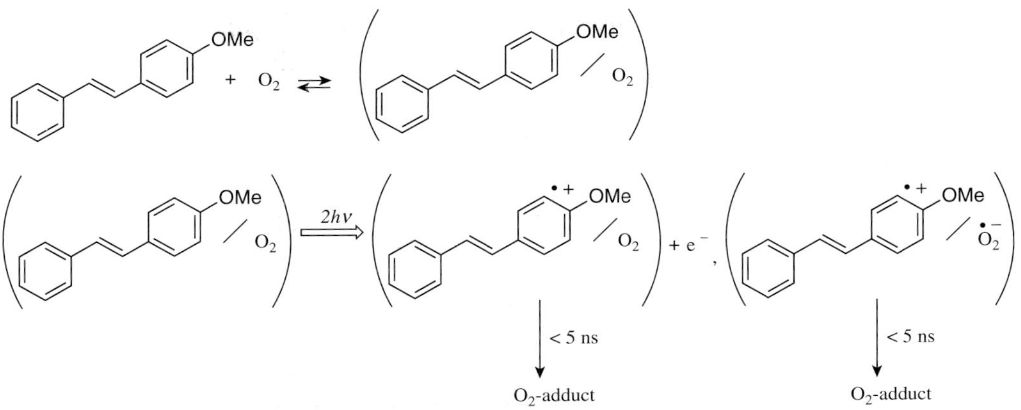

Figure 5.10 Scheme of formation of radical cations of trans-stilbene and p-substituted trans-stilbenes (S.bul.+) during the resonant two-photon ionization of S in acetonitrile in the presence of O_2 [49]. (Reproduced with permission.)

nanosecond duration has been examined [47]. After measuring the quantum yield of the photoreaction (γ266) of dyes under one-photon excitation (fourth harmonic Nd: YAG laser 266 nm) by absorption method, the photochemical decomposition of the stilbenes investigated at two-photon excitation cross section was detected. Paracyclophane two-photon-absorbing chromophores having stilbenoid groups were used as photopolymerization initiators [48].

Formation and decay kinetics of radical cations of trans-stilbene and p-substituted trans-stilbenes (S.bul.+) during the resonant two-photon ionization (TPI) of trans-stilbenes in acetonitrile in the presence and absence of O_2 were studied (Figure 5.10) [49]. For ionization of stilbenes, the laser flash photolysis using a XeCl excimer laser (308 nm, fwhm 25 ns) was employed. It was shown that the formation quantum yield of S.bul.+ (0.06–0.29) increased with decreasing oxidation potential (E_{ox}) and increasing fluorescence lifetime (τ_f) of stilbenes, except for trans-4-methoxystilbene that has the lowest redox potential E_{ox} and longer τ_{of} among investigated stilbenes. The considerable low yield and fast decay in a few tens of nanoseconds timescale were observed for trans-4-methoxystilbene.bul.+ in the presence of O_2, but not for other S.bul.+. The authors assumed that the formation of the ground-state complex between trans-4-methoxystilbene and O_2 and localization of the positive charge on the oxygen of the p-methoxyl group and an unpaired electron on the β-olefinic carbon were responsible for the fast reaction of trans-4-methoxystilbene.bul.+ with O_2 or superoxide anion. Data on the photoinduced electron transfer in the presence of a photosensitizer such as 9,10-dicyanoanthracene and O_2 in acetonitrile were also described.

trans-4-(N-2-Hydroxyethyl-N-ethylamino)-4′-(diethylamino)stilbene and (E,E)-4-{2-[p′-(N,N-di-n-butylamino)stilben-p-yl]vinyl}pyridine (DBASVP) were used as two-photon photopolymerization initiators [50, 51]. Quantum chemical calculations showed that the new initiator possessed a large delocalized π-electron system, a large change in dipole moment on transition to the excited state, and a large transition moment. The calculated two-photon absorption cross section was as high as 881.34 × 10–50 cm^4 s/photon. The single-photon and two-photon absorption and fluorescence properties in various solvents have been investigated. The initiator exhibited outstanding solvent sensitivity, which was explained by the electron-delocalized properties of the molecule. A microstructure has been fabricated under irradiation at 800 nm using a 200 fs, 76 MHz Ti:sapphire femtosecond laser.

Nonresonant two-photon excitation of *trans*-stilbene in the presence of an excess amount of tetramethylethylene was found to induce predominantly *cis–trans* isomerization [52]. Under the same condition, one-photon excitation gave a [2 + 2] cycloadduct as the main product. Action spectra for the isomerization indicated that isomerization occurred from the excited Ag state of *trans*-stilbene. Nonresonant two-photon excitation of *trans*-stilbene in the presence of tetramethylethylene-induced predominantly *cis–trans* isomerization whereas the [2 + 2] intermolecular cycloaddition pathway was found to be suppressed [53]. In contrast, the one-photon excitation under similar conditions the cycloaddition occurred as the major pathway.

5.6
Charge Transfer Ionization

Carrying out a semiclassical Ehrenfest dynamics simulation based on the time-dependent density functional theory, the authors of the study [54] investigated the light-harvesting property of a π-conjugated dendrimer, star-shaped stilbenoid phthalocyanine (SSS1Pc) with oligo(*p*-phenylenevinylene)peripheries. The results are presented in Figures 5.11–5.13. According to these results, an electron and a hole were transferred from the periphery to the core through a π-conjugated network when an electron was selectively excited in the periphery. The one-way electron and hole transfer occurs more easily in dendrimers with a planar structure than in those

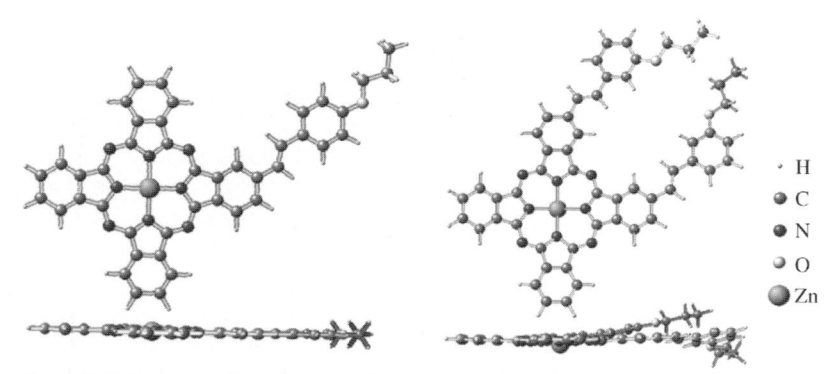

Figure 5.11 Structure of π-conjugated dendrimers, star-shaped stilbenoid phthalocyanines [54]. (Reproduced with permission.)

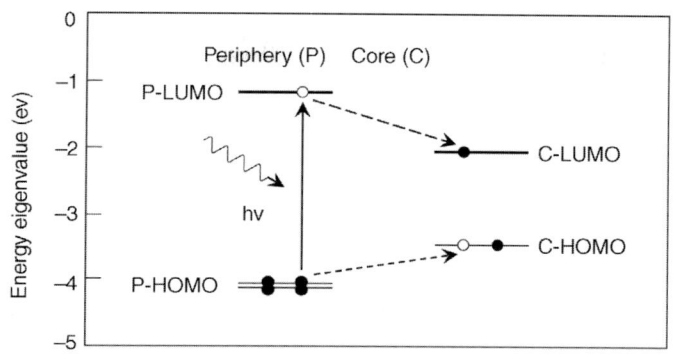

Figure 5.12 Energy of eigenvalues of π-conjugated dendrimers, star-shaped stilbenoid phthalocyanines [54]. (Reproduced with permission.)

with steric hindrance because π-conjugation was well maintained in the planar structure.

Charge transfer transitions relevant to single- and double-bond photochemical twisting have been studied [55]. The biradicaloid state theory using the AM1 method was applied to a family of donor–acceptor-substituted stilbenoids and a series of sparkle-simulated model stilbenes. Considering the varied donor–acceptor strength of the substituents, features in common and mutually interchangeable properties for the two transition types, as well as their peculiarities, were revealed. Particular attention was paid to the occurrence of S_0–S_1 state conical intersections (CIs). Photoionization of *trans*-stilbene adsorbed on the external and internal surfaces of zeolites has been studied [56]. Laser photolysis of *trans*-stilbene on zeolite NaA and in zeolite NaX produced electrons trapped in Na$_n^{n+}$ clusters, singlet- and triplet-excited states, and radical cations of the stilbene. Thermal decomposition of *trans*-stilbene in NaX produced an unknown brown product, which shortens the lifetime of

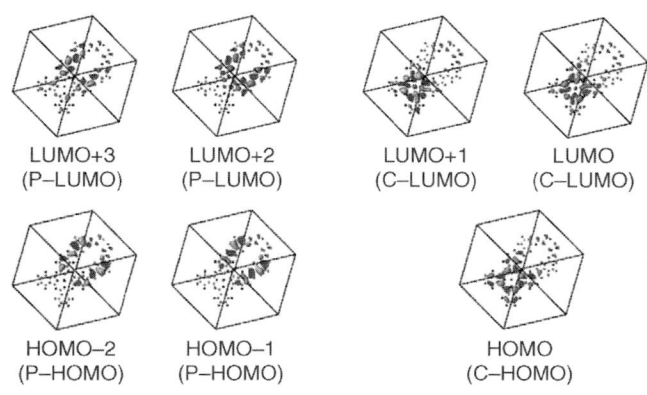

Figure 5.13 Amplitudes of the wave function of ground state of π-conjugated dendrimers, star-shaped stilbenoid phthalocyanines [54].

the electrons trapped in the sodium ionic clusters. Photoionized electrons from stilbene on NaA were trapped by sodium cation clusters inside the sodalite cage next to the surface and close to the arene radical cation. A method of time-resolved chemically induced dynamic nuclear polarization (CIDNP) was used for studying the interaction of 4,4′-dimethoxystilbene (=1,1′-[(1E)-ethane-1,2-diyl]bis[4-methoxybenzene]) with triisopropylamine or fumarodinitrile [57]. It was shown that both oxidative and reductive quenching gave almost mirror-image CIDNP spectra and photoinduced electron reverse transfer of the triplet radical ion pairs populated the stilbene triplet only, which then isomerizes.

Properties of the lowest and higher singlet excited states upon the resonant two-photon ionization of stilbenes and substituted stilbenes, using two-color two-laser irradiation, have been reported [58]. Radical cations of *trans*-stilbene (S) and substituted *trans*-stilbenes (S.bul.+) were generated from the resonant two-photon ionization in acetonitrile. The generation was performed with irradiation of one laser (266- or 355-nm laser) and with simultaneous irradiation of two-color two lasers (266- and 532-nm or 355- and 532-nm lasers) with the pulse width of 5 ns each. It was shown that the TPI proceeded through two-step two-photon excitation with the $S_0 \rightarrow S_1 \rightarrow S_n$ transition (Figure 5.14). The TPI efficiency using two-color two lasers increased compared to that when one laser was used. The authors concluded that the efficiency depends on the substituent of S, oxidation potential, molar absorption coefficient of the $S_0 \rightarrow S_1$ absorption, and properties of $S(S_1)$ and $S(S_n)$ such as lifetimes, electronic characters of $S(S_1)$ and $S(S_n)$, molar absorption coefficients of the $S_1 \rightarrow S_n$ absorption, and ionization rate of $S(S_n)$.

In order to elucidate the dendrimer effects, the two-photon ionization process (308 and 266 nm) of stilbene dendrimers having a stilbene core and benzyl ether-type dendrons has been investigated in an acetonitrile 1,2-dichloroethane mixture [59].

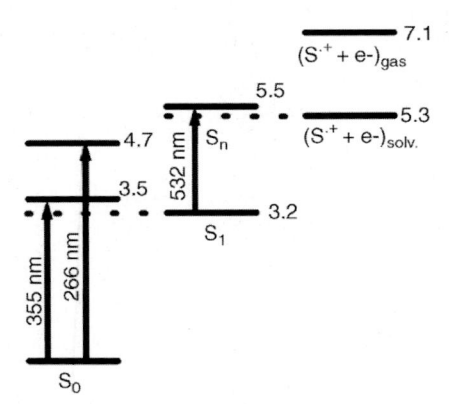

Figure 5.14 A schematic energy diagram of the TPI for the generation of $ST^{•+}$ using two-color two-laser photolysis. Numbers are energy levels in electron volts for the electronic states. One-photon energies of 266- and 355-nm light were 4.7 and 3.5 eV, respectively. IP of S is estimated to be approximately 5.3 eV. The energy level achieved by the excitation with the 266- or 355- and 532-nm two lasers is estimated to be 5.50 eV from E_{S1} (3.2 eV) and the 532-nm laser photon energy (2.30 eV) [58]. (Reproduced with permission.)

The quantum yield of the formation of stilbene core radical cation during the 308-nm TPI was independent of the dendron generation of the dendrimers, whereas a generation dependence of the quantum yield of the radical cation was observed during the 266-nm TPI, where both the stilbene core and the benzyl ether-type dendron were ionized. The authors suggested that the subsequent hole transfer occurs from the dendron to the stilbene core and that the dendron acts as a hole-harvesting antenna. The neutralization rate of the stilbene core radical cation with the chloride ion, generated from the dissociative electron capture by 1,2-dichloroethane, decreased with the increase in the dendrimer generation. Formation and decay processes of stilbene core radical cation (ST.bul. +) during the photoinduced electron transfer have been studied for a series of stilbene bearing benzyl ether-type dendrons (D) [60]. ST.bul. + and the radical cation of peripheral dendron (D.bul. +) were generated by intermolecular hole transfer from biphenyl radical cation, which was generated from photoinduced electron transfer from biphenyl to the singlet-excited 9,10-dicyanoanthracene. An intramolecular dimer radical cation of benzyl groups at the terminal of stilbene dendrimer was indicated as a hole-trapping site. It was found that D inhibits the charge recombination with 9,10-dicyanoanthracene radical anion because of the steric hindrance.

Interfacial electron hole transfer of a *trans*-stilbene radical cation photoinduced in a channel of nonacidic aluminum-rich ZSM-5 zeolite has been studied [61]. *trans*-Stilbene was incorporated as an intact molecule without solvent in the medium-size channel of nonacidic aluminum-rich $Na_{6.6}$ZSM-5 zeolite with $Na_{6.6}(SiO_2)_{89.4}(AlO_2)_{6.6}$ formula per unit cell (Figure 5.15). It was found that the interaction between Na^+ cation and stilbene occurs through one phenyl group facially coordinated to the Na^+ cation near the O atoms binding Al atoms. The similarity between Raman spectra of *trans*-stilbene in solution and occluded in $Na_{6.6}$ZSM-5 was obtained. A fast generation of primary St.bul. +-electron pair occurred by the laser UV (266 nm) photoionization. The charge carriers exhibit lifetimes about 1 h at room temperature and disappeared according to direct charge recombination and electron transfer. This subsequent electron transfer took place between the electron-deficient radical cation (*t*-St.bul. +) and the electron donor oxygen atom of the zeolite framework. The

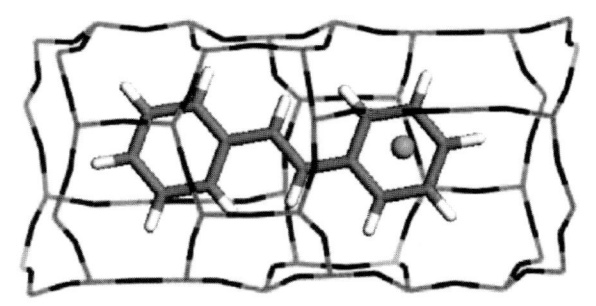

Figure 5.15 Predicted sorption site of *t*-St in straight channel of Na_4ZSM-5 (1 *t*-St/UC). The cylinders represent the H and C atoms of the *trans*-stilbene ($C_{10}H_8$) molecule and sphere represents the Na^+ cation [61]. (Reproduced with permission.)

ground state t-St@ZSM-5.bul.+.bul.− and the excited state t-St.bul.+@ZSM-5.bul.− radical cation electron pairs were characterized by the resonance Raman spectroscopy.

The mechanism and dynamics of photoinduced charge separation and charge recombination in synthetic DNA hairpins possessing donor and acceptor stilbenes separated by one–seven A:T base pairs have been investigated [62]. The application of femtosecond broadband pump-probe spectroscopy, nanosecond transient absorption spectroscopy, and picosecond fluorescence decay measurements permitted detailed analysis of the formation and decay of the stilbene acceptor singlet state and of the charge-separated intermediates. When the donor and acceptor were separated by a single A:T base pair, charge separation occurs via a single-step superexchange mechanism. However, when the donor and acceptor were separated by two or more A:T base pairs, charge separation occurs via a multistep process consisting of hole injection, hole transport, and hole trapping. In such cases, the hole arrival at the electron donor was slower than the hole injection into the bridging A-tract. Rate constants for charge separation and charge recombination depended upon the donor–acceptor distance; however, the rate constants for hole injection was independent of the donor–acceptor distance. Dynamics of ultrafast electron injection and charge recombination in DNA hairpins possessing a stilbenediether electron donor linker was observed with neighboring cytosine or thymine bases by means of femtosecond transient absorption spectroscopy [63]. It was also shown that guanine–guanine base pairs were not reduced, permitting the investigation of the distance dependence of charge injection. A scheme of dynamics of charge separation and charge recombination in (a) Hairpin **1** and (b) Hairpin **3** is shown in Figure 5.16.

A reversible bridge-mediated excited-state symmetry breaking in stilbene-linked DNA dumbbells was ascertained [64]. The excited-state behavior of synthetic DNA

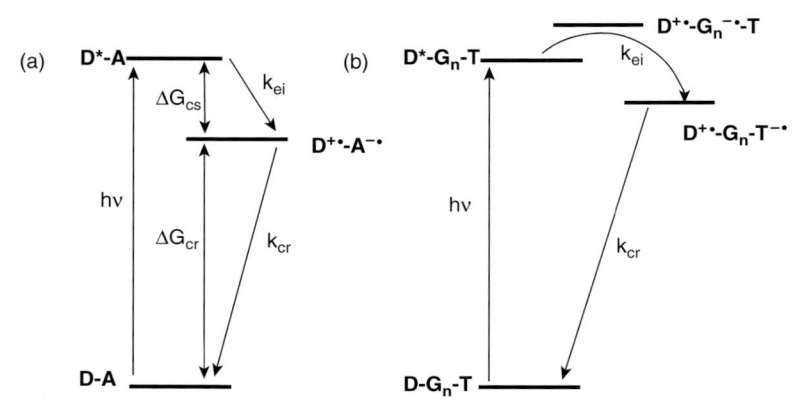

Figure 5.16 Dynamics of charge separation and charge recombination in (a) Hairpin **1** and (b) Hairpin **3** (hole on bridge shown as being localized on the middle A)[aa]. Dashed arrows indicate transient assignments [63]. (Reproduced with permission.)

dumbbells possessing stilbenedicarboxamide (Sa) linkers separated by short A-tracts or alternating A-T base-pair sequences has been investigated by means of fluorescence and transient absorption spectroscopy. Electronic excitation of Sa chromophores resulted in the conversion of a locally excited state to a charge-separated state in which one Sa is reduced and the other is oxidized. This process occurred via a multistep mechanism – hole injection followed by hole transport and hole trapping – even at short distances. Rate constants for charge separation were strongly distance-dependent at short distances. Hole trapping by Sa was highly reversible, and rapid charge recombination occurred via hole detrapping, hole transport, and charge return to regenerate the locally excited Sa singlet state. The authors suggested that neither charge separation nor charge recombination occurred via a single-step superexchange mechanism.

cis-Stilbene was excited by one photon at 270 nm in the gas phase and then probed by ionization at IR wavelengths (810 and 2100 nm), recording the delay time-dependent mass spectra [65]. A decay of 300 fs and two coherent oscillations (periods 140 and ≈600 fs) were found in the 300-fs window (Figure 5.17).

Picosecond time-resolved Raman spectroscopy has been used to study the ultrafast relaxation dynamics of *trans*-stilbene cation radicals following two-photon ionization in acetonitrile [66]. The integrated Raman intensities due to the cation radicals rise in

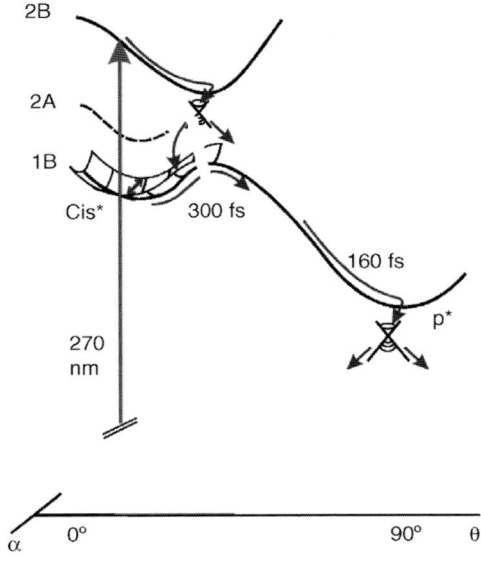

Figure 5.17 Suggested potential energy surfaces and pathways. θ is the C=C twist and * a planar CCC bend implying also some phenyl twist (ϕ). The initial motion on the 2B surface and through the first conical intersection is not time-resolved. The CI is displaced along * (outside the drawing plane), and entering from there into the 1B valley can therefore stimulate the vibration indicated. The last CI, reached from p*, is also outside the drawing plane, although the direction of displacement is not clear yet. The path toward DHP and the corresponding pericyclic minimum is not indicated [65]. (Reproduced with permission.)

tens of picoseconds and reach their maxima at a delay time of 40–60 ps from the pump pulse. It was suggested that the picosecond relaxation process increasing the cation Raman intensities occurs after the photoionization of stilbene. The rise time in the cation Raman intensity did not correlate with the dielectric relaxation time but depended on solvent polarity. The timescale of the intermolecular vibrational relaxation, obtained by the Raman spectroscopy, was found to be the same as that of the rise component of the cation Raman intensity. Photoionization and optical absorption of singlet excitons in a *trans*-stilbene crystal were studied [67]. The ionization efficiency of the singlet exciton was estimated over the energy range of 5.4–6.5 eV.

Polystyrene film doped with *trans*-stilbene has been found useful as a chemical dosimeter for ionizing radiation [68]. The dosimeter was effective for the photon energy range from 4.7 MeV to 1.25 keV (mean) and for rates above 300 rads/s. The mechanism for the dosimetry reaction, isomerization of *trans*- to *cis*-stilbene, was shown to be analogous to the mechanism proposed for radiolytic isomerization.

Chapters 3 and 4 and this chapter clearly demonstrated the rich array of photochemical and photophysical phenomena associated with stilbenes and their related substituted analogues. Because these compounds are available through their synthesis (Chapters 1 and 2), and are thermally and chemically stable, they are taking on an increasingly prominent role in the area of photochemical and biophysical investigations and multiple applications.

References

1 Whitten, D.G. (1993) *Accounts of Chemical Research*, **26**, 502–509.
2 Meier, H., Stalmach, U., Fetten, M., Seus, P., Lehmann, M., and Schnorpfeil, C. (1998) *Journal of Information Recording*, **24**, 47–60.
3 Gutlich, P., Garcia, Y., and Woike, T. (2001) *Coordination Chemistry Reviews*, **219–221**, 839–879.
4 Dou, Y. and Allen, R.E. (2004) *Journal of Modern Optics*, **51**, 2485–2491.
5 Minot, C., Roland-Gosselin, P., and Thal, C. (1980) *Tetrahedron*, **36**, 1209–1214.
6 Dyker, G., Körning, J. and Stirner, W. (1998) *European Journal of Organic Chemistry*, pp. 149–154.
7 Momotake, A., Uda, M., and Arai, T. (2003) *Journal of Photochemistry and Photobiology A: Chemistry*, **158**, 7–12.
8 Basaric, N., Marinic, Z., and Sindler-Kulyk, M. (2006) *Journal of Organic Chemistry*, **71**, 9382–9392.
9 Muszkat, K.A., Castel, N., Jakob, A., Fischer, E., Luettke, W., and Rauch, K. (1991) *Journal of Photochemistry and Photobiology A: Chemistry*, **56**, 219–226.
10 Lewis, F.D. and Reddy, G.D. (1992) *Tetrahedron Letters*, **33**, 4249–4252.
11 Hopf, H., Hucker, J., and Ernst, L. (2007) *Polish Journal of Chemistry*, **81**, 947–969.
12 Cleridou, S., Covell, C., Gadhia, A., Gilbert, A., and Kamonnawin, P. (2000) *Journal of Chemical Society, Perkin Transactions 1*, 1149–1155.
13 Maeda, H., Nishimura, K., Mizuno, K., Yamaji, M., Oshima, J., and Tobita, S. (2005) *Journal of Organic Chemistry*, **70**, 9693–9701.
14 Maeda, H., Hiranabe, R.-I., and Mizuno, K. (2006) *Tetrahedron Letters*, **47**, 7865–7869.
15 Myers, A.B., Rodier, J.-M., and Phillips, D.L. (1994) Reaction dynamics in clusters and condensed phases. 26th Jerusalem

Symposia on Quantum Chemistry and Biochemistry, pp. 261–278.
16 Ananchenko, G.S., Udachin, K.A., Ripmeester, J.A., Perrier, T., and Coleman, A.W. (2006) *Chemistry – A European Journal*, **12**, 2441–2447.
17 Kaliappan, R. and Ramamurthy. V. (2006) Abstracts of Papers, 231st ACS National Meeting, Atlanta, GA, March 26–30, 2006, ORGN-670.
18 Maeda, H., Nishimura, K.-J., Mizuno, K., Yamaji, M., Oshima, J., and Tobita, S. (2005) *Journal of Organic Chemistry*, **70**, 9693–9701.
19 Sigman, M.E., Barbas, J.T., Corbett, S., Chen, Y., Ivanov, I., and Dabestani, R. (2001) *Journal of Photochemistry and Photobiology A: Chemistry*, **138**, 269–274.
20 Amirsakis, D.G., Garcia-Garibay, M.A., Rowan, S.J., Stoddart, J.F., White, A.J.P., and Williams, D.J. (2001) *Angewandte Chemie – International Edition*, **40**, 4256–4261.
21 Saltiel, J., D'Agostino, J.T., Chapman, O.L., and Lura, R.D. (1971) *Journal of the American Chemical Society*, **93**, 2804–2805.
22 Noh, D.-Y., Lee, H.-J., Hong, J., and Underhill, A.E. (1996) *Tetrahedron Letters*, **37**, 7603–7606.
23 Kubo, Y., Suto, M., Tojo, S., and Araki, T. (1986) *Journal of the Chemical Society, Perkin Transactions, Organic and, Bio-Organic Chemistry (1972–1999)* 1, 771–779.
24 Lewis, F.D. and Crompton, E.M. (2004) SET addition of amines to alkenes, in *CRC Handbook of Organic Photochemistry and Photobiology*, 2nd edn (eds W. Horspool and F. Lenci), CRC Press LLC, Boca Raton, FL, pp. 7/1–7/18.
25 Hub, W., Schneider, S., Doerr, F., Simpson, J.T., Oxman, J.D., and Lewis, F.D. (1982) *Journal of the American Chemical Society*, **104**, 2044–2045.
26 Hub, W., Schneider, S., Doerr, F., Oxman, J.D., and Lewis, F.D. (1984) *Journal of the American Chemical Society*, **106**, 708–715.
27 Hub, W., Schneider, S., Doerr, F., Oxman, J.D., and Lewis, F.D. (1984) *Journal of the American Chemical Society*, **106**, 701–708.
28 Kaupp, G. and Grueter, H.W. (1981) *Berichte*, **114**, 2844–2858.
29 Lin, C.R., Wang, C.N., and Ho, T.I. (1991) *Journal of Organic Chemistry*, **56**, 5025–5029.
30 Kalgutkar, R.S., Yang, J.-S., and Lewis, F.D. (1998) Book of Abstracts, 216th ACS National Meeting, Boston, MA, August 23–27, 1998, PHYS-290.
31 Roberts, J.C. and Pincock, J.A. (2006) *Journal of Organic Chemistry*, **71**, 1480–1492.
32 Ho, T.I., Nozaki, K., Naito, A., Okazaki, S., and Hatano, H. (1989) *Journal of the Chemical Society, Chemical Communications*, (4), 206–208.
33 Yuan, C. and Ho, T.I. (1993) *Journal of the Chinese Chemical Society (Taipei, Taiwan)*, **40**, 47–51.
34 Lewis, F.D., Bassani, D.M., Burch, E.L., Cohen, B.E., Engleman, J.A., Reddy, G.D., Schneider, S., Jaeger, W., Gedeck, P., and Gahr, M. (1995) *Journal of the American Chemical Society*, **117**, 660–669.
35 Harder, T., Wessig, P., Bendig, J., and Stoesser, R. (1999) *Journal of the American Chemical Society*, **121**, 6580–6588.
36 Lehmann, M., Fischbach, I., Spiess, H.W., and Meier, H. (2004) *Journal of the American Chemical Society*, **126**, 772–784.
37 Soomro, S.A., Benmouna, R., Berger, R., and Meier, H. (2005) *European Journal of Organic Chemistry*, (16), 3586–3593.
38 Soomro, S.A., Schulz, A., and Meier, H. (2006) *Tetrahedron*, **62**, 8089–8094.
39 Vicinelli, V., Ceroni, P., Maestri, M., Lazzari, M., Balzani, V., Lee, S.-K., van Heyst, J., and Voegtle, F. (2004) *Organic & Biomolecular Chemistry*, **2**, 2207–2213.
40 Somlai, A.M., Creed, D., Landis, F.A., Mahadevan, S., Hoyle, C.E., and Griffin, A.C. (2000) *Polymer Preprints (American Chemical Society, Division of Polymer, Chemistry)*, **41**, 371–372.
41 Creed, D., Somlai, A.M., Hoyle, C.E., and Page, K.A. (2003) *Polymer Preprints (American Chemical Society, Division of Polymer, Chemistry)*, **44**, 84–85.

42 Meier, H., Lehmann, M., Schnorpfeil, C., and Fetten, M. (2000) *Molecular Crystals and Liquid Crystals Science and Technology, Section A: Molecular Crystals and Liquid Crystals*, **352**, 85–92.

43 Zaplo, O., Stumpe, J., Seeboth, A., and Hermel, H. (1992) *Molecular Crystals and Liquid Crystals Science and Technology, Section A: Molecular Crystals and Liquid Crystals*, **213**, 153–161.

44 Rameshbabu, K. and Kannan, P. (2007) *Journal of Applied Polymer Science*, **104**, 2760–2768.

45 Rameshbabu, K. and Kannan, P. (2006) *Polymer International*, **55**, 151–157.

46 Yoshida, M., Uchiyama, N., Aoki, N., Okoshi, M., and Iyoda, M. (1998) *Kokagaku Toronkai Koen Yoshishu*, 244.

47 Svetlichnyi, V.A. and Meshalkin, Y.P. (2007) *Proceedings of SPIE*, **6727**, 67271E/1–67271E/11.

48 Bazan, G.C., Koehler, B., Benmansour, H., Hong, J.W., Woo, H.Y., Mikhailovsky, A., Gorohmaru, H., Maeda, S., Kojima, T., and Shigeiwa, M. (2006) *US Patent Application Publication*, 2006, p. 16.

49 Hara, M., Samori, S., Xichen, C., Fujitsuka, M., and Majima, T. (2005) *Journal of Organic Chemistry*, **70**, 4370–4374.

50 Yan, Y., Tao, X., Sun, Y., Yu, W., Wang, C., Xu, G., Yang, J., Zhao, X., and Jiang, M. (2005) *Journal of Materials Science*, **40**, 597–600.

51 Ren, Y., Yu, X.-Q., Zhang, D.-J., Wang, D., Zhang, M.-L., Xu, G.-B., Zhao, X., Tian, Y.-P., Shao, Z.-S., and Jiang, M.-H. (2002) *Journal of Materials Chemistry*, **12**, 3431–3437.

52 Miyazawa, T., Liu, C., Koshihara, S.-Y., and Kira, M. (1998) *RIKEN Review*, **18**, 17–18.

53 Miyazawa, T., Liu, C., Koshihara, S.-Y., and Kira, M. (1997) *Photochemistry and Photobiology*, **66**, 566–568.

54 Kodama, Y., Ishii, S., and Ohno, K. (2007) *Journal of Physics: Condensed Matter*, **19**, 365242/1–365242/8.

55 Dekhtyar, M. and Rettig, W. (2007) *Journal of Physical Chemistry A*, **111**, 2035–2039.

56 Iu, K.-K., Liu, X., and Thomas, J.K. (1994) *Journal of Photochemistry and Photobiology A: Chemistry*, **79**, 103–107.

57 Goez, M. and Eckert, G. (2006) *Helvetica Chimica Acta*, **89**, 2183–2199.

58 Hara, M., Samori, S., Cai, X., Fujitsuka, M., and Majima, T. (2005) *Journal of Physical Chemistry A*, **109**, 9831–9835.

59 Hara, M., Samori, S., Cai, X., Tojo, S., Arai, T., Momotake, A., Hayakawa, J., Uda, M., Kawai, K., Endo, M., Fujitsuka, M., and Majima, T. (2004) *Journal of the American Chemical Society*, **126**, 14217–14223.

60 Hara, M., Samori, S., Cai, X., Tojo, S., Arai, T., Momotake, A., Hayakawa, J., Uda, M., Kawai, K., Endo, M., Fujitsuka, M., and Majima, T. (2005) *Journal of Physical Chemistry B*, **109**, 973–976.

61 Moissette, A., Bremard, C., Hureau, M., and Vezin, H. (2007) *Journal of Physical Chemistry C*, **111**, 2310–2317.

62 Lewis, F.D., Zhu, H., Daublain, P., Fiebig, T., Raytchev, M., Wang, Q., and Shafirovich, V. (2006) *Journal of the American Chemical Society*, **128**, 791–800.

63 Lewis, F.D., Liu, X., Miller, S.E., Hayes, R.T., and Wasielewski, M.R. (2002) *Journal of the American Chemical Society*, **124**, 11280–11281.

64 Lewis, F.D., Daublain, P., Zhang, L., Cohen, B., Vura-Weis, J., Wasielewski, M.R., Shafirovich, V., Wang, Q., Raytchev, M., and Fiebig, T. (2008) *Journal of Physical Chemistry B*, **112**, 3838–3843.

65 Fuss, W., Kosmidis, C., Schmid, W.E., and Trushin, S.A. (2004) *Chemical Physics Letters*, **385**, 423–430.

66 Nakabayashi, T., Kamo, S., Sakuragi, H., and Nishi, N. (2001) *Journal of Physical Chemistry A*, **105**, 8605–8614.

67 Katoh, R. and Kotani, M. (1990) *Chemical Physics Letters*, **174**, 541–545.

68 Harrah, L.A. (1969) *Radiation Research*, **39** (2), 223–229.

6
Stilbene Materials

Due to their miscellaneous photophysical and photochemical properties, stilbene compounds can be used for many applications in materials science. Data on the photochemistry of stilbenoid compounds and their role in materials science were reviewed [1–3]. In this chapter, we briefly discuss materials for lasers, nonlinear optical systems, photoconductive materials, and imaging and switching techniques.

6.1
Stilbene Lasers

A laser (light amplification by stimulated emission of radiation) is a device that emits light through a process called stimulated emission. Laser light is usually spatially coherent, which means that the light either is emitted in a narrow low-divergence beam or can be converted into one with the help of optical components such as lenses. Typically, lasers emit light with a narrow wavelength spectrum. Stimulated emission is the process by which an electron, perturbed by a photon having the correct energy, may drop to a lower energy level resulting in the creation of another photon. The second photon is created with the same phase, frequency, polarization, and direction of travel as the original. A laser consists of a gain medium inside a highly reflective optical cavity, as well as a means to supply energy to the gain medium. The gain medium is a material with properties that allow it to amplify light by stimulated emission. It can be of any state: gas, liquid, solid, or plasma.

6.1.1
Dye Lasers

A theoretical study of the excited states of stilbene and stilbenoid donor–acceptor dye systems as potential laser was performed [4]. Semiempirical calculations within the CNDO/S framework were used to characterize the nature of the "phantom-singlet" excited state P* (double-bond twisted geometry) of stilbene and stilbenoid donor–acceptor dye systems including 4-(dimethylamino)styrylpyridylmethylium and DCM laser dyes. It was shown that for stilbene, a slight geometric symmetry reduction is

Stilbenes. Applications in Chemistry, Life Sciences and Materials Science. Gertz Likhtenshtein
Copyright © 2010 WILEY-VCH Verlag GmbH & Co. KGaA, Weinheim
ISBN: 978-3-527-32388-3

necessary in order to localize the orbitals on the subunits. The results were consistent with those for methyl-substituted stilbene. The localized orbital description of twisted stilbene showed that P* contains a negligible doubly excited character and possesses a very small gap to the ground state. The planar systems were also investigated and correlated with Daehne's triad rule of polymethine systems.

Breakdown of optical power-limiting and dynamic two-photon absorption (TPA) for femtosecond laser pulses in molecular medium was reported [5]. To study the propagation of femtosecond laser pulses in a strong TPA organic molecular medium [4,4'-bis(dimethylamino)stilbene], the authors solved numerically the Maxwell–Bloch equations using an iterative predictor-corrector finite-difference time-domain technique. The hybrid density functional theory was used to calculate electronic structures of the compound. The system was described by a three-level model in an optical regime using the hybrid density functional theory. A good optical power-limiting behavior in a certain intensity region and thresholds for the breakdown of optical power was demonstrated. The dynamic two-photon absorption cross section was obtained, which was almost a linearly increasing function of the pulse width in the femtosecond time domain. Calculations showed that the propagation distance had an obvious influence on the measurement of the TPA cross section. The nonmonotonic dependence of the TPA cross section on propagation distance was observed.

A novel optical sensor for ammonia using a laser-grade dye, a 2,2'-([1,1'-biphenyl]-4,4'-diyldi-2,1-ethenediyl)-bis-benzenesulfonic acid disodium salt, was designed [6]. Fluorescence sensitization of the stilbene-3 solution in EtOH due to the presence of NH_3 in aqueous solution was observed. The authors suggested that an optical sensor for the detection of ammonia could be constructed using these properties of the dye and the sensitizer. The dynamics of pulse propagation accompanied by harmonic generation, stimulated Raman scattering, amplified spontaneous emission, and superfluorescence of the 4,4'-bis(dimethylamino)stilbene molecule was studied near the two-photon resonance [7]. Numerical solutions of the coupled Bloch–Maxwell equations for this stilbene were compared with the two-photon area theorem. In agreement with the area theorem, the authors demonstrated that the conventional dependence of the transmittance on the propagation depth was not valid for intense pulses.

6.1.2
Stilbene Solid Lasers

Analysis of the laser action of a solid-state dye laser with a thin-film ring resonator was reported [8]. The authors presented the fabrication of a compact blue solid-state dye laser and the properties of the laser action based on both experimental and numerical calculations. Stilbene-3 dye as the active medium for the laser by the sol–gel method was solidified and the medium around a 3-mm diameter glass rod was coated. The glass rod was simultaneously used as a thin-film ring resonator and a light amplifier. The dye molecules were pumped with pulsed Nd:YAG laser third harmonic generation (THG) (355 nm wavelength) and blue laser emission with the center wavelength of the laser output of 433 nm was obtained. The authors also designed a model of the thin-film ring laser and simulated the laser performance. Energy transfer solid-state dye laser with a thin-film ring resonator was invented [9]. In a suggested thin-film ring laser, the authors used organic dyes for optically active media such as stilbene-3. In order to oscillate the solid-state dye laser at any desired wavelength, two or three of the dyes were combined and doped into a xerogel thin film. The dye molecules were pumped with pulsed Nd:YAG laser THG and laser emission was obtained. By mixing dyes, the authors achieved the laser oscillation using one pumping source. The possibility for the solid-state dye laser with mixed dyes to cover the visible range of wavelength was shown.

Blue laser dye spectroscopic properties in sol–gel inorganic–organic hybrid films were described [10]. A blue solid-state laser material based on 4,4′-dibenzyl carbamide stilbene-2,2′-disulfonic acid incorporated into sol–gel zirconia and inorganic–organic hybrid matrices was presented. The absorption maxima of the dye in various matrices were around 339–361 nm, and the broad fluorescence peaks were at 411–413 nm. Optical gain measurements using the variable stripe method showed amplified spontaneous emission peaking at 437 nm. Optically active stilbene dye-doped thin-film waveguides were developed [11]. Dye-doped silica gel thin films were dip-coated on the surface of cylindrical glass rods by the sol–gel technique using stilbene-3 as laser dye. These solid-state dye thin-film ring resonators provided the feedback for laser operation. The laser with a nitrogen laser as pump source oscillated in the blue light region with an efficiency of 2% and a photostability of 70 shots. Boron-containing stilbenes as luminescent materials for organic solid lasers were patented [12]. The stilbenes were I [R_1–R_{12} = H, substituent; Q_1–Q_4 = (heteroatom-containing) C ≥ 5 aliphatic group, alicyclic group, aromatic group; n = 0–4]. The stilbenes show low threshold value, high luminescence intensity, and narrow half width of luminescence spectra.

I

6.2
Electro-Optic Materials

In one of the early works [13], a photocurrent of 10^{-10} A was measured in a system containing *trans*-stilbene irradiated with Hg high-pressure lamp HBO 500 at 20 cm and 5000 v/cm, wavelength up to 3500 Å. *trans*-α-Stilbene in which no coplanar position of the two rings had conductivity above 10^{-13} A.

To improve charge transport by reducing inhomogeneity, organic photorefractive composites (guest–host systems) consisting of charge transporting dendrimers containing eight carbazole groups highly doped (37%) with a stilbene were prepared [14]. The stilbene chromophore was prepared from N-phenyl-N,N-diethanolamine through esterification with trimethylacetyl chloride, carbonylation, and condensation with 4-nitrophenylacetic acid. The structure of the materials provided for the control of the orientation of charge transport agents and the charge transport mechanism were systematically studied. The specific photoconductivity of the photorefractive composite with stilbene chromophore was found to be 1.67×10^{-12} $(\Omega\,cm)^{-1}(W/cm^2)^{-1}$ and the linear electro-optic response was 0.29 pm/V at 66.7 kV/cm bias field, while those of the EHDNPB composite are 0.3×10^{-12} $(\Omega\,cm)^{-1}(W/cm^2)^{-1}$ and 0.22 pm/V at 66.7 kV/cm bias field. Push–pull benzoxazole-based stilbenes as electro-optic materials were proposed [15]. Photoinduced linear electro-optic effect (EOE) was discovered for three push–pull benzoxazole-based stilbenes with different electron withdrawing and donating substituents at the phenylene and methane groups. The ground-state geometry optimizations for *trans*-stilbene were performed using molecular mechanic geometry optimization within the framework of MM + force field method. Quantum chemical simulations of UV-absorption spectra were done within the framework of semiempirical RHF level (RHF) by AM1 (Austin Model 1) and PM3 (parametric method 3) methods. The maximally achieved value of photoinduced EOE coefficient was equal to 11.7 pm/V during illumination by $\lambda = 1060$ nm and $\lambda = 530$ nm coherent wavelengths.

Field-stimulated electro-optics in stilbene chromophore derivatives

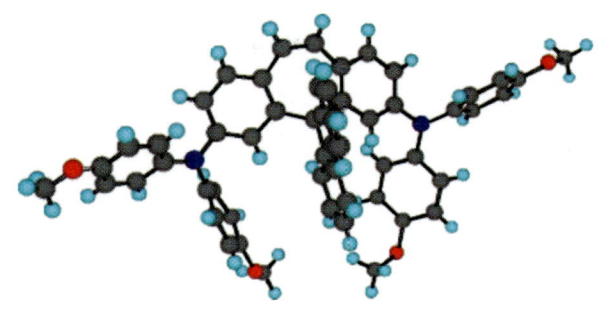

Figure 6.1 Chem-3D presentation for the X-ray crystal structure of **5c** [17].

Figure 6.2 The assembly concept of the cis-4,4′-bis(N,N-diarylamino)stilbene/fluorene optoelectronic system [17]. (Reproduced with permission.)

incorporated into photopolymer thin films were designed [16]. Optical and electric-field-induced linear electro-optic effect (PILEOE) was discovered during investigations of the stilbene chromophores embedded into polymer PMMA matrices. Doubly *ortho*-linked cis-4,4′-bis(diarylamino) stilbene/fluorene hybrids (Figures 6.1 and 6.2) as efficient nondoped, sky-blue fluorescent materials for optoelectronic applications were proposed [17]. This compound bearing a central dibenzosuberene optoelectronic unit, with functional C3 and C7, N,N-diarylamino appendages, and spiro-fluorene junction acted as blue fluorescent OLED materials. Sharp blue fluorescent (464 nm, $\Delta\lambda_{fwhm} = 47–60$ nm) OLED devices were furnished with a high η_{ext} of 7.9%, L20 of 2689 cd/m^2, η_c of 13.6 (cd/A), and η_p of 8.2 (lm/W) at 20 mA/cm^2.

The fabrication process for Mach–Zehnder modulators of poled polymer electro-optic waveguides has been developed [18]. A DANS (4-dimethylamino-4′-nitro-stilbene) polymer was suggested as the electro-optic waveguide material. It was formed of five layers that consist of an electro-optic waveguide layer, two buffer layers, and two metal electrode layers. The channel waveguide with glazed sidewall was fabricated using UV photobleaching process. Using a corona-poling setup with a tungsten wire electrode, the films were poled so that the waveguide had electro-optic property. By optimizing the fabrication process, the model device of M-Z type electro-optic intensity modulators working at 1300 nm wavelength has been developed. The half-wave voltage and modulation bandwidth of the device were found to be about

10 V and 1 GHz, respectively. A new nonconjugated polymer electroluminescent material, including stilbene derivatives, was patented [19]. The preparation process for this electroluminescent material involved monomer synthesis and polymer synthesis. The author suggested that this electroluminescent material can be used in electroluminescent devices, luminescent devices, and so on.

Stilbene-like two carbazole dimmers were synthesized by McMurry C−C coupling reaction, and their electroluminescent properties were characterized [20]. These compounds were fluorescent in blue to yellow region with moderate to good quantum yields. They were thermally stable and capable of hole transporting due to the presence of the carbazole moieties (Figure 6.3).

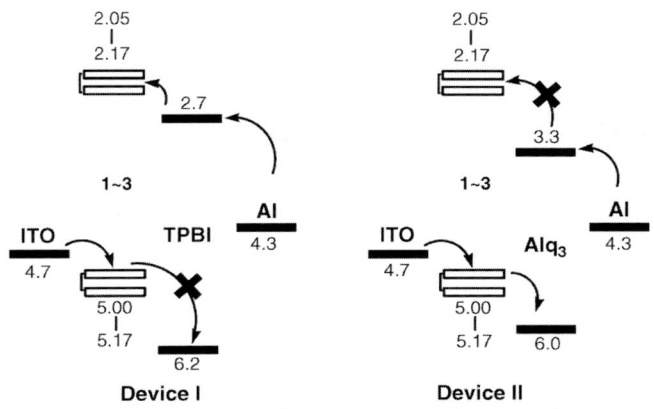

Figure 6.3 Energy alignment of the constituent in device I and device II involving stilbene-like two carbazole dimmers. See details in Ref. [20]. (Reproduced with permission.)

The electroluminescent devices fabricated using stilbene-like carbazole dimers as hole transporters/emitters with a bilayer structure ITO/Cpd/TPBI or Alq3/LiF/Al exhibited good performance (i.e., $\eta_{ext} = 1.0$–2.1%; $\eta_p = 0.9$–$1.9\,lm/W$; $\eta_c = 2.4$–4.8 cd/A at a cd of $100\,mA/cm^2$).

6.3 Electrophotographic Material

A typical electrophotographic photoconductor comprises an electroconductive substrate, a photoconductive layer formed on the electroconductive substrate, and a protective layer formed on the photoconductive layer, the protective layer comprising a binder resin and finely divided particles of at least one metal oxide with the surface. This system is coated with a coupling agent selected from the group consisting, for example, of a titanate-type coupling agent, a fluorine-containing silane-coupling agent, and an acetoalkoxyaluminum diisopropylate, dispersed in the binder resin.

A number of patents on photoconductors have recently been proposed [21–34], such as highly durable electrophotographic photoconductors with retention of image density [21], electrophotographic photoreceptor, process cartridge, image-forming apparatus and method [22], base tube for electrophotographic photoconductive members [23], electrophotographic photoconductor and image-forming apparatus [24], electrophotographic photoreceptors containing specific stilbene derivative as electron-transporting agent [25], image-forming apparatus and image-forming method [26], and so on [27].

In the recent work [24], the photoreceptor comprised a support with a photosensitive layer containing a binder, a hole-transporting agent containing stilbene derivative I with divalent organic group containing (un)substituted aromatic ring (R_{1-3} = H, halo, C1–20 alkyl, C6–30 aryl, C1–20 alkyl halide, C2–30 alkenyl, C1–25 alkoxy, C7–30 aralkyl, a, $c = 0$–4; $b = 0$–5; $d = 2$–3) and a charge-generating agent. An image was formed by using a liquid developer made of toner dispersed in a hydrocarbon solvent. The photoreceptor showed high sensitivity and good solvent resistance for a long period. An electrophotographic photoconductor, showing durability and solvent resistance, and an electrophotographic imaging apparatus for wet development were patented [28]. The photoconductor contained a stilbene derivative having 3-enamine structures represented by I containing aromatic ring-containing trivalent organic group: R_{1-7}, B_1, B_2 = H, halo, C1–20 alkyl, C1–20 haloalkyl, C1–20 alkoxy, C6–20 aryl; amino, C2–30 ethenyl, C8–20 styryl; B_1 joining together with B_2.

I

The invention of stilbene, containing polyamides for forming films for photoimaging compositions, and manufacture of semiconductor devices [29] were related to polyamides having ≥1 repeating units selected from I and II

$(R_1 = H$, alkyl; R_2, $R_3 =$ halo, alkyl, alkoxy, aryl, aralkyl; k, $l = 0$, 1; m, $n = 0$–3).

The photoreceptor has a light-sensitive layer containing a binder resin, a charge-generating agent, and a positive hole-transporting material of stilbene derivative I.

With (un)substituted C1–20 alkyl or alkoxy, (un)substituted C7–20 aralkyl, halo, (un)substituted C15–20 aryl; m, n, p, $q = 0$–3; R_3, $R_4 = H$, C1–20 aralkyl was invented.

An electrophotographic photoconductor containing a tertiary amine triphenylene derivative in a light-sensitive layer to improve antioxidant properties was reported [31]. A specified stilbene compound was used as a charge transport material. A method of purification of aminostilbenes for electrophotographic photoreceptors or electroluminescent devices was developed [32]. $Ar_1Ar_2C:CH(CH:CH)_nAr_3NR_1R_2$ [$Ar_1 =$ (un)substituted aromatic hydrocarbyl, (un)substituted aromatic heterocyclyl; $Ar_2 = H$, any group given for Ar_1; $Ar_3 =$ (un)substituted aromatic hydrocarbylene; R_1, $R_2 =$ (un)substituted aliphatic hydrocarbyl, any group given for Ar_1; $n = 0$, 1] were purified by vacuum distillation under ≤10 Pa.

Electrophotographic photoreceptors containing stilbene derivatives having triphenylamine structures have been designed. In the work [33], stilbeneamine derivatives having general structure I ($Ar_{1-4} =$ aryl, silyl-, or silyl ether-substituted aryl),

where $Ar_{1-4} =$ aryl, silyl-, or silyl ether-substituted aryl, have been used as charge-generating agents.

6.4
Light-Emitting Diodes

Light-emitting diode (LED) is a semiconductor diode that emits light when an electric current is applied in the forward direction of the device, as in a simple LED circuit. The effect is a kind of electroluminescence where incoherent and narrow-spectrum light is emitted from the p–n junction. LEDs are widely used as indicator lights on electronic devices and in higher power applications such as flashlights and area lighting.

Improvement of electroluminescence efficiency of blue polymer light-emitting diodes using polymer hole-transporting and electron-transporting layers was achieved [34]. The characteristics of blue polymer light-emitting diodes were improved by introducing hole-transporting and electron-transporting layers. Poly [(9,9-dioctylfluorene-2,7-diyl)-*alt*-(triphenylamine-4,4′-diyl)] [PF8-TPA (50%)], poly [(9,9-dioctylfluorene-2,7-diyl)-*co*-(stilbene-4,4′-diyl)] [PF8-SB (10%)], and poly[(9,9-dioctylfluorene-2,7-diyl)-*alt*-(pyridine-2,6-diyl)] [PF8-Py (50%)] were used as the hole-transporting material, blue light-emitting material, and electron-transporting material, respectively. It was shown that efficiency and external quantum efficiency at 1000 cd/m^2 can be increased up to 1.50 cd/A and 0.95%, respectively, by introducing the electron-transporting layer. These improvements of efficiencies were supposed to be achieved by a good confinement of holes and electrons in the emitting layer by the hole-transporting layer and electron-transporting layer. The authors of the work [35] have designed and synthesized new dopant materials based on the styrylamine moiety, 4-[(1,2-diphenyl)-4′-(*N*,*N*-diphenyl-4-vinylbenzenamine)]biphenyl (**4**), and 4-[(1,2-diphenyl)-4′-(*N*,*N*-diphenyl-4-vinylbenzenamine)]terphenyl (**8**). Blue OLEDs (Figure 6.4) were obtained from new styrylamine dopant materials and compared with those of blue dopant bis[4-(di-*p*-*N*,*N*-diphenylamino)styryl]stilbene (DSA-Ph) and diphenyl[4-(2-terphenyl vinyl)phenyl]amine (R-BD). The materials obtained showed efficiency of about 3.5 cd/A.

Blue organic light-emitting diode with improved color purity using 5-naphthyl-spiro[fluorene-7,9′-benzofluorene] was designed [36]. Novel spiro-type blue host

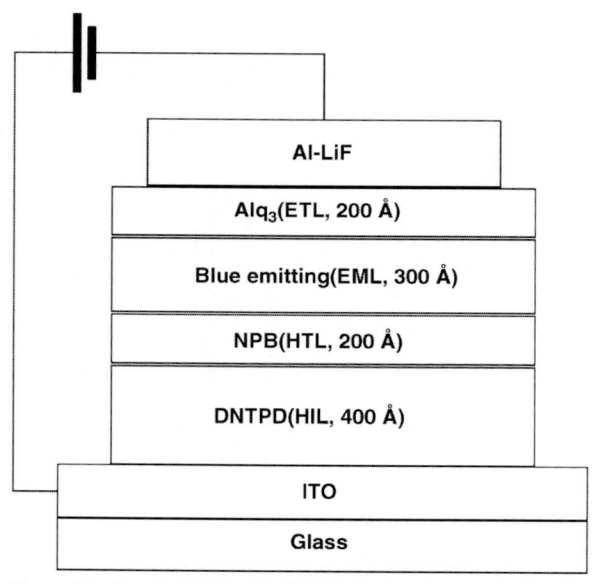

Figure 6.4 Structure of the blue OLED device [35]. (Reproduced from Ref. [35] with permission from Elsevier.)

material, 5-naphthyl-spiro[fluorene-7,9'-benzofluorene] (BH-1SN), and dopant material, 5-diphenyl amine-spiro[fluorene-7,9'-benzofluorene] (BH-1DPA), were synthesized, and a blue OLED was made from them. The structure of the blue device was ITO/DNTPD/α-NPD/BH-1SN : 5% dopant/Alq3 or ET4/Al-LiF. α-NPD was used as the hole transport layer, DNTPD as the hole injection layer, BH-1DPA or BD-1 as the blue dopant material, Alq3 or ET4 as the transporting layer, and Al as the cathode. The blue devices doped with 5% BH-1DPA and BD-1 show blue EL emissions at 444 and 448 nm at 7 V, respectively, and a high efficiency of 3.4 cd/A at 5 V for the device was obtained from BH-1SN : 5% BD-1/ET4. The CIE coordinates of the blue emission are 0.15, 0.08 at an applied voltage of 7 V for the device obtained from BH-1SN/5% BH-1DPA/Alq3.

Copolymers having aromatic amine repeating units,

their compositions, and light-emitting diodes and devices were prepared [37]. The copolymers have (A) ≥ 1 stilbenzyl units $Ar_1CR_1:CR_2Ar_2$ (Ar_1, Ar_2 = arylene, divalent heterocyclic group; R_1, R_2 = H, alkyl, alkoxy, alkylthio, alkylsilyl, alkylamino, aryl, aryloxy, arylsilyl, arylamino, arylalkyl, arylalkoxy, arylalkylsilyl, arylalkylamino, arylalkenyl, arylalkynyl, monovalent heterocyclic group, cyano) and (B) ≥ 1 aromatic amine units $Ar_3Ar_4NAr_5(NAr_6Ar_7)_n$ (Ar_3, Ar_5, Ar_7 = arylene, divalent heterocyclic group; Ar_4, Ar_6 = aryl, monovalent heterocyclic group; $n = 0–3$). The compounds comprised the copolymers and polymers, giving intense fluorescence in solid state.

6.5
Materials for Nonlinear Optics

Some principles of nonlinear optics (NLO) were described in Section 5.5.

Synthesis of novel cross-linked polyurethane containing modified stilbene and Schiff base chromophores for second-order nonlinear optics was reported [38]. The cyano and nitro groups were chosen as acceptor groups, and the substituent amino or ether groups as donor groups to the matrix. These poled polymers showed high second-order optical nonlinearity and, in authors' opinion, have potential application in frequency-doubling or electro-optical controlling devices. According to [39], the multibranched chromophores based on bis(diphenylamino) stilbene exhibited two-photon adsorption. The peripheral substituent effect and multibranched modification effected by the open aperture femtosecond Z-scan technique and the nanosecond nonlinear optical transmission were investigated. By comparing the two-photon absorptivity of (E)-4,4'-bis(diphenylamino)stilbene (BDPAS) with those of its derivatives, it was found that substituent group attached to the periphery of BDPAS has no obvious contribution to the enhancement of TPA and that the dramatic increase of effective TPA cross sections of multibranched samples in nanosecond regime suggested their larger excited-state absorption. It was shown that the one-dimensional symmetrical π-conjugated material [4,4′-bis(dimethylamino)stilbene] exhibited strong nonlinear optical properties [40]. The interaction between the ultrashort laser pulses and this compound by solving Maxwell–Bloch equations and using density functional theory on *ab initio* level was analyzed. In the case of single-photon resonance, the two-level model can describe the interaction between the small area pulse and the molecular system. For large area pulse, due to the existence of strongly secondary excitation to the higher-lying levels, the three-level model should be used.

Three-photon absorption of a series of donor–acceptor *trans*-stilbene derivatives was studied by means of density functional theory applied to the third-order response function and compared with experimental data [41]. It was shown that with a Coulomb attenuated, asymptotically corrected functional, the excitation energy of the first resonance state is much improved. A comparison with experiments indicates that this is the case for the three-photon cross section as well. The hyperpolarizabilities of nonlinear optical 138 chromophores (1–12), stilbene and heteroaromatic analogues, were investigated by *ab initio* method [42]. The results revealed a good linear relation exists between the first hyperpolarizability (P) and gas-phase substituent constants (σ + gas). The susceptibility (ρ) of the β to the donor strength was a characteristic of the conjugated bridges. Nonlinear optical properties of stilbene and azobenzene derivatives containing azaphosphane groups were reported [43]. The nitro group as the acceptor and azaphosphane ($R_3P=N-$) as the donor group were used. It was found that that both first-order polarizability and hyperpolarizabilities were larger for stilbene derivatives and maximum for the Ph substitution. Second-order polarizability was higher for Me substitution. For these molecules, the two-photon absorption cross section was obtained. Both one-photon and two-photon absorption cross sections were maximum for the first excited state in the case of stilbene and second excited state in the case of azobenzene derivatives.

Quadratic and cubic optical nonlinearities of 4-fluorophenylethynyl- and 4-nitro-(*E*)-stilbenylethynylruthenium complexes were demonstrated [44]. The complexes *trans*-[Ru(C≡C-4-C_6H_4F)X(dppe)$_2$] [X = Cl (1), C≡CPh (2), C≡C-4-$C_6H_4NO_2$ (3)], *trans*-[Ru{C≡C-4-C_6H_4-(*E*)-CH:CH-4-$C_6H_4NO_2$}X(dppe)$_2$] [X = C≡CPh (4), C≡C-4-C_6H_4C≡CPh (5)], and [C_6H_3-1,3-{C≡C-*trans*-[RuCl(dppe)$_2$]}2,5-(C≡C-4-C_6H_4F)] (6) have been synthesized and their structures confirmed by a single-crystal X-ray diffraction study, cyclic voltammetry, and hyper-Rayleigh scattering (HPS). The HPS studies at 800 nm using femtosecond pulses and amplitude modulation to remove multiphoton fluorescence contributions revealed significant fluorescence-free nonlinearities for 3–5. The frequency-independent nonlinearities calculated from the 800 nm results allowed the authors to attribute fluorescence contributions to the 1064 nm data. Z-scan studies at 820 nm revealed cubic nonlinearities that increased with the size of the π-system. Molecular geometry and atomic labeling scheme for *trans*-[Ru(C≡C-4-C_6H_4F)Cl(dppe)$_2$] is presented in (Figure 6.5).

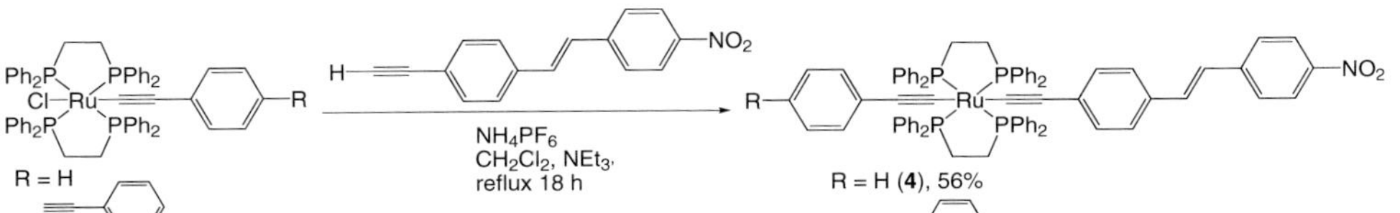

Figure 6.5 Molecular geometry and atomic labeling scheme for *trans*-[Ru(C≡C-4-C_6H_4F)Cl(dppe)$_2$] (1). Ellipsoids show 30% probability levels. Hydrogen atoms have been omitted for clarity [44]. (Reproduced with permission from Elsevier.)

Micron- and submicron-scale lateral structures of optically nonlinear organic films comprised of substituted *trans*-stilbene derivatives ($R_1 = OCH_3$, $R_2 = CN$) was characterized [45]. Second harmonic generation (SHG), optical microscopy, and atomic force microscopy (AFM) were used in this investigation. The third-order nonlinear optical properties and two-photon absorption of different types of stilbene derivatives (D-π-D, A-π-A, D-π-A) were investigated [46]. Using the INDO/CI method, the UV–vis spectra were explored and the position and strength of the two-photon absorption were predicted by sum-over-states expression. Relationships of the structures, spectra, and nonlinear optical properties have been examined. Two-photon absorption spectra (650–1000 nm) of a series of asymmetrically substituted stilbenoid chromophores

TSB

Dor

Acc

S101

N101

101

BT101

were measured via a newly developed nonlinear absorption spectral technique based on a single and powerful femtosecond white-light continuum beam [47]. The results suggested that when either an electron-donor or an electron-acceptor was attached to a *trans*-stilbene at a *para*-position, an enhancement of two-photon absorptivity was observed in both cases, particularly in the 650–800 nm region. The push–pull chromophores with both the donor and the acceptor groups showed larger overall two-photon absorption cross sections within the studied spectral region compared to their monosubstituted analogues. A large SHG excited by an Nd:YAG laser ($\lambda = 1.32\,\mu m$) has been observed [48]. The maximum output SHG was observed for the chromophore derivatives

1

2

3

possessing the highest second-order hyperpolarizabilities and corresponding dipole moments. Due to the incorporation of the chromophore into photopolymer matrices, second-order microscopic susceptibilities of the composites substantially increased (from 0.24 to 1.89 pm/V).

Synthesis and nonlinear optical characterization of a two-photon absorbing organic dye, *trans*-4-(dimethylamino)-4'-[*N*-ethyl-*N*-(2-hydroxyethyl)amino]stilbene (DMAHAS), were reported [49]. Linear absorption, single-photon-induced fluorescence, and two-photon-induced fluorescence of DMAHAS were experimentally studied. This dye showed a moderate two-photon absorption cross section of $\sigma 2 = 0.91 \times 10\text{--}46\,\text{cm}^4\,\text{s/photon}$ at 532 nm as shown by an open aperture Z-scan technique. DMAHAS also showed strong two-photon-induced blue fluorescence of 432 nm when pumped with 800 nm laser irradiation.

Quadratic and cubic hyperpolarizabilities of stilbenylethynyl–gold and –ruthenium complexes (Figure 6.6) were reported [50]. These complexes have been prepared and identified, and their electrochemistry (Ru complexes) (Figure 6.6) and nonlinear optical properties assessed. Quadratic nonlinearities at 1064 and 800 nm for the octupolar stilbenyl–Ru complexes were found to be large for compounds without strongly accepting substituents. Cubic molecular hyperpolarizabilities at 800 nm for the organic compounds and Au complexes were low. It was shown that cubic nonlinearities $\gamma 800$ and two-photon absorption cross-sections $\sigma 2$ for Ru complexes increased on proceeding from linear analogues to octupolar complexes. Cubic nonlinearities, $\text{Im}(\chi(3))/N$, for some from the first application of electroabsorption (EA) spectroscopy to organometallics were also large, scaling with the number of metal atoms.

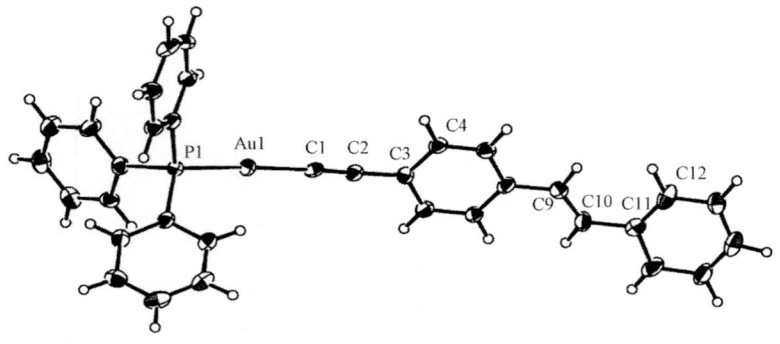

Figure 6.6 Molecular structure for (Au{E}-4-C≡CC$_6$H synthesis and properties of long conjugated organic optical limiting materials with different π-electron conjugation bridge structure were reported [50]. (Reproduced with permission from Elsevier.)

Two-, three-, and four-photon-pumped stimulated emission (cavityless lasing) properties of 10 novel stilbazolium dyes in solution phase were studied [51]. The authors demonstrated that these multiphoton active dye compounds can be used to generate highly directional stimulated emissions over a broad visible spectral range (from 490 to 618 nm) under multiphoton pump conditions. The pump source was a powerful Ti:sapphire oscillator–amplifier system associated with an optical parametric generator, which could specifically provide approximately 160 fs duration and approximately 775, 1320, and 1890 nm laser pulses for two-, three-, and four-photon excitation, respectively. The spectral, spatial, and temporal properties as well as the efficiency of multiphoton-pumped lasing output from different dye-solution samples were studied. Based on the results, two features were found: (i) the threshold pump-energy values for two-, three-, and four-photon-pumped lasing were quite close (within a factor of 3–4) and (ii) there was an obvious wavelength difference (10–30 nm) between the forward and backward lasing output under three- and four-photon pump conditions.

The dynamics of populations of the electronic states in a 4,4′-bis(dimethylamino) stilbene molecule (two-photon absorption) was studied against the frequency, intensity, and shape of the laser pulse [52]. Complete breakdown of the standard rotating wave for a two-photon absorption process was observed. An analytical solution for the interaction of a pulse with a three-level system beyond the rotating wave approximation was obtained in close agreement with the strict numerical solution of the amplitude equations. Calculations showed the strong role of the anisotropy of photoexcitation in the coherent control of populations that can affect the anisotropy of photobleaching. The two-photon absorption cross section of an ethanol solution of a *trans*-stilbene and its derivatives exposed to radiation of the second harmonic of a Nd:YAG laser (532 nm) of nanosecond duration has been detected [53]. In experiments, the method based on the measurement of the photochemical decomposition of examined molecules was used. The quantum yield of the photoreaction (γ266) of dyes under one-photon excitation (fourth harmonic Nd:YAG laser 266 nm) was detected by absorption method.

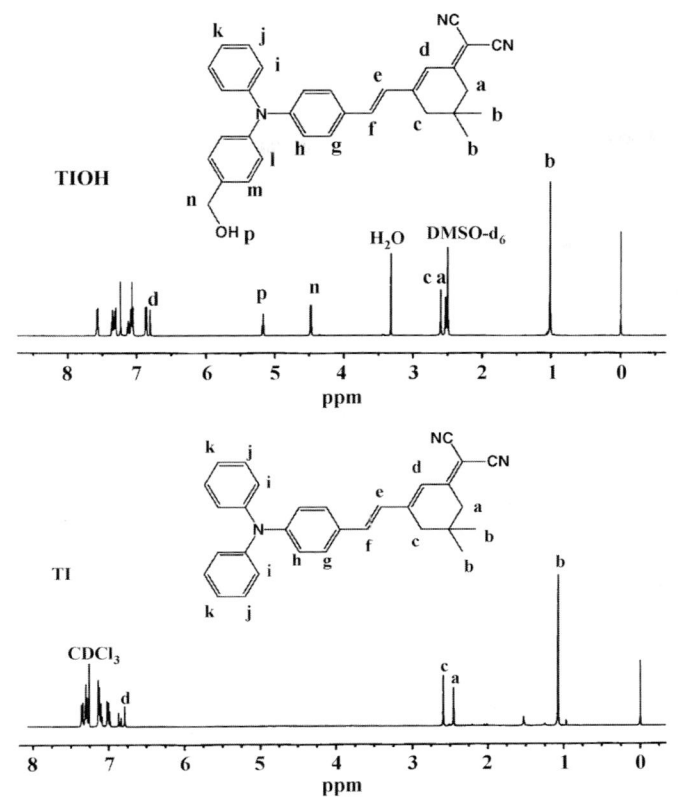

Figure 6.7 Structure and ^{1}H NMR spectra (500 MHz) and the assignment of peaks of TIOH and TI [54]. (Reproduced with permission.)

Two triphenylamino-substituted chromophores with and without hydroxyl end, named TIOH and TI (Figure 6.7), respectively, were synthesized and incorporated into hybrid organic–inorganic materials derived from 3-glydoxypropyltrimethoxysilane, tetraethoxysilane, and 3-aminopropyltriethoxysilane [54]. These stilbene-type chromophores were characterized by elemental analysis by ^{1}H NMR, FT-IR, UV–vis spectra, and TGA. The hyperpolarizabilities were characterized through solvatochromic method. Both chromophores possessed higher thermal stability and competitive hyperpolarizabilities. Second harmonic generation was observed on poled films. The nonlinear coefficient of the samples was established at 41.2 pm/V for TIOH doped film and at 24.8 pm/V for TI doped film.

6.6
Light-Emitting Materials

Synthesis and photoluminescent properties of 2,6-di(4′-bisphenylaminostilbenyl) benzo[1-2,4-5]bisoxazole were reported [55]. The compound has been synthesized with 4-(diphenylamino)benzaldehyde and 2,6-bis[4-(chloromethyl)phenylene]benzo [1-2,4-5]dioxazole through a Wittig–Horner reaction. Its chemical structure has been detected by UV–vis, IR, ^{1}H NMR and EA, and photoluminescent properties have been studied. The results indicated that the synthesized compound has *trans*-stilbene as structural character and can be used as a small-molecular blue light-emitting material. An objective of the invention [56] was to provide a novel light-emitting element material having a large energy gap and an electron transporting property. For this purpose, stilbene derivatives, represented by a general formula,

I

where *n* is an integer of 0–2 and *m* is an integer of 1 or 2 were used. Novel branched alpha-cyanostilbene fluorophores were patented [57].

α-Cyanostilbene fluorescent materials containing alkyl, C1–6 alkoxy, (un)substituted amino, (un)substituted aryl, (un)substituted heterocycle, (un)substituted aryl, or (un)substituted aryl or heterocycle were condensed at the optional site of the corresponding two benzene rings. According to the patent, the fluorescent materials may exhibit the high luminescent efficiency and are capable of tuning the fluorescent colors of red, green, and blue according to the core structure in the molecule. A higher luminescent efficiency in solid state than in solution was observed.

Three novel organic optical materials, 4'-(*N*,*N*-dihydroxyethylamino)-4-(pyridine-4-vinyl)stilbene, (*N*-((4-*N*,*N*-dihydroxyethylamino)benzylidene)-4-(pyridine-4-vinyl) aniline (a), and 4'-(*N*,*N*-dihydroxyethylamino)-4-(pyridine-4-vinyl)azobenzene (b), were synthesized and characterized by FT-IR, UV, ^{1}H NMR, and elementary analysis [58]. The results showed that these compounds possessed good optical limiting and large nonlinear optical properties. These properties were attributed to the long D-π-A-conjugated electron structure of molecules, and the π-electron-conjugated bridge structure affected the nonlinear optical and optical-limiting properties of D-π-A conjugation compounds. Compounds with C=C double bond as a conjugation bridge showed better optical limiting properties than compound **8b** with N=N double bond; under the same linear transmittance, compound **8c** with

N═N double bond as a conjugation bridge was superior in nonlinear optical properties.

Several color-tunable light-emitting copolymers containing coumarin or coumarin 1 in main chains were prepared with triphenylamine, butoxybenzene, stilbene, 9,9-bis(2-ethylhexyl)fluorene, 1,2,4,5-tetramethylbenzene, or β-methoxynaphthalene by oxidative coupling copolymerization with $FeCl_3$ [59]. The molecular structures of the polymers were characterized by ^1H NMR and FT-IR. All copolymers were amorphous and had good thermostability. The HOMO and LUMO energy levels of the polymers were calculated based on the results of voltammetry and UV–vis absorption spectroscopy. The fluorescence spectra of polymers in chloroform solution showed that the copolymer of coumarin and stilbene emitted a yellow-green or orange light. The organic polymer photoinduced micromold-type material with a core/shell structure composed of 10–90% azo-type or stilbene-type photosensitive polymer core and 10–90% colorless transparent polymer shell (acrylic ester, siloxanes, or vinylcarbazoles) was prepared [60].

Stilbene derivatives, light-emitting element, display device, and electronic device were patented [61–64]. Stilbene compounds containing a diarylamino or carbazolyl group and a double bond with substituents (Ar_1 and Ar_2 = independently selected C6–25 aryl; A11 = $-(Ar_{11})N$-α-$N(Ar_{12})Ar_{13}$, III, or IV; Ar_{11-13}, Ar_{21}, Ar_{31} = independently selected C6–25 aryl; α, β = C6–25 arylene; and R_{31}, R_{32}, R_{41}, R_{42}, R_{61}, and R_{62} = independently selected H, C1–4 alkyl, and C6–25 aryl) were described. Light-emitting devices with layers incorporating the derivatives and electronic devices with display elements employing light-emitting devices were described. Light-emitting devices containing layers incorporating the heat-resistance stilbene derivatives I

[Ar_1 = (un)substituted biphenyl or terphenyl; Ar_2 = (un)substituted Ph, biphenyl, or terphenyl; R_1–R_3 = H, C1–3 alkyl] were also patented [63].

6.7
Materials for Image-Forming Apparatuses

A method of preparing a photoreceptor for an image-forming apparatuses was patented [65]. The method included forming a photosensitive layer overlying an electroconductive substrate, coating a liquid including a radically polymerizable compound to form a protective layer, irradiating the protective layer with light to cross-link the protective layer, and then contacting the protective layer with a fluid. The latter was selected from the group consisting of fluids, which included a charge transport material. Three-dimensional (3D) optical data storage in a photobleachable dye-doped polymer using two-photon laser scanning microscopy was reported [66].

High-contrast and high-density 3D optical data storage was demonstrated using photobleaching in a polymer block in a two-photon laser scanning microscope geometry. A (hydroxyethyl methacrylate) polymer block doped with 4-[N-(2-hydroxyethyl)-N-methylaminophenyl]-4'-(6-hydroxyhexylsulfonyl)stilbene was used as the data storage medium. Digital information was written to a depth of >100 μm in the polymer block, with the established bit dimensions being 0.5 μm × 0.5 μm and 3 μm axially. This corresponds to a storage density of 1×10^{12} bits/cm^3. Photoinduced processes and holographic storage in stilbene and stilbenecarboxaldehyde in a poly(methyl methacrylate) matrix were described [67]. The authors reported a spectrophotometric investigation of the photoprocesses in PMMA films containing stilbene or stilbenecarboxaldehyde. It was shown that UV light induces *trans–cis* isomerization of the stilbene-type molecules accompanied by considerable changes in the absorption spectra. In the stilbenecarboxaldehyde/PMMA films, these changes were anisotropic. The recorded holographic gratings were stable for months.

Two-photon photobleaching three-dimensional optical storage with DVD pickup head was described [68]. Based on the objective lens and voice coil motor of the current DVD pickup head, according to photobleaching 3D optical information storage theory of two-photon absorption, a Ti:sapphire femtosecond pulse laser was used for the two-photon photobleaching of the new material consisting of stilbene derivatives. The experiments proved that proceeding with two-photon three-dimensional optical data storage with DVD pickup head indicated that two-photon photobleaching technology combined with the current CD/DVD technology provided a base for the realization of high-density and super high-density optical data storage. According to [69], photochromophores such as *cis*-stilbene can be sought out as potential candidates for media within 3D optical information storage devices. A strong molecular two-photon absorption (inducing the reversible photoisomerization) was suggested to be a necessary feature for this application due to the need for high 3D spatial resolution. The author investigated the one- and two-photon absorption (OPA and TPA) characteristics of the open- and closed-ring isomers of stilbene using time-dependent density functional theory. It was detected that the excited states populated by two-photon absorption were nearly 1 eV higher in energy than the lowest energy excited state populated by one-photon absorption. It was found that states excited by OPA had $\pi\pi^*$ character about the C—C framework associated with the bond formation/scission of the central C—C bond (Figure 6.8). In contrast, the states populated by TPA had $\pi\pi^*$ character along the C—C skeletal periphery, including Ph excitations. It was postulated that these differences in excited-state electronic structure may lead to alternative reaction pathways to photoisomerization about the central C—C bond, impacting the utility of these compounds as 3D information storage media.

Two-photon water-soluble dyes and their amine-reactive derivatives (Figure 6.9) for two-photon bioimaging applications were invented [70]. Two series of water-soluble blue fluorescent two-photon-excited (TPE) chromophores for bioimaging based on bis-stilbene structure were synthesized. Wadsworth–Emmons reaction was used to build one-dimensional D-π-aromatic core-π-D structures. The stilbene two-photon absorption cross sections in the range of 150 GM at 700 nm have been measured by

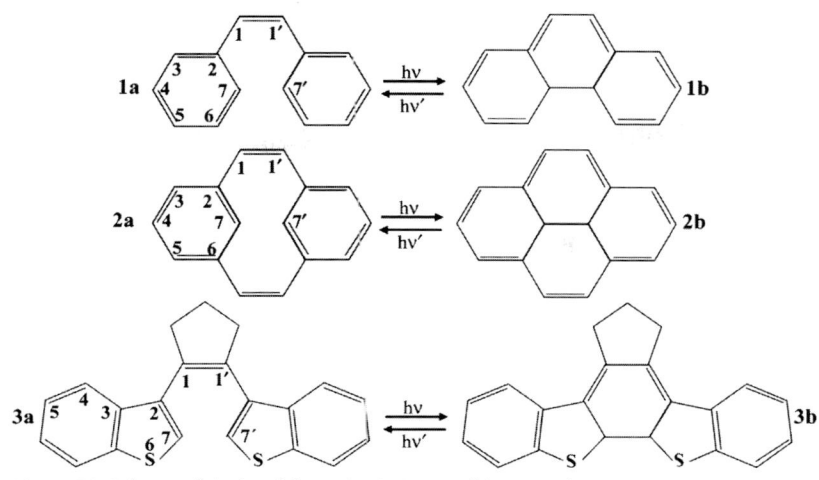

Figure 6.8 Scheme of the bond formation/scission of the central C—C bond [69]. (Reproduced with permission.)

the way of their two-photon-excited fluorescence (TPEF) properties. It was shown that it was 10–100 times higher than commercial dyes commonly used in bioimaging. In authors' opinion, this makes these new TPA chromophores good candidates for *in vivo* two-photon-excited microscopy. Synthesis and characterization of water-soluble two-photon-excited nonionic blue fluorescent chromophore analogues of GMO-4 for bioimaging were reported [70]. These compounds were specifically designed for two-photon absorption microscopy. Water solubility was induced by introducing short oligo(ethylene glycol) monomethyl ether moieties. Two-photon absorption cross sections of **7** in water and in methylene chloride are shown in Figure 6.10. This technology led to low molecular weight dyes with efficient two-photon absorption

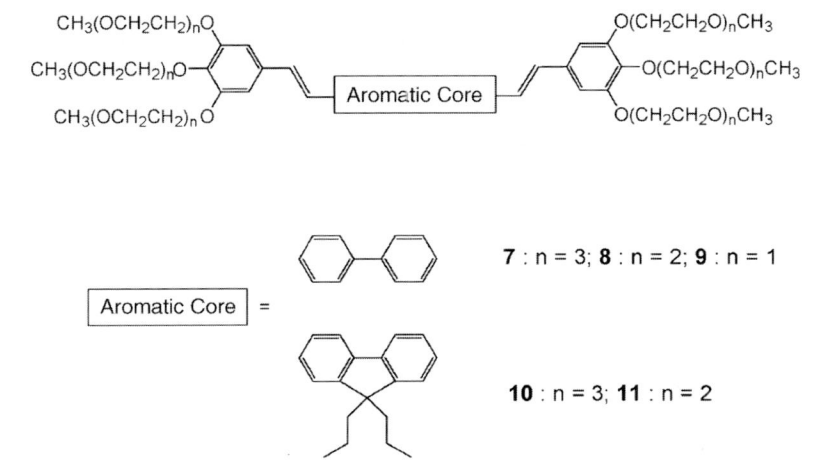

Figure 6.9 Chemical structure of compounds used in Ref. [70]. (Reproduced with permission.)

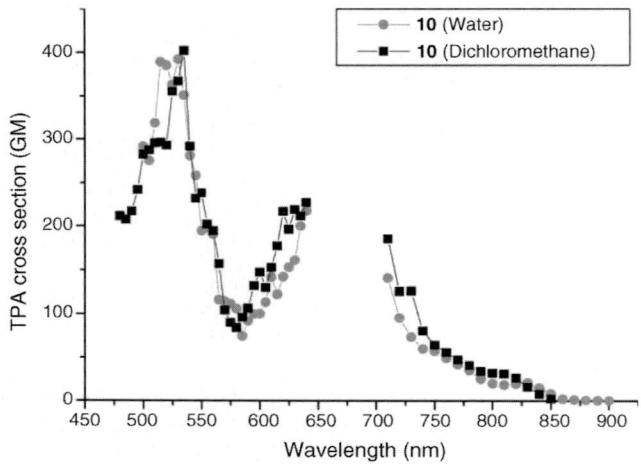

Figure 6.10 Two-photon absorption cross section of **7** in water and in methylene chloride [70]. (Reproduced with permission.)

cross sections and high fluorescence quantum yield both in organic solvents and in aqueous solution.

6.8
Radioluminescence Materials: Scintillators

A scintillator is a substance that absorbs high-energy (ionizing) electromagnetic or charged particle radiation and then, in response, fluoresces photons at a characteristic Stokes-shifted wavelength, releasing the previously absorbed energy. Scintillators are defined by their light output (number of emitted photons per unit absorbed energy), short fluorescence decay times, and optical transparency at wavelengths of their own specific emission energy. Scintillators are used in many physics research applications to detect electromagnetic waves or particles. Common scintillators used for radiation detection include inorganic crystals, organic plastics, and liquids.

A fast collimated neutron flux measurement using stilbene scintillator and flashy analogue-to-digital converter in JT-60U was described [71]. A line-integrated neutron emission profile was routinely measured using the radial neutron collimator system in JT-60U Tokamak. Stilbene neutron detectors (SNDs), which combine a stilbene organic crystal scintillation detector (SD) with an analogue neutron–gamma pulse shape discrimination (PSD) circuit, have been used to measure collimated neutron flux. A digital signal processing system (DSPS) using a flash analogue-to-digital converter (Acqiris DC252, 8 GHz, 10 bits) has been developed at Cyclotron and Radioisotope Center in Tohoku University of China. In this system, anode signals from photomultiplier of the SD were directly stored and digitized. Then, the PSD between neutrons and gamma rays was performed using software. The DSPS has been installed in the vertical neutron collimator system in JT-60U and applied to deuterium experiments. The neutron flux was successfully measured with a rate of 5×10^5 counts/s without any pileup effect of detected pulses. Stilbene crystalline powder in polymer base was prepared as a new fast neutron detector [72]. A new organic scintillation material consisting of stilbene grains in a polymer glue base was presented. The crystal grains of stilbene were obtained by mechanical grinding of stilbene single crystals. The resulting composite scintillators have been used as detectors for fast neutrons.

A study of the spectrum of scintillation amplitude generated by fast neutron in an organic scintillator was reported [73]. The features of the organic scintillator light yield as a function of the ionizing radiation energy at high specific energy losses (dE/dx), taking fast neutrons as an example, have been considered. The dependence values of light yield on neutron energy have been obtained in the experiment for stilbene single crystal and the liquid scintillator (diisopropyl naphthalene + BPO, 5 g/l) under irradiation by ^{239}Pu-Be neutron source. A method of neutron spectra reconstruction based on recoil proton spectra has been described and the spectra obtained for ^{239}Pu-Be source have been presented. Fast response neutron emission monitor for fusion reactor using stilbene scintillator and flash-ADC was demonstrated [74]. The stilbene neutron detector that has been used for neutron emission profile monitoring in JT-60U has been improved to meet the requirement of observing the high-frequency phenomena in megahertz region such as toroidicity-induced Alfven eigenmode in burning plasma as well as the spatial profile and the energy spectrum. To achieve a fast response in the stilbene detector, a flash-ADC was applied and the waveform of the anode signal was stored directly, and neutron/gamma discrimination was carried out with the help of software with a new scheme for data acquisition mode to extend the count rate limit from 1.3×10^5 neutron/s in the past to megahertz region.

Neutron measurements during trace tritium experiments at JET using a stilbene detector were carried out [75]. Neutron flux time evolution and neutron spectra measurements with a stilbene detector were performed during trace tritium experiments at JET Tokamak fusion plasma campaign evaluating fast triton energy for ion cyclotron resonance frequency (ICRF) heated plasma from the neutron spectra. It was demonstrated that with the current detector settings in JET roof laboratory, a stilbene neutron spectrometer could be used for analysis without dead-time correction and pulse shape discrimination for shots with neutron yields up to $Yn < 4 \times 10^{16}$ n/shot. A method of neutron/γ-ray digital pulse shape discrimination (DPSD) with organic scintillators was developed [76]. Neutrons and γ-rays produce light pulses with different shapes when interacting with organic scintillators. This property was used to distinguish between neutrons (n) and γ-rays (γ) in mixed n/γ fields as those encountered in radiation physics experiments. To provide data reprocessing and overcome a limit in count rate capability (typically up to 200 kHz), the performance of a n/γ DPSD system by a combination 12-bit 200 MSamples/s transient recorder card was studied. Three organic scintillators have been studied: stilbene, NE213, and anthracene. The charge comparison method was used to obtain simultaneous n/γ discrimination and pulse height analysis. Based on postexperimental simulations with acquired data, the requirements for fast digitizers to provide DPSD with organic scintillators were analyzed.

Digital pulse shape discrimination in organic scintillators for fusion applications was described [77]. Stilbene and NE213 organic scintillators were used for neutron and γ-ray detection in mixed n/γ fields due to their pulse shape discrimination properties. A system for n/γ digital pulse shape discrimination and simultaneous pulse height using a 12-bit 200 MHz transient recorder was presented. The fusion experiments employing FTU (Frascati Tokamak Upgrade) were performed. A multiparameter multichannel analyzer system for the characterization of mixed neutron–gamma field in the experimental reactor LR-0 was developed [78]. The system contained the scintillator stilbene or NE-213 scintillator. The control logic has been realized with the field programmable gate array. Measurements in a WWER-1000 type reactor pressure vessel dosimetry benchmark in the LR-0 experimental reactor have been performed. A variation of about 7% was observed in the scintillation efficiency of carbon recoils in a stilbene crystal for recoil energies of 30 keV to 1 MeV.

BTI neutron bubble detectors, a stilbene scintillation crystal, a BC501A liquid scintillator, and a silver activation counter have been used for measuring neutron emission from GOL-3 [79]. The results were in agreement with charge-exchange (CX), spectral broadening of the Dα line, and diamagnetic measurements. A neutron intensity distribution measuring apparatus, a radiation detector comprising a scintillator and a photodetector that receives fluorescence from the scintillator was patented [80]. The scintillator emitted fluorescence under irradiation with neutrons. The photodetector produced output signals from the orthogonal X- and Y-axis. The incidence position of the fluorescence was detected by the signal intensity ratio, time difference between signal pulses, and difference in width between the output pulses. Neutrons and other particles were distinguished by a pulse width of ≥ 1 output signal. The scintillator preferably contained stilbene, anthracene, poly(vinyl toluene), or another plastic. New multicomponent composite (including stilbene) scintillators in static and dynamic states were developed [81]. These scintillators comprise inorganic matrices, a set of heavy inorganic and organic scintillating fillers, and activators based on luminescent additives. The characteristics of some of the composite scintillators produced turned out to be nonadditive with respect to the characteristics of their components. The equations obtained allowed the authors to describe the scintillation process for a wide range of multicomponent media in both static and dynamically changing states.

Radioluminescence and radiation effects in metal organic framework (MOF) materials were investigated [82]. Highly fluorescent MOF materials based on stilbene dicarboxylic acid as a linker were synthesized. It was shown that the crystal structure and porosity of the product depended on synthetic conditions and choice of solvent, and a low-density cubic form has been identified by X-ray diffraction. The scintillation properties of these crystals were demonstrated. Bright proton-induced luminescence with large shifts relative to the fluorescence excitation spectra were recorded, peaking near 475 nm. Tolerance to fast proton radiation was evaluated by monitoring this radioluminescence to absorbed doses of several hundred MRAD. The PEGS4 program of EGSnrc code system was used to calculate γ and electron cross sections among stilbene, aluminum, and lead [83]. Response functions of a stilbene detector shielded by lead for monoenergy γ-rays were calculated with database cross sections, and thus a response function matrix was composed. The matrix was tested with several standard γ-ray sources. Light yield nonproportionality of organic and inorganic scintillators exposed to alpha rays of various energy was designed [84]. The influence of entrance surface for radiation state of CsI:Tl, CsI:Na, stilbene, and p-terphenyl scintillators on nonproportionality of light yield to the alpha particle energy has been studied. Formation of a disrupted near-surface layer resulted in increasing microheterogeneity of specific light yield over the crystal surface and depth. Organic and alkali-halide scintillator samples show considerable specific light output variations depending on alpha particle energy. A correlation was observed between the nonproportionality and the pulse height resolution: the less the nonproportionality was, the less was the pulse height resolution value within the range of used alpha particle energies.

6.9
Miscellaneous

Two triphenylamino-substituted stilbene-type chromophores, with different electron acceptors, were synthesized via Vilsmeier–Haack reaction and Knoevenagel condensation [85]. These compounds can be easily incorporated into polymer backbone or sol–gel matrix, and their hyperpolarizabilities were characterized through solvatochromic method. It was shown that triphenylamino-substituted chromophores possessed higher thermal stability and display better transparency in operating wavelength of electronic devices. The chlorine-containing disinfectant packing material patented in [86] consisted of CeO_2 0.01–7, TiO_2 0.01–10, ZnO 0.01–9, stilbene derivate 0.05–7, stilbene triazine 0.01–9, and polymer 60–99. This chlorine-containing disinfectant packing material has a good light ray-shading property, thermal resistance, oxidation resistance, and good antiaging property, and can increase the stability of sterilizing component in a disinfectant, prolong the product storage period for more than 1 year, and maintain the color of packing material. The influence of stilbene whitening agents on paper resistance to aging was investigated [87], which shows itself in reversion of whiteness and deterioration in strength properties. The white color and strength properties of papers whitened optically with stilbene derivatives were reduced rapidly under daylight and/or heat. A method for the detection of four stilbene-type disulfonate and one distyrylbiphenyl-type fluorescent whitening agents (FWAs) in paper materials (napkin and paper tissue) and infant clothes was developed [88]. The analytes were detected by ion-pair chromatography coupled with negative electrospray ionization-tandem mass spectrometry (HPLC–ESI-MS/MS). The method was applied to samples, showing that two stilbene-type disulfonates were predominant FWAs detected in napkin and infant cloth samples.

Linearity, precision, and the limit of detection (LOD) of *trans*-stilbene oxide and other compounds were investigated [89]. The authors investigated the second factor and evaluated two types of commonly available chiral detectors for their possible use in chiral method development and screening: polarimeters and CD detectors. It was shown that *trans*-stilbene oxide worked well across all the detectors examined, showing good linearity, precision, and low detection limits.

Stilbenes of the chemical formula

2a,b

(a) Ar =

(b) Ar =

have been used as the main building blocks for charge-transporting polymers and molecular glasses for optoelectronic applications [90]. Polymers based on 11-(4'-cyano-*trans*-4-stilbenyloxy)undecyl vinyl ether were synthesized by living cationic polymerization (LCP), photoinitiated and thermally initiated cationic polymerization using onium salts [91]. LCP resulted in a polymer of molecular mass (≤ 3600), with a uniformity index (D) of 1.2 displaying a focal conic texture indicative of smectic A (sA) phase with preserved CN group and *trans*-configuration. Polyacetylenes (1-P4) containing different stilbene groups, $-[(CH=C)-Ph-CH=CH-Ph-R]_n-$ (R = OC_mH_{2m+1} ($m = 4$ (P1), 10 (P2), 16 (P3)), or NO_2 (P4)), were synthesized using [Rh(nbd)Cl]$_2$ as a catalyst [92]. Their structures and properties were characterized and evaluated by FT-IR, ^{1}H NMR, ^{13}C NMR, GPC, UV, and PL. The optical limiting and nonlinear optical properties were investigated by using a frequency doubled, Q-switched, mode-locked continuum ns/ps Nd:YAG laser system. Experiments indicated that stilbene pendants endow polyacetylenes with a high thermal stability ($T_d \geq 270\,°C$), novel optical limiting properties, and large third-order nonlinear optical susceptibilities (up to 4.61×10^{-10} esu). The optical limiting mechanism was originated from reverse saturable absorption of molecules. It was found that the polymer with electron-accepted NO_2 moiety exhibited better optical properties than that with electron-donated alkoxy group because of larger π-electron delocalization and dipolar effect. The strong interaction between stilbene pendants and the polyene main chain results in a redshift of fluorescence emitting peak.

The unique photochemical and photophysical properties of stilbenes allow them to serve as effective materials for producing dye and solid lasers, radioluminescence, nonlinear optics, and electro-optic, electrophotographic, and light emitting devices, scintillators, image forming apparatuses, light-emitting diodes, and so on.

References

1 Meier, H., Stalmach, U., Fetten, M., Seus, P., Lehmann, M., and Schnorpfeil, C. (1998) *Journal of Information Recording*, **24**, 47–60.
2 Morrall, J.P., Dalton, G.T., Humphrey, M.G., and Samoc, M. (2008) *Advances in Organometallic Chemistry*, **55**, 61–136.
3 Fukuda, M., Takeshita, K., and Mito, K. (1999) *Japanese Journal of Applied Physics, Part 1: Regular Papers, Short Notes & Review Papers*, **38**, 6347–6350.
4 Rettig, W., Strehmel, B., and Majenz, W. (1993) *Chemical Physics*, **173**, 525–537.
5 Wang, C.-K., Liu, J.-C., Zhao, K., Sun, Y.-P., and Luo, Y. (2007) *Journal of the Optical Society of America B: Optical Physics*, **24**, 2436–2442.
6 Sahare, P.D. and Pattanaik, A. (2007) *Journal of Physics D: Applied Physics*, **40**, 7166–7171.
7 Liu, J.-C., Wang, C.-K., and Gel'mukhanov, F. (2007) *Physical Review A: Atomic, Molecular, and Optical Physics B*, **76**, 053804/1–053804/6.
8 Fukuda, M., Takeshita, K., and Mito, K. (1999) *Japanese Journal of Applied Physics, Part 1: Regular Papers, Short Notes & Review Papers*, **38**, 6347–6350.
9 Mito, K. and Fukuda, M. (2000) *MCLC S&T, Section B: Nonlinear Optics*, **24**, 151–156.
10 Saraidarov, T., Reisfeld, R., Kazes, M., and Banin, U. (2006) *Optics Letters*, **31**, 356–358.
11 Tomizawa, T., Itoh, K., Ishii, K., Mito, K., and Sasaki, K. (1995) *MCLC S&T, Section B: Nonlinear Optics*, **14**, 321–326.
12 Tsuzuki, T., Ito, Y., Fujimoto, T., Yamamoto, I., and Taniguchi, A. (2005) Jpn. Kokai Tokkyo Koho (Shinshu University, Japan), pp. 13.
13 Drefahl, D. and Henkel, H.J. (1955) *Die Naturwissenschaften*, **42**, 624–629.
14 Bai, J., Ducharme, S., Leonov, A.G., Lu, L., and Takacs, J.M. (1999) *Proceedings of SPIE*, **3799**, 22–30.
15 Fuks-Janczarek, I., Kityk, I.V., Miedzinski, R., Gondek, E., Ebothe, J., Nzoghe-Mendome, L., and Danel, A. (2007) *Journal of Materials Science: Materials in Electronics*, **18**, 519–526.
16 Fuks-Janczarek, I., Miedzinski, R., Gondek, E., Szlachcic, P., and Kityk, I.V. (2008) *Journal of Materials Science: Materials in Electronics*, **19**, 434–441.
17 Wei, Y. and Chen, C.-T. (2007) *Journal of the American Chemical Society*, **129**, 7478–7479.
18 Gao, F., Du, G., Zhang, P., and Xing, R. (2006) *Guangzi Xuebao*, **35**, 646–649.
19 Xu, L., Li, X., and Gao, Y. (2005) Faming Zhuanli Shenqing Gongkai Shuomingshu, pp. 25.
20 Chen, C.-H., Lin, J.T., and Yeh, M.-C.P. (2006) *Tetrahedron*, **62**, 8564–8570.
21 Ikegami, T., Shimada, T., and Suzuki, Y. (2008) Jpn. Kokai Tokkyo Koho, pp. 57.
22 Tamura, H., Suzuki, T., Nagai, K., Horiuchi, T., and Lee, H.G. (2008) Jpn. Kokai Tokkyo Koho, pp. 43.
23 Takemoto, H. and Saito, S. (2008) *US Patent Application Publication*, pp. 23.
24 Obata, T. and Kondoh, A. (2007) *US Patent Application Publication*, pp. 30.
25 Okada, H. (2007) Jpn. Kokai Tokkyo Koho, pp. 15.
26 Shishido, M. and Okawa, S. (2007) *US Patent Application Publication*, pp. 20.
27 Obata, T. and Kondoh, A. (2007) *US Patent Application Publication*, pp. 30.
28 Azuma, J. (2005) Jpn. Kokai Tokkyo Koho, pp. 34.

29 Sato, K., Mizutani, K., and Yamanaka, T. (2007) Jpn. Kokai Tokkyo Koho, pp. 35.
30 Sugai, A. and Inagaki, Y. (2005) Jpn. Kokai Tokkyo Koho, pp. 20.
31 Tanaka, Y., Shimada, T., Ikegami, T., and Suzuki, Y. (2005) Jpn. Kokai Tokkyo Koho, pp. 53.
32 Takei, A., Abe, K., Okubo, M., and Anzai, M. (2005) Jpn. Kokai Tokkyo Koho, pp. 11.
33 Ichiguchi, T. (2005) Jpn. Kokai Tokkyo Koho, pp. 20.
34 Li, J., Sano, T., Hirayama, Y., Tomita, T., Fujii, H., and Wakisaka, K. (2006) *Japanese Journal of Applied Physics, Part 1: Regular Papers, Brief Communications & Review Papers, 4B*, **45**, 3746–3749.
35 Jeon, S.-O., Jeon, Y.-M., Kim, J.-W., Lee, C.-W., and Gong, M.-S. (2007) *Synthetic Metals*, **157**, 558–563.
36 Jeon, S.-O., Jeon, Y.-M., Kim, J.-W., Lee, C.-W., and Gong, M.-S. (2008) *Organic Electronics*, **9**, 522–532.
37 Suzuki, T., Doi, S., and Noguchi, K. (2003) Jpn. Kokai Tokkyo Koho, pp. 16.
38 Ye, M., Xu, L., Ji, L., Liu, L., and Wang, W. (2006) *Journal of Nonlinear Optical Physics & Materials*, **15**, 275–285.
39 Huang, Z., Wang, X., Li, B., Lv, C., Xu, J., Jiang, W., Tao, X., Qian, S., Chui, Y., and Yang, P. (2007) *Optical Materials (Amsterdam, Netherlands)*, **29**, 1084–1090.
40 Liu, J.-C., Ke, Z., Song, Y.-Z., and Wang, C.-K. (2006) *Wuli Xuebao*, **55**, 1803–1808.
41 Salek, P., Aagren, H., Baev, A., and Prasad, P.N. (2005) *Journal of Physical Chemistry A*, **109**, 11037–11042.
42 Park, G., Jung, W.S., and Ra, C.S. (2004) *Bulletin of the Korean Chemical Society*, **25**, 1427–1429.
43 Jha, P.C., Anusooya, P.Y., and Ramasesha, S. (2005) *Molecular Physics*, **103**, 1859–1873.
44 Morrall, J.P.L., Cifuentes, M.P., Humphrey, M.G., Kellens, R., Robijns, E., Asselberghs, I., Clays, K., Persoons, A., Samoc, M., and Willis, A.C. (2006) *Inorganica Chimica Acta*, **359**, 998–1005.
45 Maslyanitsyn, I.A., Shigorin, V.D., Todorova, L., Marinov, Y., and Petrov, A.G. (2004) *Molecular Crystals and Liquid Crystals*, **408**, 71–81.
46 Ren, A.-M., Feng, J.-K., and Liu, X.-J. (2004) *Chinese Journal of Chemistry*, **22**, 243–251.
47 Lin, T.-C., He, G.S., Prasad, P.N., and Tan, L.-S. (2004) *Journal of Materials Chemistry*, **14**, 982–991.
48 Kityk, I.V., Makowska-Janusik, M., Gondek, E., Krzeminska, L., Danel, A., Plucinski, K.J., Benet, S., and Sahraoui, B. (2004) *Journal of Physics: Condensed Matter Process*, **16**, 231–239.
49 Yan, Y.-X., Wang, D., Zhao, X., Tao, X.-T., and Jiang, M.-H. (2003) *Chinese Journal of Chemistry*, **21**, 626–629.
50 Hurst, S.K., Lucas, N.T., Humphrey, M.G., Isoshima, T., Wostyn, K., Asselberghs, I., Clays, K., Persoons, A., Samoc, M., and Luther-Davies, B. (2003) *Inorganica Chimica Acta*, **350**, 62–76.
51 He, G.S., Lin, Tzu.-C., Chung, S.-J., Zheng, Q., Lu, C., Cui, Y., and Prasad, P.N. (2005) *Journal of the Optical Society of America B: Optical Physics*, **22**, 2219–2228.
52 Liu, J.-C., Felicissimo, V.C., Guimaraes, F.F., Wang, C.-K., and Gel'mukhanov, F. (2008) *Journal of Physics B: Atomic, Molecular and Optical Physics*, **41**, 074016/1–074016/11.
53 Svetlichnyi, V.A. and Meshalkin, Y.P. (2007) *Proceedings of SPIE*, **6727**, 67271E/1–67271E/11.
54 Chen, L., Zhong, Q., Cui, Y., Qian, G., and Wang, M. (2008) *Dyes and Pigments*, **76**, 195–201.
55 Zhang, T.-L., Ma, J.-J., and Xiong, J.-J. (2006) *Ganguang Kexue Yu Guang Huaxue*, **24**, 133–139.
56 Egawa, M. and Seo, S. (2007) US Patent Appl. Publ., pp. 39.
57 Park, S., Kwon, S., and An, B. (2006) PCT Int. Appl., pp. 23.
58 Guang, S., Yin, S., Xu, H., Zhu, W., Gao, Y., and Song, Y. (2007) *Dyes and Pigments*, **73**, 285–291.
59 Zheng, J., Xu, Z., Feng, X., and Zhan, C. (2005) *Gaofenzi Tongbao*, (3), 82–88.

60 Wu, S., Yao, S., Tang, T., Zeng, F., and Zhu, H. (2003) Faming Zhuanli Shenqing Gongkai Shuomingshu, pp. 8.

61 Egawa, M., Osaka, H., Kawakami, S., Ohsawa, N., Seo, S., and Nomura, R. (2008) US Patent Appl. Publ., pp. 82.

62 Egawa, M. (2008) US Patent Appl. Publ., pp. 54.

63 Egawa, M. (2008) Faming Zhuanli Shenqing Gongkai Shuomingshu, 71 pp.

64 Guang, S., Yin, S., Xu, H., Zhu, W., Gao, Y., and Song, Y. (2006) *Dyes and Pigments*, **73**, 285–291.

65 Tomoyuki, T.C. (2008) US Patent Appl. Publ., pp. 73.

66 Bhawalkar, J.D., Kumar, N.D., Swiatkiewicz, J., and Prasad, P.N. (1998) *MCLC S&T, Section B: Nonlinear Optics*, **19**, 249–257.

67 Ilieva, D., Nedelchev, L., Petrova, T., Dragostinova, V., Todorov, T., Nikolova, L., and Ramanujam, P.S. (2006) *Journal of Optics A: Pure and Applied Optics*, **8**, 221–224.

68 Cai, J., Shen, Z., Jiang, B., Tang, H., Xing, H., Huang, W., Xia, A., Guo, F., and Zhang, Q. (2005) *Guangxue Xuebao*, **25**, 1401–1405.

69 Clark, A.E. (2006) *Journal of Physical Chemistry A*, **110**, 3790–3796.

70 Hayek, A., Bolze, F., Nicoud, J.-F., Duperray, A., Grichine, A., Baldeck, P.L., and Vial, J.-C. (2006) *Photochemical and Photobiological Sciences*, **5**, 102–106.

71 Ishikawa, M., Itoga, T., Okuji, T., Nakhostin, M., Shinohara, K., Hayashi, T., Sukegawa, A., Baba, M., and Nishitani, T. (2006) *Review of Scientific Instruments*, **77**, 10E706/1–10E706/3.

72 Budakovsky, S.V., Galunov, N.Z., Grinyov, B.V., Karavaeva, N.L., Kim, J.K., Kim, Y.-K., Pogorelova, N.V., and Tarasenko, O.A. (2007) *Radiation Measurements*, **42**, 565–568.

73 Martynenko, E.V. (2007) *National Functional Materials*, **14**, 238–242.

74 Itoga, T., Ishikawa, M., Baba, M., Okuji, T., Oishi, T., Nakhostin, M., and Nishitani, T. (2007) *Radiation Protection Dosimetry*, **126**, 380–383.

75 Kaschuck, Yu.A., Popovichev, S., Trykov, L.A., Bertalot, L., Oleynikov, A.A., Murari, A., and Krasilnikov, A.V. (2004) European Fusion Development Agreement (Conference Papers), EFDA-JET-CP (EFDA-JET-CP(04)03-(19-39). EFDA-JET Papers presented at the 31st EPS Conference, 2004, 2, pp. 9/1–9/8.

76 Kaschuck, Y. and Esposito, B. (2005) *Nuclear Instruments & Methods in Physics Research, Section A: Accelerators, Spectrometers, Detectors, and Associated Equipment*, **551**, 420–428.

77 Esposito, B., Kaschuck, Y., Rizzo, A., Bertalot, L., and Pensa, A. (2004) *Nuclear Instruments & Methods in Physics Research, Section A: Accelerators, Spectrometers, Detectors, and Associated Equipment*, **518**, 626–628.

78 Bures, Z., Cvachovec, J., Cvachovec, F., Celeda, P., and Osmera, B. (2003) Reactor dosimetry in the 21st century. Proceedings of the 11th International Symposium on Reactor Dosimetry, Brussels, Belgium, August 18–23, 2002 (ed. J. Wagemans), pp. 194–201.

79 Burdakov, A.V., England, A.C., Kim, C.S., Koidan, V.S., Kwon, M., Postupaev, V.V., Rovenskikh, A.F., and Sulyaev, Yu.S. (2005) *Fusion Science and Technology*, **47**, 333–335.

80 Yamaguchi, H. and Yahagi, Y. (2008) Jpn. Kokai Tokkyo Koho, pp. 11.

81 Vasil'chenko, V.G. and Solov'ev, A.S. (2004) *Instruments and Experimental Techniques (Translation of Pribory i Tekhnika Eksperimenta)*, **47**, 602–610.

82 Doty, F.P., Bauer, C.A., Grant, P.G., Simmons, B.A., Skulan, A.J., and Allendorf, M.D. (2007) *Proceedings of SPIE*, **6707**, 67070F/1–67070F/8.

83 Wang, X.-H., Chen, Y., Guo, H.-P., Mou, Y.-F., An, L., and Zhu, C.-X. (2005) *Yuanzineng Kexue Jishu*, **39**, 202–204.

84 Grinyov, B.V., Tarasov, V.A., Vydaj, Yu.T., Kudin, A.M., Andryushchenko, L.A.,

Kilimchuk, I.V., Ananenko, A.A., and Gordienko, L.S. (2006) *Functional Materials*, **13**, 355–358.

85 Chen, L., Cui, Y., Mei, X., Qian, G., and Wang, M. (2006) *Dyes and Pigments*, **72**, 293–298.

86 Shen, K. (2006) Faming Zhuanli Shenqing Gongkai Shuomingshu, pp. 13.

87 Drzewinska, E. and Wysocka-Robak, A. (2006) *Polish Journal of Chemical Technology*, **8**, 91–94.

88 Chen, H.-C. and Ding, W.-H. (2006) *Journal of Chromatography A*, **1108**, 202–207.

89 Kott, L., Holzheuer, W.B., Wong, M.M., and Webster, G.K. (2007) *Journal of Pharmaceutical and Biomedical Analysis*, **43**, 57–65.

90 Grazulevicius, J.V. (2006) *Polymers for Advanced Technologies*, **17**, 694–696.

91 Hellermark, C. and Gedde, U.W. (1992) *Polymer Bulletin*, **28**, 267–274.

92 Xinyan, S., Hongyao, X., Quanzhen, G., Guang, S., Junyi, Y., Yinglin, S., and Xiangyang, L. (2008) *Journal of Polymer Science Part A: Polymer Chemistry*, **46**, 4529–4541.

7
Bioactive Stilbenes

Many stilbene derivates (stilbenoids) are naturally present in plants. The most widely distributed stilbenoids are resveratrol, combretastatins, and pterostilbene. These agents are derived from terrestrial plants, microorganisms, marine organisms, and animals [1, 2]. This chapter is a brief review on diverse biological activities of stilbenes in cells, organs, and animals.

7.1
Resveratrol

7.1.1
General

Resveratrol (*trans*-3,5,4'-trihydroxystilbene; 3,4',5-stilbenetriol; *trans*-3,5,4'-trihydroxystilbene; (*E*)-5-(*p*-hydroxystyryl)resorcinol)

is a phytoalexin produced naturally by several plants when under attack by pathogens such as bacteria or fungi and is found in the skin of red grapes and is a constituent of red wine. Resveratrol was originally isolated by Takaoka from the roots of *white hellebore* in 1940, and later, in 1963, from the roots of *Japanese knotweed*. Resveratrol was detected in grape, cranberry, and wine samples in concentrations ranging from 1.56 to 1042 nmol/g [3] and can be synthesized ([4] and references therein and Chapter 1).

A number of the resveratrol potential beneficial health effects have been reported in *in vitro* experiments on yeast, worms, and fruit flies, and this compound also has

Stilbenes. Applications in Chemistry, Life Sciences and Materials Science. Gertz Likhtenshtein
Copyright © 2010 WILEY-VCH Verlag GmbH & Co. KGaA, Weinheim
ISBN: 978-3-527-32388-3

positive anticancer effects on the cells of breast, skin, gastric, colon, esophageal, and prostate, and pancreatic cancer and leukemia ([5] and references therein). The "French paradox" that the incidence of coronary heart disease is relatively low in southern France despite high dietary intake of saturated fats is explained by the presence of resveratrol in wine and grapes [6]. Nevertheless, benefits of resveratrol to humans are still unproven. It was reported that resveratrol significantly extends the lifespan of yeasts, worms, and fruit fly [7]. Recently, however, contradictory data on stilbenes' effect on the lifespan were published.

Various aspects of biochemistry and biomedicine of resveratrol were recently reviewed [5, 8–16]. In this chapter, we briefly describe recent publications on the biological activity of resveratrol, its derivatives, and chemical analogues.

7.1.2
Resveratrol Content in Biological Objects

The paper [15] reviewed the resveratrol content in red wine. Levels of resveratrol in grapes and grape products, including wine, varies from region to region and from year to year. Red wine contains on an average 1.9 ± 1.7 mg *trans*-resveratrol/l (8.2 ± 7.5 µM), ranging from nondetectable levels to 14.3 mg/l (62.7 µM) *trans*-resveratrol. Wines made from grapes of the Pinot Noir and St. Laurent varieties showed the highest level of *trans*-resveratrol. Levels of *cis*-resveratrol follow the same trend as *trans*-resveratrol. The average level of *trans*-resveratrol–glucoside (*trans*-piceid) in a red wine may be as much as 29.2 mg/l (128.1 µM), that is, three times that of *trans*-resveratrol. The authors concluded that no region can be said to produce wines with significantly higher level of *trans*-resveratrol than all other regions. A liquid chromatography–mass spectrometry method was described to analyze total resveratrol (including free resveratrol and resveratrol from piceid) in fruit products and wine [17]. Samples were extracted using methanol, enzymatically hydrolyzed, and analyzed using reversed-phase HPLC with chemical ionization (CI) mass spectrometric detection. Following CI, an abundance of protonated molecules was recorded using selected ion monitoring (SIM) of m/z 229. Resveratrol was detected in grape, cranberry, and wine samples. The compound concentration ranged from 1.56 to 1042 nmol/g in Concord grape products and from 8.63 to 24.84 µmol/l in Italian red wine. Concentrations of resveratrol were similar in cranberry and grape juice at 1.07 and 1.56 nmol/g, respectively. The aim of the study [18] was to detect the effect of representative minor components of wine and olive oil on reactive oxygen species (ROS) and eicosanoid synthesis induced by oxLDL-stimulated macrophages. The authors suggested that a synergistic action of polyphenols of olive oil and wine and β-sitosterol of olive oil led to the modulation of the effects of oxLDL on oxidative stress and PGE2 synthesis.

The presence of and relationship between *trans*- and *cis*-resveratrol monomers in the varietal wines from Dalmatia (Croatia), produced according to the Croatian appellation of origin system, were reported [19]. Standard methods of analysis for general wine components were used for a preliminary control of the selected wines. Resveratrol monomers in wine were measured by HPLC. Significant differences were found in the phenolic components even between the wines produced from same grape varieties but from different localities. The mean concentration of free resveratrol monomers was 0.43 mg/l (range 0.11–1.04) in white wines and 2.98 mg/l (range 0.5–8.57) in red wines. Evolution of *trans-* and *cis*-resveratrol content in red grapes (*Vitis vinifera* L. cv Mencia, Albarello, and Merenzao) during ripening was examined [20]. Concentrations of *cis-* and *trans*-resveratrol were analyzed in grapes from three red Galician varieties (Mencia, Albarello, and Merenzao). Free resveratrol was quantified using an HPLC method with fluorescence detection. Grapes were sampled weekly during the last 3 weeks before the optimal stage of maturation and skin, pulp, and seed were separated. The content in skin and seed varied (7–24 mg/l). A rapid and sensitive capillary electrophoretic method for analyzing resveratrol in wine was established [21]. The protocol consists of sample preparation using a C-18 solid-phase extraction cartridge. The limits of detection for *trans-* and *cis*-resveratrol were 0.1 and 0.15 µmol/l, respectively. These procedures were used to analyze the *trans-* and *cis*-resveratrol levels in 26 wines. It was found that the concentration of *trans*-resveratrol ranged from 0.987 to 25.4 µmol/l, whereas the concentration of *cis*-resveratrol was much lower.

Analysis of *trans*-resveratrol in Iranian grape cultivars by LC was reported [22]. *trans*-Resveratrol levels were detected in 147 Iranian grape cultivars by using a modified extraction and gradient HPLC procedure with photodiode array detection. It was found that 41 out of 147 cultivars contained significant levels of *trans*-resveratrol. The detected amounts ranged from 0.98 to 6.25 mg/kg fresh weight with a mean value of 3.59 (white grapes) and 3.08 mg/kg (red grapes), respectively. Resveratrol concentration in muscadine berries, juice, pomace, purees, seeds, and wines was measured [23]. A reversed-phase HPLC method for the detection of astilbin and resveratrol in *Smilax glabra* Roxb. from different sources was developed [24]. Average recoveries of astilbin and resveratrol were found to be 99.6 and 99.2%, respectively. Resveratrol in the fruits of bilberry (*Vaccinium myrtillus* L.), the lowbush "wild" blueberry (*V. angustifolium* Aiton), the rabbiteye blueberry (*V. ashei* Reade), and the highbush blueberry (*V. corymbosum* L.) were measured using assay based on high performance liquid chromatography–tandem mass spectrometry (LC–MS/MS) [25]. Recoveries of resveratrol from blueberries spiked with 1.8, 3.6, or 36 ng/g were 91.5 ± 4.5, 95.6 ± 6.5, and $88.0 \pm 3.6\%$, respectively. The highest levels of *trans*-resveratrol in blueberry and bilberry specimens (140.0 ± 30) pmol/g in highbush blueberries from Michigan and 71.0 ± 15.0 pmol/g in bilberries from Poland were reported. The level of this chemoprotective compounds in these fruits was found to be <10% than reported for grapes. Cooking or heat processing of these berries will contribute to the degradation of resveratrol.

The extraction of resveratrol, piceid, emodin, and other effective components from *Polygonum*, a genus of the buckwheat family *Polygonaceae*, was performed [26]. The yield of effective components was detected by UV spectrometry after separation by using organic solvent extraction. The yield of resveratrol was 5.57%. The antioxidant effect of components was detected by pyrogallol method. The isolation procedure of resveratrol-producing fungus and its accumulating characteristics were described [27]. A fungus was separated from the tissue culture processes of *Polygonum cuspidatum*, and the zymotic extracts were analyzed by TLC and HPLC. Three millimoles per liter of Phe with Mg^{2+} and Zn^{2+} increased the growth of the fungus B-39 and the content of resveratrol. The resveratrol content in the fungus was detected at 8.7 µg/100 ml. A patented invention [28] provided an *in vitro* resveratrol-rich callus tissue of *V. thunbergii* Sieb. et Zucc. The *in vitro* resveratrol-rich callus tissue of *V. thunbergii* was characterized by itcontaining at least about 1000–10 000 mg/kg of dried weight of resveratrol, predominantly in the form of *trans*-resveratrol and/or resveratrol-*O*-glucoside, and ready for harvest or subculture in about 30 days. A method for cultivating the *in vitro* resveratrol-rich callus tissue, for extracting the resveratrol from the *in vitro* resveratrol-rich callus tissue, and for detecting the resveratrol amounts in the *in vitro* resveratrol-rich callus tissue by HPLC was developed.

Ultrasensitive assay for three polyphenols (including resveratrol) and their conjugates in biological fluids using gas chromatography with mass selective detection was developed [29]. The concentration of three polyphenols ((+), catechin, quercetin, and *trans*-resveratrol) in blood serum, plasma and urine, as well as whole blood, have been measured after their oral and intragastric administration, respectively, to humans and rats. The method involved ethyl acetate extraction of 100 µl samples and their derivatization with bis(trimethylsilyl)trifluoroacetamide (BSTFA) followed by gas chromatography analysis on a DB-5 column followed by mass selective detection employing two target ions and one qualifier ion for each compounds. The limits of detection (LOD) and quantitation (LOQ) were found to be 0.01 and 0.1 µg/l, respectively, for all compounds. A method for increasing resveratrol content in peanut kernel by incorporating resveratrol synthase into peanut was invented [30]. The method involved cloning full-length resveratrol synthase gene with PCR method, obtaining Arah1P promoter specifically expressing in peanut cotyledon with PCR method, replacing constitutive promoter CaMV35S of vector pCAMBIA1301 with Arah1P promoter, ligating the resveratrol synthase gene into the vector to obtain plant binary expression vector pHT711, transforming into peanut with gene gun, electroporation, pollen tube pathway, or agrobacterium-mediated method, culturing to obtain transgenic peanut kernel, screening peanut kernel containing resveratrol synthase gene, and testing resveratrol content in the transgenic peanut kernel.

7.1.3
Metabolism and Pharmacokinetics

Pharmacometrics of stilbenes and its relationship with clinical studies were recently reviewed [9]. The aim of the research [31] was to study the sulfation of resveratrol in the human liver and duodenum. A simple and reproducible radiometric assay for resveratrol sulfation was developed. It employed 3′-phosphoadenosine-5′-phosphosulfate-[35S] as the sulfate donor, and the rate of resveratrol sulfation (pmol/min/mg cytosolic protein) were found to be 90 (liver) and 74 (duodenum). Resveratrol sulfotransferase followed Michaelis–Menten kinetics, and K_m (µM) was 0.63 (liver) and 0.50 (duodenum) and V_{max} (pmol/min/mg cytosolic protein) were 125 (liver) and 129 (duodenum), respectively. Resveratrol sulfation was inhibited by the flavonoid quercetin, mefenamic acid, and salicylic acid. Potent inhibition of resveratrol sulfation by quercetin, a flavonoid present in wine, fruits, and vegetables, was suggested. In authors' opinion, quercetin present in the diet may inhibit the sulfation of resveratrol, thus improving its bioavailability.

It was demonstrated that carotenoid cleavage oxygenases from *Novosphingobium aromaticivorans* cleaved the interphenyl α–β double bond of stilbenes that had an oxygen functional group at the 4′ carbon atom (i.e., resveratrol, piceatannol, and rhaponticin) to the corresponding aldehyde products [32]. Labeling studies showed that the double bond cleavage of stilbenes occurred via a monooxygenase reaction mechanism (Figure 7.1).

Metabolism and disposition of resveratrol in rats were investigated [33]. Pharmacokinetics of *trans*-resveratrol in its aglycon (RES(AGL)) and glucuronide (RES(GLU)) forms was studied following intravenous (15 mg/kg) and oral (50 mg/kg) administration of *trans*-resveratrol in a solution of β-cyclodextrin to intact rats. The enterohepatic recirculation of RES(AGL) and RES(GLU) was assessed in a linked-rat model. Multiple plasma and urine samples were collected and concentrations of RES(AGL) and RES(GLU) were determined using an electrospray ionization–liquid chromatography/tandem mass spectrometry method. Results showed that RES(AGL) is bioavailable and undergoes extensive first-pass glucuronidation, and that enterohepatic recirculation significantly contributes to the exposure of RES(AGL) and RES(GLU) in rats. The purpose of the study [34] was to investigate the implications of selected chemopreventive parameters and metabolic conversion of resveratrol *in vivo*. In two 8-week long feeding experiments with rats, a low-resveratrol diet containing 50 mg resveratrol per kg body weight (bw) and day and a high-resveratrol diet with 300 mg/kg bw and day were administered. For chemopreventive evaluation, selected phase I and phase II enzymes of the biotransformation system, the total antioxidant activity, and the vitamin E status of the animals were detected. The level of resveratrol and its metabolites in the feces, urine, plasma, liver, and kidneys was identified and quantitated by high-performance liquid chromatography–diode array detection. The formation of *trans*-resveratrol-3-sulfate, *trans*-resveratrol-4′-sulfate, *trans*-resveratrol-3,5-disulfate, *trans*-resveratrol-3,4′-disulfate, *trans*-resveratrol-3,4′,5-trisulfate, *trans*-resveratrol-3-O-β-D-glucuronide, and resveratrol aglycon was detected by HPLC analysis.

The feasibility of the topical/transdermal delivery of resveratrol was examined [35]. Effects of vehicles on the *in vitro* permeation and skin deposition from saturated solution such as aqueous buffers and soybean oil were investigated. The general trend for the delivery from solution was pH 6 buffer = pH 8 buffer > 10% glycerol formal in pH 6 buffer > pH 9.9 buffer > pH 10.8 buffer > soybean oil. A linear relationship was established between the permeability coefficient (K_p) and drug accumulation in the skin reservoir.

Absorption of *trans*-resveratrol and other related compounds in rats was investigated [36]. Male Wistar rats (350 g) were used in a bioavailability study of [^{3}H]*trans*-resveratrol administered by gavage together with (+)-catechin, quercetin, and unlabeled *trans*-resveratrol in matrices of 10% ethanol, V8 vegetable homogenate mixture, and white grape juice. Whole blood, blood serum, urine, feces, and tissue

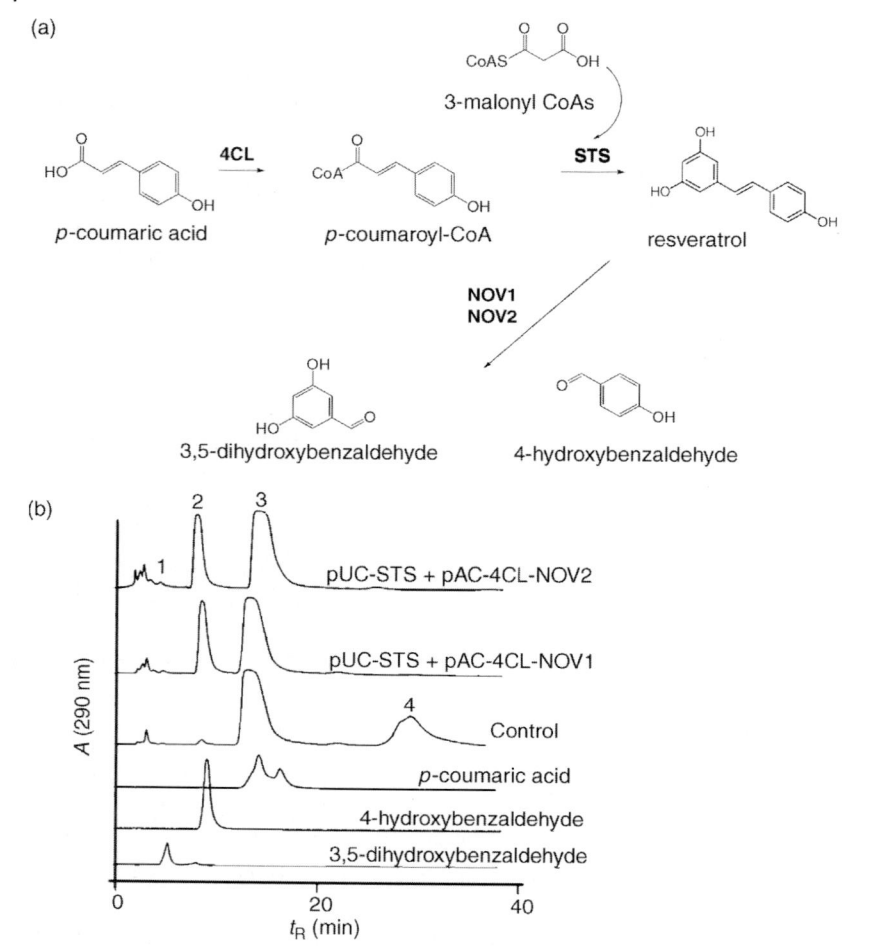

Figure 7.1 *In vivo* cleavage of resveratrol by NOV1 and NOV2. (a) Engineered pathway in *E. coli* for resveratrol biosynthesis from fed coumaric acid and cleavage of synthesized resveratrol in *E. coli* by coexpressed NOV enzymes. The enzymes that are shown are 4-coumaroyl-CoA ligase (4CL; EC 6.2.1.12), stilbene synthase (STS; EC 2.3.1.95), and NOV oxygenases (NOV1, NOV2). (b) HPLC analysis of extracts from coumaric acid-fed recombinant *E. coli* cultures that coexpressed the stilbene biosynthetic genes and NOV1 (pUC-STS + pAC-4CL + NOV1) or NOV2 (pUC-STS + pAC-4CL + NOV2). The control culture contained only stilbene biosynthesis genes (pUC-STS + pAC-4CL). HPLC traces of culture extracts and of authentic standard compounds are shown. Peaks are (1) 3,5-dihydroxybenzaldehyde (m/z 137.0), (2) 4-hydroxybenzaldehyde (m/z 121.0), (3) p-coumaric acid (m/z 163.2), (4) resveratrol (m/z 227.1). Control cultures converted coumaric acid to resveratrol. Addition of the oxygenase enzymes NOV1 and NOV1 resulted in a decrease in resveratrol and the appearance of 4-hydroxybenzaldeyde and 3,5-dihydroxybenaldehyde. The products were confirmed by standards and mass spectral analysis. 3,5-Dihydroxybenzaldehyde was not extracted from the medium in stoichiometric amounts [32]. (Reproduced with permission.)

(liver, kidney, heart, and spleen) levels of *trans*-resveratrol were detected by GC–MS and radioactivity counting. The time profiles of *trans*-resveratrol levels, interactive effects of (+)-catechin and quercetin, and matrix effects were evaluated. Pharmacokinetics of *trans*- and *cis*-resveratrol (3,4′,5-trihydroxystilbene) after red wine oral administration in rats was monitored [37]. Kinetics of *trans*- and *cis*-resveratrol has been evaluated in *rats* after oral administration of red wine. Resveratrol concentration were measured in plasma, heart, liver, and kidneys. Tissue concentration showed a significant cardiac bioavailability and strong affinity for liver and kidneys.

7.1.4
Antioxidant Activity

Five new stilbene oligomers, laetevirenol A(I)–E (4–8), were isolated from *Parthenocissus laetevirens*, together with three known stilbene oligomers (**2**, **3**, and **9**) [38]. The absolute configurations of the new compounds were elucidated by spectroscopic analysis, including 1D and 2D NMR experiments. Biomimetic transformations revealed a possible biogenetic route, where stilbene trimers were enzymatically synthesized for the first time. The oligomers' antioxidant activities were evaluated by 1,1-diphenyl-2-picrylhydrazyl assay. Results showed that stilbene oligomers with an unusual phenanthrene moiety exhibited much stronger antioxidant activities. ESR spectroscopy in combination with 5-(diethoxyphosphoryl)-5-methylpyrroline-*N*-oxide (DEPMPO)-spin trapping technique was used to detect the ability of resveratrol in scavenging superoxide anions generated from both potassium superoxide and the xanthine oxidase/xanthine system [39]. It was demonstrated that the presence of resveratrol resulted in decreased formation of DEPMPO-superoxide adduct (DEPMPO-OOH) in both the potassium superoxide and the xanthine oxidase/xanthine systems, indicating that resveratrol could directly scavenge superoxide anions. The inhibition of DEPMPO-OOH in the xanthine oxidase/xanthine system was found to be much potent compared to that observed in potassium superoxide system. Resveratrol could also directly inhibit xanthine oxidase activity as assessed by oxygen consumption and formation of uric acid. The dual role of resveratrol in directly scavenging superoxide and inhibiting its generation via xanthine oxidase in protection against oxidative injury in various disease processes was stressed.

The study in Ref. [40] examined the effects of phenolic compounds and food sources on cytokine and antioxidant production by A549 cells. The effects of resveratrol, raspberry juice, and quercetin 3′-sulfate on basal and interleukin (IL)-1-stimulated release of IL-8, IL-6, and reduced glutathione (GSH) were examined. Resveratrol at concentration $\geq 50\,\mu mol/ml$ significantly inhibited IL-8 and IL-6 production. Similar findings were made with raspberry juice at concentration $\geq 25\,\mu l/ml$. Molecular mechanisms of oxidative stress resistance induced by resveratrol via specific and progressive induction of MnSOD were investigated [41]. Effects of the long-term exposure to micromolar concentrations of resveratrol on antioxidant and DNA repair enzyme activities in a human cell line (MRC-5) were established. This stilbenoid dramatically and progressively induced mitochondrial MnSOD expression and activity. A two-week exposure to resveratrol increased MnSOD protein level by 6-fold and activity by 14-fold. The long-term exposure of human cells to RES results in a highly specific upregulation of MnSOD and was suggested to be an important mechanism by which it elicits its effects on human cells.

It was reported that resveratrol induces glutathione synthesis by activating Nrf2 and protects against cigarette smoke-mediated oxidative stress in human lung epithelium [42]. Treatment of human primary small airway epithelial and human alveolar epithelial (A549) cells with CS (CSE) dose dependently decreased GSH levels and GCL activity, effects that were associated with enhanced production of reactive oxygen species. Resveratrol restored CSE-depleted GSH levels by upregulation of GCL via activation of Nrf2 and also quenched CSE-induced release of reactive oxygen species. Nrf2 was localized in the cytosol of alveolar and airway epithelial cells due to CSE-mediated posttranslational modifications such as aldehyde/carbonyl adduct formation and nitration. Resveratrol attenuated CSE-mediated Nrf2 modifications, thereby inducing its nuclear translocation associated with GCL gene transcription, as was demonstrated by GCL-promoter reporter and Nrf2 small interfering RNA approaches. The authors stressed that these data may have implications in dietary modulation of antioxidants in the treatment of chronic obstructive pulmonary disease (Figure 7.2).

Comparative studies of the antioxidant effects of a naturally occurring resveratrol analogue – *trans*-3,3′,5,5′-tetrahydroxy-4′-methoxystilbene and resveratrol – against oxidation and nitration of biomolecules in blood platelets were carried out [43]. The action of these two compounds isolated from the bark of *Yucca schidigera* on oxidative/nitrative stress induced by peroxynitrite (ONOO−, which is a strong physiological oxidant and inflammatory mediator) in human blood platelets was compared. The *trans*-3,3′,5,5′-tetrahydroxy-4′-methoxystilbene, like resveratrol, significantly inhibited protein carbonylation and nitration (measured by ELISA) in blood platelets

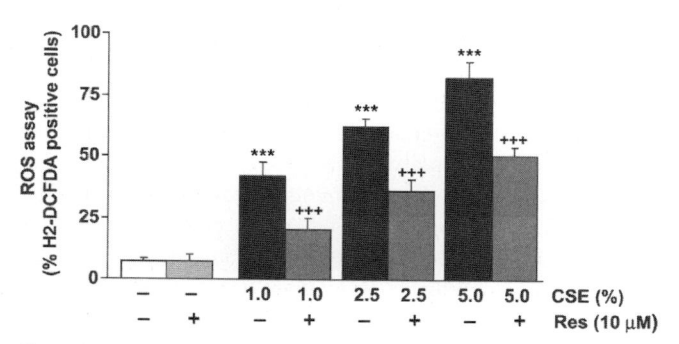

Figure 7.2 (a) Cigarette smoke extract (CSE)-induced reactive oxygen species were quenched by resveratrol (Res). Human alveolar epithelial (A549) cells were treated with 1.0–5.0% CSE with or without 10 μM resveratrol for 24 h, and ROS were measured by flow cytometry using 5- (and 6)-carboxy-2′,7′-dichlorodihydrofluorescein diacetate (H$_2$DCFDA). CSE significantly increased ROS levels in a dose-dependent manner. In cells treated with CSE + resveratrol, ROS production was significantly decreased compared to cells treated with CSE alone. Values are mean ± SE ($n = 3$). ***$P < 0.001$ versus control. +++$P < 0.001$ versus CSE [42]. (Reproduced with permission.)

treated with peroxynitrite (0.1 mM) and markedly reduced the oxidation of thiol groups of proteins or glutathione in these cells. This compound also caused a distinct reduction of platelet lipid peroxidation induced by peroxynitrite. Results obtained indicate that *in vitro* trans-3,3′,5,5′-tetrahydroxy-4′-methoxystilbene and resveratrol have very similar protective effects against peroxynitrite-induced oxidative/nitrative damage to the human platelet proteins and lipids, and *trans*-3,3′,5,5′-tetrahydroxy-4′-methoxystilbene proved to be even more potent than resveratrol in antioxidative tests.

The prooxidant effect of resveratrol and its synthetic analogues (ArOH), that is, 3,4,4′-trihydroxy-*trans*-stilbene (3,4,4′-THS), 3,4,5-trihydroxy-*trans*-stilbene (3,4,5-THS), 3,4-dihydroxy-*trans*-stilbene (3,4-DHS), 4,4′-dihydroxy-*trans*-stilbene (4,4′-DHS), 2,4-dihydroxy-*trans*-stilbene (2,4-DHS), 3,5-dihydroxy-*trans*-stilbene (3,5-DHS), and 3,4′,5-trimethoxy-*trans*-stilbene (3,5,4′-TMS), on supercoiled *p*BR322 plasmid DNA strand breakage and calf thymus DNA damage in the presence of Cu(II) ions has been studied [44]. It was found that the compounds bearing *ortho*-dihydroxyl groups (3,4-DHS, 3,4,4′-THS, and 3,4,5-THS) or bearing 4-hydroxyl groups (2,4-DHS, 4,4′-DHS, and resveratrol) exhibited remarkably high activity in the DNA damage. Kinetic analysis by UV–vis spectra demonstrated that the formation of ArOH–Cu(II) complexes, and the stabilization of oxidative intermediate derived from ArOH and Cu(II)/Cu(I) redox cycles, might be responsible for the DNA damage. A good correlation between antioxidant and prooxidant activity, as well as cytotoxicity against human leukemia (HL-60 and Jurkat) cell lines has been revealed. An increase in antioxidant enzyme activities following exposure to *trans*-resveratrol, including upregulation of mitochondrial superoxide dismutase, an enzyme in the mouse brain that is capable of reducing both oxidative stress and cell death, was observed [45]. Three separate modes of this stilbenoid delivery were utilized in a high-fat diet and through an osmotic minipump. Resveratrol given in a high-fat diet proved to be effective in elevating antioxidant capacity in the brain resulting in an increase in both MnSOD protein level (140%) and activity (75%). The potential neuroprotective properties of MnSOD have been well established, and it was suggested that a dietary delivery of RES may be able to increase the expression and activity of this enzyme *in vivo*. Antioxidative activities of resveratrol and anthocyanins leading to lipid peroxidation in various tissues of experimental animals were evaluated [46]. It was observed that application of resveratrol and anthocyanins to rats by means of gastric tube for 3 weeks protected the tissues from lipid peroxidation. Resveratrol antioxidative activity was found to be stronger than that of anthocyanins.

7.1.5
Resveratrol and Apoptosis

To explore the contribution of resveratrol to the proliferation and apoptosis of lens epithelial cells, the thiazolyl tetrazolium (MTT) assay was conducted [47]. The stilbene effects on the proliferation of lens epithelial cells cultured with different concentrations of this compound were measured. HE staining, transmission electron microscopy (TEM), TUNEL fluorescence staining, and the annexin V assay by flow cytometer (FCM) were applied to observe the CEU morphological change to detect the apoptosis induced by resveratrol. It was found that this stilbenoid suppressed the proliferation and induced the apoptosis of lens epithelial cells. The authors suggested that resveratrol might be considered as a possible treatment strategy for after-cataract. The aim of the study [48] was to analyze the protective effect of a diet rich in resveratrol against the dopaminergic neurotoxin 1-methyl-4-phenyl-1,2,3,6-tetrahydropyridine (MPTP)-induced neuronal death. Male mice were kept on a phytoestrogen-free diet, supplemented with, or not with, 50 or 100 mg/kg/day of resveratrol for 1 or 2 weeks, after which MPTP was intraperitoneally injected. It was observed that daily administration of resveratrol prevented MPTP-induced depletion of striatal DA, and maintained striatal tyrosine hydroxylase (TH) protein levels. Mice treated with resveratrol prior to MPTP administration showed more abundant TH-immunopositive neurons than mice given only MPTP, indicating that resveratrol protects nigral neurons from MPTP insults.

Effects of resveratrol on the MCL cell line Jeko-1 using a combination of flow cytometry have been investigated [49]. Western blotting and two-dimensional electrophoresis to identify the molecule involved in the induction of apoptosis and cell growth regulation were used. It was shown that resveratrol induces apoptosis in Jeko-1 cells and modulates several key molecules. By high-resolution 2D-PAGE and nanoreversed-phase high-performance liquid chromatography coupled with tandem mass spectrometry, 32 differentially expressed proteins in response to resveratrol treatment that belong to important cell death-related networks were identified. The author suggested that these findings form the basis for its potential use as a therapeutic agent. Resveratrol treatment of human LY8 follicular lymphoma cells led to an accumulation of LY8 cell in G0/G1 phase and apoptosis [50]. Resveratrol decreased the expression of BCL6 protein, concomitant with increased expression of several BCL6-regulated gene products and reduces Myc expression in LY8 cells. It was suggested that the use of resveratrol to treat aggressive lymphomas with BCL6 and/or MYC translocations may prove useful as an effective therapy.

It was shown that *trans*- and *cis*-stilbene polyphenols induced rapid perinuclear mitochondrial clustering and p53-independent apoptosis in cancer cells but not normal cells [51]. 3,4,5,4'-Tetramethoxy-*trans*-stilbene (MR-4) was found to induce p53 and perinuclear mitochondrial clustering in cancer cells and a methoxy derivative of resveratrol analogue selectively induced activation of the mitochondrial apoptotic pathway in transformed fibroblasts. The study of over 20 *trans*-stilbene derivatives and their *cis*-isomers to explore structure–activity relationship was expanded. Among them, 3,4,5,4'-tetramethoxy-*cis*-stilbene (MC-4), the *cis*-isomer of MR-4, was most potent, with IC_{50} of 20 nM for growth inhibition. MC-4 induced a rapid perinuclear mitochondrial clustering, membrane permeability transition, cytochrome *c* release, and DNA fragmentation. According to authors, these findings suggested that MC-4 and MR-4 may share a common mechanism whereby the perinuclear mitochondrial clustering, rather than p53, p21, or microtubule depolymerization, is critical for their proapoptotic action (Figure 7.3).

Alpha-tocopherol ether-linked acetic acid analogue [2,5,7,8-tetramethyl-2R-(4R,8R-12-trimethyltridecyl)chroman-6-yloxyacetic acid, α-TEA] alone and together with methylseleninic acid (MSA) and *trans*-resveratrol were investigated for the ability to induce apoptosis, DNA synthesis arrest, and cellular differentiation and inhibit colony formation in human MDA-MB-435-F-L breast cancer cells in culture [52]. The three agents alone were effective in inhibiting cell growth by each of the four different assays, and three-way combination treatments synergistically inhibited cell proliferation in each assay in comparison to individual treatments. Combinations of α-TEA, *trans*-resveratrol, and MSA significantly enhanced levels of apoptosis in human breast (MDA-MB-231, MCF7, and T47D) and prostate (LnCaP, PC-3, and DU-145) cancer cell lines as well as in immortalized but nontumorigenic MCF10A cells. Western immunoblotting confirmed the induction of apoptosis in that the three agents induced poly(ADP-ribose) polymerase cleavage. Mechanistic studies showed combination treatments to inhibit cell proliferation via downregulation of cyclin D1 and induced apoptosis via activation of caspases 8 and 9 and downregulation of prosurvival proteins FLIP and survivin. To determine the molecular mechanisms by which resveratrol induces retinoblastoma tumor cell death, after resveratrol treatment, Y79 tumor cell viability was measured using a fluorescence-based assay, and proapoptotic and antiproliferative effects were characterized by Hoechst stain and flow cytometry, respectively [53]. Mitochondrial transmembrane potential (DeltaP-sim) was measured as a function of drug treatment using 5,5′,6,6′-tetrachloro-1,1′,3,3′-tetraethyl-benzamidazolocarbocyanin iodide (JC-1), whereas the release of cytochrome *c* from mitochondria was assayed by immunoblotting and caspase activities were determined by monitoring the cleavage of fluorogenic peptide substrates. Resveratrol induced a dose- and time-dependent decrease in Y79 tumor cell viability and inhibited proliferation by inducing S-phase growth arrest and apoptotic cell death. The release of cytochrome *c* into the cytoplasm, and a substantial increase in the activities of caspase-9 and caspase-3, was detected. In a cell-free system, resveratrol directly induced the depolarization of isolated mitochondria. The authors concluded that obtained data may warrant further exploration as an adjuvant to conventional anticancer therapies for retinoblastoma.

To enhance biological effects of resveratrol, the molecule was modified by introducing additional methoxyl and hydroxyl groups [54]. The resulting novel RV analogues, M5 (3,4′,5-trimethoxy-*trans*-stilbene), M5A (3,3′,4,5′-tetramethoxy-*trans*-stilbene), and M8 (3,3′,4,4′,5,5′-hexahydroxy-*trans*-stilbene), were investigated in HT29 human colon cancer cells. Cytotoxicity was evaluated by clonogenic assays and induction of apoptosis was detected using a specific Hoechst/propidium iodide double staining method. The influence of M8 on the concentration of deoxyribonucleoside triphosphates (dNTPs), products of ribonucleotide reductase (RR), was evaluated by high-performance liquid chromatography. It was shown that M5 and M5A caused a dose-dependent induction of apoptosis and led to changes in the cell cycle distribution. After treatment with M5, growth arrest occurred mainly in the G2-M-phase, whereas incubation with M5A resulted in arrest in the G0–G1 phase of the cell cycle. Incubation of HT29 cells with M8 produced a significant imbalance

200 | 7 Bioactive Stilbenes

of intracellular dNTP pools, being synonymous with the inhibition of RR activity. The dATP pools were abolished, whereas the dCTP and dTTP pools increased. It was suggested that due to these results, the investigated RV analogues deserve further preclinic and *in vivo* testing.

Resveratrol-induced apoptosis in 7,12-dimethylbenz[*a*]anthracene (DMBA)-initiated and 12-*O*-tetradecanoylphorbol-13-acetate (TPA) promoted mouse skin tumors was investigated [55]. The chemopreventive effects of resveratrol in terms of delayed onset of tumorigenesis, cumulative number of tumors, and average number of tumors/mouse were recorded. Resveratrol treatment resulted in the regression of tumors (28%) after withdrawal of the TPA treatment. Induction of apoptosis by resveratrol in DMBA–TPA-induced skin tumors was observed by the appearance of a sub-G1 fraction (30%) using flow cytometry and an increase in the number of apoptotic cells by terminal deoxynucleotidyl transferase-mediated dUTP nick end labeling (TUNEL) assay. Western blot analysis combined with multivariable flow cytometry showed that resveratrol application induces the expression of the p53 and proapoptotic Bax, with concomitant decrease in antiapoptotic protein Bcl-2. The experimental findings demonstrated that resveratrol induces apoptosis through the activation of p53 activity in mouse skin tumors, thereupon suggesting its chemopreventive activity, through the modulation of proteins involved in mitochondrial pathway of apoptosis.

The chemotherapeutic effects and mode of action of resveratrol on K562 (CML) cells was studied [56]. Resveratrol induced apoptosis in K562 cells in a time-dependent manner. The increased annexin V binding, corroborated with an enhanced caspase-3 activity and a rise in the sub-G0/G1 population, was established. Resveratrol treatment also caused suppression of Hsp70 both in mRNA and protein levels. High endogenous levels of Hsp70 have been found to be a deterrent for sensitivity to chemotherapy. The authors concluded that resveratrol significantly downregulated Hsp70 levels through inhibition of HSF1 transcriptional activity and appreciably augmented the proapoptotic effects of 17-allylamino-17-demethoxygeldanamycin. The proapoptotic ability of (Z)-3,5,4′-tri-O-methyl-resveratrol (R3) was investigated *in vitro* on the human lymphoblastoid cell line TK6 and its p53-knockout counterpart (NH32) [57]. In both cell lines, R3 induced the stimulation of caspase-3. The experimental results showed that the proapoptotic effects of R3 against tumor cells were independent of their p53 status.

The mechanism of action of resveratrol in imatinib-sensitive (IM-S) and -resistant (IM-R) CML cell lines was investigated [58]. Resveratrol induced the loss of viability and apoptosis in IM-S and IM-R in a time- and dose-dependent fashion. Inhibition of

Figure 7.3 Structure and activity of *cis*- and *trans*-stilbene polyphenol analogues. Resveratrol (**2.1**) was used as the prototype compound to synthesize *trans*- and *cis*-stilbene analogues. The compounds are grouped according to their cytotoxicity. The IC$_{50}$ for compounds in each group is: Group 1, >100 µM; Group 2, 20–80 µM; Group 3, 2–10 µM; Group 4, 0.1–1 µM; Group 5, 20–50 nM. The compounds are referred by a numbering system with the first digit indicating the group that the compound belongs to and the second digit identifying the compound within the group. Compound 1.1, however, is not a bona fide stilbene due to the presence of a triple bond instead of a double bond that connects the two benzene rings [51]. (Reproduced with permission from Elsevier.)

cell viability was detected for concentration of resveratrol as low as 5 μM, and the IC_{50} values for viability, clonogenic assays, apoptosis, and erythroid differentiation were in the 10–25 μM range. The effect of imatinib and resveratrol was additive in IM-S but not in IM-R clones in which the resveratrol effect was already maximal. Resveratrol action was independent of BCR-ABL expression and phosphorylation, and in agreement was additive to BCR-ABL silencing. The phytoalexin also inhibited the growth of BaF3 cells expressing mutant BCR-ABL proteins found in resistant patients, including the multiresistant T315I mutation. The author suggested that resveratrol should provide therapeutic benefits in IM-R patients and in other hematopoietic malignancies (Figure 7.4).

7.1.6
Biochemical Effect

7.1.6.1 Enzymes

In the study [59], effect of several resveratrol structural analogues on (AMP-activated protein kinase (AMPK) activity in HepG2 cells was analyzed, and combretastatin A-4 (CA-4) was identified as an activator of AMPK detection by its phosphorylation. The AMPK activation was further confirmed by the phosphorylation of downstream acetyl-CoA carboxylase (ACC) and the decrease of upstream ATP level. In addition, it was shown that CA-4 activated the peroxisome proliferator-activated receptor (PPAR) transcriptional activity *in vitro* with the luciferase reporter assay, AMPK and downregulated gluconeogenic enzyme mRNA levels in liver, and improved the fasting blood glucose level in diabetic db/db mice. Inhibition of cholinesterase and amyloid-β aggregation by resveratrol oligomers from *V. amurensis* was investigated [60]. In the course of screening for acetylcholinesterase and butyrylcholinesterase inhibitors from natural products by an *in vitro* Ellman method, the experiments on the roots of *V. amurensis* Rupr. (Vitaceae) showed a significant cholinesterase inhibitory activity. Employing a bioassay-linked HPLC method, followed by a semipreparative HPLC method, two compounds of interest were isolated and characterized as vitisin A and heyneanol A. They inhibited effectively both acetylcholinesterase and butyrylcholinesterase in a dose-dependent manner.

Figure 7.4 Resveratrol (Res) induces loss of cell viability in IM-S and IM-R cells. IM-S (a) and IM-R (b) K562 cells (10^5/ml) were incubated for 48 h (filled bars) or 72 h (open bars) at 37 °C with increasing concentrations of resveratrol in 100 μl of RPMI 1640 medium containing 5% FCS in 96-well plates. Cell viability was measured by the XTT assay. Results are mean ± SD of four different determinations. Error bars = 95% confidence intervals. Resveratrol in the 1–100 μM range or 1 μM imatinib (Ima) was added to IM-S (c) or IM-R (d) CML cell lines growing in semisolid methyl cellulose medium (0.5×10^3 cells/ml). Colonies were detected after 10 days of culture by adding 1 mg/ml of MTT reagent and were scored by Image J quantification software. Results are expressed as the percentage of colony forming cells after drug treatment in comparison with the untreated control cells (CT). Results are mean ± SD of three different determinations. Error bars = 95% confidence intervals. Photographs of IM-S and IM-R cultures treated for 10 days with various concentrations of resveratrol or 1 μM imatinib are also shown [58]. (Reproduced with permission.)

7.1 Resveratrol

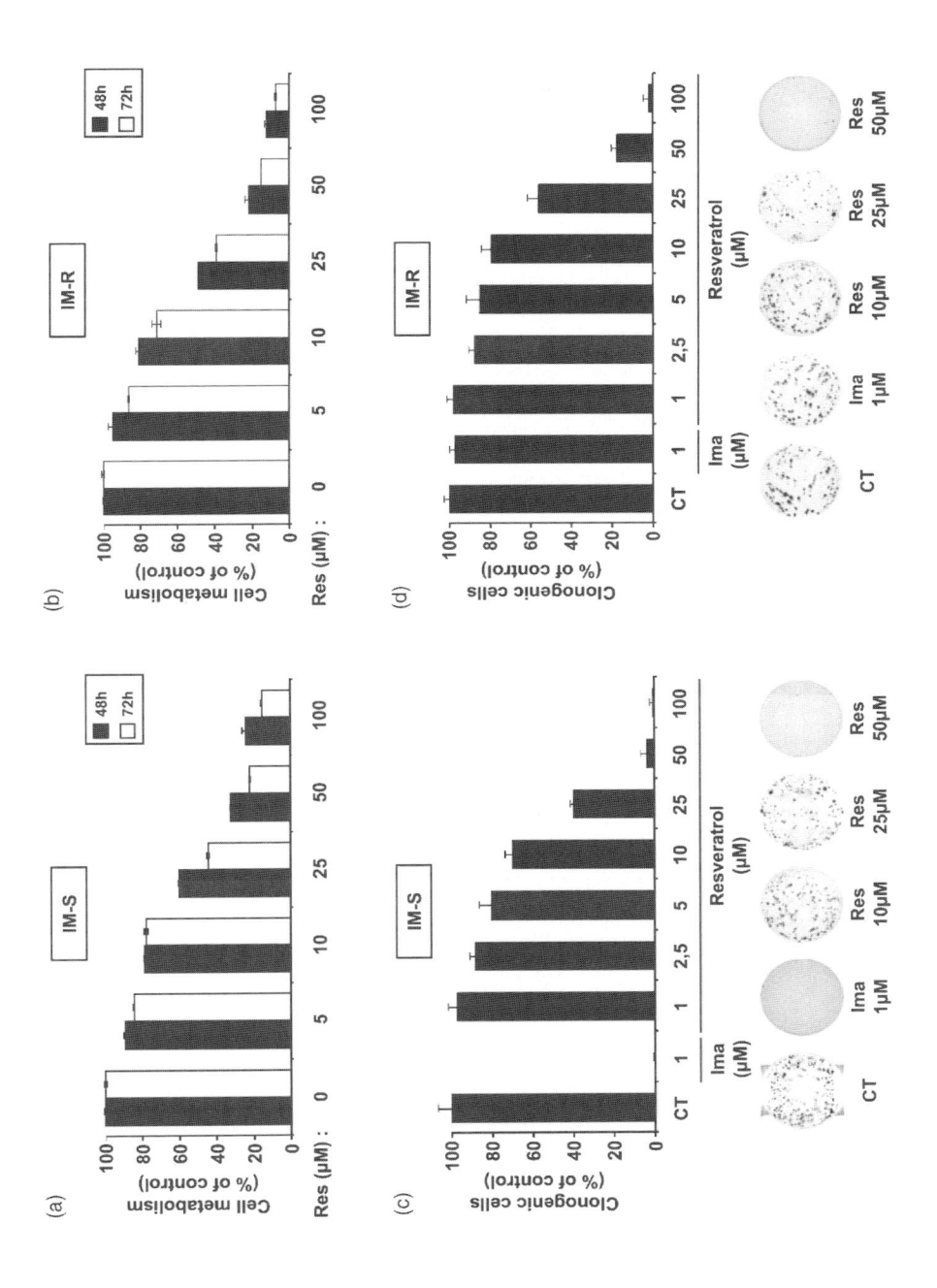

The aim of the study [61] was to investigate the effects of resveratrol, extracted from the Chinese medicinal herb *Polygonum cuspidatum* Sieb et Zucc, on the platelet activation induced by ADP and its possible mechanism. The percentage of platelet aggregation and surface P-selectin-positive platelets and the activity of protein kinase C (PKC) of platelet were observed with platelet aggregometer, flow cytometry, and phosphor imaging system, respectively. Resveratrol showed antiplatelet aggregation and inhibition of surface P-selectin-positive platelets, inhibited the activity of PKC in the membrane fraction of platelets. It was shown that cryptococcal production of both PGE2 and PGF2α can be inhibited by resveratrol, caffeic acid, resveratrol, and nordihydroguaiaretic acid [62]. These polyphenolic molecules acted as inhibitors of lipoxygenase enzymes.

7.1.6.2 Cells and Animals

Modulation of peroxisome proliferator-activated receptor γ stability and transcriptional activity in adipocytes by resveratrol was reported [63]. It was shown that resveratrol modulated PPARγ protein levels in 3T3-L1 adipocytes via inhibition of PPARγ gene expression coupled with increased ubiquitin proteasome-dependent degradation of PPARγ proteins. Resveratrol-mediated decreases in PPARγ expression were associated with the repression of PPARγ transcriptional activity when assayed using a panel of PPARγ target genes in adipocytes. It was also demonstrated that resveratrol inhibits insulin-dependent changes in glucose uptake and glycogen levels and decreases insulin receptor substrate 1 and glucose transporter 4 protein levels, indicating that resveratrol represses insulin sensitivity in adipocytes. An objective of the work [64] was to investigate the resveratrol-inhibited cell proliferation and prolactin (PRL) synthesis induced by 17[β]-estradiol (E2). Pituitary tumor CH3 cell (GH3) were treated with 17[β]-estradiol or a combination of 17[β]-estradiol and RE in serum-free and phenol red-free media. The cell proliferation was assessed by methyl thiazolyl tetrazolium (MTT) assay. GH3 cell proliferation and PRL expression were examined using immunocytochemistry and Western blot analysis. The experiments showed that resveratrol inhibited cell proliferation and PRL synthesis induced by E2, and its inhibition of E2-induced PRL synthesis was more sensitive than its suppression of E2-induced proliferation. Effects of resveratrol on the potassium current in guinea pig ventricular myocytes were evaluated [65]. Enzyme digestion was performed to isolate single ventricular myocyte of guinea pig and whole-cell patch-clamp technique was used to record potassium current. In ventricular myocytes of guinea pig, the slow component of the cardiac delayed rectified potassium current (IKS) was enhanced by resveratrol in a concentration-dependent manner; whereas the rapid component of the cardiac delayed rectified potassium current (IKS) was not changed by resveratrol.

It was shown that resveratrol lowered the levels of secreted and intracellular amyloid-β (Aβ) peptides produced from different cell lines [66]. Resveratrol did not inhibit Aβ production because it has no effect on the Aβ-producing enzymes β- and γ-secretases but promotes intracellular degradation of Aβ via a mechanism that involves the proteasome. The resveratrol-induced decrease of Aβ was shown to be prevented by several selective proteasome inhibitors and by siRNA-directed silencing of the proteasome subunit β5. These findings demonstrated a proteasome-dependent antiamyloidogenic activity of resveratrol. The authors suggested that this natural compound has a therapeutic potential in Alzheimer's disease. It was demonstrated that resveratrol improves mitochondrial function and protects against metabolic disease by activating SIRT1 and PGC-1α [67]. Treatment of mice with resveratrol significantly increased their aerobic capacity, as evident by their increased running time and consumption of oxygen in muscle fibers. Resveratrol effects were presumably associated with the induction of genes for oxidative phosphorylation and mitochondrial biogenesis and were explained by an RSV-mediated decrease in PGC-1α acetylation and an increase in PGC-1α activity. Resveratrol treatment protected mice against diet-induced obesity and insulin resistance (Figure 7.5). The authors suggested that these pharmacological effects of RSV combined with the association of three Sirt1 SNPs and energy homeostasis in Finnish subjects implicated SIRT1 as a key regulator of energy and metabolic homeostasis. Sir2 activation through increased sir2.1 dosage or treatment with the sirtuin activator resveratrol specifically rescuing early neuronal dysfunction phenotypes induced by mutant polyglutamines in transgenic *Caenorhabditis elegans* was reported [68]. These effects depended on daf-16 (Forkhead).

Resveratrol rescued mutant polyglutamine-specific cell death in neuronal cells derived from HdhQ111 knock-in mice. Effects of resveratrol and its derivatives on lipopolysaccharide-induced microglial activation and their structure–activity relationships were studied [69]. The inhibitory effects of 21 resveratrol derivatives on lipopolysaccharide (LPS)-induced nitric oxide (NO) production in microglia and their structure–activity relationships were evaluated. It was found that resveratrol derivative that have 3,5-dimethoxyl groups in the A-ring, such as (E)-4-(3,5-dimethoxystyryl)phenol (pterostilbene, **2**) or have substituted the B-ring of resveratrol with quinolyl, such as (E)-5-[2-(quinolin-4-yl)vinyl]benzene-1,3-diol (**18**) and (E)-4-(3,5-dimethoxystyryl)quinoline (**19**), strongly inhibited NO production. Compounds **2**, **18**, and **19** reduced LPS-induced protein and mRNA expression of inducible NO synthase (iNOS). These compounds also significantly inhibited the production of TNF-α by LPS-activated microglia. According to authors, the potent inhibitory effects of compounds **2**, **18**, and **19** on microglial activation suggested their potential for treatment of neurodegenerative diseases accompanied by microglial activation.

The effects of *trans*-resveratrol (4.38–438 μM/implant) in the vasculogenesis of yolk-sac membranes and its capacity to improve chick embryo growth were studied [70]. High concentration of the stilbene (43.8–438 μM) significantly inhibited early vessel formation, decreasing the percentage of vitelline vessels of 3.5-day embryos by 50% compared to the control. Basic fibroblast growth factor-stimulated vasculogenesis was partially reversed by resveratrol. Treatments with *t*-resveratrol (4.38–43.8 μM/implant) significantly increased the body length of embryos incubated *in vitro* uncoupled from any impairment in the body shape or detectable embryotoxic effect. The authors suggested that resveratrol not only can be useful as a reliable functional nutriment but is also useful for the development of prophylactic and/or therapeutic agent.

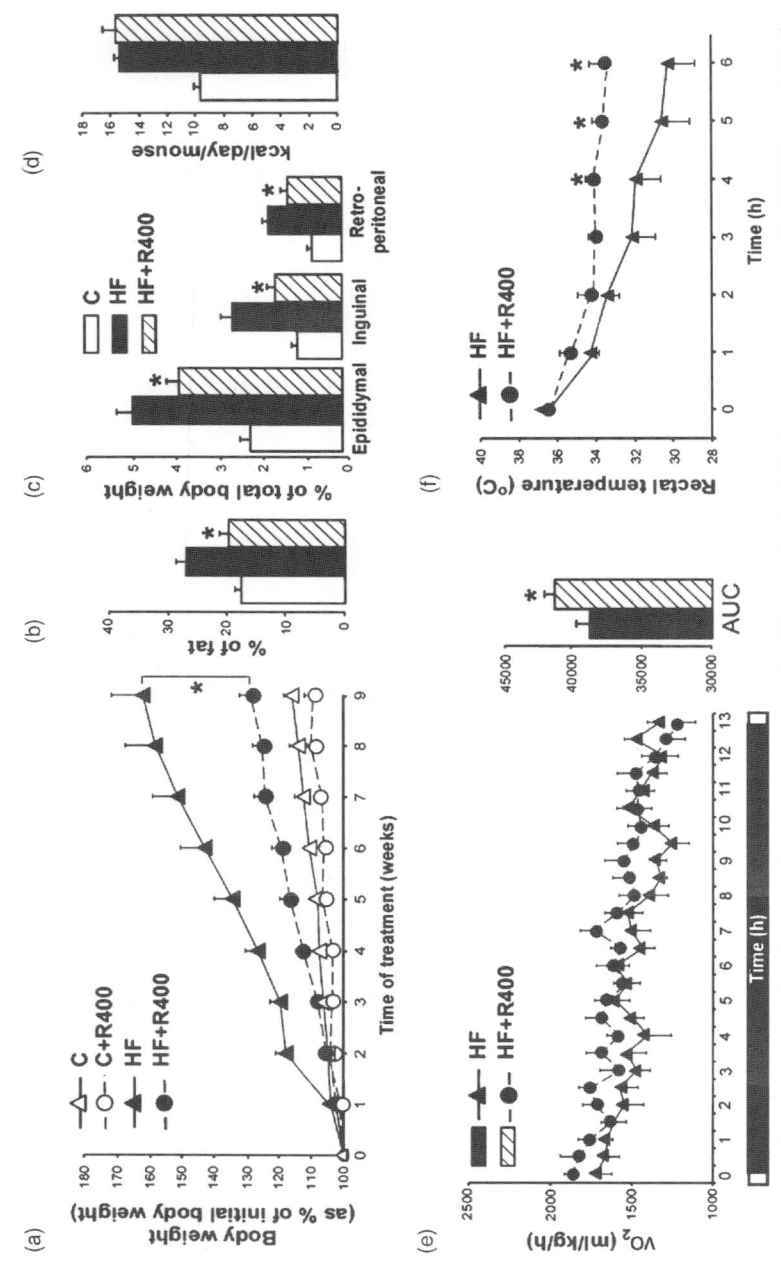

Figure 7.5 RSV prevents diet-induced obesity. 6J mice were fed a chow diet (C) or high-fat diet (HF) alone or supplemented with RSV (400 mpk, R400) for 15 weeks. (a) Evolution of body weight gain expressed as percentage of initial body weight. (b) Body fat content expressed as percentage of total body mass as analyzed by DEXA. (c) Weight of the WAT depots, expressed as percentage of total body weight. (d) Average food intake expressed as kcal/mouse/day. (e) EE as measured by changes in VO$_2$ consumption in indirect calorimetry during 13 h (time 0 is 7 p.m.). The mean areas under the curves (AUCs) are shown in the right side graph ($n = 7$). (f) The evolution of the body temperature during a cold test (4 °C for 6 h). *$P < 0.05$ and $n = 10$ animals/group unless stated otherwise. Values represent means ± SEM [67]. (Reproduced with permission from Elsevier.)

7.1.6.3 Effects on Metabolism of Estrogens

The effects of resveratrol on the metabolism of estrogens in normal breast epithelial cells (MCF-10F) treated with 4-hydroxyestradiol (4-OHE2) or estradiol-3,4-quinone (E2-3,4-Q) was presented in [71]. Ultraperformance liquid chromatography/tandem mass spectrometry was used to detect the effects of resveratrol on estrogen metabolism. Resveratrol induced NQO1 in a dose- and time-dependent manner but did not affect the expression of catechol-O-methyltransferase. Preincubation of the cells with resveratrol for 48 h decreased the formation of depurinating estrogen-DNA adducts from 4-OHE2 or E2-3,4-Q and increased the formation of methoxycatechol estrogens. The authors concluded that resveratrol can protect breast cells from carcinogenic estrogen metabolites, suggesting that it could be used in breast cancer prevention. It was reported that resveratrol acted as a mixed agonist/antagonist for estrogen receptors alpha and beta [72]. It was shown that resveratrol bounded ERβ and ERα with comparable affinity and acted as an estrogen agonist and stimulates ERE-driven reporter gene activity in CHO-K1 cells expressing either ERα or ERβ. The estrogen agonist activity of resveratrol depends on the ERE sequence and the type of ER. Resveratrol-liganded ERβ has higher transcriptional activity than E2-liganded ERβ at a single palindromic ERE. The authors concluded that these data indicate that resveratrol differentially affected the transcriptional activity of ERα and ERβ in an ERE sequence-dependent manner. The objective of the study [73] was to characterize the estrogen-modulatory effects of resveratrol in a variety of *in vitro* and *in vivo* mammary models. The effect of resveratrol alone and in combination with 17β-estradiol (E2) was assessed with MCF-7, T47D, LY2, and S30 mammary cancer cell lines. With cells transfected with reporter gene systems, the activation of estrogen response element-luciferase was studied, and using Western blot analysis, the expression of E2-responsive progesterone receptor (PR) and presnelin 2 protein was monitored. The effect of resveratrol on the formation of preneoplastic lesions (induced by 7,12-dimethylbenz(*a*)anthracene) and PR expression (with or without E2) was evaluated with mammary glands of BALB/c mice placed in organ culture. The effect of orally administered resveratrol on N-methyl-N-nitrosourea-induced mammary tumors was studied in female Sprague-Dawley rats. When resveratrol was combined with E2 (1 nM), a clear dose-dependent antagonism was observed. It was suggested that resveratrol may have beneficial effects if used as a chemopreventive agent for breast cancer.

7.1.6.4 Signaling Pathway

Statuses of Notch1 and Notch2 signaling systems linked with medulloblastoma (MB) formation in three MB human cell lines with and without resveratrol treatment were investigated [74]. Notch1 and Notch2 were detected in the cytoplasm of three cell lines under normal condition, which were upregulated by resveratrol along with differentiation, apoptosis, and enhanced Hes1 nuclear translocation. Results demonstrated that Notch signaling had little relevance to resveratrol-induced differentiation and apoptosis and may not be a universal critical factor for MB cells. It was demonstrated that brief resveratrol pretreatment conferred neuroprotection against cerebral ischemia via SIRT1 activation [75]. This neuroprotective effect produced by resveratrol was similar to the ischemic preconditioning-induced neuroprotection, which protects against lethal ischemic insults in the brain and other organ systems. Inhibition of SIRT1 abolished ischemic preconditioning-induced neuroprotection in CA1 region of the hippocampus. Since resveratrol and ischemic preconditioning-induced neuroprotection require activation of SIRT1, this common *signaling pathway* may provide targeted therapeutic treatment modalities as it relates to stroke and other brain pathologies.

7.1.7
Resveratrol in Genetics

The regulatory effect of resveratrol on the expression of CC-chemokine receptor-5 (CCR5) in human peripheral monocyte/macrophage was studied [76]. Mononuclear cells were obtained from human peripheral blood by the Ficoll-Hypaque density gradient centrifugation method. IFN-χ (1×10^5 U/l) were added into the medium to induce the cells expressing CCR5. Different concentrations of resveratrol (0.5–100 µmol/l) were applied to the cells at the same time. Monocytes/macrophages were collected after culturing for 24 h, the expression level of CCR5 mRNA was deleted by RT-PCR, and CCR5 cell rate was assayed by flow cytometry. CCR5 reporter genes were transfected into each group, the relative luciferase activity of CCR5 reporter gene was tested. The expression of CCR5, rates of CCR5 cells, and relative activity of CCR5 reporter gene significantly decreased in resveratrol treated group compared to control group. Medium and high concentrations of resveratrol inhibit the expression of CCR5 in human peripheral monocyte/macrophage. The effects of the *cis*-isomer of *cis*-resveratrol (c-RESV) on CCR5, whose expression is controlled by nuclear factor kappa B (NF-κB), were investigated [77]. In inflammatory peritoneal macrophages stimulated with lipopolysaccharide and gamma interferon (IFN-γ), *cis*-resveratrol significantly blocked the expression of genes related to the REL/NF-κB/IκB family, adhesion molecules, and acute-phase proteins. The greatest modulatory effect was obtained on the expression of genes related to the proinflammatory cytokines. C-resveratrol downregulated the nuclear factor of kappa light chain gene enhancer in B-cells 1 (NFκBL1) gene product p105 and upregulated the nuclear factor of kappa light chain gene enhancer in B-cells inhibitor alpha (IκBα) gene. The authors concluded that *cis*-resveratrol has a significant modulatory effect on the NF-κB signaling pathway and, consequently, an important antioxidant role that may partially explain the cardioprotective effects attributed to long-term moderate red wine consumption.

The effect of resveratrol on the expressions of matrix metalloproteinase and tissue inhibitors of metalloproteinase in cervical cancer HeLa cells was investigated [78]. The effect of resveratrol on the invasive ability of HeLa cells was observed with a transwell cell culture chamber. Activities of MMP-2 and MMP-9 were analyzed using gelatin zymog. Activities of TIMP-1 and TIMP-2 were analyzed using reverse zymog. mRNA expressions of MMP-2, MMP-9, TIMP-1, and TIMP-2 in HeLa cells were detected by RT-PCR. Their protein expressions were detected by Western blotting. Activities of MMP-2 and MMP-9 were found to be markedly inhibited by resveratrol, and the levels of mRNA and the protein of MMP-2 and MMP-9 were also decreased by resveratrol. Meanwhile the ratios of MMP-2/TIMP-2 and MMP-9/TIMP-1 were reduced. Resveratrol can effectively inhibit the migration and invasion of HeLa cells *in vitro*. The authors suggested that the mechanism of the resveratrol action related to the downregulation of the expressions and the activities of MMP-2 and MMP-9, increased the expressions and the activities of TIMP-1 and TIMP-2, reduced the ratios of MMP-2/TIMP-2 and MMP-9/TIMP-1, and increased the balance between them. The effect of the phytocompound resveratrol on this Promoter I.1-controlled expression of aromatase (cytochrome P 450 (CYP) 19 enzyme) was investigated [79]. Results indicated that resveratrol reduced the estradiol-induced mRNA abundance in SK-BR-3 cells expressing ERα. Luciferase reporter gene assays revealed that resveratrol could also repress the transcriptional control dictated by Promoter I.1. According to the authors' suggestion, the phytochemistry reduced the amount of ERK activated by estradiol, which could be the pathway responsible for Promoter I.1 transactivation and the induced CYP19 expression. Because nuclear factor-κB (NF-κB) plays a key role in cell survival and proliferation of human MM cells, the effect of resveratrol in NF-κB expression by Western blot analysis and immunofluorescence was tested [80]. NF-κB was constitutively active in all human MM cell lines examined, and resveratrol downregulated NF-κB expressions in all cell lines. Resveratrol also down-regulated the expression of NF-κB–regulated gene.

7.1.8
Effect on Aging

The effect of resveratrol on a cell model of human aging was investigated [81]. It was found that resveratrol inhibited expression of replicative senescence marker INK4a in human dermal fibroblasts, and 47 of 19 000 genes. These included genes for growth, cell division, cell signaling, apoptosis, and transcription. Genes involved in Ras and ubiquitin pathways, Ras-GRF1, RAC3, and UBE2D3, were downregulated. These data suggested that resveratrol might alter sirtuin-regulated downstream pathways, rather than sirtuin activity and that resveratrol's actions might cause FOXO recruitment to the nucleus. The authors of the work [7] used the short-lived seasonal fish *Nothobranchius furzeri* with a maximum recorded lifespan of 13 weeks in captivity. Resveratrol was added to the food starting in early adulthood and caused a dose-dependent increase of median and maximum lifespan. In addition, resveratrol delays the age-dependent decay of locomotor activity and cognitive performances and reduces the expression of neurofibrillary degeneration in the brain.

Effects of a low dose of dietary resveratrol that partially mimics caloric restriction and retards aging parameters in mice were investigated [82]. Mice from middle age (14-months) to old age (30-months) were fed with control diet (84 kcal/mouse/week), diet with low-dose *trans*-resveratrol (4.9 mg/kg feed/day), or energy restricted (CR; 63 kcal/mouse/week) diet and their genome-wide transcriptional profiles were evaluated. A transcriptional overlap of CR and resveratrol was found in the heart, skeletal muscle, and brain neocortex. Both dietary interventions inhibited gene expression profiles associated with cardiac and skeletal muscle aging and prevented age-related cardiac dysfunction. Dietary resveratrol also mimicked the effects of CR in insulin-mediated glucose uptake in the muscle. Gene expression profiles suggested that both CR and resveratrol may retard some aspects of animal aging through alterations in chromatin structure and transcription. Resveratrol, at doses readily achieved in humans, meets the definition of dietary compounds that mimic some aspects of CR.

The study [83] investigated whether resveratrol can improve nonalcoholic fatty liver disease (NAFLD) and evaluated the possible mechanism. Rats fed a high-fat diet were treated with resveratrol and the liver histology was observed. Hyperinsulinemic euglycemic clamp was performed to assess insulin sensitivity. Fat accumulation was induced in HepG2 cells, and the cells were treated with RSV. AMP-activated protein kinase phosphorylation levels were detected both in the animal study and cell study. Rats fed a high-fat diet developed abdominal obesity, NAFLD, and insulin resistance, which were markedly improved by 10 week.

7.1.9
Miscellaneous

Evidence that the Ca^{2+}-induced Ca^{2+}-release (CICR) mechanism that was expressed in A7r5 and 16HBE14o-cells was strongly activated by suramin and 4,4′-diisothio-cyanatostilbene-2,2′-disulfonic acid (DIDS) was presented [84]. Effects of the stilbene derivatives DIDS, SITS, and DNDS on intracellular Ca^{2+} release in A7r5 cells is shown in Figure 7.6. Suramin/DIDS-induced Ca^{2+} release was only detected in cells that displayed the CICR mechanism, and cell types that do not express this type of CICR mechanism did not exhibit suramin/DIDS-induced Ca^{2+} release. It was shown that the suramin-stimulated Ca^{2+} release was regulated by Ca^{2+} and CaM in a similar way as the CICR mechanism. The pharmacological characterization of the suramin/DIDS-induced Ca^{2+} release further confirmed its properties as a novel CaM-regulated Ca^{2+}-release mechanism. The authors also investigated the effects of disulfonated stilbene derivatives on IP_3-induced Ca^{2+} release and found, in contrast to the effect on CICR, a strong inhibition by DIDS and 4′-acetoamido-4′-isothiocyanostilbene-2′,2′-disulfonic acid.

Figure 7.6 Effects of the stilbene derivatives DIDS, SITS, and DNDS on intracellular Ca^{2+} release in A7r5 cells. (a) Chemical structure of suramin, DIDS, SITS, and DNDS. (b) Intracellular Ca^{2+} release induced by stilbene derivatives. Permeabilized A7r5 cells were loaded during 45 min in 150 nM $^{45}Ca^{2+}$. From time 0 onward, cells were incubated in efflux medium. The traces illustrate how the $^{45}Ca^{2+}$ content of the stores decreased during the efflux (■) and how this Ca^{2+} content was affected by a 2-min application (arrow) of 100 μM suramin (●), 100 μM DIDS (□), 100 μM SITS (○), or 100 μM DNDS (△). A23 187 was applied to measure the total releasable Ca^{2+} (▲). Results represent the means ± SEM for three wells. (c) Intracellular Ca^{2+} release by suramin (●) and the stilbene derivatives, DIDS (□), SITS (○), and DNDS (△) was plotted as a function of their concentration. Permeabilized A7r5 cells were loaded during 45 min in 150 nM $^{45}Ca^{2+}$. Cells were then incubated in efflux medium, and after 10 min an increased concentration of the stilbene derivative was added to the cells for 2 min. A23 187 (5 μM) was applied to measure the total releasable Ca^{2+}. Results represent the mean ± SEM of three independent experiments each performed in twofold [84]. (Reproduced with permission.)

(a) suramin

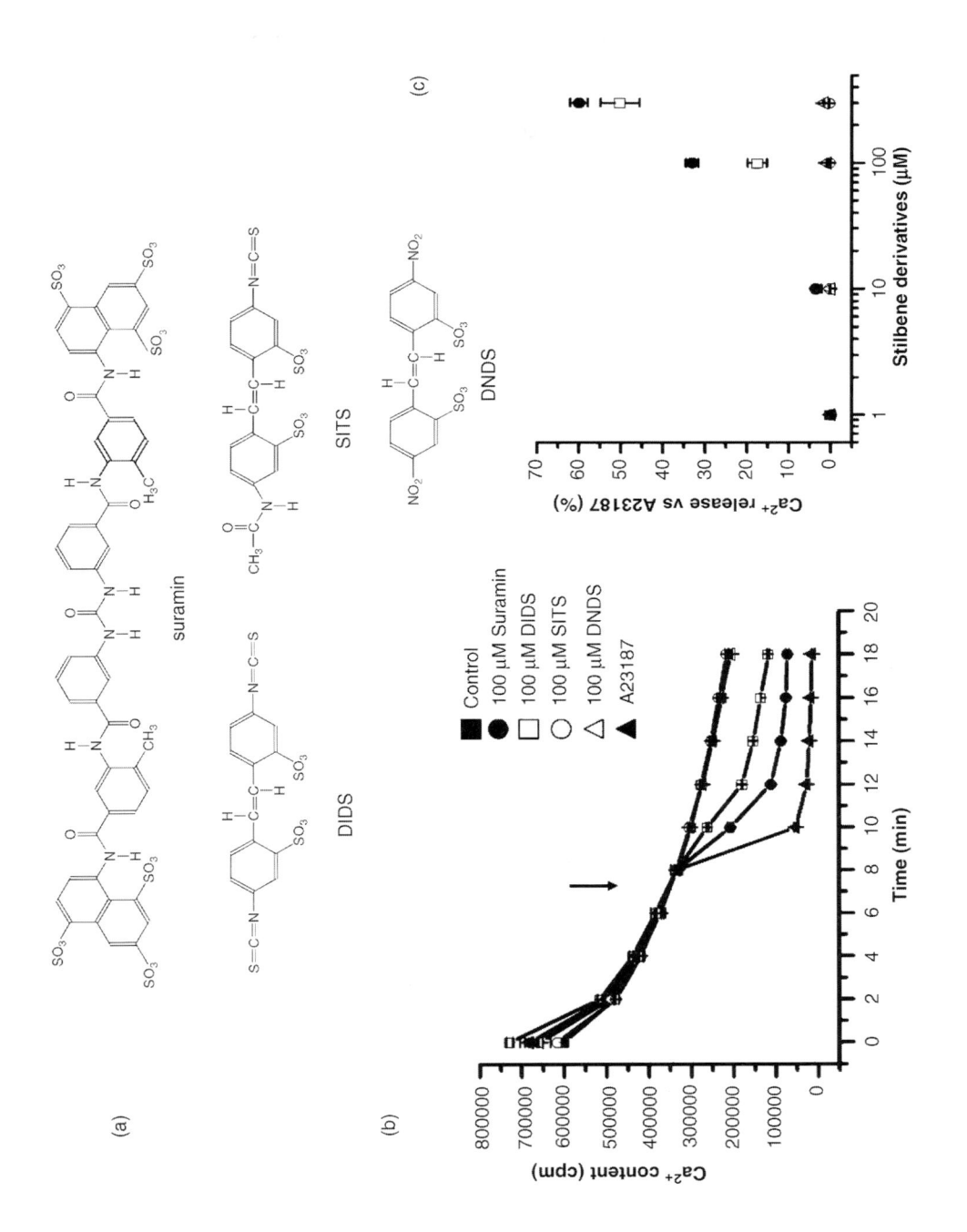

SITS

DIDS

DNDS

(b)
Control
100 μM Suramin
100 μM DIDS
100 μM SITS
100 μM DNDS
A23187

(c)

The synthesis of several novel aza-stilbene derivatives was carried out [85]. The compounds were tested for their c-RAF enzyme inhibition.

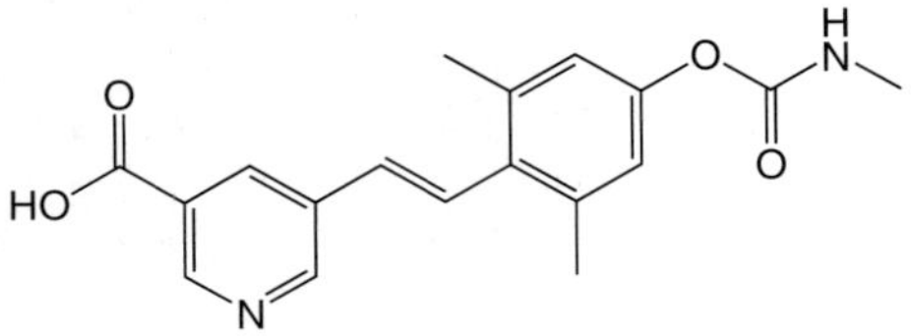

possessed significant potency against c-RAF and demonstrated selectivity over other protein kinases. A hypothesis for the binding mode, activity, and selectivity was proposed (Figure 7.7).

A system for producing "unnatural" flavonoids and stilbenes in *Escherichia coli* was developed [86]. The artificial biosynthetic pathway included a substrate synthesis step for the synthesis of CoA esters from carboxylic acids by 4-coumarate:CoA ligase, a polyketide synthesis step for the conversion of CoA esters into flavanones by chalcone synthase and chalcone isomerase, and into stilbenes by stilbene synthase, and a modification step for modification of flavanones by flavone synthase, flavanone 3β-hydroxylase, and flavonol synthase. Incubation of the recombinant *E. coli* with exogenously supplied carboxylic acids led to the production of 87 polyketides, including 36 unnatural flavonoids and stilbenes. A process for the preparation of resveratrol nanocrystal and its application in beautifying and skin-nursing cosmetics was patented [87]. This invention pertained to a process for producing resveratrol nanocrystal, which comprises treating stearic acid (or lauric acid, palmitic acid, linoleic acid, oleic acid, and synthetic fatty acid) with alcohol under refluxing to obtain 1–8% stearic acid solution. The invention involves the application of resveratrol nanocrystal in beauti-

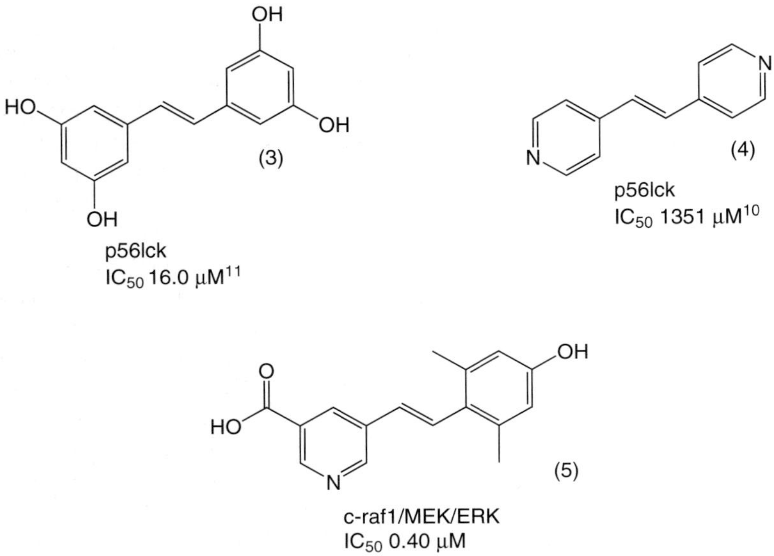

(3)
p56lck
IC_{50} 16.0 μM[11]

(4)
p56lck
IC_{50} 1351 μM[10]

(5)
c-raf1/MEK/ERK
IC_{50} 0.40 μM

Figure 7.7 Related stilbene derivatives that inhibit kinase activity [85]. (Reproduced with permission from Elsevier.)

fying and skin-nursing cosmetics for resisting skin crease, aging, regulating immunity, inhibiting bacteria, diminishing inflammation, preventing allergy, desensitizing, resisting radiation, and withstanding full-wave UV ray. The authors claimed that the invention provided a process for producing the resveratrol nanocrystal with simplified procedure, good stability and water solubility, high utilization rate and bioavailability, small dosage of raw materials, low cost, and convenient use.

Effects of resveratrol on cigarette smoke-induced vascular oxidative stress and inflammation, which in authors' opinion is a clinically highly relevant model of accelerated vascular aging, were elucidated [88]. It was demonstrated that smoking and *in vitro* treatment with cigarette smoke experiments (CSEs) increased reactive oxygen species production in rat arteries and cultured coronary arterial endothelial cells (CAECs), respectively, which was attenuated by resveratrol treatment. The smoking-induced upregulation of inflammatory markers (ICAM-1, inducible nitric oxide synthase, IL-6, and TNF-α) in rat arteries was abrogated by resveratrol treatment. Resveratrol also inhibited CSE-induced NF-κB activation and inflammatory gene expression in CAECs. In CAECs, the aforementioned protective effects of resveratrol were abolished by knockdown of SIRT1, whereas the overexpression of SIRT1 mimicked the effects of resveratrol. Resveratrol treatment of rats protected aortic endothelial cells against cigarette smoking-induced apoptotic cell death and exerted antiapoptotic effects in CSE-treated CAECs, which could be abrogated by knockdown of SIRT1. Resveratrol treatment also attenuated CSE-induced DNA damage in CAECs. According to the authors, resveratrol and SIRT1 exerted antioxidant, anti-inflammatory, and antiapoptotic effects, which protect the endothelial cells against the adverse effects of cigarette smoking-induced oxidative stress. This stilbenoid can contribute to its antiaging action in mammals and may be beneficial in pathophysiological conditions associated with accelerated vascular aging (Figure 7.8).

7.2
Combretastatin and Its Analogues

Combretastatins (1-(3,4,5-trimethoxyphenyl)-2-(3′-hydroxy-4′-methoxyphenyl)ethane, 3,4,5-trimethoxy-3′-hydroxy-4′-methoxystilbene)

are a class of natural stilbenoid phenols. Molecules of the combretastatin family generally share three common structural features: a trimethoxy "A"-ring, a "B"-ring containing substituents often at C3′ and C4′, and an ethene bridge between the two rings that provides necessary structural rigidity. Several reviews on biological activity of combretastatin and its analogues were recently published [89–99].

In this chapter, we concentrate on recent results in this rapidly developing area.

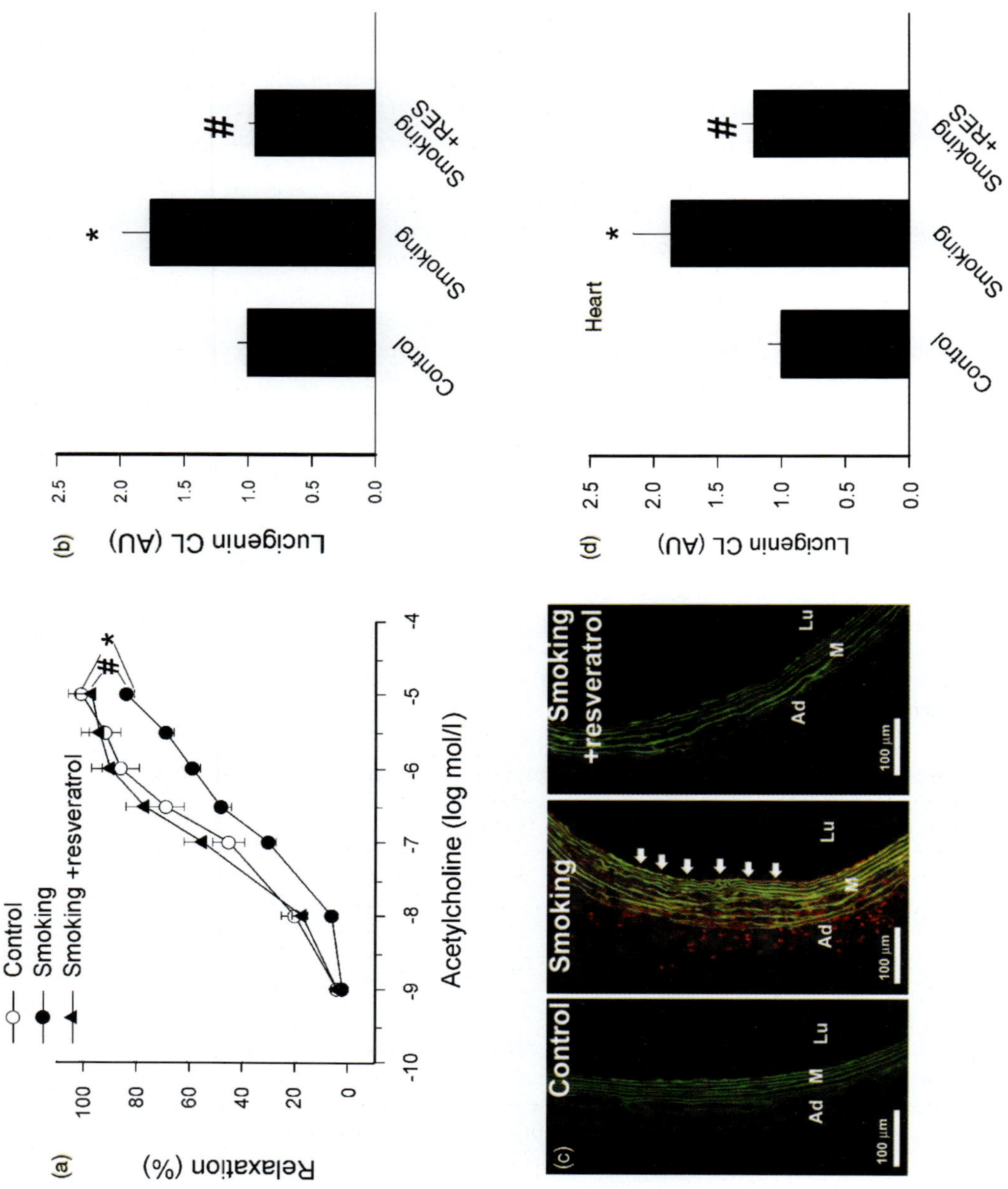
(a)
Control
Smoking
Smoking +resveratrol
Relaxation (%)
100
80
60
40
20
0
-10
-9
-8
-7
-6
-5
-4
Acetylcholine (log mol/l)
#
*
(b)
Lucigenin CL (AU)
2.5
2.0
1.5
1.0
0.5
0.0
Control
Smoking
Smoking +RES
*
#
(c)
Control
Smoking
Smoking +resveratrol
Ad
M
Lu
100 μm
(d)
Heart
Lucigenin CL (AU)
2.5
2.0
1.5
1.0
0.5
0.0
Control
Smoking
Smoking +RES
*
#

7.2.1
Effect on Tubulin Polymerization

It was shown [100] that combretastatin A-4 (I), a potent inhibitor of tubulin polymerization (II), possessed growth inhibitory properties.

A series of aryl- and aroyl-substituted chalcone analogues of the tubulin binding agent combretastatin A-4 (1) were prepared, using a one-pot palladium-mediated hydrostannylation-coupling reaction sequence [101]. These chalcones were converted to indanones by Nazarov cyclization, followed by oxidation to give the corresponding indenones. Indenones were also prepared using a palladium-mediated formal [3 + 2]-cycloaddition process between *ortho*-halobenzaldehydes and diarylpropynones. All compounds were assessed as inhibitors of tubulin polymerization, but only I had activity similar to that of 1.

Fourteen N-acetylated and nonacetylated 3,4,5-tri- or 2,5-dimethoxypyrazoline analogues with the same substituents as CA-4 was the most active compound in the series [102]. A cell-based assay indicated that compound I

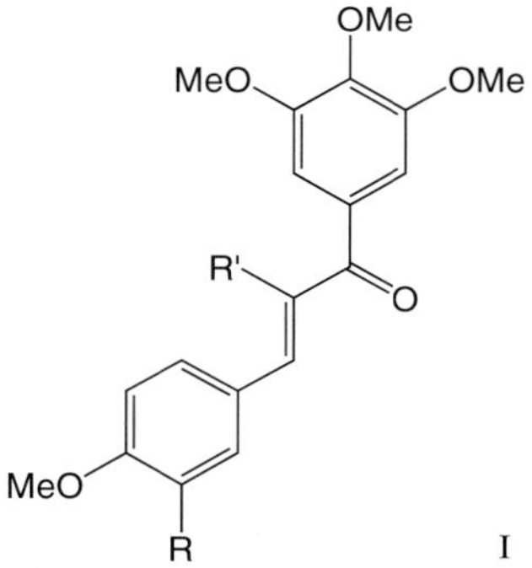

caused extensive microtubule depolymerization with an EC_{50} value of 7.1 μM in A-10 cells.

Figure 7.8 (a) Effects of resveratrol treatment on relaxation responses to acetylcholine in ring preparations of carotid arteries of control rats and rats exposed to cigarette smoke. Data are mean ± SE; $n = 6$ animals for each group. (b) Effects of Res treatment on $O_2^{\bullet-}$ generation in vessels of control rats and rats exposed to cigarette smoke. $O_2^{\bullet-}$ generation was determined by the lucigenin (5 μmol/l) chemiluminescence (CL) method. Data are normalized to the mean value of the untreated control group. Data are mean ± SE. (c) Representative fluorescent photomicrographs showing increased nuclear ethidium bromide (EB) fluorescence in endothelial cells (arrows) in sections of carotid arteries of rats exposed to cigarette smoke compared to vessels of control rats. Vessels were incubated with the dye dihydroethidium, which produces a red nuclear fluorescence when oxidized to EB by $O_2^{\bullet-}$. Res treatment prevented smoking-induced increases in vascular $O_2^{\bullet-}$ production. Green autofluorescence of elastic laminae is shown for orientation purposes. Lu, lumen; M, media; Ad, adventitia. (d) Effects of resveratrol treatment on $O_2^{\bullet-}$ generation by myocardium of control rats and rats exposed to cigarette smoke (with or without Res treatment). $O_2^{\bullet-}$ generation was determined by the lucigenin (5 μmol/l) CL method. Data are mean ± SE. *$P < 0.05$ versus untreated; #$P < 0.05$ versus no resveratrol AU [88]. (Reproduced with permission.)

Conformationally restricted macrocyclic analogues of combretastatins have been evaluated as inhibitors of tubulin polymerization [103]. These compounds present a macrocyclic structure, in which the *para* positions of the aromatic moieties have been linked by a 5- or 6-atoms chain, in order to produce a conformational restriction. This could contribute to detect the active conformation for these ligands. Such a conformational restriction and/or the steric hindrance made them less potent inhibitors than the model compound CA-4.

7.2.2
Miscellaneous

The activation ability of combretastatin A-4 in HepG2 cells was detected by the enzyme phosphorylation [104]. AMP-activated protein kinase activation was confirmed by the phosphorylation of downstream acetyl-CoA carboxylase and the decrease of upstream ATP level. It was also shown that combretastatin A-4 activated PPAR transcriptional activity *in vitro* with the luciferase reporter assay, activated AMPK and downregulated gluconeogenic enzyme mRNA levels in liver, and improved the fasting blood glucose level in diabetic db/db mice. The quantitative structure–activity relationship (QSAR) of a series of combretastatin analogues with ring B modification was studied [105]. The two-dimensional structure–activity relationship was carried out by genetic function analytical (GFA) method. The results indicate that Apol, PMI-mag, Dipole-mag, Hbond donor, and RadOfGyration descriptors contributed significantly to the activities. A three-dimensional structure–activity was performed via comparative molecular field analysis (CoMFA) and comparative molecular similarity indices analysis (CoMSIA). An analysis of CoMFA and CoMSIA models resulted in a cross-validated coefficient ($q2$) of 0.630 and 0.634, respectively, which showed a strong predictive ability. A molecular docking was used to analyze and validate QSAR models. The authors suggested that those results provide a useful information to design novel tubulin inhibitors.

The phase I biotransformation of combretastatin A-4, was studied using rat and human liver subcellular fractions [106]. The metabolites were separated by high-performance liquid chromatography and detected with simultaneous UV and electrospray ionization (ESI) mass spectrometry. The assignment of metabolite structures was based on ESI-tandem mass spectrometry experiments. *O*-Demethylation and aromatic hydroxylation were the two major phase I biotransformation pathways, the latter being regioselective for Ph ring B of CA-4. Incubation with rat and human microsomal fractions led to the formation of a number of metabolites, eight of which were identified. The regioselectivity of microsomal oxidation was also demonstrated by the lack of metabolites arising from stilbenic double bond epoxidation. When CA-4 was incubated with a cytosolic fraction, metabolites were not observed. A series of boronic acid containing *cis*-stilbenes as potent inhibitors of tubulin polymerization was synthesized by the introduction of boronic acid as an acceptor-type functional group into the aromatic ring B of the combretastatin framework [107]. High cell-growth inhibition was observed with boron compounds in which a hydroxyl group on the aromatic ring B of combretastatin

A-4 was replaced with boronic acid; IC_{50} values toward B-16 and 1–87 cell lines were 0.48–2.1 µM. These compounds exhibited a significant inhibitory activity against tubulin polymerization ($IC_{50} = 21$–22 µM). According to the FACScan analysis using Jurkat cells, apoptosis was induced after incubation for 8 h with 13c in which a hydroxy group on the aromatic ring B of combretastatin A-4 was replaced with boronic acid at a concentration of $>10^{-8}$ M.

An image-processing-based method to quantify the rate of extravasation of fluorescent contrast agents from tumor microvessels was developed, and the effect of combretastatin A-4-P

R = OH, combretastatin A-4
R = OPO_3Na_2, CA-4P

on apparent tumor vascular permeability to 40 kDa fluorescein isothiocyanate (FITC) labeled dextran was investigated [108]. Extravasation of FITC-dextran was imaged in three dimensions over time within P22 rat sarcomas growing in dorsal skin flap window chambers in BDIX rats using multiphoton fluorescence microscopy. Image processing techniques were used to segment the data into intra- and extravascular regions or classes. Quantitative estimation of the tissue influx rate constant, K_i, was obtained from the time courses of the fluorescence intensities in the two classes and apparent permeability, P, was calculated. A methodology was developed that provided evidence for a combretastatin A-4-P-induced increase in tumor macromolecular vascular permeability, likely to be central to its anticancer activity. The objective of study carried out in the work [109] was to develop a targeted liposome delivery system for combretastatin A-4 with high loading and stable drug encapsulation. Liposomes composed of hydrogenated soybean phosphatidylcholine (HSPC), cholesterol, and distearoyl phosphoethanolamine-PEG-2000 conjugate (DSPE-PEG) were prepared by the lipid film hydration and extrusion process. Cyclic arginine–glycine–aspartic acid (RGD) peptides with affinity for αvβ3-integrins overexpressed in tumor vascular endothelial cells were coupled to the distal end of polyethylene glycol (PEG) on the liposomes sterically stabilized with PEG (nontargeted liposomes; LCLs). The effect of lipid concentration, drug-to-lipid ratio, cholesterol, and DSPE-PEG content in the formulation on CA-4 loading and its release from the liposomes was studied. Ligand coupling to the liposome surface increased drug leakage as a function of ligand. Liposomes, with measured size of 123.84 ± 41.23 nm, released no significant amount of the encapsulated drug over 48 h at 37 °C. It was shown that combretastatin A-4-P modulates hypoxia inducible factor-1 and gene expression [110]. The effect of factors on the upstream and downstream signaling pathway of HIF-1 *in vitro* was investigated. Combretastatin A-4-P treatment under hypoxia tended to reduce HIF-1 accumulation in a concentration-dependent manner and increased HIF-1 accumulation under aerobic conditions *in vitro*. At these concentrations of combretastatin A-4-P under aerobic conditions, nuclear factor κB was activated via the small GTPase RhoA, and the expression of the HIF-1 downstream angiogenic effector gene, vascular endothelial growth factor (VEGF-A), was increased.

Ophthalmic preparation containing combretastatin A-4 for treating diabetic retinopathy was patented [111]. The ophthalmic preparation was composed of combretastatin A-4 and other auxiliary materials acceptable for treating eye diseases. The authors claimed the preparation could be used as eye drop, ointment, and gel for treating diabetic retinopathy by inhibiting angiogenesis in a dose-dependent manner without affecting the development of retina vascular system.

7.3
Pterostilbene

Pterostilbene (3,5-dimethoxy-4'-hydroxy-*trans*-stilbene)

is a stilbenoid chemically related to resveratrol. It is found predominantly in blueberries and grapes that exhibit anticancer, antihypercholesterolemia, antihypertriglyceridemia properties, as well as fight off and reverse cognitive decline. It was suggested that the compound also has antidiabetic and antifungal properties, but so far very little has been studied on this issue.

The inhibitory effects of pterostilbene on the induction of NO synthase (NOS) and cyclooxygenase-2 (COX-2) in murine RAW 264.7 cells activated with lipopolysaccharide was investigated [112]. Western blotting and real-time polymerase chain reaction analyses demonstrated that pterostilbene significantly blocked the protein and mRNA expression of iNOS and COX-2 in LPS-induced macrophages and inhibited the LPS-induced activation of PI3K/Akt, extracellular signal-regulated kinase 1/2 and p38 MAPK. The aim of the work [113] was to assess the inhibitory effect of a series of naturally occurring *trans*-resveratrol analogues on cytochromes P 450, namely, CYP1A2 and CYP2E1, *in vitro* in order to analyze structure–activity relationships. 3,5-Pterostilbene, 3,4',5-trimethoxy-*trans*-stilbene (TMS), 3,4'-dihydroxy-5-methoxy-*trans*-stilbene (3,4'-DH-5-MS), and 3,5-dihydroxy-4'-methoxy-*trans*-stilbene (3,5-DH-4'-MS) inhibited the activity of CYP1A2, with $K_i = 0.39$, 0.79, 0.94, and 1.04 µM, respectively. Structure–activity relationship analysis led to the conclusion that the substitution of hydroxyl groups of resveratrol with methoxy groups increases the inhibition of CYP1A2. It was presumed that the 4'-hydroxyl group in *trans*-resveratrol and its analogues may play an important role in the interaction with a binding site of CYP2E1. It was shown that resveratrol, pterostilbene, piceatannol, and resveratrol tri-Me ether) activated the peroxisome proliferator-activated receptor alpha (PPARα) isoform [114]. These stilbenoids were evaluated along with ciprofibrate at 1, 10, 100, 300 μM concentrations, for the activation of endogenous PPARα in H4IIEC3 cells. Cells were transfected with a peroxisome proliferator response element-AB (rat fatty acyl CoA β-oxidase response element) – the luciferase gene reporter construct. Of the four compounds, pterostilbene demonstrated the highest induction of PPARα showing 7- and 9–14-fold increases in luciferase activity at 100 and 300 μM, respectively, compared to the control. The maximal luciferase activity responses to pterostilbene at 100 μM were found to be similar to those obtained with the hypolipidemic drug, ciprofibrate. Design, synthesis, biological evaluation, and docking studies of pterostilbene analogues inside PPARα were reported [115]. Pterostilbene and its analogues exposed PPARα activation in H4IIEC3 cells and was found to decrease cholesterol levels in animals. Among analogues that were synthesized, (E)-4-(3,5-dimethoxystyryl)phenyl dihydrogen phosphate showed activity higher than pterostilbene and control drug ciprofibrate. Docking of the stilbenes inside PPARα showed the presence of important hydrogen bond interactions for PPARα activation.

The peroxyl-radical scavenging activity of pterostilbene was found to be the same as that of resveratrol, with total reactive antioxidant potentials of 237 ± 58 and 253 ± 53 μM, respectively [116]. Using a plant system, pterostilbene was also shown to be as effective as resveratrol in inhibiting electrolyte leakage caused by herbicide-induced oxidative damage, and both compounds had the same activity as α-tocopherol. Using a mouse mammary organ culture model, it was found that carcinogen-induced preneoplastic lesions were significantly inhibited by pterostilbene ($ED_{50} = 4.8$ μM).

Stilbenes such as resveratrol, combretastatin, and pterostilbene and their analogues possess a wide spectrum of biological activities. These compounds have been proved to be antioxidant reagents that cause cell apoptosis, suppress growing cancer cells, inhibit and activate specific enzymes, effect the animal aging and metabolisms of estrogens, and so on.

References

1 Newman, D.J. and Cragg, G.M. (2004) *Journal of Natural Products*, **67**, 1216–1238.

2 Schwartsmann, G., Da Rocha, A.B., Mattei, J., and Lopes, R. (2003) *Expert Opinion on Investigational Drugs*, **12**, 1367–1383.

3 Jang, M., Cai, L., Udeani, G.O., Slowing, K.V., Thomas, C.F., Beecher, C.W., Fong, H.H., Farnsworth, N.R., Kinghorn, A.D., Mehta, R.G., Moon, R.C., and Pezzuto, J.M. (1997) *Science*, **275**, 218–220.

4 Farina, A., Ferranti, C., and Marra, C. (2006) *Natural Products Research*, **20**, 247–252, and references therein.

5 Baur, J.A. and Sinclair, D.A. (2006) *Nature Reviews. Drug Discovery*, **5**, 493–506, and references therein.

6 Renaud, S. and Ruf, J.C. (1994) *Circulation*, **90**, 3118–3119.

7 Valenzano, D.R., Terzibasi, E., Genade, T., Cattaneo, A., Domenici, L., and Cellerino, A. (2006) *Current Biology*, **16**, 296–300.

8 Athar, M., Back, J.H., Tang, X., Kim, K.H., Kopelovich, L., Bickers, D.R., and Kim,

A.L. (2007) *Toxicology and Applied Pharmacology*, **224**, 274–283.
9 Roupe, K.A., Remsberg, C.M., Yanez, J.A., and Davies, N.M. (2006) *Current Clinical Pharmacology*, **1**, 81–101.
10 Bastianetto, S. and Quirion, R. (2008) *Agro Food Industry Hi-Tech*, **19**, 20–21.
11 Soleas, G.J., Diamandis, E.P. and Goldberg, D.M. (1997) *Clinical Biochemistry*, **30**, 91–113.
12 Pezzuto, J.M. (2008) *Pharmaceutical Biology*, **46**, 443–573.
13 Raval, A.P., Lin, H.W., Dave, K.R., DeFazio, R.A., Della Morte, D., Kim, E.J., and Perez-Pinzon, M.A. (2008) *Current Medicinal Chemistry*, **15**, 1545–1551.
14 Harikumar, K.B. and Aggarwal, B.B. (2008) *Cell Cycle*, **7**, 1020–1035.
15 Stervbo, U., Vang, O., and Bonnesen, C. (2006) *Food Chemistry*, **101**, 449–457.
16 Shankar, S., Singh, G., and Srivastava, R.K. (2007) *Frontiers in Bioscience*, **12**, 4839–4854.
17 Wang, Y., Catana, F., Yang, Y., Roderick, R., and van Breemen, R.B. (2002) *Journal of Agricultural and Food Chemistry*, **50**, 431–435.
18 Vivancos, M. and Moreno, J.J. (2008) *British Journal of Nutrition*, **99**, 1199–1207.
19 Katalinic, V., Ljubenkov, I., Pezo, I., Generalic, I., Stricevic, O., Milos, M., Modun, D., and Boban, M. (2008) *Periodicum Biologorum*, **110**, 77–83 AN 2.
20 Moreno, A., Castro, M., and Falque, E. (2008) *European Food Research and Technology*, **227**, 667–674.
21 Gu, X., Creasy, L., Kester, A., and Zeece, M. (1999) *Journal of Agricultural and Food Chemistry*, **47**, 3223–3227.
22 Esna-Ashari, M., Gholami, M., Zolfigol, M.A., Shiri, M., Mahmoodi-Pour, A., and Hesari, M. (2008) *Chromatographia*, **67**, 1017–1020.
23 Ector, B.J., Magee, J.B., Hegwood, C.P., and Coign, M.J. (1996) *American Journal of Enology and Viticulture*, **47**, 57–62.
24 Li, L., Zhang, H., and Qiao, Y. (2007) *Yaowu Fenxi Zazhi*, **27**, 654–656.
25 Lyons, M.M., Yu, C., Toma, R.B., Cho, S.Y., Reiboldt, W., Lee, J., and Van Breemen, R.B. (2003) *Journal of Agricultural and Food Chemistry*, **51**, 5867–5870.
26 Nie, Y., Li, M.-Q., Zhang, J., and Guo, C.-Y. (2008) *Hebei Gongye Daxue Xuebao*, **37**, 60–65.
27 Cao, Y., Tang, Y., Lu, C., and Huang, Z. (2007) *Shipin Kexue*, **28**, 245–248.
28 Ho, C.-W. and Kuo, H.-S. (2008) *US Patent Application Publication*, p. 17.
29 Soleas, G.J., Yan, J., and Goldberg, D.M. (2001) *Journal of Chromatography B: Biomedical Sciences and Applications*, **757**, 161–172.
30 Liu, W., Bi, Y., Han, J., Wang, X., Shan, L., Liu, Z., Li, Z., Wang, Q., and Zhang, M. (2008) Faming Zhuanli Shenqing Gongkai Shuomingshu, pp. 14.
31 De Santi, C., Pietrabissa, A., Spisni, R., Mosca, F., and Pacifici, G.M. (2000) *Xenobiotica*, **30**, 609–617.
32 Marasco, E.K. and Schmidt-Dannert, C. (2008) *ChemBioChem*, **9**, 1450–1461.
33 Marier, J.-F., Vachon, P., Gritsas, A., Zhang, J., Moreau, J.-P., and Ducharme, M.P. (2002) *The Journal of Pharmacology and Experimental Therapeutics*, **302**, 369–373.
34 Wenzel, E., Soldo, T., Erbersdobler, H., and Somoza, V. (2005) *Molecular Nutrition & Food Research*, **49**, 482–494.
35 Hung, C.-F., Lin, Y.-K., Huang, Z.-R., and Fang, J.-Y. (2008) *Biological & Pharmaceutical Bulletin*, **31**, 955–962.
36 Soleas, G.J., Angelini, M., Grass, L., Diamandis, E.P., and Goldberg, D.M. (2001) *Methods in Enzymology (Flavonoids and Other Polyphenols)*, **335**, 145–154.
37 Bertelli, A.A.E., Giovannini, L., Stradi, R., Urien, S., Tillement, J.-P., and Bertelli, A. (1996) *International Journal of Clinical Pharmacology Research*, **16**, 77–81.
38 He, S., Wu, B., Pan, Y., and Jiang, L. (2008) *Journal of Organic Chemistry*, **73** (14), 5233–5241.
39 Jia, Z., Zhu, H., Misra, B.R., Mahaney, J.E., Li, Y., and Misra, H.P. (2008)

Molecular and Cellular Biochemistry, **313**, 187–194.

40 Gauliard, B., Grieve, D., Wilson, R., Crozier, A., Jenkins, C., Mullen, W.D., and Lean, M. (2008) *Journal of Medicinal Food*, **11**, 382–384.

41 Robb, E.L., Page, M.M., Wiens, B.E., and Stuart, J.A. (2008) *Biochemical and Biophysical Research Communications*, **367**, 406–412.

42 Kode, A., Rajendrasozhan, S., Caito, S., Yang, Se.-R., Megson, I.L., and Rahman, I. (2008) *American Journal of Physiology*, **294**, L478–L488.

43 Olas, B., Wachowicz, B., Nowak, P., Stochmal, A., Oleszek, W., Glowacki, R., and Bald, E. (2008) *Cell Biology and Toxicology*, **24**, 331–340.

44 Zheng, L.-F., Wei, Q.-Y., Cai, Y.-J., Fang, J.-G., Zhou, B., Yang, L., and Liu, Z.-L. (2006) *Free radical Biology & Medicine*, **41**, 1807–1816.

45 Robb, E.L., Winkelmolen, L., Visanji, N., and Brotchie, J.J.A. (2008) *Biochemical and Biophysical Research Communications*, **372**, 254–259.

46 Kowalczyk, E., Fijalkowski, P., Blaszczyk-Suszynska, J., Blaszczyk, J., Popinska, M., Andryskowski, G., Kopff, A., Kopff, M., and Rapacka, E. (2007) *Zywienie Czlowieka i Metabolizm*, **34**, 1285–1288.

47 Feng, Q., Wu, Q., Lu, B., and Chen, Q.-J. (2008) *Zhongguo Jiceng Yiyao*, **15**, 264–266.

48 Blanchet, J., Longpre, F., Bureau, G., Morissette, M., DiPaolo, T., Bronchti, G., and Martinoli, M.-G. (2008) *Progress in Neuro-Psychopharmacology & Biological Psychiatry*, **32**, 1243–1250.

49 Cecconi, D., Zamo, A., Parisi, A., Bianchi, E., Parolini, C., Timperio F A.M., Zolla, L., and Chilosi F M. (2008) *Journal of Proteome Research*, **7**, 2670–2680.

50 Faber, A.C. and Chiles, T.C. (2006) *International Journal of Oncology*, **29**, 1561–1566.

51 Gosslau, A., Pabbaraja, S., Knapp, S., and Chen, K.Y. (2008) *European Journal of Pharmacology*, **587**, 25–34.

52 Snyder, R.M., Yu, W., Li, J., Sanders, B.G., and Kline, K. (2008) *Nutrition and Cancer*, **60**, 401–411.

53 Sareen, D., van Ginkel, P.R., Takach, J.C., Mohiuddin, A., Darjatmoko, S.R., Albert, D.M., and Polans, A.S. (2006) *Investigative Ophthalmology & Visual Science*, **47**, 3708–3716.

54 Saiko, P., Pemberger, M., Horvath, Z., Savinc, I., Grusch, M., Handler, N., Erker, T., Jaeger, W., Fritzer-Szekeres, M., and Szekeres, T. (2008) *Oncology Reports*, **19**, 1621–1626.

55 Kalra, N., Roy, P., Prasad, S., and Shukla, Y. (2008) *Life Sciences*, **82**, 348–358.

56 Chakraborty, P.K., Mustafi, S.B., Ganguly, S., Chatterjee, M., and Raha, S. (2008) *Cancer Science*, **99**, 1109–1116.

57 Schneider, Y., Fischer, B., Coelho, D., Roussi, S., Gosse, F., and Bischoff, P. (2004) *Cancer Letters*, **211**, 155–161.

58 Puissant, A., Grosso, S., Jacquel, A., Belhacene, N., Colosetti, P., Cassuto, J.-P., and Auberger, P. (2008) *FASEB Journal*, **22**, 1894–1904, doi: 10.1096/fj.07–101394.

59 Zhang, F., Sun, C., Wu, J., He, C., Ge, X., Huang, W., Zou, Y., Chen, X., Qi, W., and Zhai, Q. (2008) *Pharmacological Research*, **57**, 318–332.

60 Jang, M.H., Piao, X.L., Kim, J.M., Kwon, S.W., and Park, J.H. (2008) *Phytotherapy Research*, **22**, 544–549.

61 Yang, Y.-M., Wang, X.-X., Chen, J.-Z., Wang, S.-J., Hu, H., and Wang, H.-Q. (2008) *American Journal of Chinese Medicine*, **36**, 603–613.

62 Erb-Downward, J.R., Noggle, R.M., Williamson, P.R., and Huffnagle, G.B. (2008) *Molecular Microbiology*, **68**, 1428–1437.

63 Floyd, Z.E., Wang, Z.Q., Kilroy, G., and Cefalu, W.T. (2008) *Metabolism, Clinical and Experimental*, **57**, S32–S38

64 Hu, Z., Wang, C., Chu, M., Zhang, Y., Zhu, G., and Guan, F. (2008) *Zhonghua Shenjing Waike Jibing Yanjiu Zazhi*, **7**, 107–110.

65. Wang, T., Bai, Y., Liu, Y., Zhao, W., Zhang, Y., Li, B., Lu, Y., and Yang, B. (2007) *Harbin Yike Daxue Xuebao*, **41**, 189–191.
66. Marambaud, P., Zhao, H., and Davies, P. (2005) *Journal of Biological Chemistry*, **280**, 37377–37382.
67. Lagouge, M., Argmann, C., Gerhart-Hines, Z., Meziane, H., Lerin, C., Daussin, F., Messadeq, N., Milne, J., Lambert, P., Elliott, P., Geny, B., Laakso, M., Puigserver, P., and Auwerx F J. (2006) *Cell*, **127**, 1109–1122.
68. Parker, J.A., Arango, M., Abderrahmane, S., Lambert, E., Tourette, C., Catoire, H., and Neri, C. (2005) *Nature Genetics*, **37**, 349–350.
69. Meng, X.-L., Yang, J.-Y., Chen, G.-L., Wang, Li.-H., Zhang, L.-J., Wang, S., Li, J., and Wu, C.-F. (2008) *Chemico-Biological Interactions*, **174**, 51–59.
70. Dias, P.F., Berti, F.V., Siqueira, J.M., Jr., Maraschin, M., Gagliardi, A.R., and Ribeiro-do-Valle, R.M. (2008) *Journal of Pharmacological Sciences*, **107**, 118–127.
71. Zahid, M., Gaikwad, N.W., Ali, M.F., Lu, F., Saeed, M., Yang, L., Rogan, E.G., and Cavalieri, E.L. (2008) *Free Radical Biology & Medicine*, **45**, 136–145.
72. Bowers, J.L., Tyulmenkov, V.V., Jernigan, S.C., and Klinge, C.M. (2000) *Endocrinology*, **141**, 3657–3667.
73. Bhat, K.P.L., Lantvit, D., Christov, K., Mehta, R.G., Moon, R.C., and Pezzuto, J.M. (2001) *Cancer Research*, **61**, 7456–7463.
74. Wang, Q., Li, H., Liu, N., Chen, X.-Y., Wu, M.-L., Zhang, K.-L., Kong, Q.-Y., and Liu, J. (2008) *Neuroscienc Letters*, **438**, 168–173.
75. Raval, A.P., Lin, H.W., Dave, K.R., DeFazio, R.A., Della Morte, D., Kim, E.J., and Perez-Pinzon, M.A. (2008) *Current Medicinal Chemistry*, **15**, 1545–1551.
76. Guo, J. and Sun, A. (2008) *Jiangsu Daxue Xuebao, Yixueban*, **18**, 102–106.
77. Leiro, J., Arranz, J.A., Fraiz, N., Sanmartin, M.L., Quezada, E., and Orallo, F. (2005) *International Immunopharmacology*, **5**, 393–406.
78. Dong, D., Guo, E., Zhang, Y., Lu, Y., and Zhang, S. (2007) *Zhonghua Zhongliu Fangzhi Zazhi*, **14**, 489–493.
79. Wang, Y., Ye, L., and Leung, L.K. (2008) *Toxicology*, **248**, 130–135.
80. Sun, C., Hu, Y., Liu, X., Wu, T., Wang, Y., He, W. and Wei, W. (2006) *Cancer Genetics and Cytogenetics*, **165**, 9–19.
81. Stefani, M., Markus, M.A., Lin, R.C.Y., Pinese, M., Dawes, I.W., and Morris, B.J. (2007) *Annals of the New York Academy of Sciences*, **1114** (Healthy Aging and Longevity), 407–418.
82. Barger, J.L., Kayo, T., Vann, J.M., Arias, E.B., Wang, J., Hacker, T.A., Wang, Y., Raederstorff, D., Morrow, J.D., Leeuwenburgh, C., Allison, D.B., Saupe, K.W., Cartee, G.D., Weindruch S R., and Prolla F T.A. (2008) *PLoS One*, **3** (6).
83. Shang, J., Chen, L.-L., Xiao, F.-X., Sun, H., Ding, H.-C., and Xiao, H. (2008) *Acta Pharmacologica Sinica*, **29**, 698–706.
84. Kasri, N.N., Bultynck, G., Parys, J.B., Callewaert, G., Missiaen, L., and De Smedt, H. (2005) *Molecular Pharmacology*, **68**, 241–250.
85. McDonald, O., Lackey, K., Ward, R.D., Wood, E., Samano, V., Maloney, P., Deanda, F., and Hunter, R. (2006) *Bioorganic & Medicinal Chemistry Letters*, **16**, 5378–5383.
86. Katsuyama, Y., Funa, N., Miyahisa, I., and Horinouchi, S. (2007) *Chemistry & Biology*, **14**, 613–621.
87. Wang, Y. (2008) Faming Zhuanli Shenqing Gongkai Shuomingshu, pp. 10.
88. Csiszar, A., Labinskyy, N., Podlutsky, A., Kaminski, P.M., Wolin, M.S., Zhang, C., Mukhopadhyay, P., Pacher, P., Hu, F., de Cabo, R., Ballabh, P., and Ungvari, Z. (2008) *American Journal of Physiology*, **294**, H2721–H2735.
89. Chaudhary, A., Pandeya, S.N., Kumar, P., Sharma, P.P., Gupta, S., Soni, N., Verma, K.K., and Bhardwaj, G. (2007) *Mini-Reviews in Medicinal Chemistry*, **7**, 1186–1205.

90 Patterson, D.M. and Rustin, G.J.S. (2007) *Drugs of the Future*, **32**, 1025–1032.

91 Brown, T., Holt, H., Jr., and Lee, M. (2006) *Topics in Heterocyclic Chemistry (Heterocyclic Antitumor Antibiotics)*, **2**, 1–51.

92 Sajan, D., Abraham, J.P., Hubert, J.I., Jayakumar, V.S., Aubard, J., and Faurskov, N.O. (2008) *Journal of Molecular Structure*, **889**, 129–143.

93 Griggs, J., Metcalfe, J.C., and Hesketh, R. (2001) *The Lancet Oncology*, **2**, 82–87.

94 Hinnen, P. (2007) *British Journal of Cancer*, **96**, 1159–1165.

95 Jordan, M.A. (2004) *Nature Reviews Cancer*, **4**, 253–265.

96 Salmon, B. (2007) *International Journal of Radiation Oncology, Biology, Physics*, **68**, 211–217.

97 Tozer, G. (2005) *Nature Reviews Cancer*, **5**, 432–435.

98 Tron, G., Pirali, T., Sorba, G., Pagliai, F., Busacca, S., and Genazzani, A. (2006) *Journal of Medicinal Chemistry*, **49**, 3033–3044.

99 Patterson, D. (2007) *Clinical Oncology*, **19**, 443–456;

100 Salmon, B. (2007) *International Journal of Radiation Oncology, Biology, Physics*, **68**, 211–217.

101 Robinson, J.E. and Taylor, R.J.K. (2007) *Chemical Communications*, (16), 1617–1619.

102 Kerr, D.J., Hamel, E., Jung, M.K. and Flynn, B. (2007) *Bioorganic & Medicinal Chemistry*, **15**, 3290–3298.

103 Johnson, M., Younglove, B., Lee, L., LeBlanc, R., Holt, H., Jr., Hills, P., Mackay, H., Brown, T., Mooberry, S.L., and Lee, M. (2007) *Bioorganic & Medicinal Chemistry Letters*, **17**, 5897–5901.

104 Mateo, C., Alvarez, R., Perez-Melero, C., Pelaez, R., and Medarde, M. (2007) *Bioorganic & Medicinal Chemistry Letters*, **17**, 6316–6320.

105 Zhang, F., Sun, C., Wu, J., He, C., Ge, X., Huang, W., Zou, Y., Chen, X., Qi, W., and Zhai, Q. (2008) *Pharmacological Research*, **57**, 318–332.

106 Tian, R., Liu, Z.-M., Jin, H.-W., Zhang, L.-R., and Lin, W.-H. (2007) *Gaodeng Xuexiao Huaxue Xuebao*, **28**, 2150–2155.

107 Aprile, S., Del Grosso, E., Tron, G.C., and Grosa, G. (2007) *Drug Metabolism and Disposition*, **35**, 2252–2261.

108 Nakamura, H., Kuroda, H., Saito, H., Suzuki, R., Yamori, T., Maruyama, K., and Haga, T. (2006) *ChemMedChem*, **1**, 729–740.

109 Reyes-Aldasoro, C.C., Wilson, I., Prise, V.E., Barber, P.R., Ameer-Beg, M., Vojnovic, B., Cunningham, V.J., and Tozer, G.M. (2008) *Microcirculation*, **15**, 65–79.

110 Nallamothu, R., Wood, G.C., Kiani, M.F., Moore, B.M., Horton, F.P., and Thoma, L.A. (2006) *Journal of Pharmaceutical Science and Technology*, **60**, 144–155.

111 Dachs, G.U., Steele, A.J., Coralli, C., Kanthou, C., Brooks, A.C., Gunningham, S.P., Currie, M.J., Watson, A.I., Robinson, B.A., and Tozer, G.M. (2006) *BMC Cancer*, **6**.

112 He, Y., Xu, X., and Bao, X. (2007) *Faming Zhuanli Shenqing Gongkai Shuomingshu*, pp. 11.

113 Pan, M.-H., Chang, Y.-H., Tsai, M.-L., Lai, C.-S., Ho, S.-Y., Badmaev, V., and Ho, C.-T. (2008) *Journal of Agricultural and Food Chemistry*, **56**, 7502–7509.

114 Mikstacka, R., Rimando, A.M., Szalaty, K., Stasik, K., and Baer-Dubowska, W. (2006) *Xenobiotica*, **36**, 269–285.

115 Rimando, A.M., Yokoyama, W., and Feller, D.R. (2006) PCT Int. Appl., pp. 17.

116 Mizuno, C.S., Ma, G., Khan, S., Patny, A., Avery, M.A., and Rimando, A.M. (2008) *Bioorganic & Medicinal Chemistry*, **16**, 3800–3808.

117 Rimando, A.M., Cuendet, M., Desmarchelier, C., Mehta, R.G., Pezzuto, J.M., and Duke, S.O. (2002) *Journal of Agricultural and Food Chemistry*, **50**, 3453–3457

8
Preclinic Effects of Stilbenes

Preclinical research on cells, organs, and animals using stilbenes occupy a great deal of attention as a necessary step for further clinical investigation and invention of new drugs. Data on cancer and other chemopreventive activities of resveratrol (RES) and other stilbenes were summarized in reviews [1–22].

In this chapter, we present a number of recent developments in this boundless area.

8.1
Resveratrol

8.1.1
Cancer Protection in Animal

Data on the cancer chemopreventive activity of resveratrol were summarized in reviews [6, 22, 25]. The role of resveratrol in prevention and therapy of cancer revealed in preclinical and clinical studies was reviewed [22]. The authors demonstrated that resveratrol exhibits cardioprotective effects, and anticancer properties, as suggested by its ability to suppress the proliferation of a wide variety of tumor cells, including lymphoid and myeloid cancers; multiple myeloma; cancers of the breast, prostate, stomach, colon, pancreas, and thyroid; melanoma; head and neck squamous cell carcinoma; ovarian carcinoma; and cervical carcinoma. Pharmacokinetic studies revealed that the target organs of resveratrol are liver and kidney, where it is concentrated after absorption and is mainly converted to a sulfated form and a glucuronide conjugate. *In vivo*, resveratrol blocks the multistep process of carcinogenesis at various stages: it blocks carcinogen activation by inhibiting aryl hydrocarbon-induced CYP1A1 expression and activity, and suppresses tumor initiation, promotion, and progression. Limited data on humans have revealed that resveratrol is pharmacologically quite safe. The authors concluded that structural analogues of resveratrol with improved bioavailability are being pursued as potential therapeutic agents for cancer.

Authors of the work [23] investigated whether resveratrol would inhibit human melanoma xenograft growth. Athymic mice received control diets or diets containing

Stilbenes. Applications in Chemistry, Life Sciences and Materials Science. Gertz Likhtenshtein
Copyright © 2010 WILEY-VCH Verlag GmbH & Co. KGaA, Weinheim
ISBN: 978-3-527-32388-3

110 or 263 µM I, 2 week prior to injection of the tumor cells. Tumor growth was measured during a 3-week period. The effect of a major resveratrol metabolite, piceatannol (II), on experimental lung metastasis I, and its major metabolites, resveratrol glucuronide and II, on serum, liver, skin, and tumor tissue was also detected. The relationship between resveratrol bioavailability and its effect on tumor growth was investigated. As it was shown in the work, the most efficient way of administering resveratrol in humans appears to be buccal delivery, that is, without swallowing, by direct absorption through the inside of the mouth. Tissue levels of RES were studied after intervenous and oral administration of *trans*-resveratrol (*t*-RES) to rabbits, rats, and mice. When 1 mg of resveratrol in 50 ml solution was retained in the mouth for 1 min before swallowing, 37 ng/ml of free resveratrol was measured in plasma 2 min later. This level of unchanged resveratrol in blood can be achieved only with 250 mg of resveratrol taken in a pill form. Figure 8.1 shows the plasma levels and half-life of resveratrol after its intravenous administration to rabbits. The authors suggested the antimetastatic mechanism involving a *t*-RES (1 µM)-induced inhibition of vascular adhesion molecule 1 (VCAM-1) expression in the hepatic sinusoidal endothelium (HSE), which consequently decreased *in vitro* B16M cell adhesion to the endothelium via very late activation antigen 4 (VLA-4).

Suppression of prostate cancer progression in transgenic mice by resveratrol was reported [24]. Male mice with transgenic adenocarcinoma of the prostate were fed resveratrol (625 mg resveratrol/kg AIN-76A diet) and phytoestrogen-free control diet (AIN-76A) starting at 5 weeks of age. Mechanisms of action and histopathologic studies were conducted at 12 and 28 weeks of age, respectively. Resveratrol in the diet significantly reduced the incidence of poorly differentiated prostatic adenocarcinoma

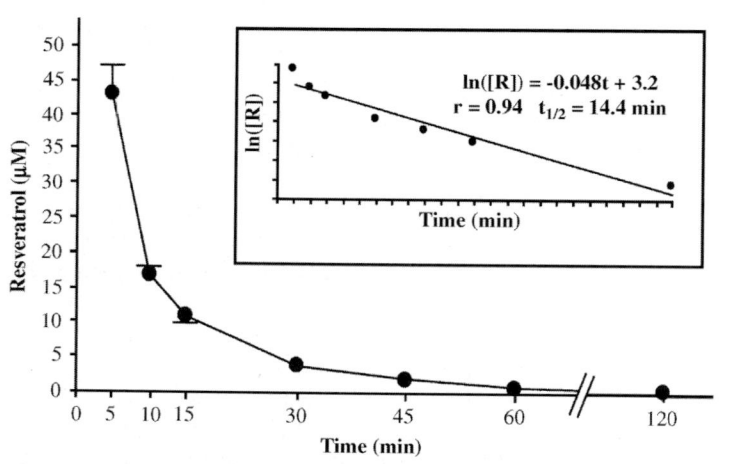

Figure 8.1 Plasma levels and half-life of resveratrol after intravenous administration to rabbits. Animals were treated with 20 mg *t*-RES per kg of bulk weight. All time points were determined in the same animal. Results are means ± SD for six different rabbits [18]. (Reproduced with permission from Elsevier.)

Inset equation: $\ln([R]) = -0.048t + 3.2$, $r = 0.94$, $t_{1/2} = 14.4$ min

by 7.7-fold. The authors claimed that the decrease in cell proliferation and the potent growth factor, IGF-1, the downregulation of downstream effectors, phospho-ERKs 1 and 2, and the increase in the putative tumor suppressor, estrogen receptor-β, provided a biochemical basis for resveratrol suppressing the development of prostate cancer.

Biological and preclinical activity of resveratrol has been recently reviewed [25].

8.1.2
Cell Cancer Protection

The antiproliferative and proapoptotic effects of *trans*-resveratrol on the human colorectal carcinoma HT-29 cells, as well as the mechanisms underlying these effects, were examined [26]. Proliferation, cytotoxicity, and apoptosis were measured by fluorescence-based techniques. Studies of dose-dependent effects of *trans*-resveratrol showed antiproliferative activity with an EC_{50} value of $78.9 \pm 5.4\,\mu M$. Caspase-3 was activated in a dose-dependent manner after incubation for 24 h giving an EC_{50} value of $276.1 \pm 1.7\,\mu M$. Apoptosis was also confirmed with microscopic observation of changes in membrane permeability and detection of DNA fragmentation. The activity of *trans*-resveratrol on the mitochondria apoptosis pathway was evident in the production of superoxide anions in the mitochondria of cells undergoing apoptosis. The chemopreventive/antiproliferative potential of resveratrol against prostate cancer and its mechanism of action were evaluated [27]. Treatment with resveratrol (0–50 μmol/l for 24 h) resulted in a significant (i) decrease in cell viability, (ii) decrease in clonogenic cell survival, (iii) inhibition of androgen (R1881)-stimulated growth, and (iv) induction of apoptosis in androgen-responsive human prostate carcinoma (LNCaP) cells. Treatment with resveratrol also resulted in a significant dose-dependent inhibition of the constitutive expression of phosphatidylinositol 3′-kinase, phosphorylated (active) Akt in LNCaP cells, and a significant (i) loss of mitochondrial membrane potential, (ii) decrease in the protein level of antiapoptotic Bcl-2, and (iii) increase in proapoptotic members of the Bcl-2 family. On the basis of these studies, the authors suggested that resveratrol could be developed as an agent for the management of prostate cancer.

Potential antitumor agents, aminoalkyl phosphonate derivatives of resveratrol, were described [28]. These derivatives were prepared by partial synthesis of resveratrol. Antitumor activities of the synthesized compounds were detected against a human nasopharyngeal epidermoid tumor cell line KB and a human normal cell line L02. The results indicated that these compounds showed good cytotoxic activity against KB with IC_{50} of 0.4–0.9 M but weak cytotoxic activity against L02. In authors' opinion, the potent antitumor activities shown by the compounds make these resveratrol phosphonate derivatives more interesting for further investigation. Inhibitory effects of *trans*-resveratrol analogue molecules on the proliferation and the cell cycle progression of human colon tumoral cells were reported [29]. Effects of resveratrol, ε-viniferin, their acetylated forms (resveratrol triacetate, ε-viniferin pentaacetate), and vineatrol (a wine grape extract) on human adenocarcinoma colon cells were studied. Resveratrol and resveratrol triacetate inhibited cell proliferation and arrested cell cycle. Vineatrol inhibits cell proliferation and favors accumulation in the S-phase of the cell cycle. Consequently, resveratrol triacetate and vineatrol could constitute new putative anticancer agents against colon carcinoma. It was shown [30] that resveratrol did not induce apoptosis in primary cultures of normal prostate epithelial cells but arrested cells at the G1-S-phase transition of the cell cycle associated with increased expression of p21 and decreased expression of cyclin D1 and cyclin-dependent kinase 4 proteins.

According to [31], introduction of additional hydroxyl groups into the stilbene structure increased the biological activity of resveratrol and the activity of 3,3′,4,4′,5,5′-hexahydroxystilbene (**M8**) in ZR-75-1, MDA-MB-231, and T47D human breast cancer cells. For the evaluation of cytotoxic activity of **M8**, clonogenic and cell proliferation assays were used. Compound **M8** caused the activation of caspase-8 in MDA-MB-231 cells, while activities of caspase-9 and caspase-3 increased in all three tested cell lines. Activation of caspase-9 and caspase-3 was associated with the loss of mitochondrial potential and increase in p53, which could have an impact on downregulation of mitochondrial superoxide dismutase (MnSOD). An increase in oxidative stress conditions was suggested by the loss of reduced glutathione in tested cells. Since cancer cells are usually under permanent oxidative stress, an additional increase in the generation of reactive oxygen species (ROS) as a result of the interaction of **M8** with the mitochondrial respiratory chain and a decrease in oxidative defense were suggested to be a promising method for selective elimination of cancer cells. To investigate the mechanism of anticarcinogenic activities of resveratrol, the effects on cytochrome P 450 were examined in human liver microsomes and *Escherichia coli* membranes coexpressing human P 450 1A1 or P 450 1A2 with human NADPH-P 450 reductase [32]. Resveratrol exhibited potent inhibition of human P 450 1A1 in a dose-dependent manner with IC_{50} of 23 µM for EROD and IC_{50} of 11 µM for methoxyresorufin O-demethylation (MROD). Resveratrol showed over 50-fold selectivity for P 450 1A1 over P 450 1A2. In a human P 450 1A1/reductase bicistronic system, resveratrol inhibited human P 450 1A1 activity in a mixed-type inhibition with K_i values of 9 and 89 µM. These results suggested that reservatrol may be considered for use as a strong cancer chemopreventive agent in humans. Alpha-tocopherol ether-linked acetic acid analogue [2,5,7,8-tetramethyl-2*R*-(4*R*, 8*R*-12-trimethyltridecyl) chroman-6-yloxyacetic acid (α-TEA)] alone and together with methylseleninic acid (MSA) and *t*-RES was investigated for their ability to induce apoptosis, DNA synthesis arrest, and cellular differentiation and inhibit colony formation in human MDA-MB-435-F-L breast cancer cells in culture [33]. The three agents alone were effective in inhibiting cell growth by each of the four different assays, and three-way combination treatments synergistically inhibited cell proliferation. Combinations of α-TEA, resveratrol, and MSA significantly enhanced levels of apoptosis in human breast (MDA-MB-231, MCF7, and T47D) and prostate (LnCaP, PC-3, and DU-145) cancer cell lines. Western immunoblotting confirmed the induction of apoptosis in that the three agents induced poly(ADP-ribose) polymerase cleavage [32].

Four phenolic compounds were isolated from the seeds by solvent fractionation Sephadex LJ-20 column chromatography and preparative HPLC, and three of them showed strong inhibitory activity against soybean liposygenase (SLO) and were characterized as *trans*-resveratrol, ε-viniferin, and luteolin by UV, IR, ^{1}H NMR, ^{13}C NMR, and MS spectrometry [34]. The following activity has been established: *trans*-resveratrol (IC$_{50}$ = 1.02 µM), ε-viniferin (IC$_{50}$ = 0.81 µM), and luteolin (IC$_{50}$ = 10.01 µM).

Mechanistic studies [35] showed combination treatments to inhibit cell proliferation via downregulation of cyclin D1 and induce apoptosis via activation of caspases-8 and -9 and downregulation of prosurvival proteins, FLIP and survivin [35]. Several stilbenes, related to resveratrol, have been synthesized and tested for their anticancer effect on HL-60 leukemia cell line, taking particular care of the cell cycle analysis. Figure 8.2 shows synthesis of stilbenes and its effects on cell cycle. A scheme of flow cytometry analysis of cell cycle is presented in Figure 8.3. The most potent compound was found to be (Z)-3,4′,5-trimethoxystilbene(I)

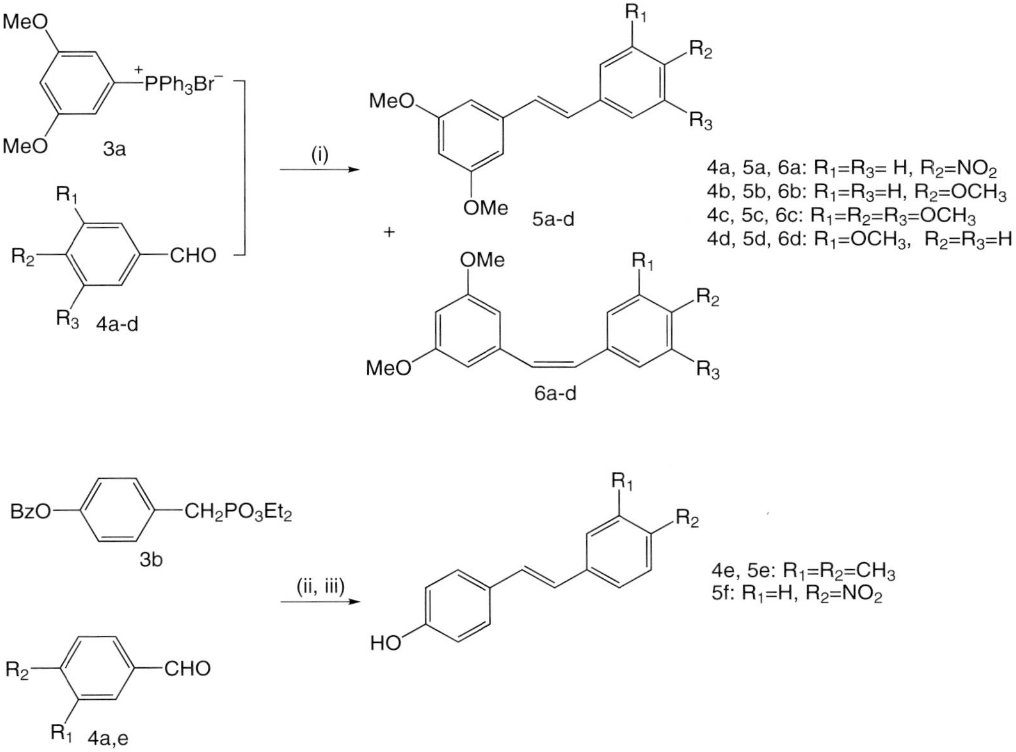

Figure 8.2 Synthesis of stilbenes. Reagents and conditions: (i) **3a** (1.1 equiv), **4a–d** (1 equiv), NaH (1.2 equiv), THF, 4–16 h, room or reflux temperature, 30–50% yield, 1:1 to 1:3 Z/E isomer ratio; (ii) same conditions as (i), but only E-isomers were recovered; (iii) AlCl$_3$, N,N-dimethylaniline, CH$_2$Cl$_2$, rt, 60% yield. Effects on cell cycle of the most active *cis* (**6b, 6c**) and *trans* (**5b, 5f**) were examined by flow cytometry after cells were stained with propidium iodide. HL-60 cells were exposed to each compound at the concentrations reported in Ref. [35]. (Reproduced with permission from Elsevier.)

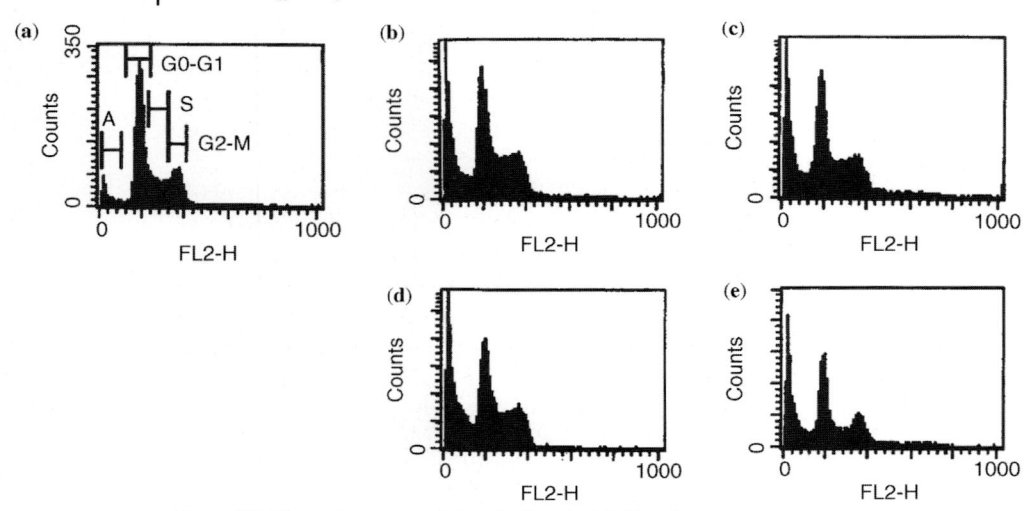

Figure 8.3 Flow cytometry analysis of cell cycle. HL-60 cells were exposed for 24 h to 3.5 µM **5b** (b), 2.5 µM **6c** (c), 4 µM **5f** (d), and 0.2 µM **6b** (e). (a) Control; A, sub-G0–G1 peak [35]. (Reproduced with permission from Elsevier.)

which was active as an apoptotic agent at 0.24 µM.

Similar to resveratrol, compounds **5b**, **5f**, and **6c** induced a partial block of cells in S-phase and an apoptotic sub-G0–G1 peak corresponding to about 20%, suggesting that these compounds act on HL-60 cells as phase-specific cytotoxic agents. In contrast, compound **6b** caused an evident sub-G0–G1 peak increase but no modification in cell cycle distribution (phases G0–G1, S, and G2–M) compared to the control group.

8.1.3
Miscellaneous Effects

The objective of the study [36] was to examine the cardioprotective effect of resveratrol in the rat after ischemia and ischemia–reperfusion (I–R). The left main coronary artery was occluded for 30 or 5 min followed by a 30-min reperfusion in anesthetized rats. Animals were preinfused with and without resveratrol before occlusion, and the severity of ischemia- and I–R-induced arrhythmias and mortality was compared. Resveratrol pretreatment had any effect neither on ischemia-induced arrhythmias nor on mortality. In contrast, a dramatic protective effect was observed against I–R-induced arrhythmias and mortality. Resveratrol pretreatment reduced both the incidence and the duration of ventricular tachycardia (VT) and ventricular fibrillation (VF). During the same period, resveratrol pretreatment also increased nitric oxide (NO) and decreased lactate dehydrogenase levels in the carotid blood. Resveratrol was suggested to be a potent antiarrhythmic agent with cardioprotective properties in I–R rats. It was suggested that the cardioprotective effects of resveratrol in I–R rats may be correlated with its antioxidant activity and upregulation of NO production.

To elucidate the role of a high-fat diet on vascular dysfunction and cardiac fibrosis in the absence of overt obesity and hyperlipidemia, normal female rats were fed a high-fat diet for 8 weeks [37]. This was associated with a modest increase in the body weight and a normal plasma lipid profile. In rats fed a high-fat diet, systolic (171 ± 7 mmHg) and diastolic blood pressures (109 ± 3) increased compared to a standard diet (systolic blood pressure, 134 ± 8; diastolic blood pressure, 96 ± 5 mmHg), and the acetylcholine-dependent relaxation of isolated aortic rings significantly reduced. Perivascular fibrosis was detected in the heart of rats fed a high-fat diet. The exogenous addition of resveratrol (*trans*-3,5,4′-trihydroxystilbene) (0.1 µM) to aortic rings isolated from rats fed a high-fat diet restored the acetylcholine-mediated relaxation. The administration of resveratrol (20 mg/kg/day for 8 weeks) to rats fed a high-fat diet prevented the increase in blood pressure and preserved acetylcholine-dependent relaxation of isolated aortic rings. The authors suggested that resveratrol therapy can prevent the hypertensive response in female rats fed a high-fat diet but is without effect on the progression of perivascular fibrosis.

Effects of chronic resveratrol treatment on vascular responsiveness of streptozotocin-induced diabetic rats were examined [38]. Resveratrol (5 mg/kg/day, intraperitoneal) was administered for 42 days to streptozotocin (STZ) (60 mg/kg)-induced diabetic rats. Loss of weight, hyperglycemia, and elevated levels of plasma malondialdehyde (MDA) were observed in diabetic rats. Resveratrol treatment was significantly effective for these metabolic and biochemical abnormalities. The contractile responses of the aorta were recorded. Compared to control subjects, the aorta showed significantly enhanced contractile responses to noradrenaline, but not to potassium chloride, in diabetic rats. Treatment of diabetic rats with resveratrol significantly reversed the increase in responsiveness and sensitivity of aorta to noradrenaline. In diabetic aorta, the relaxation response to acetylcholine was found to be significantly decreased compared to control subjects. The effect of resveratrol on early ovarian follicle development and oocyte apoptosis in rats was reported [39]. The female neonatal rats of adult SD rats were divided into control group, intraperitoneal injection group, which was injected daily with 25 mg/kg resveratrol within 12 h of birth, and intragastric perfusion group, in which pregnant rats were treated with resveratrol (25 mg/kg daily) by intragastric administration on day 12 after pregnancy till delivery. Ovaries were collected from the rats of each group on postnatal days 2 and 4. The development of ovarian follicles (the ratio of oocytes + primordial follicle and developed follicles) was examined by HE staining, and the oocyte apoptosis was detected by TUNEL staining. It was shown that resveratrol could delay the rupture of oocyte nests and inhibit the development of primordial follicle; however, it had no effect on oocyte apoptosis.

According to Ref. [40], resveratrol attenuated early pyramidal neuron excitability impairment and death in acute rat hippocampal slices caused by oxygen-glucose deprivation (OGD). Ischemic conditions by applying OGD to acute rat hippocampal slices were simulated and the effect of the resveratrol analyzed. The stilbene on OGD-induced pyramidal neuron excitability impairment was examined using whole-cell patch clamp recording. A 100 μM of resveratrol largely inhibited the 15 min OGD-induced progressive membrane potential (V_m) depolarization and the reduction in evoked action potential frequency and amplitude in pyramidal neurons. In a parallel neuronal viability study using TO-PRO-3 iodide staining, a 20-min OGD induced irreversible combretastatin A-1 (CA-1) pyramidal neuronal death that was significantly reduced by 100 μM of resveratrol. Resveratrol also markedly reduced the frequency and amplitude of AMPA-mediated spontaneous excitatory postsynaptic currents (sEPSCs) in pyramidal neurons. Resveratrol effects on neuronal excitability were inhibited by the large-conductance potassium channel (BK channel) inhibitor paxilline. Together, these studies demonstrated that resveratrol attenuates OGD-induced neuronal impairment occurring early on in the simulated ischemia slice model by enhancing the activation of BK channel and reducing the OGD-enhanced AMPA/NMDA receptor-mediated neuronal EPSCs. It was showed that resveratrol improved mitochondrial function and protected against metabolic disease by activating SIRT1 and PGC-1α (Section 7.1.6.2 and Figure 7.5) [41].

The peroxisome proliferator-activated receptors (PPARs) are a group of nuclear receptor proteins that function as transcription factors regulating the expression of genes. PPARs play essential roles in the regulation of cellular differentiation, development, and metabolism (carbohydrate, lipid, and protein) of higher organisms. Activation of PPARα with resveratrol was investigated [42]. Docking of resveratrol natural analogues was performed in PPAR-alpha ligand-binding domain. The proposed binding pose of these compounds in PPARα appeared similar for the active as well as the inactive compounds. Hypercholesterolemic hamsters were fed with blueberry skins and these animals showed lower levels of lipids compared to those fed ciprofibrate. Resveratrol and its three analogues (pterostilbene, piceatannol, and resveratrol trimethyl ether) were evaluated at 1, 10, 100, and 300 μM for the activation of endogenous PPARα in H4IIEC3 cells [43]. Cells were transfected with a peroxisome proliferator response element-AB (rat fatty acyl CoA β-oxidase response element)-luciferase gene reporter construct. Pterostilbene demonstrated the highest induction of PPARα, showing an 8- and 14-fold increase in luciferase activity at 100 and 300 μM, respectively compared to the control group. Hypercholesterolemic hamsters fed with pterostilbene at 25 ppm of the diet showed 29% lower plasma low-density lipoprotein (LDL) cholesterol, 7% higher plasma high-density lipoprotein (HDL) cholesterol, and 14% lower plasma glucose compared to the control group. The LDL/HDL ratio was also lower for pterostilbene, compared to that of the control animals, at this diet concentration. *In vivo* studies demonstrated that pterostilbene possesses lipid- and glucose-lowering effects.

The aim of the research [44] was to study the sulfation of resveratrol in the human liver and duodenum. A simple and reproducible radiometric assay for resveratrol sulfation was developed. It employed 3′-phosphoadenosine-5′-phosphosulfate-[35S] as the sulfate donor and the rates of resveratrol sulfation (pmol/min/mg cytosolic protein) were 90 (liver) and 74 (duodenum). Resveratrol sulfotransferase followed Michaelis–Menten kinetics, and K_m (µM) was 0.63 (liver) and 0.50 (duodenum) and V_{max} (pmol/min/mg cytosolic protein) was 125 (liver) and 129 (duodenum). Resveratrol sulfation was inhibited by flavonoid quercetin, by mefenamic acid, and by salicylic acid. The potent inhibition of resveratrol sulfation by quercetin, a flavonoid present in wine, fruits, and vegetables, suggested that compounds present in the diet may inhibit the sulfation of resveratrol, thus improving its bioavailability.

8.2
Combretastatin

8.2.1
Effects on Cancer Cells

It was demonstrated that low and nontoxic doses of microtubule-destabilizing agent combretastatin A-4-phosphate (CA-4-P) inhibited leukemic cell proliferation *in vitro* and induced mitotic arrest and cell death [45]. Treating acute myeloid leukemias (AMLs) with CA-4-P led to the disruption of mitochondrial membrane potential, release of proapoptotic mitochondrial membrane proteins, and DNA fragmentation, resulting in cell death in part in a caspase-dependent manner. CA-4-P increases intracellular reactive oxygen species, and antioxidant treatment imparts partial protection from cell death, suggesting that ROS accumulation contributes to CA-4-P-induced cytotoxicity in AML. A series of *cis*-restricted 1,5-disubstituted 1,2,3-triazole analogues of combretastatin A-4 (1) have been prepared [46]. The triazole 12f, 2-methoxy-5-(1-(3,4,5-trimethoxyphenyl)-1*H*-1,2,3-triazol-5-yl)-aniline, displayed potent cytotoxic activity against several cancer cell lines with IC_{50} values in the nanomolar range. The ability of triazoles to inhibit tubulin polymerization with IC_{50} about 5 µM has been evaluated. Molecular modeling involving the stilbene analogue and the colchicine binding site of α,β-tubulin showed that the triazole moiety interacts with β-tubulin via hydrogen bonding with several amino acids (Figure 8.4).

The relationship between microtubular dynamics, dismantling of pericentriolar components, and induction of apoptosis was analyzed after exposure of H460 nonsmall lung cancer cells to combretastatin A-4, as an antimitotic drug [47]. This compound led to microtubular array disorganization, arrest in mitosis, and abnormal metaphases, accompanied by the presence of numerous centrosome-independent "star-like" structures containing tubulin and aggregates of pericentrosomal matrix components such as γ-tubulin, pericentrin, and ninein. The structural integrity of centrioles was not affected by treatment. Treatment with combretastatin A-4 (7.5 nM), which produced high-frequency "star-like" aggregates, was accompanied by mitotic catastrophe commitment characterized by proapoptotic translocation. High drug concentrations, which fail to block cells at mitosis, were also unable to activate apoptosis. It was suggested that the maintenance of microtubular integrity plays a relevant role in stabilizing the pericentriolar matrix, whose dismantling can be associated with apoptosis after exposure to microtubule depolymerizing agents. Biological assay of new potent inhibitory activity of combretastatin D-4 showed such an activity against cellular proliferation of human HT-29 colon carcinoma cells [48]. Preparation of fluoro-containing combretastatin derivatives as microtubule polymerization inhibitors was patented [49]. The fluoro-containing combretastatin derivatives with R = (un)substituted amino, hydroxy, NO_2, halo, alkoxy, fluoro-containing alkyl, or pharmaceutically acceptable salts were prepared. These compounds acted as microtubule polymerization inhibitors for the treatment of tumor and neovascularization. For example, a compound prepared from Wittig reaction of 4-(2,2,2-trifluoroethoxy)-3-hydroxybenzaldehyde with (3,4,5-trimethoxybenzyl)triphenylphosphonium bromide (79.5%) showed inhibitory activity with IC_{50} of 0.005 µM against SGC-7901 lung cancer cells.

It was demonstrated that mitogen-activated protein kinases (MAPKs) were critically involved in the cytotoxicity of CA-4 [50]. CA-4 stimulated both extracellular signal-regulated kinases (ERK1/2) and p38 MAPK in the BEL-7402 hepatocellular carcinoma cell line in a time- and dose-dependent manner. This stimulation was a result of CA-4-induced microtubule disassembly, a reversible process. In authors' opinion, these data indicated that p38 MAPK is a potential anticancer target and that the combination of CA-4 with p38 MAPK inhibitors may be a novel and promising strategy for cancer therapy.

It was found that naphthalene was a good surrogate for the isovanillin moiety (3-hydroxy-4-methoxyphenyl) of combretastatin A-4 (Figure 8.4), generating highly cytotoxic analogues when combined with 3,4,5-trimethoxyphenyl or related systems [51]. The most cytotoxic naphthalene analogues of combretastatins, which also produce inhibition of tubulin polymerization, exerted their antimitotic effects through microtubule network disruption and subsequent G2/M arrest of the cell cycle in human cancer cells.

The effect of combretastatin A-4 on the growth and metastasis of gastric cancer cells at clinically achievable concentration and the associated antitumor mechanisms

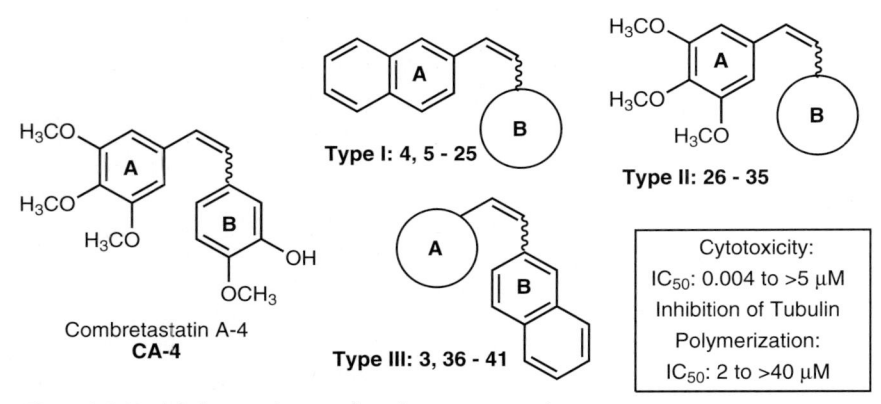

Figure 8.4 Naphthalene analogues of combretastatins used in Ref. [51]. (Reproduced with permission.)

were investigated [52]. Nine human gastric cancer cell lines, including two metastatic gastric cancer cell lines (AGS-GFPM1/2), constitutively expressing green fluorescence protein (GFP) were used. These metastatic AGS-GFPM1/2 cells expressed a higher level of phosphorylated serine 473 on AKT (p-AKT). Results showed that CA-4 (0.02–20 µM) has significant *in vitro* effects on reducing cell attachment, migration, and invasiveness, as well as cell G2/M disturbance on p-AKT-positive gastric cancer cells. A phosphoinositide 3-kinase inhibitor, LY294 002 [2-(4-morpholinyl)-8-phenyl-1(4H)-benzopyran-4-one hydrochloride], a specific AKT inhibitor, and 0.2–20 µM CA-4 displayed a similar response to p-AKT-positive cells, suggesting that the CA-4-induced effect was mediated by the inhibition of the PI3 kinase/AKT pathway. CA-4-phosphate (CA-4-P; 200 mg/kg) significantly inhibited the intraabdominal growth of xenotransplanted AGS-GFPM2 cells in nude mice. CA-4-P treatment showed an ability to inhibit gastric tumor metastasis and attenuate p-AKT expression.

Compounds structurally similar to combretastatin A-4 were represented by the formula,

which can be used as tubulin polymerization inhibitor and antitumor agent, wherein R is –OH or –OPO(ONa)2, have been prepared [53]. A novel series of 15 pyrazoline analogues of combretastatin A-4 have been synthesized and the biological activity of these compounds assessed using murine B16 and L1210 cell lines (melanoma and leukemia, respectively) [54]. The compounds possessed a pyrazoline, acetyl-pyrazoline, or a thiourea group in the central core of the molecule. The most potent compound in L1210 cells ($IC_{50} = 0.5$ µM) was a pyrazoline derivative with the same substituents as CA-4. This compound inhibited tubulin polymerization with an EC_{50} of 46 µM in L1210 cells. A thiourea compound with the same substituents as CA-4 also showed potent cyctotoxicity ($IC_{50} = 2.6$ and 1.4 µM in B16 and L1210 cells, respectively) and tubulin inhibitory properties. Ten 1,2,3,4-tetrahydro-2-thioxopyrimidine analogues of combretastatin A-4 designed to have improved water solubility over CA-4 were synthesized and their cytotoxicity against the growth of two murine cancer cell lines (B16 melanoma and L1210 leukemia) in culture was detected using an MTT assay [55]. Two of the 2-thioxopyrimidine analogues exhibited significant activity (IC_{50} 1 M, L1210, and B16 cells). Molecular modeling studies using Macspartan indicated that the two active 2-thioxopyrimidine analogues preferably adopt a twisted conformation, similar to CA-4, suggesting that conformation and structure are associated with compound activity.

A series of benzil derivatives related to combretastatin A-4 have been synthesized by the oxidation of diarylalkynes promoted by PdI_2 in DMSO (Figure 8.5) [56]. Several benzils exhibited antiproliferative activity and inhibited cell growth of four

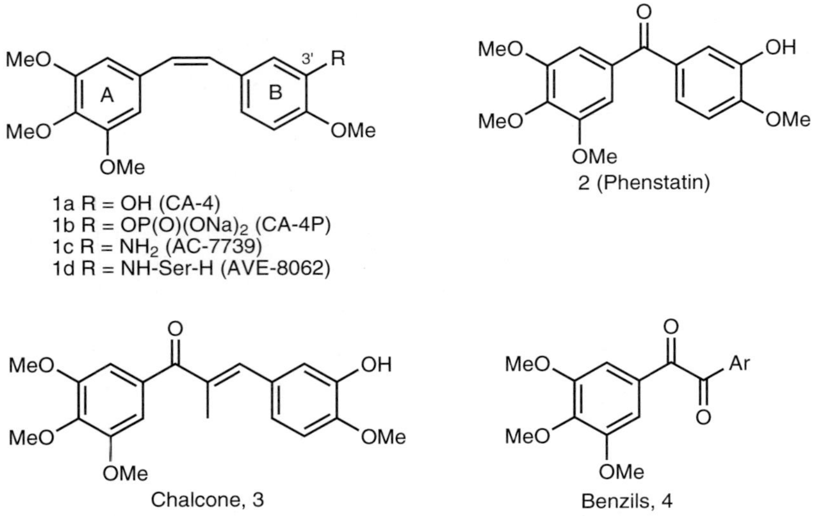
MeO
MeO
OMe
A
B
3'
R
OMe
1a R = OH (CA-4)
1b R = OP(O)(ONa)2 (CA-4P)
1c R = NH2 (AC-7739)
1d R = NH-Ser-H (AVE-8062)
O
MeO
MeO
OMe
OH
OMe
2 (Phenstatin)
O
MeO
MeO
OMe
OH
OMe
Chalcone, 3
O
MeO
MeO
OMe
Ar
O
Benzils, 4

Figure 8.5 Representative tubulin binding agents and general structure of the synthesized benzils **4** [56]. (Reproduced with permission from Elsevier.)

human tumor cell lines at the nanomolar level (20–50 nM). Flow cytometric analysis indicated that these compounds acted as antimitotics and arrested the cell cycle in G2/M phase (Figure 8.6). A cell-based assay indicated that these compounds displayed a similar inhibition of tubulin assembly with an IC_{50} value similar to CA-4. These results demonstrated that the *Z*-double bond of CA-4 can be replaced by a 1,2-diketone unit without any significant loss of cytotoxicity and inhibition of tubulin assembly potency.

Sixteen 2-cyclohexenone and 6-(ethoxycarbonyl)-2-cyclohexenone analogues of combretastatin A-4 were synthesized, and their ability to inhibit the growth of two murine cancer cell lines (B16 melanoma and L1210 leukemia) was detected using an MTT assay [57]. The cyclohexenone analogues, which contain the same substituents as CA-4 after exposure of A-10 aortic cells to cyclohexenone, produced significant reduction in cellular microtubules, with EC_{50}. The synthesis and biological evaluation of a series of tubulin polymerization inhibitors, with substituents R = 4-$MeOC_6H_4$, 3-HO-4-$MeOC_6H_3$, *N*-methyl-5-indolyl and containing the 1,2,4-triazole ring to retain the bioactive configuration afforded by the *cis*-double bond in combretastatin A-4, were described [58]. Several compounds exhibited potent tubulin polymerization inhibitory activity as well as cytotoxicity against a variety of cancer cells including multidrug-resistant cancer cell lines. Attachment of the *N*-methyl-5-indolyl moiety to the 1,2,4-triazole core, as exemplified by R = *N*-methyl-5-indolyl, conferred optimal properties on this series. Computer docking and molecular simulations inside the colchicine binding site of tubulin enabled identification of residues most likely to strongly interact with these inhibitors and explain their potent antitubulin activity and cytotoxicity.

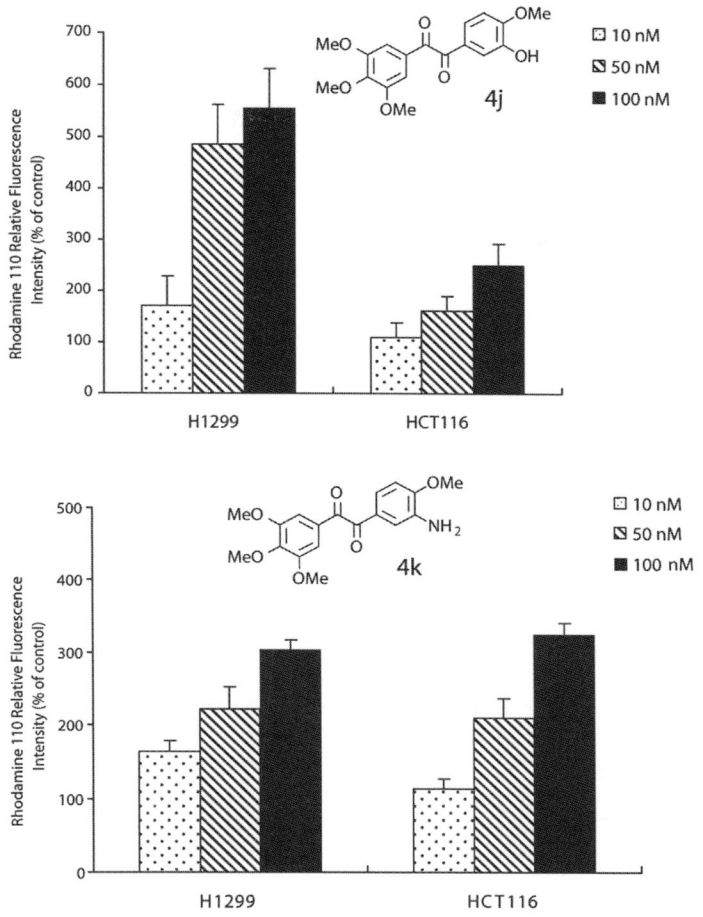

Figure 8.6 Apoptotic effects of benzils **4j** and **4k** on HCT116 and H1299 cells. Percentage of apoptotic cells induced by different concentrations of **4j** and **4k** (evaluation after 48 h of treatment) [56]. (Reproduced with permission from Elsevier.)

Several stilbenoid compounds having structural similarity to the combretastatin group of natural products and characterized by the incorporation of two N-bearing groups (amine, nitro, and serinamide) were prepared by chemical synthesis and evaluated in terms of their biochemistry and biological activities [59]. The 2′,3′-diamino B-ring analogue (Z)-2-(2,3-diamino-4-methoxyphenyl)-1-(3,4,5-trimethoxyphenyl)ethene demonstrated cytotoxicity against selected human cancer cell lines in vitro ($GI_{50} = 13.9$ nM) and also showed good activity with regard to inhibition of tubulin assembly ($IC_{50} = 2.8$ μM). The potential value of this compound and its corresponding salt formulations as new vascular disrupting agents (VDA) was discussed.

Molecular mechanisms leading to cell death in nonsmall lung cancer H460 cells induced by natural (CA-4) and synthetic stilbenoids (ST2151), structurally related to CA-4, were investigated [60]. It was found that both compounds induced depolymerization and rearrangement of spindle microtubules, as well as an increasingly aberrant organization of metaphase chromosomes in a dose- and time-dependent manner. Prolonged exposure to ST2151 led cells to organize multiple sites of tubulin repolymerization whereas tubulin repolymerization was observed only after CA-4 washout. H460 cells were arrested at a prometaphase stage, with condensed chromosomes and a triggered spindle assembly checkpoint, as evaluated by kinetochore localization of Bub1 and Mad1 antibodies. A persistent checkpoint activation led to mitochondrial membrane permeabilization (MMP) alterations, cytochrome c release, activation of caspase-9 and -3, PARP cleavage, and DNA fragmentation. In authors' opinion, the ability of cells to reassemble tubulin in the presence of an activated checkpoint may be responsible for ST2151-induced multinucleation, a recognized sign of mitotic catastrophe.

A variety of derivatives of the most common anticancer drugs including combretastatin A-4, DOPA carrying an accessible COOH group, convenient for coupling with peptide-NH_2, under standard solid-phase synthesis conditions were prepared [61]. The choice of linkers between the tetrabranched NT peptide and the functional unit was found to be crucial for the release of the cytotoxic molecules inside tumor cells. Ester or disulfide linkages as suitable covalent bonds able to safely release the drug inside the cell either by hydrolysis or by exchange reactions with cytoplasmic thiols (e.g., glutathione, GSH) were exploited. A synthesis of combretastatin A-4 and a small library of analogues led to the discovery of some new cooperative *ortho* effects allowing (Z)-stilbenes to be prepared in high yield and diastereomeric ratio [62]. Combretastatin A-4 and a dibromo analogue exhibited cell growth inhibitory activity on breast cancer cells. Synthesis and biological properties of bioreductively targeted nitrothienyl prodrugs of combretastatin A-4

15 R_1=H, R_2=H
16 R_1=H, R_2=Me
17 R_1=Me, R_2=Me
18 R_1=Me, R_2=CO_2Et

exhibited cell growth inhibitory activity on breast cancer cells were reported [63]. Nitrothienylprop-2-yl ether formation on the 3′-phenolic position of combretastatin A-4 (1) abolished the cytotoxicity and tubulin polymerization inhibitory effects of the drug. 5-Nitrothiophene derivatives of **1** were synthesized, and compound (I) represented a new lead in bioreductively targeted cytotoxic anticancer therapies. In this compound, optimized *gem*-dimethyl α-carbon substitution enhanced both the aerobic metabolic stability and the efficiency of hypoxia-mediated drug release. Only the *gem*-substituted derivative released **1** under anoxia in either *in vitro* whole-cell

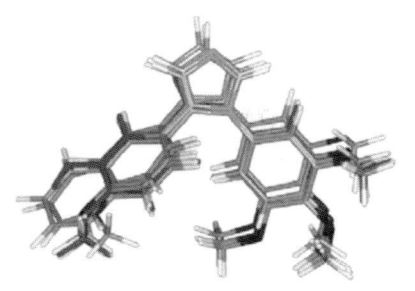

Figure 8.7 Arrangement of **1** diaryl-1*H*-imidazoles inside the colchicine binding site [64]. (Reproduced with permission from Elsevier.)

experiments or supersomal suspensions. The rate of release of **1** from the radical anions of these prodrugs was enhanced by more methyl-substitution on the α-carbon. Cellular and supersomal studies showed that this α-substitution pattern controls the useful range of oxygen concentration over which **1** can be effectively released by the prodrug. The *in vitro* antitumor activity of novel combretastatin-like 1,5- and 1,2-diaryl-1*H*-imidazoles was evaluated against the NCI 60 human tumor cell lines panel [64]. These compounds proved to be more cytotoxic than CA-4 in tests involving their evaluation over a 10-4–10-8 range. Docking experiments showed a good correlation between the MG-ID log GI_{50} values of all these compounds and their calculated interaction energies with the colchicine binding site of α,β-tubulin. Arrangement of **1** diaryl-1*H*-imidazoles inside the colchicine binding site is shown in Figure 8.7.

A new series of aryl-substituted imidazol-2-one derivatives structurally related to combretastatin A-4 were synthesized and evaluated for their cytotoxic activities *in vitro* against various human cancer cell lines including MDR cell line [65]. The highly active compounds also exhibited inhibitory activity against tumor growth *in vivo*.

8.2.2
Xenografts and Tumors

It was shown that CA-4-P (**4**) prodrug has underwent extensive preclinical evaluation in human tumor xenografts and orthotopically transplanted tumors in murine models, demonstrating that the prodrug caused profound disruption of the tumor blood vessel network [66–74].

To examine the pathophysiological impact of treatment with combretastatin A-4-phosphate on the regions of tumors that ultimately either necrose or survive treatment with this agent, proliferation, perfusion, and expression of vascular endothelial growth factor (VEGF) were analyzed in the KHT tumor model after treatment with CA-4-P [2]. Analyses were conducted on the whole tumor and the tumor periphery. It was shown that perfusion in the tumor periphery decreased 4 h after treatment but returned to baseline 20 h later. Whole-tumor perfusion also decreased 4 h after treatment but did not return to baseline. The decrease in perfusion could have negative affect on therapies using the combination of CA-4-P and conventional anticancer agents by decreasing drug delivery and tissue oxygenation. These findings suggested that the timing of CA-4-P treatments when used in conjunction with conventional anticancer therapies should be considered carefully. The authors of the works [75] showed that CA-4-P selectively targeted endothelial cells, but not smooth muscle cells, and induced the regression of unstable nascent tumor neovessels by rapidly disrupting the molecular engagement of the endothelial cell-specific junctional vascular endothelial-cadherin (VE-cadherin) *in vitro* and *in vivo* in mice. CA-4-P increases endothelial cell permeability while inhibiting endothelial cell migration and capillary tube formation predominantly through disruption of VE-cadherin/β-catenin/Akt signaling pathway, thereby leading to rapid vascular collapse and tumor necrosis. Stabilization of VE-cadherin signaling in endothelial cells with adenovirus E4 gene or ensheathment with smooth muscle cells conferred resistance to CA-4-P. CA-4-P synergized with low and nontoxic doses of neutralizing mAbs to VE-cadherin by blocking the assembly of neovessels, thereby inhibiting tumor growth. The authors suggested that combined treatment with anti-VE-cadherin agents in conjunction with microtubule-disrupting agents provides a novel synergistic strategy to selectively disrupt the assembly and induce regression of nascent tumor neovessels, with minimal toxicity and without affecting normal stabilized vasculature. Figure 8.8 illustrates the CA-4-P inhibition of growth factor-induced endothelial cell proliferation and migration.

A combination chemotherapy including combretastatin A-4-phosphate and paclitaxel was found to be effective against anaplastic thyroid cancer in a nude mouse xenograft model [76]. The nude mouse xenograft model with ARO and KAT-4 cells was treated by the first combination consisting of CA-4-P, paclitaxel, and manumycin A (a farnesyltransferase inhibitor), and the second, CA-4-P, paclitaxel, and carboplatin. Main outcome measures included tumor growth curves and tumor weights. The tumor growth curve analysis demonstrated that both triple-drug combinations were significantly better than placebo for both cell lines. CA-4-P decreased the depth of the viable outer rim of tumor cells on xenograft sections. It was demonstrated that radioimmunotherapy using 131I-A5B7, an anti-CEA antibody, in combination with combretastatin A-4-phosphate (200 mg/kg), produced tumor cures in SW1222 colorectal xenografts [77]. CA-4-P caused acute tumor blood vessel shutdown, which can be monitored in clinical trials using dynamic contrast-enhanced magnetic resonance imaging (DCE-MRI). The magnitude of the antivascular effect of CA-4-P in the SW1222 tumor, at 200 mg/kg and at lower doses, was detected using conventional assays. In addition, related effects of changes in DCE-MRI parameters and the corresponding effects on tumor retention of 131I-A5B7 were evaluated. A significant reduction in tumor DCE-MRI kinetic parameters, the initial area under the contrast agent concentration–time curve (IAUGC) and the transfer constant (K_{trans}), was demonstrated 4 h after CA-4-P. The authors concluded that moderate tumor blood flow reduction following antibody administration is sufficient to improve tumor antibody retention and this is encouraging for the combination of CA-4-P and 131I-A5B7 in clinical trials.

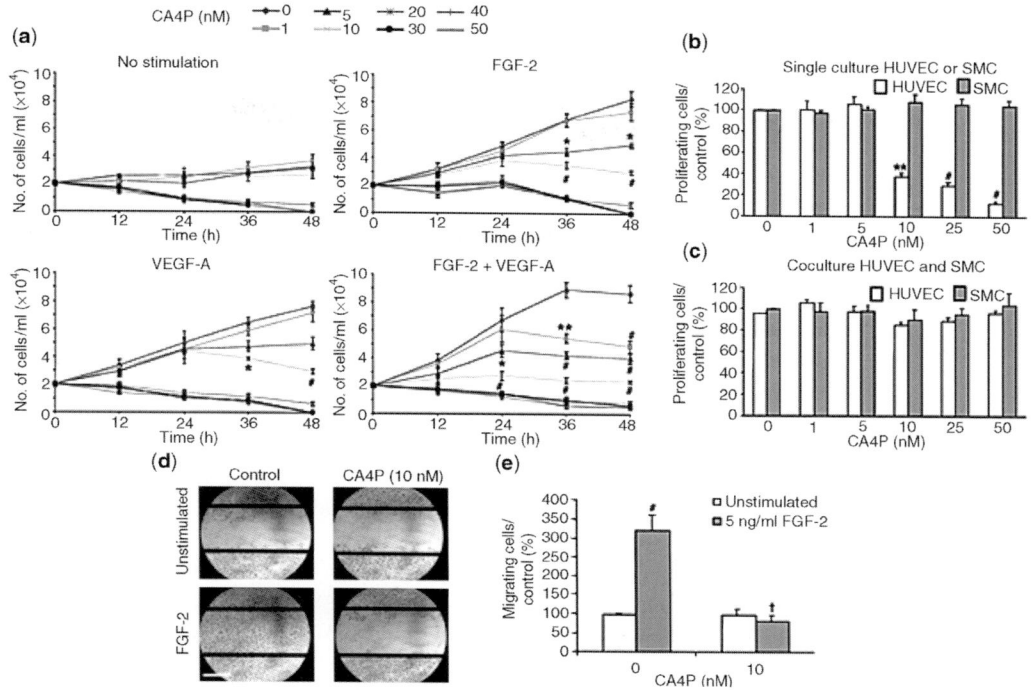

Figure 8.8 Data on the CA-4-P inhibition of growth factor-induced endothelial cell proliferation and migration. (a) CA-4-P inhibition of HUVEC proliferation. (b) Absence SMCs sensitivity to CA-4-P. (c) Resistance of HUVECs to CA-4-P. (d) CA-4-P inhibition of HUVEC migration. (e) Quantification of recovery of each denuded area after CA-4-P treatment [75]. (Reproduced with permission from Elsevier.)

For the evaluation of the effect of the vascular targeting agent, combretastatin A-4-phosphate, on tumor oxygenation compared to vascular perfusion/permeability, ^{19}F MRI oximetry and DCE-MRI were used to monitor tumor oxygenation and perfusion/permeability in syngeneic 13762NF rat breast carcinoma [78]. A significant drop was found in the mean tumor pO_2 (from 23 to 9 mmHg) within 90 min after treatment (30 mg/kg of combretastatin A-4-phosphate) and a further decrease was observed at 2 h (2 mmHg). The initial area under signal intensity curve was shown to be fully recovered after 24 h in a thin peripheral region but not in the tumor center. The response observed by DCE-MRI, indicating vascular shutdown, paralleled the pO_2 measurements. It was concluded that quantitative pO_2 measurements are potentially important for optimizing the therapeutic combination of vascular targeting agents with radiotherapy. It was demonstrated that vascular targeting agents enhanced chemotherapeutic agent activities in solid tumor therapy [79]. The utility of combining the vascular targeting agents 5,6-dimethyl-xanthenone-4-acetic acid (DMXAA) and combretastatin A-4 disodium phosphate (CA-4-DP) with the anticancer drugs cisplatin and cyclophosphamide (CP) was evaluated in experimental rodent (KHT sarcoma), human breast (SKBR3), and ovarian (OW-1) tumor models. Doses of the vascular targeting agents that led to rapid vascular shutdown and subsequent extensive central tumor necrosis were identified. A histological evaluation showed morphological damage of tumor cells within a few hours after treatment, followed by extensive hemorrhagic necrosis and dose-dependent neoplastic cell death as a result of prolonged ischemia. DMXAA also enhanced the tumor cell killing of cisplatin, but doses >15 mg/kg were required. In contrast, CA-4-DP increased cisplatin-induced tumor cell killing at all doses studied.

The transverse magnetic resonance imaging relaxation rate $R2^*$ was used as a biomarker of tumor vascular response to monitor vascular disrupting agent therapy [80]. Multigradient echo MRI was used to quantify $R2^*$ in rat GH3 prolactinomas. R2 is a sensitive index of deoxyHb in the blood and therefore was used to give an index of tissue oxygenation. Tumor $R2^*$ was measured before and up to 35 min after treatment, and 24 h after treatment with either 350 mg/kg 5,6-dimethylxanthenone-4-acetic acid or 100 mg/kg CA-4-P. After acquisition of the MRI data, functional tumor blood vessels remaining after VDA treatment were quantified using fluorescence microscopy of the perfusion marker Hoechst 33342. DMXAA induced a transient, significant increase in tumor $R2^*$ 7 min after treatment, whereas CA-4-P induced no significant changes in tumor $R2^*$ over the first 35 min. These results suggest that DMXAA was less effective than CA-4-P in this rat tumor model. Acute effects of the antivascular drug, combretastatin A-4-phosphate, on tumor energy status and perfusion were assessed using MRI and spectroscopy (MRS) [81]. Localized ^{31}P magnetic resonance spectroscopy showed that LoVo and RIF-1 tumors responded well to drug treatment, with significant increase in the Pi/nucleoside triphosphate ratio within 3 h, whereas SaS, SaF, and HT29 tumors did not respond to the same extent. This variable response was also seen in MRI experiments in which tumor perfusion was assessed by monitoring the kinetics of inflow of the contrast agent, gadolinium diethylenetriaminepentaacetate (GdDTPA). These data were analyzed to give the initial rate and time constant for the inflow of contrast agent and the integral under the inflow curve. The differential susceptibility of the tumors to combretastatin A-4-phosphate showed a correlation with prior MRI measurements of tumor vascular permeability, which was detected by measuring the inflow of a macromolecular contrast agent, BSA-GdDTPA.

A purpose of the study [82] was to detect how combretastatin A-4 disodium phosphate dose-dependent changes in radiation response of a C3H mouse mammary carcinoma relate to measurements of DCE-MRI parameters. C3H mammary carcinomas grown in female CDF1 mice were treated when at 200 mm^3 in size. Groups of mice were given graded radiation doses, either alone or followed 30 min later by an intraperitoneal injection of CA-4-DP, administered at doses of 10–250 mg/kg. The radiation dose producing local tumor control in 50% of treated animals in 90 days (TCD50) was calculated for each CA-4-DP dose. DCE-MRI was performed before and 3 h after CA-4-DP administration, and parameters describing vascularity and interstitial volume were estimated. TCD50 showed a dose-dependent decrease reaching a significant level at 25 mg/kg. DCE-MRI data predicted the CA-4-DP enhancement of the tumor radiation response and suggested clinical CA-4-DP doses necessary to improve the radiation response in patients.

Factors that influence tumor susceptibility were evaluated [83]. Mouse fibrosarcoma cell lines that are capable of expressing all vascular endothelial growth factor isoforms (control) or only single isoforms of VEGF (VEGF120, VEGF164, or VEGF188) were developed under endogenous VEGF promoter control. It was shown that VEGF188 was uniquely associated with tumor vascular maturity, resistance to hemorrhage, and resistance to CA-4-P. Intravital microscopy measurements of vascular length and RBC velocity showed that CA-4-P produced significantly more vascular damage in VEGF120 and VEGF164 tumors than in VEGF188 and control tumors. It was suggested that VEGF isoforms might be useful in vascular disrupting cancer therapy to predict tumor susceptibility. Using nuclear magnetic resonance imaging (MRI) and spectroscopy (MRS) the effects of combretastatin A-4 prodrug on perfusion and the levels of 31P metabolites in an implanted were examined for 3 h after drug treatment [84]. The area of regions of low signal intensity in spin-echo images of tumors increased slightly after treatment with the drug. These regions of low signal intensity corresponded to necrosis seen in histological sections, whereas the expanding regions surrounding them corresponded to hemorrhage. Tumor perfusion was assessed before and 160 min after drug treatment using dynamic MRI measurements of gadolinium diethylenetriaminepentaacetate uptake and washout. Perfusion decreased significantly in central regions of the tumor after treatment, which was attributed to the disruption of the vasculature and was consistent with the hemorrhage seen in histological sections. The mean apparent diffusion coefficient of water within the tumor did not change, indicating that there was no expansion of necrotic regions 3 h after drug treatment. Localized 31P-MRS showed that there was decline in cellular energy status in the tumor after treatment with the drug. Concentrations of nucleoside triphosphates within the tumor fell, the inorganic phosphate concentration increased, and there was a significant decrease in tumor pH for 80 min after drug treatment. The rapid, selective, and extensive damage caused to these tumors by combretastatin A-4 prodrug had highlighted the potential of the agent as a novel cancer chemotherapeutic agent. It was concluded that the response of tumors to treatment with the drug may be monitored noninvasively using MRI and MRS experiments that are appropriate for use in a clinical setting.

It was established [85] that combretastatin induced extensive blood flow shutdown in the tumor compared to normal tissues. A histological assessment of vascular shutdown showed that over 90% of vessels were rendered nonfunctional 6 h posttreatment with 100 mg/kg bul.i.p. Measurement of blood flow using a diffusible tracer 86RbCl indicated an overall reduction in perfusion by only 50–60%. Results showed that combretastatin can significantly enhance tumor response to both *cis*-platinum and radiation. The studies confirmed combretastatin A-4-phosphate as an agent that targets and damages tumor vasculature and indicated its potential therapeutic usefulness as an adjuvant to conventional cytotoxic approaches.

8.2.3
Animals

Preclinical studies to predict the efficacy of vascular changes induced by combretastatin A-4 disodium phosphate in patients were performed [86]. The dose-dependent effect of CA-4-DP on changes in radiation response of a C3H mouse mammary carcinoma by measuring dynamic contrast-enhanced magnetic resonance imaging parameters was investigated. The DCE-MRI results were compared with published clinical DCE-MRI data. C3H mammary carcinomas grown in female CDF1 mice were treated when at 200 mm^3 in size. Groups of mice were given graded radiation doses, either alone or followed 30 min after by an intraperitoneal injection of CA-4-DP, administered at doses of 10–250 mg/kg. The radiation dose producing local tumor control in 50% of treated animals in 90 days (TCD50) showed a dose-dependent decrease reaching a significant level at 25 mg/kg. The authors concluded that DCE-MRI data could predict the CA-4-DP enhancement of the tumor radiation response and suggest the clinical CA-4-DP doses necessary to improve the radiation response in patients. The *in vivo/in vitro* inhibitory effects of compound combretastatin A-4-phosphate on human leukemia HL-60 cell line and on HL-60 cell inoculated with transplant tumor nude mice were investigated [87]. The inhibitory effect of CA-4-P on human leukemia cell line HL-60 was assayed by MTT analysis, cell growth curve, clone forming, and cell dying experiments with different concentrations of CA-4-P. The inhibitory effect of CA-4-P on nude mice inoculated with HL-60 cells *in vivo* was demonstrated by the prolongation of survival times in mice carrying tumor. The dose–effect relationship was also observed in the experiment of HL-60 cell inoculated with transplant tumor nude mice (Figure 8.9).

It was demonstrated that *in vivo* CA-4-P inhibited proliferation and circulation of leukemic cells and diminished the extent of perivascular leukemic infiltrates, prolonging survival of mice that underwent xenotransplantation without inducing hematologic toxicity [88]. CA-4-P decreases the interaction of leukemic cells with neovessels by downregulating the expression of the adhesion molecular VCAM-1, thereby augmenting leukemic cell death. The authors suggested that CA-4-P targets both circulating and vascular-adherent leukemic cells through mitochondrial damage and downregulation of VCAM-1 without incurring hematologic toxicities and provides an effective means to treat refractory organ-infiltrating leukemias. With an objective to provide a novel derivative of a combretastatin that has water solubility and is capable of releasing a drug independent of biological enzymes and whose effective therapeutic effect can be expected, a polymer conjugate of combretastatin was prepared [89]. A polymer conjugate of a combretastatin, characterized by a structure in which a hydroxy group of a combretastatin was linked by an ester bond to a carboxylic acid group of the following polymer moiety in a block copolymer of a polyethylene glycol structure moiety with a polymer moiety having two of more carboxylic acid groups such as polyaspartic acid or polyglutamic acid, was provided. A conjugate of combretastatin A-4 with methoxypolyethylene glycol-polyaspartic acid block copolymer was synthesized. The conjugates showed the drug release property and antitumor effect in mice.

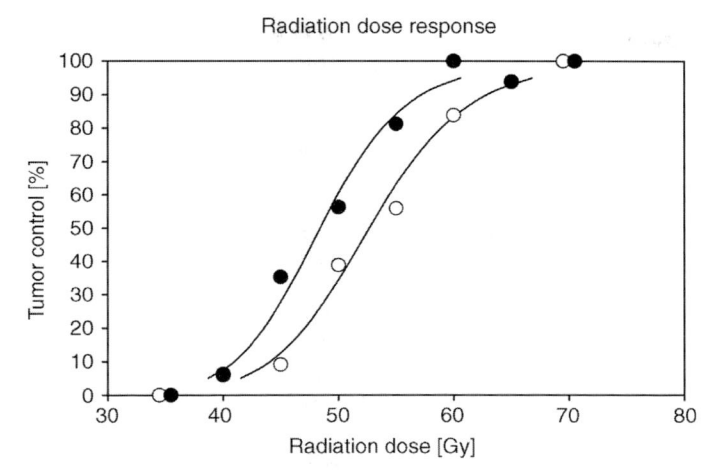

Figure 8.9 Effect of combretastatin A-4 disodium phosphate on radiation response of C3H mammary carcinoma. Tumors locally irradiated with single graded doses of radiation in control animals (open circles) or 30 min before injection of CA-4-DP at dose of 25 mg/kg (filled circles). Tumor response expressed as percentage of animals in each treatment group showing local tumor control 90 days after treatment. Each treatment group consisted of an average of 12 mice. Lines were fitted using logit analysis [86]. (Reproduced with permission of Elsevier.)

The anticancer effect of the novel vascular disrupting agent, combretastatin A-1 disodium phosphate (OXi4503), when combined with mild hyperthermia and/or radiation was described [90]. C3H mammary carcinoma was grown in the rear right foot of female CDF1 mice and treated when a volume of 200 mm^3 was reached. OXi4503 was administered intraperitoneally at variable doses. Hyperthermia was administered locally to the tumor-bearing foot using a thermostat-controlled water bath. Radiation treatment was performed locally using a conventional X-ray machine. Tumor response was assessed with either a tumor growth time or a tumor control assay. The authors conceded that OXi4503 was capable of significantly enhancing the anticancer effect of mild hyperthermia, and the mild temperature of diosensitization was enhanced.

It was shown that oxidation of CA-1 by peroxidase, tyrosinase, or Fe(III) generated a species with mass characteristics of the corresponding *ortho*-quinone Q1 [91]. After the administration of CA-1-bis(phosphate) to mice, the hydroquinone-thioether conjugate Q1H2-SG, formed from the nucleophilic addition of GSH to Q1, was detected in liver. Electrocyclic ring closure of Q1, over a few minutes at pH 7.4, led to a second *ortho*-quinone product Q2, characterized by an exact mass and NMR. This product was also generated by human promyelocytic leukemia (HL-60) cells *in vitro* provided that superoxide dismutase was added. Free radical intermediates formed during autoxidation of CA-1 were characterized by EPR, and the effects of GSH and ascorbate on the signals were studied. Pulse radiolysis was used to initiate selective one-electron oxidation or reduction and provided further evidence, from the differing absorption spectra of the radicals formed on oxidation of CA-1 or reduction of Q2, that two different quinones were formed on oxidation of CA-1. The results demonstrated fundamental differences between the pharmacological properties of CA-1 and CA-4. In the study [4], the authors investigated whether novel water-soluble combretastatin A-4 derivative, (Z)-N-[2-methoxy-5-[2-(3,4,5-trimethoxyphenyl)vinyl] phenyl]-L-serinamide hydrochloride, AC7700, acts in the same way against solid tumors growing in the liver, stomach, kidney, muscle, and lymph nodes. Tumor blood flow and the change in tumor blood flow induced by AC7700 were measured by the hydrogen clearance method. In a model of cancer chemotherapy against metastases, LY80 cells (2×10^6) were injected into the lateral tail vein, and AC7700 at 10 mg/kg was injected intravenously five times at 2-day intervals, starting on day 7 after tumor cell injection. The number and size of tumors were compared with those in the control group. The change in tumor blood flow and the therapeutic effect of AC7700 on microtumors were observed directly by using Sato lung carcinoma implanted in a rat transparent chamber. AC7700 caused a marked decrease in the tumor blood flow of all LY80 tumors developing in various tissues and organs, and the growth of all tumors including lymph node metastases and microtumors was inhibited. In every tumor, tumor blood flow began to decrease immediately after AC7700 administration and reached a minimum at approximately 30 min after injection. In many tumor capillaries, blood flow completely stopped within 3 min after AC7700 administration. These results demonstrated that AC7700 is effective for tumors growing in various tissues and organs and for metastases. It was concluded that tumor blood flow stanching induced by AC7700 may become an effective therapeutic strategy for all cancers, including refractory cancers, because the therapeutic effect is independent of tumor site and specific type of cancer.

The potential of targeting the tumor vascular flow by using the tubulin destabilizing agent disodium combretastatin A-4 3-O-phosphate (CA-4-P) was assessed in a rat system [92]. The early vascular effects of CA-4-P were assessed in the implanted P22 carcinosarcoma and in a range of normal tissues. Blood flow was measured by the uptake of radiolabeled iodoantipyrine (IAP), and quantitative autoradiography was used to measure spatial heterogeneity of blood flow in tumor sections. CA-4-P (100 mg/kg i.p.) caused a significant increase in the mean arterial blood pressure (MABP) at 1 and 6 h after treatment and a very large decrease in tumor blood flow, which by 6 h was reduced approximately 100-fold. The spleen was the most affected normal tissue with a sevenfold reduction in blood flow at 6 h. Calculations of vascular resistance revealed some vascular changes in the heart and the kidney for which there were no significant changes in blood flow. Quantitative autoradiography showed that CA-4-P increased the spatial heterogeneity in tumor blood flow. The drug affected peripheral tumor regions less than central regions. Administering CA-4-P (30 mg/kg) in the presence of the nitric oxide synthase (NOS) inhibitor, N(omega)-nitro-L-arginine methyl ester, showed that tissue production of nitric oxide protects against the damaging vascular effects of CA-4-P. The action of CA-4-P includes mechanisms other than those involving red cell viscosity, intravascular coagulation, and neutrophil adhesion. The uptake of CA-4-P and CA-4 was more efficient in tumor than in skeletal muscle tissue and dephosphorylation of CA-4-P to CA-4 was faster in the former. The authors claimed that these results are promising for the use of CA-4-P as a tumor vascular-targeting agent.

An antitumor combination including combretastatin derivatives and an anticancer compound selected from VEGF inhibitors, more particularly VEGF-Trap, was patented [93]. The invention also disclosed methods of using these pharmaceutical preparations for the treatment of solid and similar carcinomas and particularly for modulating tumor growth or metastasis in animal and human breast carcinoma (MX-1). The immediate effects of combretastatin A-4 disodium phosphate on tumor IFP in C3H mammary carcinoma in mice were examined [94]. Animals were treated with 100 mg/kg CA-4-DP by i.p. injection. Tumor perfusion was recorded by laser Doppler flowmetry at separate time points, and IFP was recorded continuously by the wick-in-needle method. It was found that CA-4-DP treatment resulted in a rapid reduction in tumor perfusion, followed by a decrease in IFP; no increase in IFP was observed. This suggested that CA-4-DP-induced reduction in tumor perfusion does not depend on the increase in IFP. Combining tumor cell immunization with the vascular-targeting drug CA-4-P, which enhanced tumor retardation and/or affected the antitumor immune response, was investigated [95]. Rats with intrahepatic colon carcinoma were immunized weekly with IL-18/IFN-γ-transfected tumor cells, starting on day 9, and were treated with a low-dose CA-4-P (2 mg/kg, 5 days a week starting on day 7). The effect of CA-4-P on tumor growth and on immune reactivity was studied *in vitro*. Rats with pre-existing tumor, immunized and treated with low-dose CA-4-P, had a significantly retarded tumor growth compared to rats receiving CA-4-P or immunization alone. Splenocytes from rats treated with this combination had a significantly enhanced antitumor immune response compared to splenocytes from control rats. Combining the nitric oxide synthase inhibitor N-nitro-L-arginine methyl ester with CA-4-P and immunization further retarded tumor growth. Concomitant treatment of rats with progressively growing tumor with immunization and low-dose CA-4-P significantly enhanced the therapeutic effect compared to either treatment alone and resulted in an enhanced antitumor immune reactivity.

The antivascular actions of disodium combretastatin A-4 3-*O*-phosphate (CA-4-P) were investigated in the rat P22 carcinosarcoma after single doses of 10 or 30 mg/kg [96]. Pharmacokinetic data showed that 10 mg/kg in the rat gave a plasma exposure similar to that achieved in the clinic. Blood flow rate to the tumor and normal tissues was measured using the uptake of radiolabeled iodoantipyrine. Quantitative autoradiography was used to detect changes in spatial distribution of tumor blood flow. Both doses caused an increase in MABP and a reduction in heart rate 1 h after treatment. In authors' opinion, the data obtained provide an insight into the mechanisms underlying tissue blood flow changes occurring after clinically relevant doses of CA-4-P are administered and may help interpret pharmacodynamic data obtained from phase I/II clinical trials of CA-4-P and are relevant for future drug development in this area.

The ability of combretastatin A-4 disodium phosphate to induce vascular damage and enhance the radiation response of murine tumors was investigated [97]. A C3H mouse mammary carcinoma transplanted in the foot of CDF1 mice and the KHT mouse sarcoma growing in the leg muscle of C3H/HeJ mice were used. CA-4-DP was dissolved in saline and injected intraperitoneally. Tumor blood perfusion was estimated using 86RbCl extraction and Hoechst 33342 fluorescent labeling. Necrotic fraction was detected from histological sections. Tumors were locally irradiated in nonanesthetized mice and the response was assessed by local tumor control for the C3H mammary carcinoma and *in vivo/in vitro* clonogenic cell survival for the KHT sarcoma. CA-4-DP decreased tumor blood perfusion and increased necrosis in a dose-dependent fashion in the C3H mammary carcinoma, which was maximal at 250 mg/kg. CA-4-DP enhanced radiation damage in the two tumor models without enhancing normal tissue damage. These radiation effects were clearly consistent with the antivascular action of CA-4-DP. The effect of a single intraperitoneal combretastatin A-4-phosphate combreAp injection on the growth of rhabdomyosarcomas syngeneic in WAG/Rij rats was evaluated [98]. Different tumor volume groups, ranging 0.1–27 cm^3, were selected to assess the relationship between the size at treatment time and the response to combreAp, and a double combreAp treatment (25 mg/kg, twice) was investigated. The systemic administration of combreAp induced a clear-cut differential growth delay in the solid rat rhabdomyosarcomas: with very large tumors ($\geq$14 cm^3), a 17.6-fold stronger effect was measured than with very small tumors (<1 cm^3). It was demonstrated that growth delay was related to an early (within 3–6 h) and extensive breakdown of tumor blood vessels. The differential volume–response obtained with "selective" vascular targeting, stronger in larger tumors than smaller ones, suggested the potential of broadening the therapeutic window.

The toxicity of combretastatin A-4 disodium phosphate and its vascular effects on the BT4An rat glioma, and the tumor response of CA-4 combined with hyperthermia, were investigated [99]. To assess drug toxicity, rats were given 50, 75, or 100 mg/kg CA-4 and followed by daily registration of weight and side effects. Interstitial tumor blood flow was detected by laser Doppler flowmetry in rats injected with 50 mg/kg CA-4. It was found that CA-4, at a well-tolerated dose of 50 mg/kg, induced a considerable time-dependent decrease in the tumor blood flow. CA-4 induces a gradual reduction in tumor blood flow that, in authors' opinion, can be exploited to sensitize the BT4An tumor for hyperthermia. Hemodynamic changes after CA-4 is injected are shown in Figure 8.10.

The efficacy of the tumor vascular-targeting agent combretastatin A-4-P at clinically relevant doses was demonstrated [100]. Clinically relevant doses of the tumor vascular-targeting agent, combretastatin A-4-phosphate, were investigated in the rat P22 tumor system. Single intraperitoneal bolus doses of 3 and 10 mg/kg CA-4-P were as effective as much higher doses, in terms of tumor blood flow response up to several hours. Tumor necrosis induction, at 24 h following drug treatment, was suboptimal. The combination of 3 mg/kg CA-4-P with systemic nitric oxide synthase inhibition, using L-NAME, significantly increased the early tumor vascular effects of CA-4-P alone. CA-4-P (3 mg/kg) had less effect in normal tissues than in tumor and addition of L-NAME did not increase the effects of CA-4-P alone in critical normal tissues. The authors suggested that NOS inhibition has potential for increasing the therapeutic efficacy of drugs such as CA-4-P.

It was shown [101] that CA-4-P selectively targeted endothelial cells, but not smooth muscle cells, and induced the regression of unstable nascent tumor neovessels

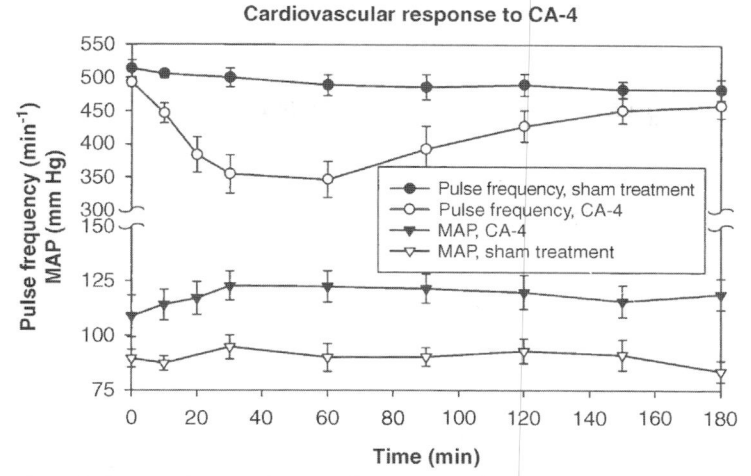

Figure 8.10 Hemodynamic changes after injection of CA-4 50 mg/kg intraperitoneal or 0.9% NaCl 1 ml/kg intraperitoneal (sham treatment). Values are means ± SEM of six tumors (CA-4) or four tumors (sham treatment). MAP: mean arterial pressure [99]. (Reproduced with permission of Elsevier.)

by rapidly disrupting the molecular engagement of the endothelial cell-specific junctional molecular vascular endothelial-cadherin (VE-cadherin) *in vitro* and *in vivo* in mice. CA-4-P increased endothelial cell permeability while inhibiting endothelial cell migration and capillary tube formation predominantly through the disruption of VE-cadherin/β-catenin/Akt signaling pathway, thereby leading to a rapid vascular collapse and tumor necrosis. Stabilization of VE-cadherin signaling in endothelial cells with adenovirus E4 gene or ensheathment with smooth muscle cells conferred resistance to CA-4-P. CA-4-P synergizes with low and nontoxic doses of neutralizing mAbs to VE-cadherin by blocking assembly of neovessels, thereby inhibiting tumor growth. The authors claimed that combined treatment with anti-VE-cadherin agents in conjunction with microtubule-disrupting agents provided a novel synergistic strategy to selectively disrupt assembly and induce the regression of nascent tumor neovessels, with minimal toxicity and without affecting normal stabilized vasculature.

Preclinical studies to predict the efficacy of vascular changes induced by combretastatin A-4 disodium phosphate in patients were performed [102]. To detect how combretastatin A-4 disodium phosphate dose-dependent changes in radiation response of a C3H mouse mammary carcinoma are related to measurements of dynamic contrast-enhanced magnetic resonance imaging parameters and how those results compare with published clinical DCE-MRI data. C3H mammary carcinomas grown in female CDF1 mice were treated when at 200 mm^3 in size. Groups of mice were given graded radiation doses, either alone or followed 30 min later by an intraperitoneal injection of CA-4-DP, administered at doses of 10–250 mg/kg. The radiation dose producing local tumor control in 50% of treated animals in 90 days

(TCD50) was calculated for each CA-4-DP dose. TCD50 showed a dose-dependent decrease reaching significance at 25 mg/kg. The authors concluded that the preclinical DCE-MRI data could predict the CA-4-DP enhancement of the tumor radiation response and suggest the clinical CA-4-DP doses necessary for improving the radiation response in patients.

In a pilot study [103], the ability of combretastatin A-4 to modify retinal neovascularization, which results in an altered retinal vessel blood flow and retinal permeability, was evaluated in aphakic long-term galactose-fed beagles, an animal model that develops diabetes-like retinal neovascularization. Two groups of aphakic dogs, each group comprising four galactose-fed dogs and two age-matched control dogs, were used. Each group initially received the combretastatin A-4-phosphate prodrug as either sub-Tenon's injections, administered at the corneoscleral junction, or intravitreal injections. Six weeks after this treatment, all dogs also received systemic (i.v.) injections of CA-4-P. Retinal vascular changes were monitored at 2-week intervals by fluorescein angiogenesis. All galactose-fed dogs demonstrated the presence of retinal neovascular lesions by fluorescein angiograms. The authors suggested that the failure of CA-4-P to ameliorate neovascularization suggests that chronic, long-term administration may be required to destroy the slowly growing retinal endothelial cells. Proposed model for CA-4-P-mediated angiogenesis inhibition is presented in Figure 8.11.

8.3
Pterostilbene

8.3.1
Cells

The inhibitory effect of pterostilbene on the induction of NO synthase and cyclooxygenase-2 (COX-2) in murine RAW 264.7 cells activated with lipopolysaccharide (LPS) was investigated [104, 105]. Western blotting and real-time polymerase chain reaction (PCR) analyses demonstrated that pterostilbene significantly blocked the protein and mRNA expression of iNOS and COX-2 in LPS-induced macrophages. Treatment with pterostilbene resulted in the reduction of LPS-induced nuclear translocation of the nuclear factor-kappaB (NF-κB) subunit and the dependent transcriptional activity of NF-κB by blocking phosphorylation of inhibitor kappaB (IκB)alpha and p65 and subsequent degradation of IκBα. Transient transfection experiments using NF-κB reporter constructs indicated that pterostilbene inhibited the transcriptional activity of NF-κB in LPS-stimulated mouse macrophages. It was found that pterostilbene also inhibited LPS-induced activation of PI3K/Akt, extracellular signal-regulated kinase 1/2, and p38 MAPK. Taken together, these results showed that pterostilbene downregulated inflammatory iNOS and COX-2 gene expression in macrophages by inhibiting the activation of NF-κB by interfering with the activation of PI3K/Akt/IKK and MAPK. The authors concluded that these results have an important implication for using pterostilbene toward the development of an effective anti-inflammatory agent [105].

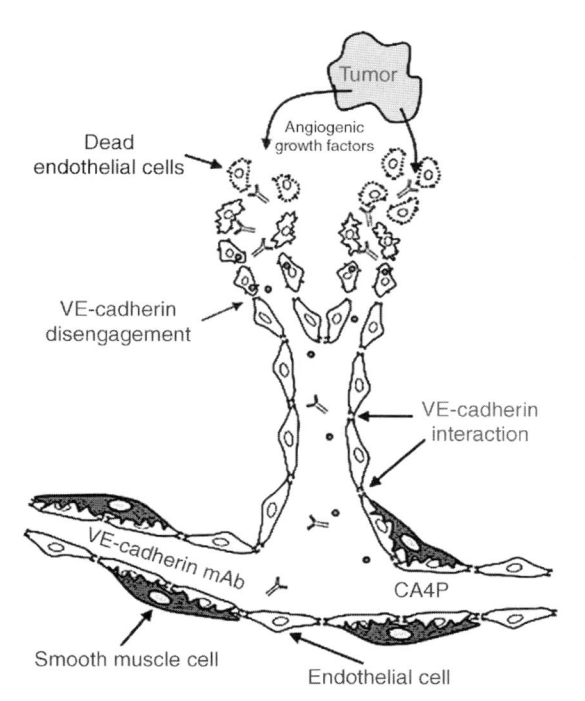

Figure 8.11 Proposed model for CA-4-P-mediated angiogenesis inhibition. Pre-existing vessels are invested with SMCs protecting endothelial cells against CA-4-P-induced cell death. Because nascent unstable tumor neovessels are not ensheathed by periendothelial mural cells, CA-4-P selectively destabilizes neovessels by inducing VE-cadherin disengagement, therefore increasing the antiangiogenic effects of the neutralizing mAb against VE-cadherin without increasing toxicity to the normal vasculature [103]. (Reproduced with permission.)

Pterostilbene and 3′-hydroxypterostilbene were found to be effective apoptosis-inducing agents in MDR and BCR-ABL-expressing leukemia cells (Figure 8.12) [106, 107]. Both compounds were able to induce apoptosis in the two Fas-ligand-resistant lymphoma cell lines, HUT78B1 and HUT78B3, and the multidrug-resistant leukemia cell lines, HL-60-R and K562-ADR (a BCR-ABL-expressing cell line resistant to imatinib mesylate). Pterostilbene and 3′-hydroxypterostilbene, when used at concentrations that elicit significant apoptotic effects in tumor cell lines, did not show any cytotoxicity in normal hemopoietic stem cells. The authors concluded that these data showed that pterostilbene and particularly 3′-hydroxypterostilbene may be useful in the treatment of resistant hematologic malignancies, including imatinib, nonresponsive neoplasms.

8.3.2
Animals

The invention [108] related to the combined use of pterostilbene and quercetin for the production of cancer treatment medicaments. The *in vitro* growth of melanoma

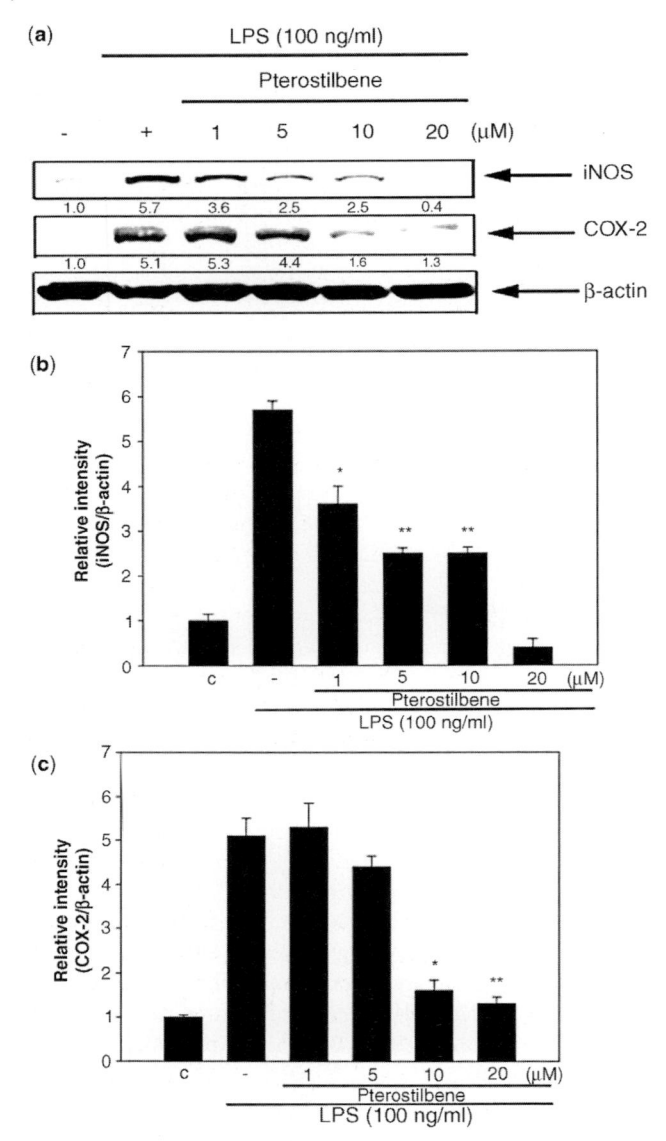

Figure 8.12 Effect of pterostilbene on LPS-induced iNOS and COX-2 protein expression in RAW 264 cells. (a) The cells were treated with different concentrations of pterostilbene for 24 h. Equal amounts of total proteins (50 μg) were subjected to 10% SDS-PAGE. The expression of iNOS, COX-2, and β-actin protein was detected by Western blot using specific antibodies. This experiment was repeated three times with similar results. (b and c) Real-time PCR analyses of the expression of iNOS and COX-2 mRNA. Cells were treated with LPS (100 ng/ml) and pterostilbene (1–20 μM) for 5 h; the mRNA expression of *iNOS* and *COX-2* genes was performed using the LightCycler System and TaqMan probe real-time PCR. The values are expressed as means ± standard error of triplicate tests. *$P<0.05$ and **$P<0.01$, indicating statistically significant differences from the LPS-treated group [106]. (Reproduced with permission.)

cells B16-F10 (B16M-F10) was inhibited (56%) by combined exposures of short-duration (60 min/day) to PTER (40 μM) and QUER (20 μM). The combined intravenous administration of PTER and QUER (20 mg/kg × day) to mice inhibited (73%) the metastatic growth of melanoma B16M-F10 in the liver, a common site for metastasis development. Association between pterostilbene and quercetin that inhibited the metastatic activity of B16 melanoma was investigated [109]. *In vitro* growth of highly malignant B16 melanoma F10 cells (B16M-F10) was inhibited (56%) by short-time exposure (60 min/day) to PTER (40 μM) and QUER (20 μM) for the first hour after intravenous administration of 20 mg/kg of each polyphenol. Intravenous administration of PTER and QUER (20 mg/kg per day) to mice inhibited (73%) metastatic growth of B16M-F10 cells in the liver, a common site for metastasis development. These findings demonstrated that the association of PTER and QUER inhibits metastatic melanoma growth and extends host survival. The antimetastatic mechanism in metastatic cells was discussed.

Preclinical pharmacokinetics and metabolism, anticancer, anti-inflammatory, antioxidant, and analgesic activity of pterostilbene were evaluated [110]. Right jugular vein cannulated male Sprague-Dawley rats were dosed intravenously with 20 mg/kg of pterostilbene and samples were analyzed by the reversed-phase HPLC method. A pterostilbene glucuronidated metabolite was detected in both serum and urine. The *in vitro* metabolism in rat liver microsomes furthermore suggests phase II metabolism of pterostilbene. Pterostilbene demonstrated concentration-dependent anticancer activity in five cancer cell lines (1–100 μg/ml) and suppression of PGE2 production in the media of HT-29 cells *in vitro* colitis and decreased levels of MMP-3, sGAG, and TNF-α. Pterostilbene also exhibited antioxidant capacity measured by the ABTS method.

The effect of pterostilbene on lipids and lipid profiles in streptozotocin-nicotinamide-induced type 2 diabetes mellitus was reported [111]. It was shown that oral administration of pterostilbene (40 mg/kg bodyweight) to streptozotocin-nicotinamide-induced diabetic rats for 6 weeks significantly reduced the elevated serum very low-density lipoprotein (VLDL) and LDL-cholesterol levels and significantly increased the serum HDL-cholesterol level and lowered the levels of triglycerides, phospholipids, free fatty acids, and total cholesterol in the serum, liver, and kidney of diabetic rats. The effect of pterostilbene on hepatic key enzymes of glucose metabolism in streptozotocin-nicotinamide-induced diabetic rats was evaluated [112]. Diabetic rats were orally administered with pterostilbene (10, 20, and 40 mg/kg) for 2, 4, and 6 weeks on glucose. Administration of pterostilbene at 40 mg/kg significantly decreased plasma glucose. Effects of the oral administration of pterostilbene for 6 weeks on glucose, insulin levels, and hepatic enzymes in normal and streptozotocin-nicotinamide-induced diabetic rats were studied. A significant decrease in glucose and significant increase in plasma insulin levels were observed in normal and diabetic rats treated with pterostilbene. Treatment with pterostilbene resulted in a significant reduction in glycosylated Hb and an increase in total Hb level. The activity of hepatic enzymes such as hexokinase increased whereas the activities of glucose-6-phosphatase and fructose-1,6-bisphosphatase were decreased by the administration of pterostilbene in diabetic rats.

The antioxidant role of pterostilbene in streptozotocin-nicotinamide-induced type 2 diabetes mellitus in Wistar rats was evaluated [113]. The activity of superoxide dismutase, catalase, glutathione peroxidase, glutathione-S-transferase and reduced glutathione significantly decreased in liver and kidney of diabetic animals compared to normal control. The increased levels of lipid peroxidation measured as thiobarbituric acid reactive substances in liver and kidney of diabetic rats were also normalized by treatment with pterostilbene. Chronic treatment of pterostilbene remarkably reduced the pathological changes observed in the liver and the kidney of diabetic rats.

It was shown that intravenous administration of *trans*-pterostilbene and quercetin (QUER; 3,3′,4′,5,6-pentahydroxyflavone) to mice inhibited metastatic growth of highly malignant B16 melanoma F10 (B16M-F10) cells [114]. *trans*-Pterostilbene and QUER also inhibited bcl-2 expression in metastatic cells, which sensitizes them to vascular endothelium-induced cytotoxicity. A 60 min/day exposure to *trans*-pterostilbene (40 µM) and QUER (20 µM) within the first hour after intravenous administration of 20 mg of each polyphenol/kg downregulated the inducible NO synthetase in B16M-F10 cells and upregulated the endothelial NO synthetase in the vascular endothelium and thereby facilitated endothelium-induced tumor cytotoxicity. Very low and high NO levels downregulated bcl-2 expression in B16M-F10 cells; *trans*-pterostilbene and QUER induced a NO shortage-dependent decrease in cAMP-response element-binding protein phosphorylation, a regulator of bcl-2 expression, in B16M-F10 cells.

Pterostilbene, from blueberries, was tested for its preventive activity against aberrant crypt foci formation in the azoxymethane-induced colon carcinogenesis model (Figure 8.13) [115]. Experiments were designed to study the inhibitory effect of pterostilbene in male F344 rats (Figure 8.13). Beginning at 7 weeks of age, rats were treated with azoxymethane (15 mg/kg body weight once a week for 2 weeks). Administration of pterostilbene for 8 weeks significantly suppressed azoxymethane-induced formation of ACF (57% inhibition) and multiple clusters of aberrant crypts (29% inhibition) and azoxymethane-induced colonic cell proliferation and iNOS expression.

The preclinic pharmacokinetics and pharmacodynamics of *trans*-pterostilbene, a constituent of some plants, were investigated [116]. Right jugular vein-cannulated male Sprague-Dawley rats were dosed intravenously with 20 mg/kg of pterostilbene and samples were analyzed by the reversed-phase HPLC method. A pterostilbene glucuronidated metabolite was detected in both serum and urine. The *in vitro* metabolism in rat liver microsomes furthermore suggested phase II metabolism of pterostilbene. Pterostilbene demonstrated concentration-dependent anticancer activity against five cancer cell lines (1–100 µg/ml). An *in vitro* colitis model showed concentration-dependent suppression of PGE2 production in the media of HT-29 cells. Anti-inflammatory activity was examined by inducing inflammation in canine chondrocytes followed by treatment with pterostilbene (1–100 µg/ml). The results showed decreased levels of MMP-3, sGAG, and TNF-α compared to control levels. Pterostilbene exhibited concentration-dependent antioxidant capacity measured by the ABTS method. Pterostilbene increased the latency period of response in both tail-flick and hot-plate analgesic tests. A novel and simple high-performance liquid

8.3 Pterostilbene

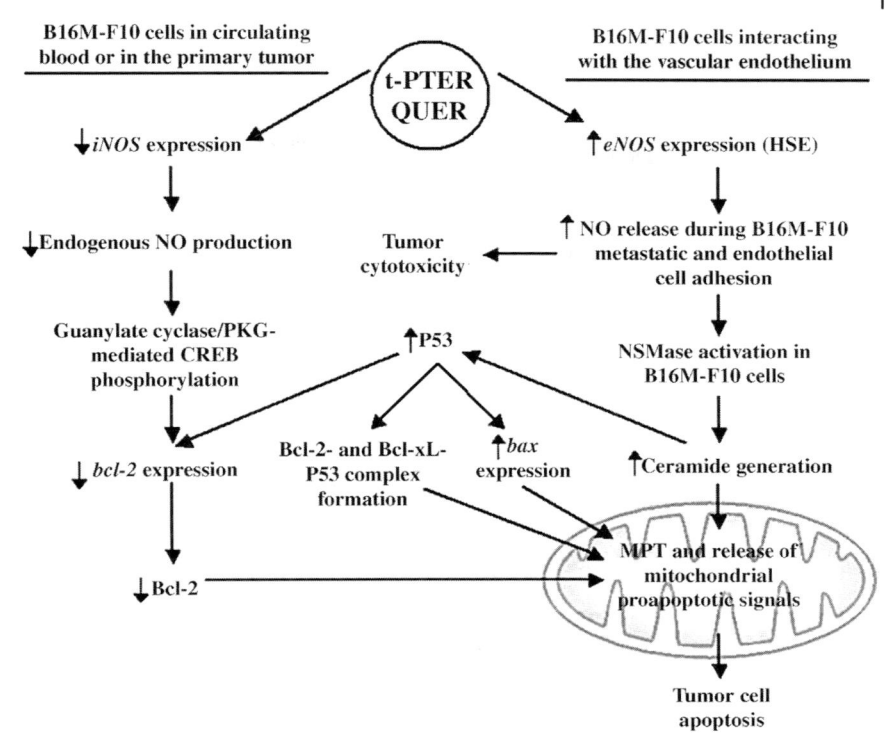

Figure 8.13 Inhibition of iNOS protein by pterostilbene in HT-29 colon carcinoma cells. HT-29 human colon carcinoma cells were grown in complete medium (DMEM supplemented with 10% fetal bovine serum and 1% penicillin/streptomycin) at 37 °C, 5% CO_2. At day 0, HT-29 cells were plated in 100 mm dish (2×10^6 cells per dish). Cells were then treated with pterostilbene (1, 10, or 30 μmol/l) together with a cytokine mixture (IFN-γ, tumor necrosis factor-α, and lipopolysaccharide, each at 10 ng/ml) for 15 h and cell lysates were harvested and subjected to Western blot analysis [114].

chromatography method was developed for detecting pterostilbene in rat serum [117]. Separation of pterostilbene was achieved on a Phenomenex C18 column (250 mm × 4.60 mm) with fluorescence excitation at 330 nm and emission at 374 nm. The calibration curves were linear ranging from 0.5 to 100 μg/ml. The assay was applied to the study of pterostilbene pharmacokinetics in rats. Pterostilbene and synthetic analogues of resveratrol were synthesized and their ability to activate peroxisome proliferator-activated receptor alpha was investigated [117]. Docking of resveratrol natural analogues was performed in PPARα ligand-binding domain. The proposed binding pose of these compounds in PPARα was similar for both the active and the inactive compounds. Blueberry skins were fed to hypercholesterolemic hamsters and these animals showed lower levels of lipids compared to those fed ciprofibrate.

Data presented in this chapter unequivocally indicate that stilbenoids possess a wide spectrum of properties, which allow one to consider these compounds as potentially important drugs with a broad therapeutic window. *trans*-Resveratrol may be considered for use as a cancer chemopreventive agent showing antioxidant activity and upregulation of NO production. Resveratrol therapy may prevent the hypertensive response and the relaxation response to acetylcholine. Combretastatins are potential new vascular disrupting agents and show a remarkable ability to inhibit gastric tumor metastasis and enhanced antitumor immune reactivity. These compounds may provide an effective means to treat refractory organ-infiltrating leukemias and are potentially important for optimizing the therapeutic combination of vascular-targeting agents with radiotherapy. Pterostilbene possess lipid- and glucose-lowering effects useful in the treatment of resistant hematology malignancies, exhibit antioxidant capacity, and demonstrate concentration-dependent anticancer activity.

References

1 Patterson, D. (2007) *Clinical Oncology*, **19**, 443–456.
2 Salmon, B. and Siemann, D.W. (2007) *International Journal of Radiation Oncology, Biology, Physics*, **68**, 211–217.
3 Hinnen, P. and Eskens, F.A.L.M. (2007) *British Journal of Cancer*, **96**, 1159–1165.
4 Hori, K., Saito, S., and Kubota, K. (2002) *British Journal of Cancer*, **86**, 1604–1614.
5 Athar, M., Back, J.H., Tang, X., Kim, K.H., Kopelovich, L., Bickers, D.R., and Kim, A.L. (2007) *Toxicology and Applied Pharmacology*, **224**, 274–283.
6 Jang, M., Cai, L., Udeani, G.O., Slowing, K.V., Thomas, C.F., Beecher, C.W., Fong, H.H., Farnsworth, N.R., Kinghorn, A.D., Mehta, R.G., Moon, R.C., and Pezzuto, J.M. (1997) *Science*, **275**, 218–222.
7 Shankar, S., Singh, G., and Srivastava, R.K. (2007) *Frontiers in Bioscience*, **12**, 4839–4854.
8 Patterson, D.M. and Rustin, G.J.S. (2007) *Drugs of the Future*, **32**, 1025–1032.
9 Bissery, M.C., Vrignaud, P., Demers, B., and Chiron, M. (2007) Combination Comprising Combretastatin and Anticancer Agents. French Patent PCT/FR2006/002771, filed Dec. 18, 2006 and issued July 12, 2007.
10 Johnson, M., Younglove, B., Lee, L., LeBlanc, R., Holt, H., Jr., Hills, P., Mackay, H., Brown, T., Mooberry, S.L., and Lee, M. (2007) *Bioorganic & Medicinal Chemistry Letters*, **17**, 5897–5901.
11 Tozer, G.M. (2003) *British Journal of Radiology*, **76**, S23–S35.
12 Jordan, M.A. and Wilson, L. (2004) *Nature Reviews Cancer*, **4**, 253–226.
13 Banerjee, S., Wang, Z., Mohammad, M., Sarkar, F.H., and Mohammad, R.M. (2008) *Journal of Natural Products*, **71**, 492–496.
14 Chaudhary, A., Pandeya, S.N., Kumar, P., Sharma, P.P., Gupta, S., Soni, N., Verma, K.K., and Bhardwaj, G. (2007) *Mini-Reviews in Medicinal Chemistry*, **7**, 1186–1205.
15 Bavaresco, L. and Vezzulli, S. (2006) *Recent Progress in Medicinal Plants*, **11**, 389–410.
16 Bavaresco, L., Fregoni, C., Cantu, E., and Trevisan, M. (1999) *Drugs under Experimental and Clinical Research*, **25**, 57–63.
17 Roupe, K.A., Remsberg, C.M., Yanez, J.A., and Davies, N.M. (2006) *Current Clinical Pharmacology*, **1**, 81–101.
18 Asensi, M., Medina, I., Ortega, A., Carretero, J., Bano, M.C., Obrador, E., and Estrela, J.M. (2002) *Free Radical Biology & Medicine*, **33**, 387–398.
19 Wenzel, E. and Somoza, V. (2005) *Molecular Nutrition & Food Research*, **49**, 472–481.

20 Tozer, G. (2005) *Nature Reviews Cancer*, **5**, 432–435.
21 Aggarwal, B.B., Bhardwaj, A., Aggarwal, R.S., Seeram, N.P., Shishodia, S., and Takada, Y. (2004) *Anticancer Research*, **24**, 2783–2840.
22 Niles, R.M., Cook, C.P., Meadows, G.G., Fu, Y.-M., McLaughlin, J.L., and Rankin, G.O. (2006) *Journal of Nutrition*, **136**, 2542–2546.
23 Harper, C.E., Patel, B.B., Wang, J., Arabshahi, A., Eltoum, I.A., and Lamartiniere, C.A. (2007) *Carcinogenesis*, **28**, 1946–1953.
24 Shankar, S., Singh, G., and Srivastava, R.K. (2007) *Frontiers in Bioscience*, **12**, 4839–4854.
25 Juan, M.E., Wenzel, U., Daniel, H., and Planas, J.M. (2008) *Journal of Agricultural and Food Chemistry*, **56**, 4813–4818.
26 Aziz, M.H., Nihal, M., Fu, V.X., Jarrard, D.F., and Ahmad, N. (2006) *Molecular Cancer Therapeutics*, **5**, 1335–1341.
27 Shi, L., Huang, X.-F., Zhu, Z.-W., Li, H.-Q., Xue, J.-Y., Zhu, H.-L., and Liu, C.-H. (2008) *Australian Journal of Chemistry*, **61**, 472–475.
28 Marel, A.-K., Lizard, G., Izard, J.-C., Latruffe, N., and Delmas, D. (2008) *Molecular Nutrition & Food Research*, **52**, 538–548.
29 Hudson, T.S., Hartle, D.K., Hursting, S.D., Nunez, N.P., Wang, T.T.Y., Young, H.A., Arany, P., and Green, J.E. (2007) *Cancer Research*, **67**, 8396–8405.
30 Murias, M., Luczak, M.W., Niepsuj, A., Krajka-Kuzniak, V., Zielinska-Przyjemska, M., Jagodzinski, P.P., Jaeger, W., Szekeres, T., and Jodynis-Liebert, J. (2008) *Toxicology in Vitro*, **22**, 1361–1370.
31 Kim, H.-J., Chung, S.-K., and Park, S.-W. (1999) *Journal of Food Science and Nutrition*, **4**, 163–166.
32 Snyder, R.M., Yu, W., Li, J., Sanders, B.G., and Kline, K. (2008) *Nutrition and Cancer*, **60**, 401–411.
33 Chun, Y.J., Kim, M.Y., and Guengerich, F.P. (1999) *Biochemical and Biophysical Research Communications*, **262**, 20–24.
34 Simoni, D., Roberti, M., Invidiata, F.P., Aiello, E., Aiello, S., Marchetti, P., Baruchello, R., Eleopra, M., Di Cristina, A., Grimaudo, S., Gebbia, N., Crosta, L., Dieli, F., and Tolomeo, M. (2006) *Bioorganic & Medicinal Chemistry Letters*, **16**, 3245–3248.
35 Hung, L.-M., Chen, J.-K., Huang, S.-S., Lee, R.-S., and Su, M.-J. (2000) *Cardiovascular Research*, **47**, 549–555.
36 Aubin, M.-C., Lajoie, C., Clement, R., Gosselin, H., Calderone, A., and Perrault, L.P. (2008) *Journal of Pharmacology and Experimental Therapeutics*, **325**, 961–968.
37 Silan, C. (2008) *Biological & Pharmaceutical Bulletin*, **31**, 897–902.
38 Zhang, L., Luo, L., Fu, Y., Zhang, R., Qian, Y., and Xu, J. (2007) *Zhongguo Zuzhi Gongcheng Yanjiu Yu Linchuang Kangfu*, **11**, 3764–3767.
39 Zhang, H., Schools, G.P., Lei, T., Wang, W., Kimelberg, H.K., and Zhou, M. (2008) *Experimental Neurology*, **212**, 44–52.
40 Lagouge, M., Argmann, C., Gerhart-Hines, Z., Meziane, H., Lerin, C., Daussin, F., Messadeq, N., Milne, J., Lambert, P., Elliott, P., Geny, B., Laakso, M., Puigserver, P., and Auwerx, J. (2006) *Cell (Cambridge, MA, United States)*, **127**, 1109–1122.
41 Mizuno, C.S., Patny, A., Avery, M.A., Yokoyama, W.H., and Rimando, A.M. (2007) Abstracts of Papers, 233rd ACS National Meeting, Chicago, IL, March 25–29, 2007, MEDI-014.
42 Rimando, A.M., Nagmani, R., Feller, D.R., and Yokoyama, W. (2005) *Journal of Agricultural and Food Chemistry*, **53**, 3403–3407.
43 De Santi, C., Pietrabissa, A., Spisni, R., Mosca, F., and Pacifici, G.M. (2000) *Xenobiotica*, **30**, 609–617.
44 Petit, I., Karajannis, M.A., Vincent, L., Young, L., Butler, J., Hooper, A.T., Shido, K., Steller, H., Chaplin, D.J., Feldman, E., and Rafii, S. (2008) *Blood*, **111**, 1951–1961.
45 Odlo, K., Hentzen, J., dit Chabert, J.F., Ducki, S., Gani, O.A.B.S.M., Sylte, I., Skrede, M., Florenes, V.A., and

Hansen, T.V. (2008) *Bioorganic & Medicinal Chemistry*, **16**, 4829–4838.

46 Cenciarelli, C., Tanzarella, C., Vitale, I., Pisano, C., Crateri, P., Meschini, S., Arancia, G., and Antoccia, A. (2008) *Apoptosis*, **13**, 659–669.

47 Uno, K., Tanabe, T., Ogamino, T., Okada, R., Imoto, M., and Nishiyama, S. (2008) *Heterocycles*, **75**, 291–292.

48 Shen, W., Wang, J., Wang, J., Jin, H., Qian, F., and Wang, F. (2007) *PCT International Patent Application*, pp. 31.

49 Quan, H., Xu, Y., and Lou, L. (2008) *International Journal of Cancer*, **122**, 1730–1737.

50 Maya, A.B.S., Perez-Melero, C., Mateo, C., Alonso, D., Fernandez, J.L., Gajate, C., Mollinedo, F., Pelaez, R., Caballero, E., and Medarde, M. (2005) *Journal of Medicinal Chemistry*, **48**, 556–568.

51 Lin, H.-L., Chiou, S.-H., Wu, C.-W., Lin, W.-B., Chen, L.-H., Yang, Y.-P., Tsai, M.-L., Uen, Y.-H., Liou, J.-P., and Chi, C.-W. (2007) *Journal of Pharmacology and Experimental Therapeutics*, **323**, 365–373.

52 Xian, L., Zou, Y., Cai, Y., and Wang, Z. (2007) Faming Zhuanli Shenqing Gongkai Shuomingshu, pp. 14.

53 Younglove, B., Lee, L., Mackay, H., Mooberry, S.L., Hills, P., Brown, T., and Lee, M. (2007) Abstracts of Papers, 233rd ACS National Meeting, Chicago, IL, March 25–29, 2007, CHED-1315.

54 Lee, L.E., Davis, R., VanderHam, J., Mackay, H., Brown, T., Mooberry, S., and Lee, M. (2007) Abstracts, 59th Southeast Regional Meeting of the American Chemical Society, Greenville, SC, October 24–27, 2007, GEN-356.

55 Mousset, C., Giraud, A., Provot, O., Hamze, A., Bignon, J., Liu, J.-M., Thoret, S., Dubois, J., Brion, J.-D., and Alami, M. (2008) *Bioorganic & Medicinal Chemistry Letters*, **18**, 3266–3271.

56 Ruprich, J., Prout, A., Dickson, J., Younglove, B., Nolan, L., Baxi, K., LeBlanc, R., Forrest, L., Hills, P., Holt, H., Jr., Mackay, H., Brown, T., Mooberry, S.L., and Lee, L.M. (2007) *Letters in Drug Design & Discovery*, **4**, 144–148.

57 Zhang, Q., Peng, Y., Wang, X.I., Keenan, S.M., Arora, S., and Welsh, W.J. (2007) *Journal of Medicinal Chemistry*, **50**, 749–754.

58 Siles, R., Ackley, J.F., Hadimani, M.B., Hall, J.J., Mugabe, B.E., Guddneppanavar, R., Monk, K.A., Chapuis, J.-C., Pettit, G.R., Chaplin, D.J., Edvardsen, K., Trawick, M.L., Garner, C.M., and Pinney, K.G. (2008) *Journal of Natural Products*, **71**, 313–320.

59 Vitale, I., Antoccia, A., Cenciarelli, C., Crateri, P., Meschini, S., Arancia, G., Pisano, C., and Tanzarella, C. (2007) *Apoptosis*, **12**, 155–166.

60 Pagliuca, C., Menichetti, S., Vitellozzi, L., Bracci, L., and Falciani, C. (2007) Frontiers in CNS and oncology medicinal chemistry. ACS-EFMC, Siena, Italy, October 7–9, 2007, COMC-018.

61 Harrowven, D.C., Guy, I.L., Howell, M., and Packham, G. (2006) *Synlett*, (18), 2977–2298.

62 Thomson, P., Naylor, M.A., Everett, S.A., Stratford, M.R.L., Lewis, G., Hill, S., Patel, K.B., Wardman, P., and Davis, P.D. (2006) *Molecular Cancer Therapeutics*, **5**, 2886–2894.

63 Bellina, F., Cauteruccio, S., Monti, S., and Rossi, R. (2006) *Bioorganic & Medicinal Chemistry Letters*, **16**, 5757–5762.

64 Xue, N., Yang, X., Wu, R., Chen, J., He, Q., Yang, B., Lu, X., and Hu, Y. (2008) *Bioorganic & Medicinal Chemistry*, **16**, 2550–2557.

65 Siemann, D.W., Mercer, E., Lepler, S., and Rojiani, A.M. (2002) *International Journal of Cancer*, **99**, 1–6.

66 Galbraith, S.M., Maxwell, R.J., Lodge, M.A., Tozer, G.M., Wilson, J., Taylor, N.J., Stirling, J.J., Sena, L., Padhani, A.R., and Rustin, G.J. (2003) *Journal of Clinical Oncology*, **21**, 2831–2842.

67 Horsman, M.R., Murata, R., Breidahl, T., Nielsen, F.U., Maxwell, R.J., Stodkiled-Jorgensen, H., and Overgaard, J. (2000)

Advances in Experimental Medicine and Biology, **476**, 311–323.
68 Li, L., Rojiani, A., and Siemann, D.W. (1998) *International Journal of Radiation Oncology, Biology, Physics*, **42**, 899–903.
69 Nelkin, B.D. and Ball, D.W. (2001) *Oncology Reports*, **8**, 157–160.
70 Li, L., Rojiani, A.M., and Siemann, D.W. (2002) *Acta Oncologica*, **41**, 91–9778.
71 Yeung, S.C., She, M., Yang, H., Pan, J., Sun, L., and Chaplin, D. (2007) *Journal of Clinical Endocrinology and Metabolism*, **92**, 2902–2909.
72 Nabha, S.M., Wall, N.R., Mohammad, R.M., Pettit, G.R., and Al-Katib, A.M. (2000) *Anticancer Drugs*, **11**, 385–392.
73 Vincent, L., Kermani, P., Young, L.M., Cheng, J., Zhang, F., Shido, K., Lam, G., Bompais-Vincent, H., Zhu, Z., Hicklin, D.J., Bohlen, P., Chaplin, D.J., May, C., and Rafii, S. (2005) *Journal of Clinical Investigation*, **115**, 2992–3006.
74 Yeung, S.-C.J., She, M., Yang, H., Pan, J., Sun, L., and Chaplin, D. (2007) *Journal of Clinical Endocrinology and Metabolism*, **92**, 2902–2909.
75 Lankester, K.J., Maxwell, R.J., Pedley, R.B., Dearling, J.L., Qureshi, U.A., El-Emir, E., Hill, S.A., and Tozer, G.M. (2007) *International Journal of Oncology*, **30**, 453–460.
76 Zhao, D., Jiang, L., Hahn, E.W., and Mason, R.P. (2005) *International Journal of Radiation Oncology, Biology, Physics*, **62**, 872–880.
77 Siemann, D.W., Mercer, E., Lepler, S., and Rojiani, A.M. (2002) *International Journal of Cancer*, **99**, 1–6.
78 McPhail, L.D., Griffiths, J.R., and Robinson, S.P. (2007) *International Journal of Radiation Oncology, Biology, Physics*, **69**, 1238–1245.
79 Beauregard, D.A., Hill, S.A., Chaplin, D.J., and Brindle, K.M. (2001) *Cancer Research*, **61**, 6811–6815.
80 Nielsen, T., Murata, R., Maxwell, R.J., Stodkilde-Jorgensen, H., Ostergaard, L., and Horsman, M.R. (2008) *International Journal of Radiation Oncology, Biology, Physics*, **70**, 859–866.
81 Reyes-Aldasoro, C.C., Ruhrberg, C., Shima, D.T., and Kanthou, C. (2008) *Cancer Research*, **68**, 2301–2311.
82 Beauregard, D.A., Thelwall, P.E., Chaplin, D.J., Hill, S.A., Adams, G.E., and Brindle, K.M. (1998) *British Journal of Cancer*, **77**, 1761–1767.
83 Chaplin, D.J., Pettit, G.R., and Hill, S.A. (1999) *Anticancer Research*, **19**, 189–196.
84 Nielsen, T., Murata, R., Maxwell, R.J., Stodkilde-Jorgensen, H., Ostergaard, L., and Horsman, M.R. (2008) *International Journal of Radiation Oncology, Biology, Physics*, **70**, 859–866.
85 Niu, B., Fang, Su-H., Bao, X., Qiu, G., Yin, H.-L., Bao, D.-Y., Wang, H.-X., and Xu, X.-P. (2007) *Huaxi Yaoxue Zazhi*, **22**, 632–635.
86 Petit, I., Karajannis, M.A., Vincent, L., Young, L., Butler, J., Hooper, A.T., Shido, K., Steller, H., Chaplin, D.J., Feldman, E., and Rafii, S. (2008) *Blood*, **111**, 1951–1961.
87 Kitagawa, M., Ishikawa, K., Masuda, A., and Takashio, K. (2008) *PCT International Patent Application*, pp. 29.
88 Hokland, S.L., and Horsman, M.R. (2007) *International Journal of Hyperthermia*, **23**, 599–606.
89 Folkes, L.K., Christlieb, M., Madej, E., Stratford, M.R.L., and Wardman, P. (2007) *Chemical Research in Toxicology*, **20**, 1885–1894.
90 Tozer, G.M., Prise, V.E., Wilson, J., Locke, R.J., Vojnovic, B., Stratford, M.R., Dennis, M.F., and Chaplin, D.J. (1999) *Cancer Research*, **59**, 1626–1634.
91 Tao, C., Wang, Q., De, T., Desai, N.P., and Soon-Shiong, P. (2007) *PCT International Patent Application*, pp. 71.
92 Ley, C.D., Horsman, M.R., and Kristjansen, P.E.G. (2007) *Neoplasia*, **9**, 108–112.
93 Badn, W., Kalliomaeki, S., Widegren, B., and Sjoegren, H.O. (2006) *Clinical Cancer Research*, **12**, 4714–4719.
94 Prise, V.E., Honess, D.J., Stratford, M.R.L., Wilson, J., and Tozer, G.M. (2002)

International Journal of Oncology, **21**, 717–726.
95 Murata, R., Siemann, D.W., Overgaard, J., and Horsman, M.R. (2001) *Radiotherapy and Oncology*, **60**, 155–161.
96 Landuyt, W., Verdoes, O., Darius, D.O., Drijkoningen, M., Nuyts, S., Theys, J., Stockx, L., Wynendaele, W., Fowler, J.F., Maleux, G., Van den Bogaert, W., Anne, J., van Oosterom, A., and Lambin, P. (2000) *European Journal of Cancer*, **36**, 1833–1843.
97 Eikesdal, H.P., Schem, B.-C., Mella, O., and Dahl, O. (2000) *International Journal of Radiation Oncology, Biology, Physics*, **46**, 645–652.
98 Tozer, G.M., Honess, D.J., Wilson, J., Hill, S.A., Hodgkiss, R.J., Ameer-Beg, S., Vojnovic, B., and Prise, V.E. (2002) Microcirculation and vascular biology. 22nd Meeting of the European Society for Microcirculation, Exeter, UK, August 28–30, 2002.
99 Vincent, L., Kermani, P., Young, L.M., Cheng, J., Zhang, F., Shido, K., Lam, G., Bompais-Vincent, H., Zhu, Z., Hicklin, D.J., Bohlen, P., Chaplin, D.J., May, C., and Rafii, S. (2005) *Journal of Clinical Investigation*, **115**, 2992–3006.
100 Nielsen, T., Murata, R., Maxwell, R.J., Stodkilde-Jorgensen, H., Ostergaard, L., and Horsman, M.R. (2008) *International Journal of Radiation Oncology, Biology, Physics*, **70**, 859–866.
101 Kador, P.F., Blessing, K., Randazzo, J., Makita, J., and Wyman, M. (2007) *Journal of Ocular Pharmacology and Therapeutics*, **23**, 132–142.
102 Pan, M.-H., Chang, Y.-H., Tsai, M.-L., Lai, C.-S., Ho, S.-Y., Badmaev, V., and Ho, C.-T. (2008) *Journal of Agricultural and Food Chemistry*.
103 Pan, M.-H., Chang, Y.-H., Badmaev, V., Nagabhushanam, K., and Ho, C.-T. (2007) *Journal of Agricultural and Food Chemistry*, **55**, 7777–7785.
104 Pan, M.-H., Chang, Y.-H., Tsai, M.-L., Lai, C.-S., Ho, S.-Y., Badmaev, V., and Ho, C.-G. (2008) *Journal of Agricultural and Food Chemistry*, **56** (16), 7502–7509.
105 Tolomeo, M., Grimaudo, S., Di Cristina, A., Roberti, M., Pizzirani, D., Meli, M., Dusonchet, L., Gebbia, N., Abbadessa, V., Crosta, L., Barucchello, R., Grisolia, G., Invidiata, F., and Simoni, D. (2005) *International Journal of Biochemistry & Cell Biology*, **37**, 1709–1726.
106 Estrela Ariquel, J.M., Asensio Aguilar, G., Asensi Miralles, M.A., Obrador Pla, E., Varea Munoz, M.T., Jorda Quilis, L., Ferrer Pastor, P., Segarra Guerrero, R., Ortega Valero, A., and Benlloch Garcia, M. (2006) *PCT International Patent Application*, pp. 41.
107 Ferrer, P., Asensi, M., Segarra, R., Ortega, A., Benlloch, M., Obrador, E., Varea, M.T., Asensio, G., Jorda, L., and Estrela, J.M. *Neoplasia*, **7**, 37–47.
108 Remsberg, C.M., Yanez, J.A., Ohgami, Y., Vega-Villa, K.R., Rimando, A.M., and Davies, N.M. (2008) *Phytotherapy Research*, **22**, 69–179.
109 Satheesh, M.A., and Pari, L. (2008) *Journal of Applied Biomedicine*, **6**, 31–37.
110 Pari, L., Satheesh, M.A., Bavaresco, L., and Vezzulli, S. (2006) *Life Sciences*, **79**, 641–645.
111 Satheesh, M.A., and Pari, L. (2006) *Journal of Pharmacy and Pharmacology*, **58**, 1483–1490.
112 Ferrer, P., Asensi, M., Priego, S., Benlloch, M., Mena, S., Ortega, A., Obrador, E., Esteve, J.M., and Estrela, J.M. (2007) *Journal of Biological Chemistry*, **282**, 2880–2890.
113 Suh, N., Paul, S., Hao, X., Simi, B., Xiao, H., Rimando, A.M., and Reddy, B.S. (2007) *Clinical Cancer Research*, **13**, 350–355.
114 Remsberg, C.M., Yanez, J.A., Ohgami, Y., Vega-Villa, K.R., Rimando, A.M., and Davies, N.M. (2008) *Phytotherapy Research*, **22**, 169–179.
115 Remsberg, C.M., Yanez, J.A., Roupe, K.A., and Davies, N.M. (2007) *Journal of Pharmaceutical and Biomedical Analysis*, **43**, 250–254.

9
Stilbenes in Clinics

9.1
General

In health care, clinical trials are conducted to collect safety and efficacy data for new drugs or devices. Depending on the type of product and the stage of its development, investigators enroll healthy volunteers and/or patients into small pilot studies initially, followed by larger scale studies in patients that often compare the new product with the currently prescribed treatment.

Clinical trials involving new drugs are commonly classified into five phases. Each phase of the drug approval process is treated as a separate clinical trial [1–5]. The drug development process will normally proceed through all four phases over many years. If the drug successfully passes through phases 0, I, II, and III, it will usually be approved by the national regulatory authority for use in the general population. Phase IV consists of "post-approval" studies.

Phase 0 is a recent designation for exploratory, first-in-human trials conducted in accordance with the US Food and Drug Administration's (FDA) 2006 Guidance on Exploratory Investigational New Drug (IND) Studies [6]. Distinctive features of phase 0 trials include the administration of single subtherapeutic doses of the study drug to a small number of subjects (10–15) to gather preliminary data on the agent's pharmacokinetics and pharmacodynamics. Phase I trials are the first stage of testing in human subjects. Normally, a small (20–80) group of healthy volunteers will be selected. This phase includes trials designed to assess the safety (pharmacovigilance), tolerability, pharmacokinetics, and pharmacodynamics of a drug. Once the initial safety of the study drug has been confirmed in phase I trials, phase II trials are performed on larger groups (20–300) and are designed to assess how well the drug works. Phase III studies are randomized controlled multicenter trials on large patient groups (300–3000 or more depending on the disease/medical condition studied) and are aimed at being the definitive assessment of how effective the drug is in comparison with current "gold standard" treatment. Phase IV trials involve the safety surveillance (pharmacovigilance) and ongoing technical support of a drug after it receives permission to be sold.

Stilbenes. Applications in Chemistry, Life Sciences and Materials Science. Gertz Likhtenshtein
Copyright © 2010 WILEY-VCH Verlag GmbH & Co. KGaA, Weinheim
ISBN: 978-3-527-32388-3

Stilbenes act as natural protective agents to defend the plant against viral and microbial attack, excessive UV exposure, and disease. These compounds have been extensively studied and have been shown to possess potent anticancer, anti-inflammatory, and antioxidant activities. Numerous studies describe different biological and clinical effects of resveratrol, combretastatin, and pterostilbene.

The following reviews on application of stilbenes in clinical trials have been recently published [7–25]. The tumor vasculature is an attractive target for therapy because of its accessibility to blood-borne anticancer agents and the reliance of most tumor cells on an intact vascular supply for their survival. The review [11] described the vascular effects of some of these agents and identified suitable end points for measuring efficacy in early clinical trials. Measurement of tumor microvascular density (MVD) from tumor biopsies is a common method for assessing the efficacy of antiangiogenic drugs. Preclinical data regarding tumor response to the antivascular agent combretastatin A-4 3-O-phosphate (CA-4-P) were discussed in the context of guiding clinical trial planning. It was shown that growth of human tumors depends on the supply of oxygen and nutrients via the surrounding vasculature [16]. Apart from angiogenesis inhibitors that compromise the formation of new blood vessels, a second class of specific anticancer drugs has been developed. These so-called vascular disrupting agents (VDAs) target the established tumor vasculature and cause an acute and pronounced shutdown of blood vessels resulting in an almost complete stop of blood flow, ultimately leading to selective tumor necrosis. The mechanism of action of a number of VDAs, which were tested in clinical studies, has been discussed. In addition, data from some considerations with regard to the future development were given.

9.2
trans-Resveratrol

Resveratrol has been extensively studied and has been shown to possess potent anticancer, anti-inflammatory, and antioxidant activities (Section 7.1). Found primarily in the skins of grapes, resveratrol is synthesized by *Vitis vinifera* grapevines in response to fungal infection or other environmental stressors. Molecular nutrition revealed information on absorption, metabolism, and the consequent bioavailability of resveratrol, *in vitro*, *ex vivo*, and *in vivo* models [7]. It was found that around 75% of this polyphenol are excreted via feces and urine. The oral bioavailability of resveratrol is almost zero due to rapid and extensive metabolism and the consequent formation of various metabolites such as resveratrol glucuronides and resveratrol sulfates. The considerable research showing resveratrol to be an attractive candidate in combating a wide variety of cancers and diseases has fueled interest in detecting the disease-fighting capabilities of other structurally similar stilbene compounds. The purpose of review [8] was to describe four structurally similar stilbene compounds, piceatannol, pinosylvin, rhapontigenin, and pterostilbene, and detail some current pharmaceutical research and highlight their potential clinical applications.

Therapeutic potential of resveratrol *in vivo* was discussed in [26]. According to the authors, despite skepticism concerning its bioavailability, a growing body of *in vivo* evidence indicates that resveratrol has protective effects in rodent models of stress and disease. A comprehensive and critical review of the *in vivo* data on resveratrol, and considering its potential as a therapeutic for humans, has been provided. The absorptive efficiency of the three of its constituents (*trans*-resveratrol, [+]-catechin, and quercetin) when given orally to healthy human subjects in three different media was tested (Figure 9.1) [27]. Twelve healthy males aged 25–45 years were randomly assigned to three different groups orally consuming one of the following polyphenols: *trans*-resveratrol, 25 mg/70 kg; [+]-catechin 25 mg/70 kg; and quercetin 10 mg/70 kg. Each polyphenol was randomly administered at 4-week intervals in three different matrices: white wine (11.5% ethanol), grape juice, and vegetable juice/homogenate. Blood was collected at zero time and at four intervals over the first 4 h after consumption; urine was collected at zero time and after 24 h. The sums of free and conjugated polyphenols were measured in blood serum and urine by a gas chromatographic method. All three polyphenols were present in serum and urine predominantly as glucuronide and sulfate conjugates, reaching peak concentrations in the former around 30 min after consumption. The free polyphenols accounted for 1.7–1.9% (*trans*-resveratrol), 1.1–6.5% ([+]-catechin), and 17.2–26.9% (quercetin) of the peak serum concentrations. The absorption of *trans*-resveratrol was the most efficient. An example of kinetics of mean total) *trans*-resveratrol in serum of four subjects is presented in Figure 9.1.

	V-8	WINE	JUICE
0	3	4	2
0.5	471	416	424
1	250	191	344
2	112	154	220
4	56	100	106

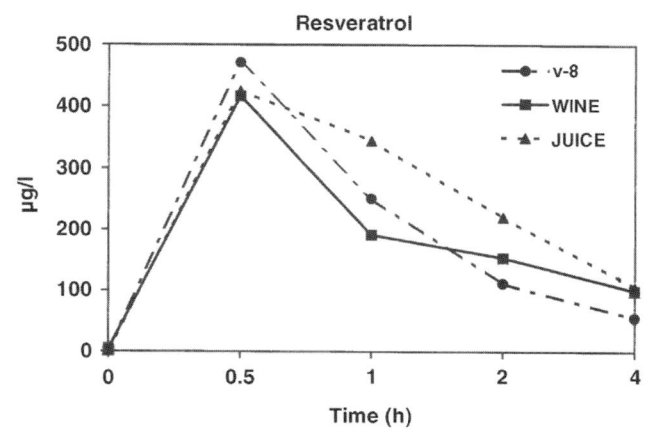

Figure 9.1 Mean total (free and conjugated) *trans*-resveratrol in serum of four subjects given 25 mg of *trans*-resveratrol in various matrices [27]. (Reproduced with permission from Elsevier.)

Application of stilbene oxygen-substituted acid or its salt to prepare the medical formulations for lowering blood lipid was patented [28]. The invention related to the application of stilbene oxygen-substituted acid to prepare the medical formulations for lowering serum triglyceride, low-density lipoprotein (LDL) cholesterol and/or total cholesterol, heightening high-density lipoprotein (HDL) cholesterol, preventing and/or treating hyperlipemia, hypercholesterolemia, fatty liver, atherosclerosis, and atherosclerosis-associated hypertension, coronary heart disease, and stroke. For example, the stilbene oxygen-substituted acid was resveratrol-3,4′,5-tris-O-acetic acid or resveratrol-3,4′,5-tris-O-dimethyl acetic acid.

Ultrasensitive assay for three polyphenols including resveratrol and their conjugates in biological fluids using gas chromatography with mass selective detection was developed [29]. The concentration of three polyphenols ((+)-catechin, quercetin, and *trans*-resveratrol) in blood serum, plasma, and urine, as well as whole blood, has been measured after their oral and intragastric administration, respectively, to humans and rats. The method used ethyl acetate extraction of 100 µl samples and their derivatization with bis(trimethylsilyl)trifluoroacetamide (BSTFA) followed by gas chromatography analysis on a DB-5 column followed by mass selective detection employing two target ions and one qualifier ion for each compound. The limits of detection (LOD) and quantitation (LOQ) was found to be 0.01 and 0.1 µg/l, respectively, for all compounds. After oral administration of the three polyphenols to humans, their conjugates vastly exceeded the concentration of the aglycons in both plasma and urine. The concentration peaked within 0.5–1.0 h in plasma and within 8 h in urine. During the first 24 h, 5.1% of the (+)-catechin and 24.6% of the *trans*-resveratrol given were recovered in the urine. This method can be proposed as the method of choice to assay these polyphenols and their conjugates in biological fluids.

Phase I dose escalation pharmacokinetic study in healthy volunteers of resveratrol was described [30]. A phase I study of oral resveratrol (single doses of 0.5, 1, 2.5, or 5 g) was conducted in 10 healthy volunteers per dose level. Resveratrol and its metabolites were identified in plasma and urine by high-performance liquid chromatography–tandem mass spectrometry and quantitated by high-performance liquid chromatography–UV. Resveratrol and six metabolites were recovered from plasma and urine. The area under the plasma concentration values for resveratrol-3-sulfate and resveratrol monoglucuronides was up to 23 times greater than that of resveratrol. Cancer chemopreventive effects of resveratrol in cells *in vitro* require levels of at least 5 µmol/l. The authors concluded that consumption of high-dose resveratrol might be insufficient to elicit systemic levels commensurate with cancer chemopreventive efficacy. However, the high systemic levels of resveratrol conjugate metabolites suggested that their cancer chemopreventive properties warrant investigation. Figure 9.2 shows mean plasma concentrations of resveratrol (a), two resveratrol monoglucuronides (b and c), and resveratrol-3-sulfate (d) versus time in healthy volunteers.

High absorption but very low bioavailability of oral resveratrol in humans was described [31]. The authors examined the absorption, bioavailability, and metabolism of ^{14}C-resveratrol after oral and intravenous doses in six human volunteers. The absorption of a dietary relevant 25-mg oral dose was at least 70%, with peak plasma

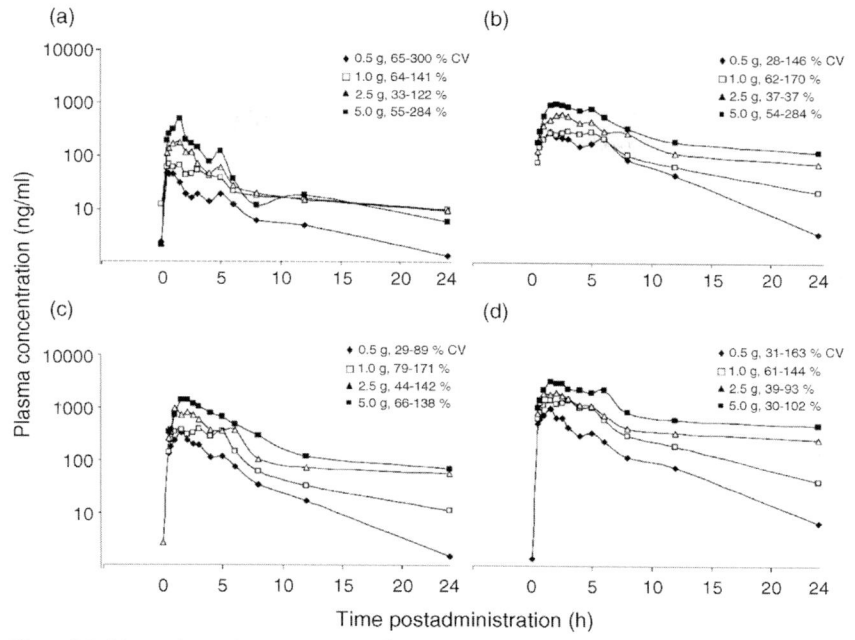

Figure 9.2 Mean plasma concentrations of resveratrol (a), two resveratrol monoglucuronides (b and c), and resveratrol-3-sulfate (d) versus time in healthy volunteers who received a single dose of resveratrol at 0.5 (♦), 1 (□), 2.5 (Δ), or 5 g (■). *Points*, mean of 10 volunteers per dose level. *Insets*, coefficients of variation [30]. (Reproduced with permission.)

levels of resveratrol and metabolites of 491 ± 90 ng/ml (about $2\,\mu M$) and a plasma half-life of 9.2 ± 0.6 h. Only trace amounts of unchanged resveratrol (<5 ng/ml) could be detected in plasma. Most of the oral dose was recovered in urine, and liquid chromatography/mass spectrometry analysis identified three metabolic pathways, which were sulfate and glucuronic acid conjugation of the phenolic groups and hydrogenation of the aliphatic double bond, the latter likely produced by the intestinal microflora. An extremely rapid sulfate conjugation by the intestine/liver appears to be the rate-limiting step in resveratrol's bioavailability. It was suggested that although the systemic bioavailability of resveratrol is very low, accumulation of resveratrol in epithelial cells along the aerodigestive tract and potentially active resveratrol metabolites may still produce cancer preventive and other effects.

Sirtuins represent a novel family of enzymes that are collectively well situated to help regulate nutrient sensing and utilization, metabolic rate, and ultimately metabolic disease. It was shown that resveratrol activated one of these enzymes, SIRT1 [32]. The activation of SIRT1 leads to enhanced activity of multiple proteins, including peroxisome proliferator-activated receptor coactivator-1α (PGC-1α), which helps mediate some of the *in vitro* and *in vivo* effects of sirtuins. Resveratrol, given in a proprietary formulation SRT-501 (3 or 5 g), reached 5–8 times higher blood levels.

SRT-501 represented the first in a novel class of SIRT1 activators that have proven to be safe and well-tolerated in humans. Clinical trials in type 2 diabetic patients were reported to be underway.

Authors of the work [7] hypothesizes that resveratrol from wine could have higher bioavailability than resveratrol from a pill. The highest level of unchanged resveratrol in the serum (7–9 ng/ml) was achieved after 30 min, and it completely disappeared from blood after 4 h. *This conclusion was not consistent with data obtained in the work* [33] *where* bioavailability of *trans*-resveratrol from red wine in humans was investigated. *trans*-Resveratrol 3- and 4′-glucuronides were synthesized, purified, and characterized as pure standards. Bioavailability data were obtained by measuring the concentration of free, 3-glucuronide and 4′-glucuronide *trans*-resveratrol by high-performance liquid chromatography, both with UV and mass spectrometry detection, in serum samples taken at different times after red wine administration. According to experiments after five men took 600 ml of red wine with a resveratrol content of 3.2 mg/l (total dose about 2 mg) before breakfast, resveratrol was found unchanged in the blood of only two of them and only in trace amounts (below 2.5 ng/ml). Resveratrol levels appeared to be slightly higher if red wine (600 ml of red wine containing 0.6 mg/ml resveratrol; total dose about 0.5 mg) was taken with meal: trace amounts (1–6 ng/ml) were found in 4 out of 10 subjects. Free *trans*-resveratrol was found in trace amounts only in some serum samples collected 30 min after red wine ingestion, while after longer times resveratrol glucuronides predominated. *trans*-Resveratrol bioavailability was shown to be independent of the meal or its lipid content. The finding in human serum of *trans*-resveratrol glucuronides, rather than the free form of the compound, with a high interindividual variability, allowed the authors to suggest that the benefits associated with red wine consumption could be probably due to the whole antioxidant pool present in red wine. The authors concluded that the trace amounts of resveratrol found in the blood were insufficient to explain the French paradox.

Quantification of free and protein-bound polyphenol *trans*-resveratrol (*t*-RES) metabolites and identification of *trans*-resveratrol-C/O-conjugated diglucuronides, two novel resveratrol metabolites in human plasma (Figure 9.3), were reported [34].

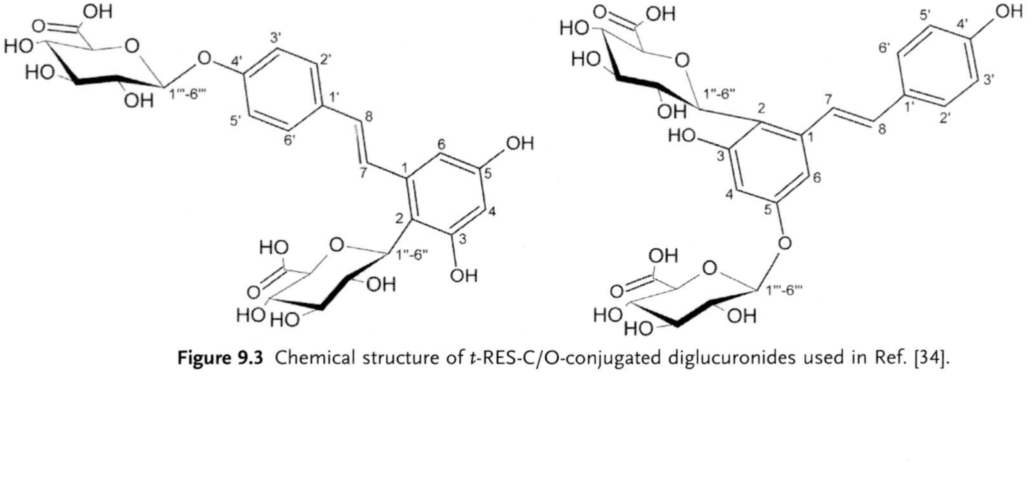

Figure 9.3 Chemical structure of *t*-RES-C/O-conjugated diglucuronides used in Ref. [34].

A 85.5 mg piceid per 70 kg body weight was taken by nine healthy men (23–41 years, BMI 21–29 kg/m^2) in a bolus dose. The t-RES metabolites formed in blood plasma and urine were identified and quantified by LC-MS/MS, NMR, and HPLC–DAD analysis using chemically synthesized t-RES conjugate standards. The amount of t-RES metabolites bound noncovalently to plasma proteins was detected. The metabolites identified and quantified were t-RES-3-sulfate, t-RES-3,4'-disulfate, t-RES-3,5-disulfate, t-RES-3-glucuronide, and t-RES-4'-glucuronide, with t-RES-sulfates being the dominant conjugates in plasma and urine. Two novel t-RES-C/O-conjugated diglucuronides were identified and quantified in plasma and urine.

9.3
Combretastatin

Combretastatin A-4 phosphate (CA-4-P) is a vascular disrupting agent that binds to tubulin and selectively damages established tumor vasculature. Preclinical studies have shown that CA-4-P causes rapid vascular shutdown, leading to central tumor necrosis, although it leaves a rim of viable cells at the periphery, which requires additional antineoplastic treatment to enhance efficacy (Sections 7.2 and 8.2). CA-4-P is also effective in disrupting ocular neovascularization [24]. Phase I monotherapy trials have shown the agent to be well tolerated, with tumor pain and mild transient cardiovascular effects being the most common side effects. Clinical trials of various CA-4-P combinations have been conducted and human proof of concept was established for intervenous drug in patients with myopic macular degeneration.

9.3.1
Vascular Damaging Agents

Comprehensive overviews of the current state of development of a novel class of anticancer drugs, the vascular disrupting agents, were provided [13–15]. Tumor vascular targeting therapy exploited differences between normal and tumor blood vessels. VDAs targeted the pre-existing vessels of tumors and caused vascular shutdown leading to tumor cell death and rapid hemorrhagic necrosis within hours. Small-molecule VDAs (Figure 9.4) worked either as tubulin binding agents or through induction of local cytokine production. VDAs killed tumor cells resistant to conventional chemotherapy and radiotherapy, combination therapy with cytotoxic chemotherapy, and external beam radiotherapy. VDAs were generally well tolerated with different side-effect profiles from current oncological therapies. The authors concluded that VDAs are a promising new class of drugs, which offer the attractive possibility of inducing responses in all tumor types with combination therapy.

A schematic representation of potential targets of CA-4 in endothelial cells is presented in Figure 9.5. Some issues related to the pharmacology of CA-4 and future directions in research were considered.

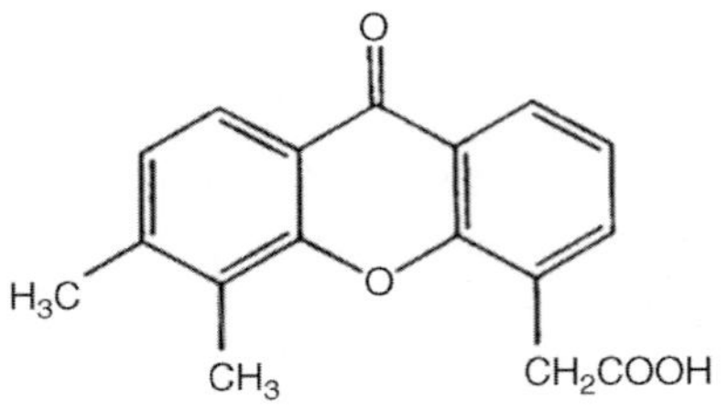

DMXAA

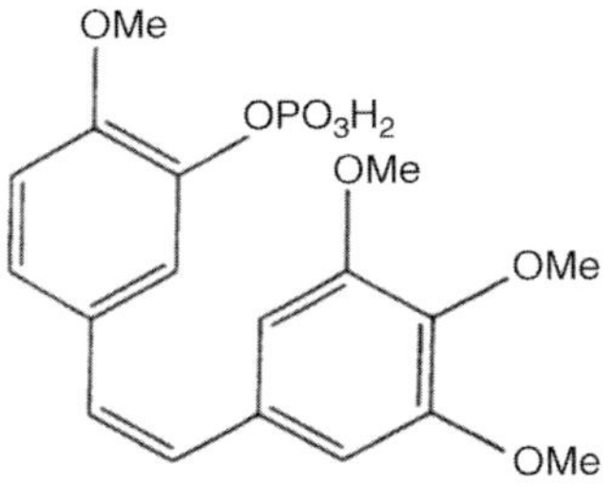

combretastatin A4 phosphate

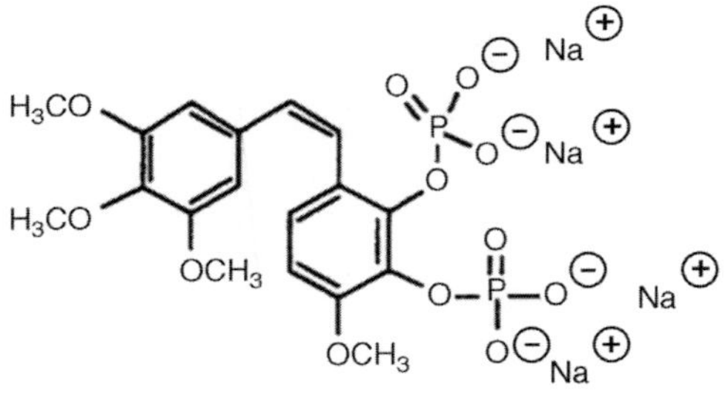

Oxi4503

Figure 9.4 Chemical structure of selected small molecule vascular disruptive agents employed in Ref. [15].

9.3.2
Pharmacometrics of Stilbenes: Seguing Toward the Clinic

The efficacy of selected natural products as therapeutic agents against cancer was reviewed [35]. The review described a few of the compounds obtained from marine and terrestrial sources [bryostatin 1 (**1**), dolastatin 10 (**2**), auristatin PE (**3**), and combretastatin A-4 (**4**)] that were investigated for their sensitization effects on other cytotoxic agents in several different site-specific tumors employing murine models or human subjects. The tumor vascular effects of radiotherapy and subsequent administration of the vascular disrupting agent combretastatin A-4 phosphate were studied in patients with advanced nonsmall-cell lung cancer using volumetric dynamic contrast-enhanced computed tomography (CT) [36]. Eight patients receiving palliative radiotherapy (27 Gy in six fractions, twice a week) also received CA-4-P (50 mg/m^2) after the second fraction of radiotherapy. Changes in dynamic CT parameters of tumor blood volume (BV) and permeability surface area product (PS) were measured for the whole-tumor volume, tumor rim, and center after radiotherapy alone and after radiotherapy in combination with CA-4-P. After the second fraction of radiotherapy, six of the eight patients showed increase in tumor PS (23.6%). Four hours after CA-4-P administration, a reduction in tumor BV (22.9%)

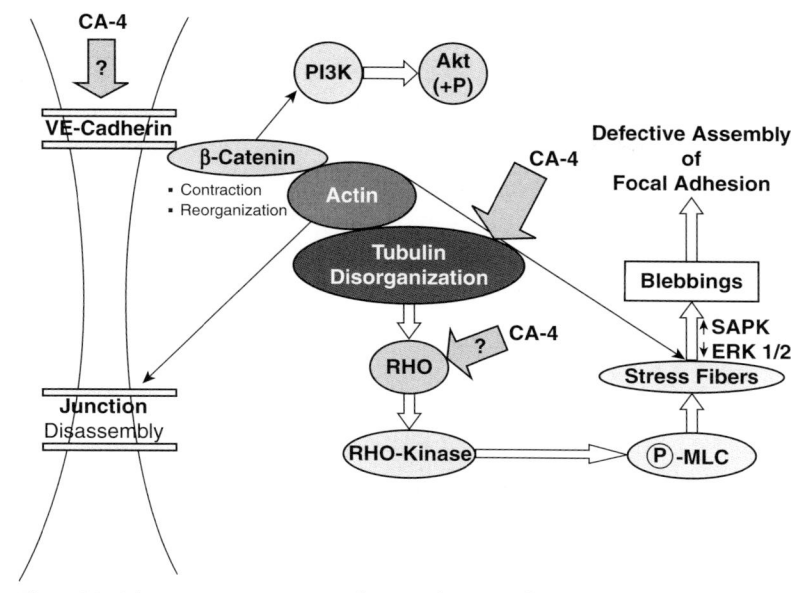

Figure 9.5 Schematic representation of potential targets of CA-4 in endothelial cells: P-MLC, phosphorylated myosin light chain; ERK, extracellular receptor kinase; SAPK, stress-activated protein kinase [15]. (Reproduced with permission from Elsevier.)

was demonstrated in the same six patients. Both increase in PS after radiotherapy and reduction in BV after CA-4-P were greater at the rim of the tumor. The BV reduction at the rim was sustained to 72 h (51.4%). Radiotherapy enhances the tumor antivascular activity of CA-4-P in human nonsmall-cell lung cancer, resulting in sustained tumor vascular shutdown. Figure 9.6 presents dynamic computed tomography images of a lung tumor from three representative axial levels. Regression plot showing percentage change in whole-tumor BV after combretastatin A-4 phosphate administration against percentage change in permeability surface area product (PS) after radiotherapy is shown in Figure 9.7.

Four phase I trials of CA-4-P (**4**) in humans have been published [37].

The purpose of the study [9] was to review and determine the cardiovascular safety profile of combretastatin A-4 phosphate in a phase I study on 25 patients with advanced solid tumors. CA-4-P was administered in a dose-escalating fashion starting at 18 mg/m^2 intervenously every 21 days, and the maximal dosage was 90 mg/m^2. Continuous evaluation included bedside blood pressure and pulse monitoring, 12-lead electrocardiogram (ECG) at fixed time points for measured QT interval determination, determination of the corrected QT interval (QTc) using Bazett's formula QTc = QT/$(R - R \text{ interval})^{1/2}$, and chart review. Pharmacodynamic correlations of CA-4-P dose, CA-4-P/CA-4 area under the curve (AUC), and C_{max} versus heart rate (HR), blood pressure, QT, and QTc intervals, over the first 4 h postdosing were analyzed. After CA-4-P administration, there were significant increases in QTc interval. Three of the 25 patients had prolonged QTc intervals at

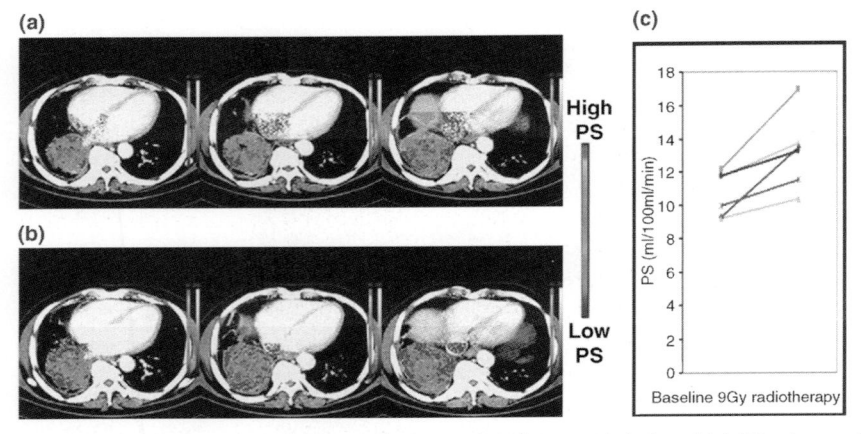

Figure 9.6 Dynamic computed tomography images of a lung tumor from three representative axial levels showing a colored parametric map of tumor vascular blood volume before combretastatin A-4 phosphate administration (a) and 4 h after a single dose of CA-4-P (b). Each pixel within the tumor map represents a vascular parameter value; the color scale indicates red pixels as high BV values and purple pixels as low BV values. An acute reduction in tumor vascular BV was seen after CA-4-P and most evident at the rim of the tumor. Tumor vascular BV decreased by 22.9% ($p < 0.001$) in six of the eight patients after CA-4-P (c) was administered [36]. (Reproduced with permission from Elsevier.)

baseline. The slope of HR and QTc increase as a function of time during the first 4 h was correlated with CA-4-P dose. Two patients had ECG changes consistent with an acute coronary syndrome within 24 h of CA-4-P infusion. There was a temporal relationship with the CA-4-P infusion and with ECG changes consistent with an acute coronary syndrome in two patients.

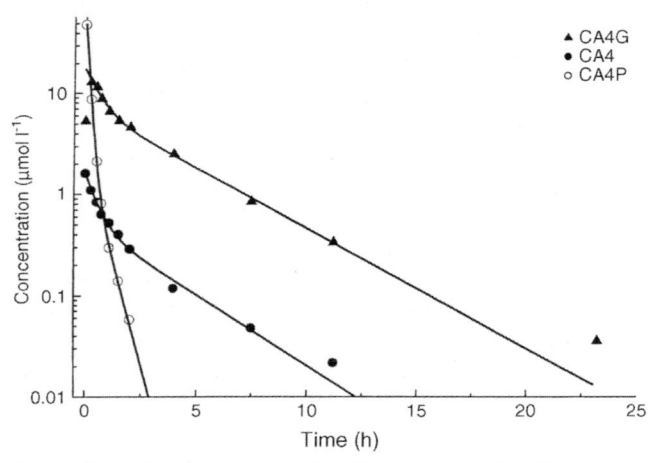

Figure 9.7 A typical plasma profile after 68 mg combretastatin A-4 phosphate, combretastatin A-4, and combretastatin A-4 glucuronide [42]. (Reproduced with permission.)

Effects of CA-4-P administered intravenously to patients with advanced cancer were investigated [38]. Patients with solid malignancies and good performance status received CA-4-P as a 10-min infusion daily for 5 days repeated every 3 weeks. Pharmacokinetic sampling was performed during cycle 1. Patients receiving ≥ 52 mg/m^2/day had serial dynamic contrast-enhanced magnetic resonance imaging (DCE-MRI) studies to measure changes in tumor perfusion with CA-4-P treatment. Thirty-seven patients received 133 treatment cycles. CA-4-P dose levels ranged from 6 to 75 mg/m^2 daily. Severe pain at sites of known tumor was dose limiting at 75 mg/m^2. Dose-limiting cardiopulmonary toxicity (syncope and dyspnea or hypoxia) was noted as well in two patients treated at 75 mg/m^2. Other toxicities included hypotension, ataxia, dyspnea, nausea or vomiting, headache, and transient sensory neuropathy. Plasma CA-4-P and CA-4 area under the concentration–time curve and maximal concentration values increased linearly with dose. Tumor perfusion, as measured by the first-order rate constant of gadolinium plasma to tissue transfer during DCE-MRI studies, was found to decrease in 8 of 10 patients. Relationships were also demonstrated between perfusion changes and pharmacokinetic indices. A partial response was observed in a patient with metastatic soft tissue sarcoma, and 14 patients exhibited disease stability for a minimum of two cycles. The authors stressed that doses of CA-4-P on a five-times-a-day schedule of 52–65 mg/m^2 were reasonably well tolerated. The 52 mg/m^2 dose was recommended for further study based on cumulative phase I experience with CA-4-P. Antitumor efficacy was observed, and the use of DCE-MRI provided a valuable.

Phase I clinical trial and pharmacokinetic results of combretastatin A-4 phosphate administered weekly were reported [39]. A phase I trial was performed with combretastatin A-4 phosphate that has been shown to rapidly reduce blood flow in animal tumors. The drug was delivered by a 10-min weekly infusion for 3 weeks followed by a week-long gap, with intrapatient dose escalation. Dose escalation was accomplished by doubling until grade 2 toxicity was seen. The starting dose was 5 mg/m^2. Thirty-four patients received 167 infusions. CA-4-P was rapidly converted to the active combretastatin A-4, which was further metabolized to glucuronide. CA-4 area under the curve increased from 0.169 at 5 mg/m^2 to 3.29 µmol h/l at 114 mg/m^2. The mean CA-4 AUC in eight patients at 68 mg/m^2 was 2.33 µmol h/l compared to 5.8 µmol h/l at 25 mg/kg (the lowest effective dose) in the mouse. The only toxicity that possibly was related to the drug dose up to 40 mg/m^2 was tumor pain. dose-limiting toxicity (DLT) was reversible ataxia at 114 mg/m^2, vasovagal syncope and motor neuropathy at 88 mg/m^2, and fatal ischemia in previously irradiated bowel at 52 mg/m^2. Other drug-related grade 2 or higher toxicities observed in more than one patient were pain, lymphopenia, fatigue, anemia, diarrhea, hypertension, hypotension, vomiting, visual disturbance, and dyspnea. One patient at 68 mg/m^2 had improvement in liver metastasis of adrenocortical carcinoma. CA-4-P was well tolerated in 14 of 16 patients at 52 or 68 mg/m^2; these are doses at which tumor blood flow reduction has been recorded.

A phase I trial to determine the maximum-tolerated dose, safety, and pharmacokinetic profile of CA-4-P on a single-dose i.v. schedule [40]. Preliminary data on its effect on tumor blood flow using DCE-MRI techniques and cell adhesion molecules at the higher-dose levels were obtained. Twenty-five assessable patients with advanced cancer received a total of 107 cycles over the following dose escalation schema: 18, 36, 60, 90 mg/m^2 as a 10-min infusion and 60 mg/m^2 as a 60-min infusion at 3-week intervals. There was no significant myelotoxicity, stomatitis, or alopecia. Tumor pain was a unique side effect, which occurred in 10% of cycles. Pharmacokinetics revealed rapid dephosphorylation of the parent compound (CA-4-P) to CA-4, with a short plasma half-life (approximately 30 min). A significant ($P < 0.03$) decline in gradient peak tumor blood flow by DCE-MRI in six of seven patients treated at 60 mg/m^2 was observed. A patient with anaplastic thyroid cancer showed a complete response and survived for 30 months after treatment.

Preclinical evidence of synergy that led to a phase I trial employing combretastatin A-4 phosphate, in combination with carboplatin, was reported [41]. Based on preclinical scheduling studies, patients were treated on day 1 of a 21-day cycle. Carboplatin was given as a 30-min intravenous infusion and CA-4-P was given 60 min later as a 10-min infusion. Sixteen patients with solid tumors received 40 cycles of therapy at CA-4-P doses of 27 and 36 mg/m^2 together with carboplatin at area under the concentration–time curve values of 4 and 5 mg min/ml. The dose-limiting toxicity of thrombocytopenia halted the dose escalation phase of the study. Four patients were treated at an amended dose level of CA-4-P of 36 mg/m^2 and carboplatin AUC of 4 mg min/ml although grade 3 neutropenia and thrombocytopenia were still observed. Three lines of evidence were adduced to suggest that a pharmacokinetic interaction between the drugs results in greater thrombocytopenia than anticipated. The carboplatin exposure (as AUC) was greater than predicted; the platelet nadirs were lower than predicted; and the deviation of the carboplatin exposure from predicted was proportional to the AUC of CA-4, the active metabolite of CA-4-P. In terms of benefits, six patients were found with stable disease lasting at least four cycles. The authors concluded that this study of CA-4-P and carboplatin given in combination showed dose-limiting thrombocytopenia. Pharmacokinetic/pharmacodynamic modeling permitted the inference that altered carboplatin pharmacokinetics caused the increment in platelet toxicity.

Phase I, pharmacokinetic, and DCE-MRI correlative study of AVE8062A, an antivascular combretastatin analogue, administered weekly for 3 weeks every 28 days, was reported [43]. Nine patients received 48 infusions of AVE8062. Cardiovascular effects consisting of asymptomatic systolic hypotension without elevation of CPK or troponin I levels or ECG changes were observed. Decreased tumor blood flow was observed by DCE-MRI at the 15.5 mg/m^2 dose level. The half-life of AVE8062 was 15 min, but an active metabolite was formed with a half-life of 7 h.

An ophthalmic preparation containing combretastatin A-4 for treating diabetic retinopathy was patented [44]. The preparation was composed of (95–100 by weight %) combretastatin A-4 and other auxiliary materials acceptable for treating eye diseases. The ophthalmic preparation can be used as eye drop, ointment, and gel for treating diabetic retinopathy by inhibiting angiogenesis in a dose-dependent manner without affecting the development of retina vascular system. The invention [45] disclosed an antitumor combination including a derivative of stilbene and an anticancer compound, particularly VEGF-Trap, and methods of use of these pharmaceutical preparation for the treatment of solid and similar carcinomas.

Methods for controlling acute hypertension and cardiotoxicity in cancer patients treated with vascular targeting agents (VTA) were designed [46]. This invention provides a method of using antihypertensive agents (AHA) to attenuate the transient increase in blood pressure observed in patients who were administered vascular targeting agents to treat diseases associated with malignant neovascularization. The invention also relates to pharmaceutical compounds comprising AHAs and VTAs, and to kits thereof. The vascular targeting agent was selected from the group consisting of combretastatin, combretastatin A-4 phosphate, a combretastatin A-1 diphosphate, combretastatin A-4 disodium phosphate, combretastatin A-1 tetrasodium phosphate, or a pharmaceutically acceptable salt thereof. The antihypertensive agent was a vasodilator selected from the group consisting of isosorbide mononitrate, isosorbide dinitrate, nitroglycerin, fenoldopam mesylate, epoprostenol sodium, milrinone lactate, and sodium nitroprusside.

Thirty-four patients with advanced solid tumors were treated with CA-4-P receiving 167 infusions [47]. The drug CA-4-P was given weekly for 3 weeks followed by a gap of 1 week. Up to 40 mg/m^2, the only drug-related toxicity was tumor pain in 35%. Tumor pain was not considered a dose-limiting toxicity because it could be controlled by analgesics. Tumor viability and tumor blood flow were assessed by positron emission therapy (PET) and DCE-MRI.

Positron emission therapy was used to measure the effects of the vascular targeting agent combretastatin A-4 phosphate on tumor and normal tissue perfusion and blood volume in patients [48]. Patients with advanced solid tumors were enrolled onto part of a phase I, accelerated titration, dose escalation study. Effects of 5–114 mg/m^2 CA-4-P on tumor, spleen, and kidney were investigated. Tissue perfusion was measured using oxygen-15 (^{15}O)-labeled water and blood volume was measured using ^{15}O-labeled carbon monoxide (C^{15}O). PET data were obtained for 13 patients with intrapatient dose escalation. Significant dose-dependent reductions in tumor blood volume were seen in tumor perfusion 30 min after CA-4-P administration. Thirty minutes after CA-4-P administration, borderline significant changes were seen in spleen perfusion, spleen blood volume, kidney perfusion, and kidney blood volume. No significant changes were seen at 24 h in spleen or kidney. The authors concluded that CA-4-P produces rapid changes in the vasculature of human tumors that can be assessed using PET measurements of tumor perfusion.

9.4
Other Stilbenoids

The purpose of the review [8] was to describe four structurally similar stilbene compounds, piceatannol, pinosylvin, rhapontigenin, and pterostilbene, and detail some current pharmaceutical research and highlight their potential clinical applications. In the study [49], an extract of *Pterocarpus marsupium* Roxb. containing pterostilbene has been evaluated for its PGE2-inhibitory activity in LPS-stimulated peripheral blood mononuclear cells (PBMC) and the COX-1/2 selective inhibitory activity of *P. marsupium* (PM) in healthy human volunteers. Pterostilbene and resveratrol inhibited PGE2 production from LPS-stimulated human PBMC with IC$_{50}$ values of 3.2 ± 1.3 μg/ml, 1.0 ± 0.6, and 3.2 ± 1.4 μM. Pterostilbene levels in serum increased, but were fivefold lower than the observed IC$_{50}$ for PGE2 inhibition in LPS-stimulated PBMC. No changes in the baseline of the safety parameters were observed. The PGE2 inhibitory activity of PM extract was related to its pterostilbene content. In humans, 450 mg PM extract resulted in elevated pterostilbene levels in serum, which were below the active concentration observed *in vitro*. The authors argued for a dose-finding study of PM extract in humans to corroborate the *in vitro* observed inhibitory activity on PGE2 production to resolve the potential use of PM extraction in inflammatory disorders and/or inflammatory pain.

Determination of glyoxal (Go), and methylglyoxal (MGo), in the serum of diabetic patients by MEKC using stilbenediamine as derivatizing reagent was performed [50] Uncoated fused silica capillary, with an effective length of 50 cm × 75 μm i.d., applied voltage 20 kV, and photodiode array detection were used. Calibration was linear within 0.02–150 μg/ml with detection limits 3.5–5.8 ng/ml. Go and MGo, observed for diabetic and healthy volunteers, were within 0.098–0.193 μg/ml Go and 0.106–0.245 μg/ml MGo with RSD 1.6–3.5 and 1.7–3.4%, respectively, in diabetics compared to 0.016–0.046 μg/ml Go and 0.021–0.06 g/ml MGo with RSD 1.5–3.5 and 1.4–3.6%, respectively, in healthy volunteers.

The active stilbene agents, reported in this chapter, derived from either marine or terrestrial sources tested in clinics have a unique chemistry that offers valuable information for their use as compounds for further chemical synthesis of more potent chemotherapeutic drugs against a variety of cancers. The stilbenes also possess the antioxidative, anti-inflammatory, and estrogenic effects and chemopreventive activities. A challenging goal is the search for more active natural products as therapeutic agents in cancer and other diseases. This search should be continued until a novel and very effective compound is found.

References

1 FDA (2007) Guidance for Institutional Review Boards and Clinical Investigators.
2 Rang, H.P., Dale, M.M., Ritter, J.M., and Moore, P.K. (2003) *Pharmacology*, 5th edn, Churchill Livingstone, Edinburgh.
3 Finn, R. (1999) *Cancer Clinical Trials: Experimental Treatments and How They Can Help You*, O'Reilly & Associates, Sebastopo.
4 (a) Pocock, S.J. (2004) *Clinical Trials: A Practical Approach*, John Wiley & Sons, Inc.; (b) Covance Inc . (2005) Periapproval Services (Phase IIIb and IV programs), Retrieved on March 27, 2007.
5 Arcangelo, V.P., Andrew, M., and Peterson, A.P. (2005) *Pharmacotherapeutics for Advanced Practice: A Practical Approach*, Lippincott Williams & Wilkin.
6 Food and Drug Administration (January 2006) Guidance for Industry, Investigators, and Reviewers Exploratory IND Studies.

7 Wenzel, E. and Somoza, V. (2005) *Molecular Nutrition & Food Research*, **49**, 472–481.

8 Roupe, K.A., Remsberg, C.M., Yanez, J.A., and Davies, N.M. (2006) *Current Clinical Pharmacology*, **1**, 81–101.

9 Cooney, M.M., Radivoyevitch, T., Dowlati, A., Overmoyer, B., Levitan, N., Robertson, K., Levine, S.L., DeCaro, K., Buchter, C., Taylor, A., Stambler, B.S., and Remick, S.C. (2004) *Clinical Cancer Research*, **10**, 96–100.

10 Patterson, D. (2007) *Clinical Oncology*, **19**, 443–456.

11 Tozer, G.M. (2003) *British Journal of Radiology*, **76** (Special Issue 1), S23–S35.

12 Chaplin, D.J. and Dougherty, G.J. (1999) *British Journal of Cancer*, **80** (Suppl. 1), 57–64.

13 Tron, G., Pirali, T., Sorba, G., Pagliai, F., Busacca, S., and Genazzani, A. (2006) *Journal of Medicinal Chemistry*, **49**, 3033–3044.

14 Cuccidla, V., Boriello, A., Oliva, A., Galletti, P., Zappia, V. and Della Ragione, F. (2007) *Cell Cycle*, **6**, 2495–2510.

15 Gaya, A.M. and Rustin, G.J.S. (2005) *Clinical Oncology*, **17**, 277–290.

16 Hinnen, P. and Eskens, F.A.L.M. (2007) *British Journal of Cancer*, **96**, 1159–1165.

17 Simmons, T.L., Andrianasolo, E., McPhail, K., Flatt, P., and Gerwick, W.H. (2005) *Molecular Cancer Therapeutics*, **4**, 333–342.

18 Thorpe, P.E. (2004) *Clinical Cancer Research*, **10**, 415–427.

19 Chaplin, D.J. (1999) *Cancer Research*, **59**, 1626–1634.

20 Li, L., Rojiani, A.M., and Siemann, D.W. (2002) *Acta Oncologica*, **41**, 91–97.

21 Yeung, S.C., She, M., Yang, H., Pan, J., Sun, L., and Chaplin, D.J. (2007) *Journal of Clinical Endocrinology and Metabolism*, **92**, 2902–2909.

22 Gumbrell, L. and Price, P.M. (2003) *Journal of Clinical Oncology*, **21**, 2815–2822.

23 Hinnen, P. (2007) *British Journal of Cancer*, **96**, 1159–1165.

24 Patterson, D.M. and Rustin, G.J.S. (2007) *Drugs of the Future*, **32**, 1025–1032.

25 Kerbel, R. and Folkman, J. (2002) *Nature Reviews. Cancer*, **2**, 727–739.

26 Baur, J.A. and Sinclair, D.A. (2005) *Nature Reviews. Drug Discovery*, **5**, 493–506.

27 Goldberg, D.M., Yan, J., and Soleas, G.J. (2003) *Clinical Biochemistry*, **36**, 79–87.

28 Wang, B., Zeng, W., Kang, H., Xue, Y., Tang, T., Shi, Z., Zhou, Y., Ye, Q., Yan, Q., Zhu, D., Chen, H., Huang, C., Zhao, J., Feng, H., and Yu, L. (2008) Faming Zhuanli Shenqing Gongkai Shuomingshu, pp. 15.

29 Soleas, G.J., Yan, J., and Goldberg, D.M. (2001) *Journal of Chromatography B: Biomedical Sciences and Applications*, **757**, 161–172.

30 Boocock, D.J., Faust, G.E.S., Patel, K.R., Schinas, A.M., Brown, V.A., Ducharme, M.P., Booth, T.D., Crowell, J.A., Perloff, M., Gescher, A.J., Steward, W.P., and Brenner, D.E. (2007) *Cancer Epidemiology, Biomarkers & Prevention*, **16**, 1246–1252.

31 Walle, T., Hsieh, F., DeLegge, M.H., Oatis, J.E., Jr, and Walle, U.K. (2004) *Drug Metabolism and Disposition*, **32**, 1377–1382.

32 Elliott, P.J. and Jirousek, M. (2008) *Current Opinion in Investigational Drugs (Thomson Scientific)*, **9**, 371–378.

33 Vitaglione, P., Sforza, S., Galaverna, G., Ghidini, C., Caporaso, N., Vescovi, P.P., Fogliano, V., and Marchelli, R. (2005) *Molecular Nutrition & Food Research*, **49**, 495–504.

34 Burkon, A. and Somoza, V. (2008) *Molecular Nutrition & Food Research*, **52**, 549–557.

35 Banerjee, S., Wang, Z., Mohammad, M., Sarkar, F.H., and Mohammad, R.M. (2008) *Journal of Natural Products*, **71**, 492–496.

36 Ng, Q.-S., Goh, V., Carnell, D., Meer, K., Padhani, A.R., Saunders, M.I., and Hoskin, P.J. (2007) *International Journal of Radiation Oncology, Biology, Physics*, **67**, 1375–1380.

37 Rustin, G.J.S., Galbraith, S.M., Anderson, H., Stratford, M., Folkes, L.K., Sena, L., Giantonio, B., Zimmer, R., Petros, W.P., Stratford, M., Chaplin, D.,

Young, S.L., Schnall, M., and O'Dwyer, P.J. (2003) *Journal of Clinical Oncology*, **21**, 4428–4438.

38 Stevenson, J.P., Rosen, M., Sun, W., Gallagher, M., Haller, D.G., Vaughn, D., Giantonio, B., Zimmer, R., Petros, R., Stratford, M.D., Chaplin, D., Young, S.L., Schnall, M., and O'Dwyer, P.J. (2003) *Journal of Clinical Oncology*, **21**, 4428–4438.

39 Rustin, G.J.S., Galbraith, S.M., Anderson, H., Stratford, M., Folkes, L.K., Sena, L., Gumbrell, L., and Price, P.M. (2003) *Journal of Clinical Oncology*, **21**, 2815–2822.

40 Dowlati, A., Robertson, K., Cooney, M., Petros, W.P., Stratford, M., Jesberger, J., Rafie, N., Overmoyer, B., Makkar, V., Stambler, B., Taylor, A., Waas, J., Lewin, J.S., McCrae, K.R., and Remick, S.C. (2002) *Cancer Research*, **62**, 3408–3416.

41 Bilenker, J.H., Flaherty, K.T., Rosen, M., Davis, L., Gallagher, M., Stevenson, J.P., Sun, W., Vaughn, D., Giantonio, B., Zimmer, R., Schnall, M., and O'Dwyer, P.J. (2005) *Clinical Cancer Research*, **11**, 1527–1533.

42 Stevenson, J.P., Rosen, M., Sun, W., Gallagher, M., Haller, D.G., Vaughn, D., Giantonio, B., Zimmer, R., Petros, W.P., Stratford, M., Chaplin, D., Young, S.L., Schnall, M., and O'Dwyer, P.J., (2003) *Journal of Clinical Oncology*, **21**, 4428–4438.

43 Tolcher, A.W., Forero, L., Celio, P., Hammond, L.A., Patnaik, A., Hill, M., Verat-Follet, C., Haacke, M., Besenva, M., and Rowinsky, E.K. (2003) *Journal of Clinical Oncology*, **22** (Suppl.), 208 (abstract 834).

44 He, Y., Xu, X. and Bao (2007) XuFaming Zhuanli Shenqing Gongkai Shuomingshu, pp. 11.

45 Bao, X., Wang, Q., De, T., Desai F N.P., and Soon-Shiong F P. (2007) *PCT International Patent Application*, pp. 71.

46 Chaplin, D. and Young, S. (2008) US Patent pp. 11.

47 Galbraith, S.M., Maxwell, R.J., and Lodge, M.A. (2003) *Journal of Clinical Oncology*, **21**, 2831–2842.

48 Anderson, H.L., Yap, J.T., Miller, M.P., Robbins, A., Jones, T., and Price, P.M. (2003) *Journal of Clinical Oncology*, **21**, 2823–2830.

49 Hougee, S., Faber, J., Sanders, A., de Jong, R.B., van den Berg, W.B., Garssen, J., Hoijer, M.A., and Smit, H.F. (2005) *Planta Medica*, **71**, 387–392.

50 Mirza, M.A., Kandhro, A.J., Memon, S.Q., Khuhawar, M.Y., and Arain, R. (2007) *Electrophoresis*, **28**, 3940–3947.

10
Stilbenes as Molecular Probes

10.1
General

A molecular probe is a group of atoms or molecules attached to other molecules or structures and used in studying the properties of these molecules and structures [1–7].

Methods of stilbene molecular probes involve a whole arsenal of photophysical and photochemical properties of these compounds (Chapters 2–5). These methods, taken in combination, allow solving a number of structural and dynamic problems on molecular level. Recently, a series of fluorescent methods of analysis and investigation of system based on the use of stilbenes and potentially important in biochemistry, biophysics, biotechnology, and biomedicine were proposed and developed [8–21]. In these methods, two new types of stilbene molecular probes have been used: (i) fluorescent photochrome molecules and (ii) supermolecules containing fluorescent and fluorescent quenching segments. These methods utilize the following photochemical and photophysical phenomena: the fluorescence quenching, photochrome photoisomerization, and triplet–triplet and singlet–singlet energy transfer. The fluorescence properties of the new probes were intensively exploited as the basis of several methodologies that include a real-time analysis of nitric oxide and trinitrotoluene, investigation of molecular dynamics of antibodies and biomembranes in a wide range characteristic times, and characterization of surface system. Information about commercial stilbene probes may be taken from Ref. [22, 23].

In this chapter, we concentrate on luminescence approaches that are most suitable for stilbenes to be molecular probes and are based on specific and nonspecific labeling, competition, solvatochromism, experimental molecular dynamics, and singlet–singlet and triplet–triplet energy transfer. A general survey is made of the physical principles and application of the fluorescence probe methods stressing on latest developments in this area.

10.2
Theoretical Grounds

10.2.1
Local Properties of Medium

10.2.1.1 Polarity

The sensitivity of the nonspecific electrostatic properties of the environment allows to use chromophores as indicators of polarity [1, 2, 4, 24, 25]. Within the framework of the Onsager model, the value of the shift of the absorption spectra at transition from the gas phase to the given medium depends on the dipole moments of the chromophore ground and excited states and the medium dielectric constant (ε_0) and the refraction index (n). The Onsager model obviously gives only a rough picture of the effects in the system under investigation. Nevertheless, in some cases this model can be a tool for approximating the relative characteristics of solvate chromic effects.

Using the fluorophores markedly increases sensitivity and accuracy of the method to polarity [1, 2, 4]. The simplest way for estimating local polarity is a previous calibration using standard liquids with well-characterized properties.

10.2.1.2 Molecular Dynamics

The excitation of a chromophore group is accompanied by a change in the electron dipole moment of the molecule. This involves a change in the interaction energy with the surrounding molecules, which manifests itself by a shift of the time-dependent frequency maximum of the fluorescence spectra (relaxation shift) [4, 25]

$$\Delta v_{max} = v_{max}(t) - v_{max}(\infty) = \left[v_{max}(0) - v_{max}(\infty) \exp\left(-\frac{t}{\tau_r}\right)\right], \quad (10.1)$$

where the indices t, ∞, and 0 are related to the maximum of the time-resolved emission spectrum at a given moment, $t \to \infty$ and $t \to 0$, and τ_r is the characteristic time of reorganization of the dipoles in the medium around the fluorophore. The value of τ_r can also be independently derived from the analysis of the temperature (T) dependencies of the relaxation shift using the following equation:

$$\Delta v_{max}(T) = \frac{[v_{max}(0) - v_{max}(\infty)]\tau_f}{[\tau_f + \tau_r(T)]}, \quad (10.2)$$

where $\Delta v_{max}(T)$ is the relaxation shift in the steady-state fluorescence spectra and τ_f is the fluorescence lifetime. A gradual increase in temperature results in gradual decrease in τ_r. The experimental $\Delta v_{max}(T) - T$ dependence can be used to estimate $\tau_r(T)$ in each temperature if τ_f is known. In real systems (viscous liquids, polymers, proteins, membranes, and so on), there as a rule is a set of τ_r values, relaxation energy and entropy activation, and other parameters. Analysis of relaxation shifts in such systems requires special approaches. For instance, if one assumes a Gaussian distribution over the free activation energies of the reorientation of surrounding particles ($\Delta F^{\#}$), it is possible to find an expression to relate the energy activation of relaxation in the distribution maximum (E_{max}) to the second moment of the distribution curve (ΔF_0^2)

$$E_{app}(T) = E_{max} - \frac{\Delta F_0^2}{RT}, \tag{10.3}$$

where $E_{app}(T)$ is the experimental value of apparent energy activation derived from the Arrhenius plot, $\log \Delta v_{max}(T) - 1/T$. Equation 10.3 allows the estimation of E_{max} and (ΔF_0^2) plotting $E_{app}(T)$ versus $1/T$ [4].

A pump-damp-probe method (PDPM), which allows the characterization of solvation dynamics of a fluorescence probe not only in excited states but also in the ground state has been developed recently ([26] and references therein). In PDPM, a pump produces a nonequilibrium population of the probe excited, which, after media relaxation, is stimulated back to the ground state. The solvent relaxation of the nonequilibrium ground state is probed by monitoring with absorption technique. In the pump-damp-probe experiments, part of a series of laser output pulses was frequency doubled and softer beams were used as the probe. The delay of the probe with respect to the pump was fixed at 500 ps.

Asymmetry of chromophores suggests anisotropy of their transition dipole moments. The rotation of an asymmetric chromophore with correlation time τ_R comparable to the molecular excited singlet or triplet excited state, t^*, may be accompanied by depolarization. The value of τ_R depends on the microviscosity and microstructure of the medium. The theory of the process and the experimental data have been described in Refs [1, 2].

10.2.2
Excited Energy Transfer

10.2.2.1 Fluorescence Resonance Energy Transfer

Fluorescence resonance energy transfer (FRET), a phenomenon first described by Förster in 1948 [27], involves dipole–dipole energy transfer from the emitting fluorophore moiety (donor) to the absorbing moiety (acceptor). The rate of energy transfer (k_T) for any specific donor (D) and acceptor (A) pair is given by

$$k_T = \frac{R_0^6}{r^6 \tau_D} = \frac{9000(\ln 10)k^2 \varphi_D}{128\pi^5 n^4 N r^6 \tau_D} \int_0^\infty \frac{F_D(v)\varepsilon_A(v)}{v^4} dv$$

$$= \left(r^{-6} J k^2 n^{-4} \frac{\varphi_D}{\tau_D} \right) \times 8.71 \times 10^{23} \text{ s}^{-1} \tag{10.4}$$

and depends upon (a) the degree of overlapping of the emission spectrum of donor and absorption spectrum of acceptor (overlap integral, J), (b) relative orientation of the donor and acceptor excited-state dipoles (orientational factor, k^2), and (c) distance between donor and acceptor (r). In the formula above, τ_D is the lifetime of donor in the absence of acceptor, R_0 is Förster distance at which the efficiency of transfer is 50%, φ_D is the quantum yield of the donor in the absence of acceptor; n is the refractive index of the medium, N is Avogadro number, $F_D(v)$ is the corrected fluorescence intensity of donor at wavenumber ν with total intensity normalized to unity, and $\varepsilon_A(\nu)$ is the extinction coefficient of the acceptor.

The efficiency of energy transfer is the proportion of photons absorbed by the donor that are transferred to the acceptor and it is equal to relative fluorescence yield in the presence (F_{DA}) and the absence (F_D) of acceptor [1, 2]:

$$E = 1 - \frac{F_{DA}}{F_D} = \frac{R_0^6}{R_0^6 + r^6}. \tag{10.5}$$

Equations 10.4 and 10.5 are used to calculate distances between the donor and the acceptor.

The rate constant of TTET

$$k_{TT} = \frac{2\pi}{h} J_{TT} FC, \tag{10.6}$$

where J_{TT} is the TT exchange integral. The Hamiltonian of the exchange interaction (SE) between spins with operators S_1 and S_2 is described by the equation

$$H_{SE} = -2 J_{SE} S_1 S_2, \tag{10.7}$$

where J_{SE} is the SE exchange integral.

A vast literature pertains to the quantitative investigation of exchange processes (see, e.g., [4, 28–30] and references therein).

As seen in Figure 10.1, experimental data on the dependence of k_{TT} and J_{SE} on the distance between the centers (ΔR) lie on two curves, which are approximated by the following equation [30]:

$$k_{TT}, J_{SE} \propto \exp(-\beta \Delta R). \tag{10.8}$$

For systems in which the centers are separated by a "nonconductive" medium (molecules or groups with saturated chemical bonds), β_{TT} is equal to 2.6 Å^{-1}. For systems in which the radical centers are linked by "conducting" conjugated bonds, β_{SE} is 0.3 Å^{-1}.

These dependences can be used for distance estimation. Measurements of TTET at encounters between the donor and the acceptor chromophores allow investigating the dynamics of the processes involved.

10.3
Experimental Methods and Their Applications

10.3.1
Probing Based on Solvatochromism

4-(N,N-Dimethylamino)stilbene (DS), 4-(1-aza-15-Crown-5)stilbene (DS-Crown) and regioselectively bridged 4-(N,N-dialkylamino)stilbene derivatives (DS-B2, DS-B4, DS-B24, DS-B34) have been synthesized and their solvatochromism was measured [31]. In ethanol, the fluorescence quantum yields were obtained at room

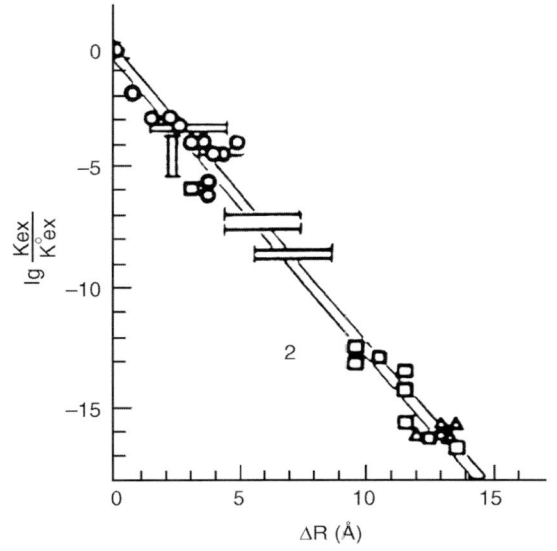

Figure 10.1 Dependence of the logarithm of relative parameters of the exchange interaction on the distance between the interacting centers (ΔR). k_{TT} is the rate constant of triplet–triplet electron transfer and JSE is the spin-exchange integral. Index 0 is related to van der Waals' contact [30]. (Reproduced with permission from Se.)

temperature (RT), while the lifetimes were measured from 77 K to RT. When the single bond connecting the dialkyanilino group to the double bond was bridged, a strong fluorescence quenching was observed. When this bond was flexible, the fluorescence quenching was strongly reduced and the lifetime maxima at intermediate temperature indicated the involvement of a further emitting state (TICT). The resulting kinetic scheme led to the design of fluorescence probes for calcium sensing (DS-Crown).

The fluorescent probes, 4-(dimethylamino)-4'-(methylsulfonyl)stilbene, 4-(dimethylamino)-4'-(methylsulfonyl)diphenylbutadiene, and 4-(dimethylamino)-4'-nitrostilbene, were incorporated in polymeric networks formed by photopolymerization of dimethacrylates of different molecular sizes and polarities [32]. The response of the probe's emission to changes in its environment during photopolymerization was detected and compared with the emissions in solvents of low viscosity of differing polarities. A comparison between these data revealed that the blueshifts observed during polymerization were roughly proportional to the solvatochromism of each probe.

The sensitive solvatochromism of 2,5-dicyano-4-methyl-4'-(9-carbazolyl)stilbene (**1**) and 2,5-dicyano-4-methyl-4'-bis-(hydroxyethyl) aminostilbene (**2**) was described [33]. The emission maximum and two-photon absorption cross section (δTPA) of **1** and **2** vary from 412 and 453 nm in cyclohexane to 534 and 629 nm in DMSO and from 6930 GM in cyclohexane and 2880 GM in 1,4-dioxane to 1050 and 130 GM in

DMF, respectively. Both **1** and **2** with remarkably large δTPA exhibited very strong polarity, viscosity, and temperature dependence of fluorescence and can be used to detect polarities, viscosities, and temperature. The covalent attachment of two cyano groups in the single aromatic ring and bis-(hydroxyethyl)amino group (or 9-carbazolyl group) to stilbene made the compound a highly sensitive two-photon thermo-solvatochromic probe with large two-photon absorption cross sections and pretty high fluorescence quantum yield that can be used to detect analytes with high π^*. The results provide a prototype for the development of effective two-photon fluorescence probes.

Stilbene-like fluorescent probes 2-(p-N,N-dimethylaminostyryl)benzoxazole (OS), 2-(p-N,N-dimethylaminostyryl)-benzothiazole (SS), and 2-(p-N,N-dimethylaminostyryl)naphthiazole (PS) were prepared and their absorption and fluorescence spectra were measured in various solvents at room temperature [34]. The normalized absorption and fluorescence spectra of PS in ethyl acetate (AcOEt), butyronitrile (BuCN), and N,N-dimethylformamide (DMF) as solvents are presented in Figure 10.2. On the basis of the solvatochromic behavior, the ground state (μ_g) and excited state (μ_e) dipole moments of these p-N,N-dimethylaminostyryl derivatives were evaluated. The dipole moments (μ_g and μ_e) were estimated from solvatochromic shifts of absorption and fluorescence spectra as function of the dielectric constant and refractive index (n) of applied solvents. The absorption spectra were only slightly affected by the solvent polarity in contrast to the fluorescence spectra that are highly solvatochromic and display a large Stokes shift. The analysis of the solvatochromic behavior of the fluorescence spectra revealed that the emission occurred from a high polarity excited state. The large changes in dipole moments along with the

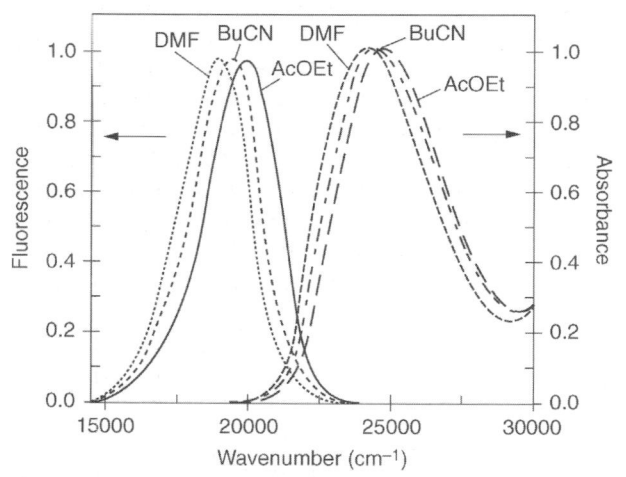

Figure 10.2 Normalized absorption and fluorescence spectra of PS in ethyl acetate (AcOEt), butyronitrile (BuCN), and N,N-dimethylformamide as solvents [34]. (Reproduced with permission.)

strongly redshifted fluorescence, as the solvent polarity was increased, demonstrated the formation of an intramolecular charge transfer state. Compounds under study were used as fluorescence probes for monitoring the kinetics of thermally initiated polymerization of methyl methacrylate and photoinitiated polymerization of 2-ethyl-2-(hydroxymethyl)propane-1,3-diol triacrylate.

10.3.2
Image and Structure Probing

Luminescent 4,4′-bis(2-benzoxazolyl)stilbene (BBS) was used as a molecular probe for poly(propylene) film deformation [35]. Polypropylene (PP) films containing different concentrations of BBS were prepared by melting process. It was found that the emission characteristics of the PP films depended on the BBS concentration and polymer deformation. A well-defined excimer band was observed with more than 0.2 wt% of BBS, conferring on the film a green luminescence. During drawing (130°), the PP reorganization broke the BBS excimer-type arrangement, leading to the prevalence of the blue emission of the single molecules. The photophysics of this compound was efficiently applied for the detection of tensile deformation of PP films. Figure 10.3 shows a digital image of the uniaxially oriented PP/BBS-0.5 film and fluorescence emission spectra PP/BBS-0.5.

[^{11}C]4 was prepared by ^{11}C methylation of 4-amino-4′-hydroxystilbene. The [^{11}C]4 displayed a moderate lipophilicity (log $P = 2.36$) and showed a very good brain penetration and washout from normal rat brain after an i.v. injection. *In vitro* autoradiography was performed by incubating [^{11}C]4 with brain sections from control and double mutation mice (TgCRDN8). It was an early-onset transgenic (Tg) mouse model resulting from effects of familial Alzheimer's disease (AD) mutations on amyloid beta-peptide (Aβ) biogenesis. Thioflavin S-positive Aβ amyloid deposits were present at 3 months, with dense-cored plaques and neuritic pathology becoming evident from 5 months of age. Using the brain sections from these mice, Aβ plaque labeling by [^{11}C]4 was tested. The labeled stilbene derivative, [^{11}C]4, showed an excellent binding to the plaques in the mutant mouse section while only a minimum labeling in control section was observed (Figure 10.4). Taken together, the data suggested that a relatively simple stilbene derivative, N-[^{11}C]4methylamino-4′-hydroxystilbene, synthesized by a scheme presented in Figure 10.5, may be useful as a positron emission tomography (PET) imaging agent for mapping Aβ plaques in the brain of AD patients.

Three fluorescent probes

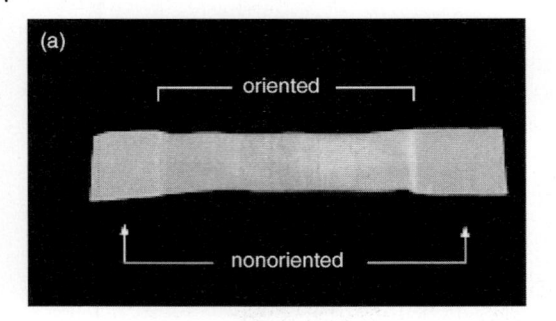

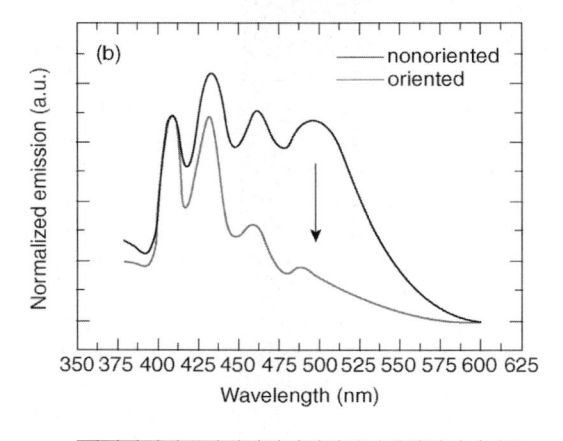

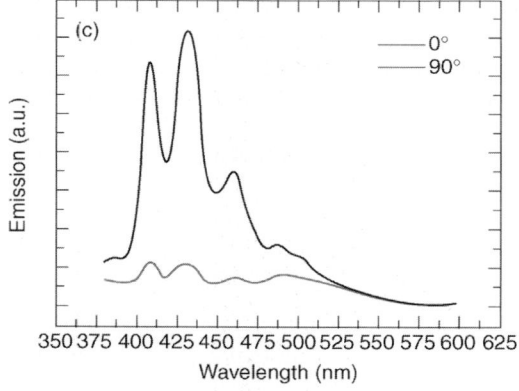

Figure 10.3 (a) Digital image of the uniaxially oriented PP/BBS-0.5 film taken under excitation with a long-range UV lamp ($\lambda = 366$ nm). (b) Fluorescence emission spectra ($\lambda_{exc} = 277$ nm) of a PP/BBS-0.5 film containing 0.5 wt% of BBS molecules, before and after solid-state drawing ($D_r = 8$). The spectra are normalized to the intensity of the isolated BBS molecules peak (409 nm). (c) Fluorescence emission spectra ($\lambda_{exc} = 277$ nm) of the uniaxially oriented PP/BBS-0.5 film recorded with polarization parallel (0°) and perpendicular (90°) to the 0.5 film recorded with polarization parallel (0°) and perpendicular (90°) to the drawing direction, respectively [35]. (Reproduced with permission.)

Figure 10.4 Scheme of the synthesis of labels 4-N-methylamino-4'-hydroxystilbene [^{11}C]4 [36]. (Reproduced with permission from Elsevier.)

were synthesized for optical imaging to detect amyloid plaques present in the patients with Alzheimer's disease [37].

These compounds were prepared via Sonogashira coupling of a well-defined fluorophore (4-bora-3a,4a-diaza-s-indacene, BODIPY) with the pharmacophore possessing either a stilbene or a diphenylacetylene moiety. Different polyethylene glycol chain lengths were used as linkers between the fluorophore and the pharmacophore to adjust the lipophilicity of these probes. The compounds exhibited strong fluorescence emission between 665 and 680 nm and had very high extinction coefficients comparable to the parent fluorophore, BODIPY dye.

For direct detection and quantification of myelin content *in vivo*, a series of stilbene derivatives as myelin imaging agents were prepared that readily enter the brain and selectively bind to myelinated regions [38]. Abnormalities and changes in myelination in the brain were seen in many neurodegenerative disorders such as multiple sclerosis (MS). Spectrophotometry-based and radioligand-based binding assays showed that these stilbene derivatives exhibited relatively high myelin binding affinities. *In vitro* myelin staining showed that the compounds selectively stained intact myelinated regions in wild-type mouse brain. *In situ* tissue staining demonstrated that the compounds readily entered the mouse brain and selectively labeled myelinated white matter regions. These studies suggested that stilbene derivatives can be used as myelin-imaging probes to monitor myelin pathology *in vivo*.

Synthesis of stilbene derivatives and their use in binding and imaging amyloid plaques were patented [39]. The invention described a method of imaging amyloid

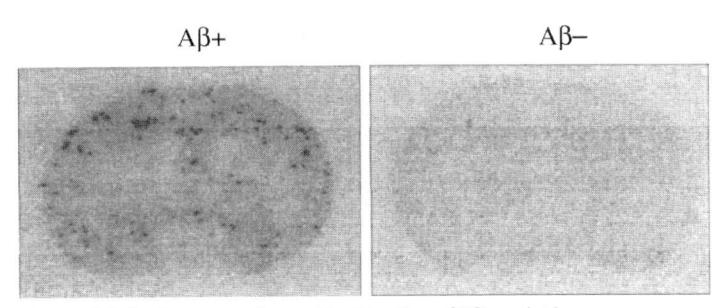

Figure 10.5 *In vitro* autoradiographic detection of Aβ amyloid deposits with [^{11}C]4 in TgCRND8 mouse brain sections. Clear differences between histochemically characterized Tg Aβ+ (left panel) and Tg Aβ− (right panel) brains are readily observable [36]. (Reproduced with permission from Elsevier.)

deposits and methods of making labeled compounds useful in imaging amyloid deposits. It also described methods of making compounds for inhibiting the aggregation of amyloid proteins to form amyloid deposits, and a method of delivering a therapeutic agent to amyloid deposits. Fluorescent molecules, p-dimethylamino-benzonitrile and *trans*-stilbene were used as probes to investigate the inner structure of the nanotubular cavities contained in a new polyorganosiloxane [40]. This polymer was prepared by the hydrosilylation coupling reaction of *cis*-isotactic ladder-like polyvinylsilsesquioxane with 1,1,3,3-tetramethyldisiloxane as a coupling agent in the presence of catalyst, dicyclopentadienyldichloroplatinum. Results from fluorescence resonance energy transfer spectra and fluorescence spectra in combination with molecular simulation revealed that the cross section of the nanotubular cavity was nearly rectangular in shape and was about 0.62 nm in width and about 0.38 nm in height.

In the study [41], authors used FRET to monitor the rapid exchange of recombinant alpha-crystallin subunits. AlphaA-crystallin was labeled with stilbene iodoacetamide (4-acetamido-4'-((iodoacetyl)amino)stilbene-2,2'-disulfonic acid), which serves as an energy donor, and with lucifer yellow iodoacetamide, which serves as an energy acceptor. Upon mixing the two populations of labeled alphaA-crystallin, a reversible, time-dependent decrease in stilbene iodoacetamide emission intensity and a concomitant increase in lucifer yellow iodoacetamide fluorescence were observed. This result was indicative of an exchange reaction that brings the fluorescent alphaA-crystallin subunits close to each other. The exchange reaction strongly depended on temperature, with a rate constant of 0.075 min^{-1} at 37 °C and an activation energy of 60 kcal/mol. The subunit exchange was independent of pH and calcium concentration but decreased at low and high ionic strength, suggesting the involvement of both ionic and hydrophobic interactions. Binding of large denatured proteins to alphaA-crystallin markedly reduced the exchange rate, indicating an association of the polypeptides with several alphaA-crystallin subunits. It was suggested that a dynamic organization of alphaA-crystallin subunits may be a key factor in preventing protein aggregation.

SecA topology was investigated using 4-acetamido-4'-maleimidylstilbene-2,2'-disulfonic acid (AMS), which is a membrane-impermeable sulfhydryl-labeling reagent [42]. In this study, the authors have used a combination of site-directed sulfhydryl labeling, proteolysis, or truncation of SecA to develop a map of the membrane topology of this protein. To detect which regions of SecA are periplasmically exposed, right-side out-membrane vesicles were prepared from strains synthesizing monocysteine SecA variants produced by mutagenesis and probed with AMS. Inverted inner membrane vesicles were subjected to proteolysis, and integral-membrane fragments of SecA were identified with region-specific antibodies. The analysis indicated that the membrane topology of SecA was complex with amino-terminal, central, and carboxyl-terminal regions of SecA integrated into the membrane where portions were periplasmically accessible. The insertion and penetration of the amino-terminal third of SecA, which included the proposed preprotein binding domain, were subjected to modulation by ATP binding.

Disulfonic stilbene anion transport inhibitor 4,4'-dibenzamido-2,2'-disulfonic stilbene (DBDS) was used for studying the radiation inactivation effect on red cell band 3 stilbene site [43]. The target sizes for binding DBDS and mercurial water transport inhibitor, p-chloromercuribenzene sulfonate (p-CMBS), to ghost membranes of human erythrocytes were measured by a fluorescence enhancement technique. The measured target size for erythrocyte ghost acetylcholinesterase was found to be 78 ± 3 kDa. It was shown that radiation (0–26 Mrad) had no effect on total membrane protein and DBDS binding affinity, whereas DBDS binding stoichiometry exponentially decreased with radiation dose, giving a target size of 59 ± 4 kDa. H2-4,4′-diisothiocyano-2,2′-disulfonic stilbene (H2-DIDS, 5 µM) blocked more than 95% of DBDS binding at all radiation doses. Results obtained supported the notion that DBDS and p-CMBS bind to the transmembrane domain of erythrocyte band 3 in NEM-treated ghosts and demonstrated that radiation inactivation may probe a target significantly smaller than a covalently linked protein subunit. It was suggested that the small target size for the band 3 stilbene binding site may correspond to the intramembrane domain of the band 3 monomer (52 kDa), which is physically distinct from the cytoplasmic domain (42 kDa). For the purpose of an effective utilization of donatable hydrogens in coal liquefaction residues (CLR), hydrogen donor properties of CLR were evaluated at 380–450 °C with a series of molecular probes including *trans*-stilbene [44].

10.3.3
Methods Based on Accessibility of Reactive Groups

A large and rigid hydrophilic sulfhydryl reagent, 4-acetamido-4′-[(iodoacetyl)amino] stilbene-2,2′-disulfonic acid (IASD) disodium salt,

was used to investigate conformations of the lipid binding region of *Escherichia coli* pyruvate oxidase (PoxB) [45]. In the absence of pyruvate, the cysteine residues of the modified PoxB proteins failed to form disulfide bonds and generally failed to react with IASD. In the presence of pyruvate, all of the C-terminal cysteine residues (except the two most distal from the C-terminus) reacted with both sulfhydryl reagents and readily formed disulfide cross-linked species, indicating conversion to a structure having a high degree of conformational freedom. In the presence of lipid activators, Triton X-100 or dipalmitoylphosphatidylglycerol, a subset of the cysteine-substituted proteins no longer reacted with the membrane-impermeable IASD reagent, indicating penetration of these protein segments into the lipid micelles. These data were discussed in terms of three distinct PoxB conformers and the known crystal structure of a highly related protein.

Structural changes in staphylococcal alpha-hemolysin (alpha HL) that occurred during oligomerization and pore formation on membranes have been examined with the hydrophilic sulfhydryl reagent, 4-acetamido-4′-((iodoacetyl)amino)stilbene-2,2′-

disulfonate [46]. A simple gel-shift assay to determine the rate of modification of key single-cysteine mutant with IASD was used. It was found that a residue in glycine-rich loop of alpha HL lining the lumen of the transmembrane channel remained accessible to IASD after assembly, in keeping with the ability of the pore to pass molecules of approximately 1000 Da. By contrast, residues near the N-terminus, which are critical for pore function, became deeply buried during oligomerization.

To analyze the accessibility and the chemical nature of functional sites of the integral enzyme protein, the effect of the photoactivated reagent 4,4′-diazidostilbene 2,2′-disulfonic (DASS) acid on rat liver microsomal glucose-6-phosphatase was investigated [47]. When native rat liver microsomes were irradiated with the photoactive reagent, the activity of glucose-6-phosphatase progressively inhibited and an intensely fluorescent adduct was formed. The adduct emission and excitation maximum corresponded with those obtained when cysteine or 3-mercaptopropionic acid was irradiated in the presence of the photolabile reagent. Data from fluorescence measurements show that p-mercuribenzoate and dithiothreitol reduced fluorescence labeling of microsomes. From these results, the authors concluded that DASS directly reacts with the integral phosphohydrolase mainly by chemical modification of essential sulfhydryl groups of the enzyme protein accessible from the cytoplasmic surface of the native microsomal membrane. New nanofibers containing poly (vinylpyrrolidone)-iodine complex (PVP-iodine) were obtained by electrospinning to prepare materials suitable for wound dressing [48]. It was shown that photo-mediated cross-linking in the presence of 4,4′-diazidostilbene-2,2′-disulfonic acid disodium salt successfully stabilized the electrospun and PEO/PVP nanofibers against water and water vapor.

10.3.4
Stilbene Probes Binding to Proteins

Inhibition of inorganic anion transport across the human red blood cell membrane by chloride-dependent association of dipyridamole with a stilbene disulfonate binding site on the band 3 protein was investigated [49]. Dipyridamole binding led to a displacement of 4,4′-dibenzoylstilbene-2,2′-disulfonate from the stilbenedisulfonate binding site of band 3. The Cl^--promoted dipyridamole binding leading to a competitive replacement of the stilbenedisulfonates was assumed. The newly described Cl^- binding site was found to be highly selective with respect to Cl^- and other monovalent anion species. It was found that stilbene disulfonates can bind specifically to proteins that are not anion transporters. For example, disulfonic acids, 4,4′-diisothiocyanatostilbene-2,2′-disulfonic acid (DIDS) and 4-acetamido-4′-isothiocyanatostilbene-2,2′-disulfonic acid (SITS), complexed specifically with the CD4 glycoprotein on T-helper lymphocytes and macrophages, blocking HIV type-1 growth at multiple stages of the virus life cycle. According to Ref. [50], the probes DIDS and SITS bounded the variable-1 immunoglobulin-like domain of CD4 on JM cell. These compounds appeared to be CD4 antagonists that block human immunodeficiency virus type-1 growth at multiple stages of the virus life cycle. 4,4′-Dinitro-2,2′-disulfonic stilbene (DNDS) and SITS were effective inhibitors of organic anion transport across the brush border membrane of the rabbit proximal tubule and Cl-oxalate and SO_4–CO_3 exchange by binding to protein band 3.

10.3.5
Depth of Immersion of a Stilbene Probe in Biomembranes

The approach proposed in Refs [21, 51] was based on quantitative investigation of the dynamic quenching of the fluorescence (phosphorescence) of a chromophore incorporated in an object of interest, by a quencher (stable nitroxide radical) freely diffused in solution. Using the concept of dynamic exchange interactions and the empirical dependence of parameters of the static exchange interactions on the distances between exchangeable centers whether by mechanisms of intersystem crossing (IC) (rate constant k_{IC}) or electron transfer (ET) (rate constant k_{ET}) (Figure 10.1, Equation 10.8), the theoretical basis of a method for the determination of the immersion depth of chromophore was developed. The theory establishes a quantitative relationship between the ratio of the experimental rate constants of dynamic quenching of fluorophore in solution (k_q^d, diffusion limit) and after the fluorophore ducking k_q^k, kinetic limit) in the matrix under investigation, on the one hand, and the depth of immersion of the fluorophore, on the other hand. The distance of closest approach R_0 derived can be taken as the depth of immersion of the fluorescent and phosphorescent chromophore in the biological matrix. The ratio between the quenching rate constants for a chromophore–quencher pair in solution and after ducking in a matrix is given by

$$k_q^k/k_q^d = \tau_c^2 10^a \exp 2[-\beta(R_0 - r_v)] \tag{10.9}$$

and the depth of immersion of the center under investigation by

$$(R_0 - r_v) = 0.5 \times \beta^{-1}\left[\ln\left(k_q^d/k_q^k\right) + \ln(\tau_c^2\, 10^a)\right], \tag{10.10}$$

where $\beta = 2$ and $1.3\,\text{Å}^{-1}$ and $a = 28$ and 26 for the intersystem crossing (ICHA) and electron transfer mechanisms, respectively, and τ_c is the value of an encounter complex lifetime τ_c. Equations 10.9 and 10.10 were used to estimate the depth of immersion of a fluorescent chromophore in a "nonconductive" matrix by experimental measurement of ratio k_q^d/k_q^k with a reasonable choice of values of an encounter complex lifetime τ_c and parameter β.

It was shown that the values of the quenching rate constant k_q^s (2.6–4.4×10^9 $M^{-1}s^{-1}$) in the solution fall within the range of values typical for fluorescence quenching in the diffusion limit. In contrast to results of fluorescence quenching in solution, the Stern–Volmer plots for these systems exhibit deviation from linearity (Figure 10.6). At the quencher concentration $[R] < 0.1\,M$ (area I), the plots can be approximated as straight lines with slopes $K_{q(I)}$ equal to 4, 13, and $18\,M^{-1}$ for liposomes, BS, and BS + CAM membranes, respectively.

Experimental data on the depth of immersion ($R_0 - r_v$) of the stilbene probe DMACS ($R_0 - r_v$) in different membranes calculated by Equations 10.9 and 10.10 for IC and ET mechanisms of the fluorescence quenching were found to range from 5.2 to $5.9\,\text{Å}$.

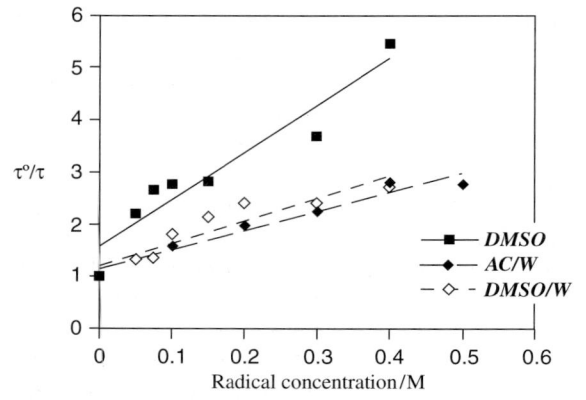

Figure 10.6 Stern–Volmer plot for the quenching of the stilbene probe DMACS ($C = 5 \times 10^{-6}$ M) incorporated in membranes [21]. (Reproduced with permissions from Elsevier.)

10.3.6
Fluorescence–Photochrome Method

10.3.6.1 General

A necessary stage in the olefinic photoisomerization process in stilbene fluorescence–fluorophore molecules, in the singlet or triplet excited state, involves twisting (about the former double bond) of stilbene fragments relative to one another (Section 4.1, Figure 10.6 [52–55]). Since *trans*-stilbene is fluorescent and *cis*-stilbene is fluorescent silent, the process can be readily monitored by a single steady-state fluorescence technique. Here, a general survey is made of the physical principles and application of the new fluorescence methods.

Theoretical considerations and existing experimental data indicate that under certain conditions, the rate of photoisomerization strongly depends on the microviscosity around the isomerized molecule and upon the effect of steric hindrance. In a viscous medium, the apparent rate constant of *trans–cis* photoisomerization k_{iso} is controlled by the reorganization rate of the process in the medium (Equations 4.2 and 4.3). This method was used for the measurement of fluidity of biological membranes and microviscosity of a specific site of a protein. On this theoretical basis, fluorescence–photochrome immunoassay (FPHIA)] were used.

10.3.6.2 Molecular Dynamics of Proteins and Biomembranes

The fluorescence–photochrome technique was first applied to study the molecular dynamics of a stilbene fluorescence–photochrome molecule, SITC, attached covalently to the terminal amino group of sperm whale myoglobin [18]. The same myoglobin residue was also labeled with a spin label, 4-iodoacetamide-TEMPO. The kinetics of the stilbene *trans–cis* photoisomerization (k_{app}) and the rotational diffusion frequency of nitroxide radicals (ν_c) was monitored by fluorescence and ESR techniques, respectively. These data on the probes in a bound state were compared with data obtained in 60% ethylene glycol/water solution. The values of k_{app} and v_c for labels bound to myoglobin were found several times less than the values for free labels indicating that microviscosity in the vicinity of labels attached to myoglobin is higher than that in the bulk solution.

The proposed fluorescence–photochrome method (FPM), based on monitoring fluorescence parameters and kinetics of photochrome photoisomerization of *para*-substituted stilbenes (PSS), was employed for studying molecular dynamics of biological membranes [8, 12–17, 21]. It was shown that PSS exhibits fluorescence characteristics that are similar to those of typical membrane fluorescence probes such as diphenylhexatriene, keeping its ability to study molecular dynamics and micropolarity of media using the steady-state and time domain fluorescence polarization and the spectral relaxation shift techniques. In addition, the study of kinetics of PSS *trans–cis* and *cis–trans* photoisomerization and measuring the rate constant (k_{iso}) makes it possible to estimate, under certain conditions, the twisting correlation time of the stilbene fragments in the excited state of PSS for the fixed angle 180°. In viscous media, this process is a rate-limiting stage. Taken together, the both techniques, fluorescence and photochrome, make it possible to establish a detailed mechanism and, under certain calibration, to measure quantitative parameters of stilbene probe mobility in a membrane.

The FPM was applied to the study of *E. coli* membrane dynamics using 4-dimethylamino-4'-aminostilbene (DMAAS) [13]. Considering the dependence of the glycerol viscosity on temperature and then the effect of viscosity on the correlation time of rotation of a nitroxide radical (τ_c), ln k_{iso} was plotted against τ_c. In the low temperature region:

$$\ln k_{iso} = 28.5 - 1.77 \ln \tau_c. \tag{10.11}$$

Because the volume of the stilbene fragments, twisted in the excited state, and the nitroxide radical are close to each other, this dependence was used for estimating the twisting correlation time of the stilbene fragments $\tau_{tw} \approx \tau_c$. It is necessary to stress that owing to its ability to integrate results of the stilbene photoisomerization, this method allows to measure k_{iso} in a wide range of values and therefore to expand values of τ_{tw} for the twisting around the single bond in the excited states for fixed angle 180°. Figure 10.8 shows the location and mobility parameters of DMAAS in *E. coli* membrane.

As seen in Figure 10.7, the fluorescence–photochrome method allows to measure the probe motion as a whole in temporal scale close to the fluorescence lifetime and in the same time and the same space to detect twisting of stilbene fragment in the singlet excited states for 180° in the microsecond scale. A separation of effects of a probe rotation frequency and angle of twisting appears to be a serious problem in fluorescence and spin probing.

Thus, the fluorescence–photochrome method may, in certain conditions, gain advantage over conventional fluorescence, and spin labeling is simpler, more sensitive, and informative.

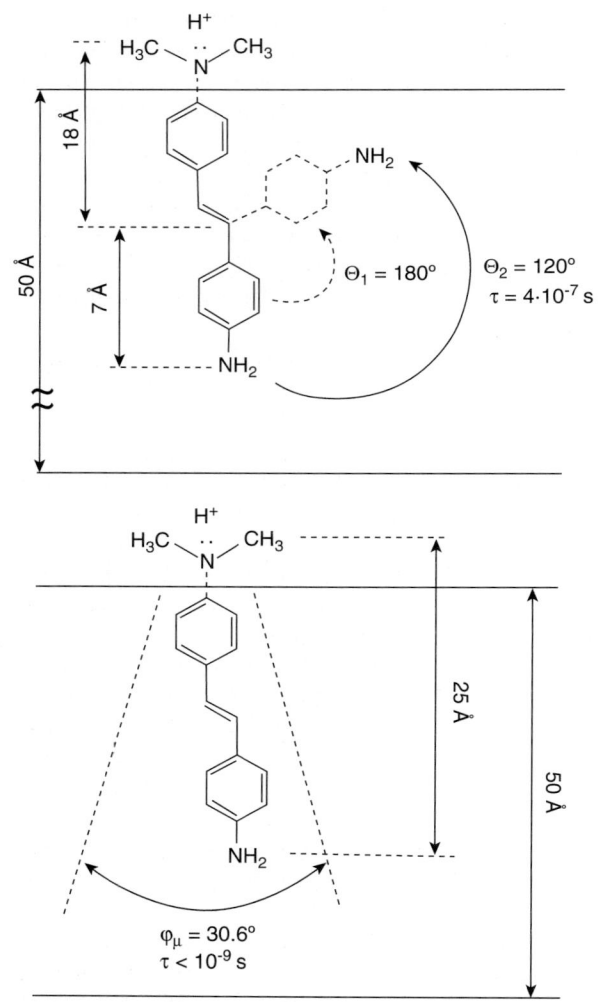

Figure 10.7 Proposed location and mobility parameters of DMAAS in *E. coli* membranes. Top panel: high-amplitude low-frequency twisting of the stilbene fragment in the excited state. Θ_1 and Θ_2 are the angles of the fragment twisting. Bottom panel: low-amplitude high-frequency wobbling of the probe as a whole [13]. (Reproduced with permission from Elsevier.)

10.3.6.3 Molecular Dynamics of anti-DNP Antibody Binding Site

A combined fluorescence–photochrome approach was used for investigating the molecular dynamics of anti-DNP antibody binding site and its cavity. A 4-(N-2,4-dinitrophenylamino)-4'-(N,N'-dimethylamino)stilbene (StDNP) fluorescence DNP analogue was incorporated into the antibody binding site [56]. This was followed by measurements of fluorescence and photochrome parameters such as the StDNP excitation and emission spectra, fluorescence lifetime, steady-state and time-resolved fluorescence polarization, kinetics of *trans–cis* and *cis–trans* photoisomerization, and fluorescence quenching by nitroxide radicals freely diffused in solution. In parallel, computational modeling studies on the location and dynamics of DNP/TEMPO spin label (Ns*l*DNP) and StDNP guests within a model of the binding site were performed. On the basis of the available experimental and computational data, the authors came to several conclusions. The StDNP probe undergoes anisotropic rotation in the 6 cP solvent solution with a correlation time $\tau_c = 1.1$ ns. This is much faster than theoretically predicted from the rotation barrier in vacuum about the C–N bond connecting the DNP and stilbene fragments ($\tau_c = 1.9$ ns). Therefore, the rigid label moves in solution as a whole.

Data on fluorescence polarization of StDNP, and analysis of the ESR spectrum of Ns*l*DNP, indicated that both labels in the binding site are involved in anisotropic wobbling with correlation times of 4 and 2.5 ns, respectively. This wobbling is significantly faster than the mobility of the anti-DNP antibody and Fab fragments as a whole and faster than theoretically calculated rotation about the C–N bond between the DNP moiety and the stilbene and nitroxide fragments in StDNP ($\tau_c = 19$ ns). Results of the photoisomerization kinetics experiments confirm the computer modeling prediction of sufficient space for twisting the StDNP stilbene fragments in the excited singlet-state without any steric hindrance. The estimated "microviscosity" parameter, the twisting correlation time ($\tau_c = 1$ ns), was found to be close to the unbound label's anisotropic rotational correlation time in solution ($\tau_c = 1.1$ ns). Therefore, the microviscosity of the medium in the binding site cavity does not differ markedly from the bulk viscosity. As was predicted by computer modeling (Figure 10.8), the StDNP fluorescence quenching experiments indicate that nitroxide quencher access to the bound stilbene fragment in the complex to afford encounter complexes is limited by an apparent steric factor of approximately 0.5. Therefore, the viscosity of the binding site cavity in the area of the stilbene label at distance approximately 5 Å from the protein tryptophanyl groups is close to the bulk viscosity.

These results showed that the combined fluorescence–photochrome approach can be used for investigating the local medium molecular dynamics in the immediate vicinity of specific sites of proteins and nucleic acids, as well as for other biologically important structures and synthetic analogues.

10.3.7
Systems Immobilized on Quartz Slides

10.3.7.1 Sensoring for Surface Microviscosity and Ascorbic Acid

With an aim to establish a basis for a fibro-optic biosensoring, a number of fluorescence molecular systems were immobilized on quartz surface and investigated.

A series of substituted *trans*-stilbene derivatives have been prepared and immobilized on a quartz surface [19]. Several immobilization methods have been tried including the silanization technique, cross-linking with cyanuric chloride, surface activation with cyanogen bromide, and surface smoothing with coating proteins.

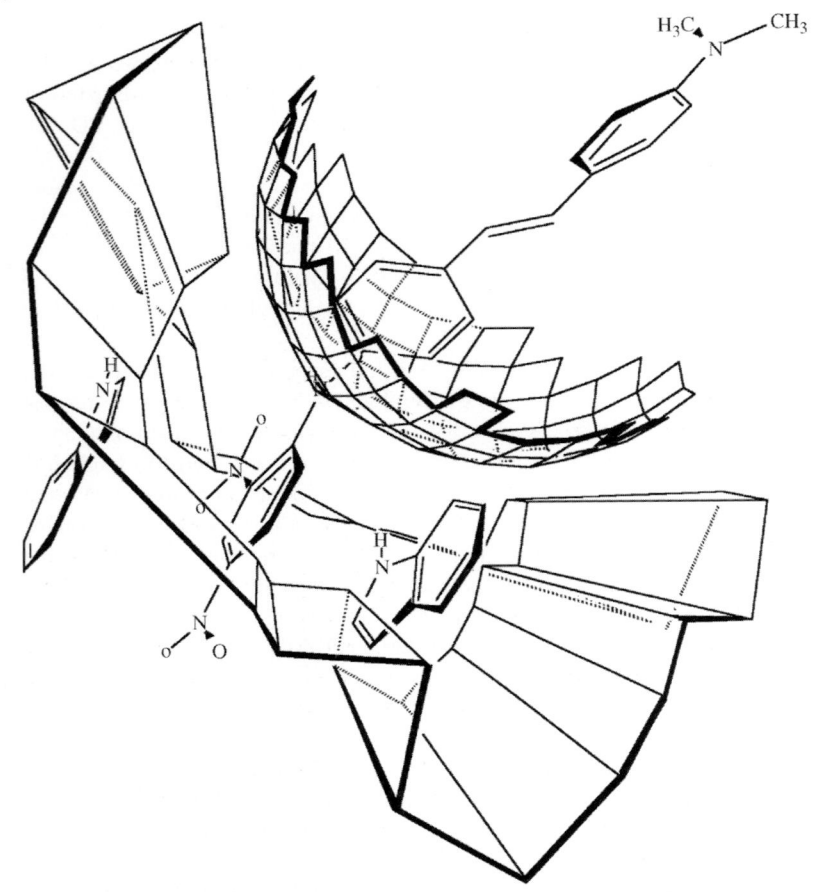

Figure 10.8 Iconic drawing of inaccessible surface area on partial sphere drawn with a 5.0 Å radius from double-bond center of mode 1 bound StDNP (**1**) in anti-DNP antibody binding site [56]. (Reproduced with permission from Elsevier.)

Studies of solvent polar effects on the fluorescence spectrum of immobilized stilbenes indicated that the maximum wavelength of the fluorescence emission is not very sensitive to solvent polarity. The apparent local polarity of the medium in the vicinity of the stilbene label was estimated. The *trans–cis* photoisomerization kinetics of the stilbene derivatives in the immobilized and free state in a medium with different viscosity was monitored by fluorescence technique at constant illumination conditions. Investigation of the microviscosity effect on the photoisomerization of the immobilized and free stilbene label was carried out by changing the relative concentration of glycerin in a glycerin–water mixture used as a solvent. With an appropriate calibration, the microviscosity in the vicinity of the stilbene label was estimated. The apparent photoisomerization rate constant of the process was found to be 3–4 times less for the immobilized label than in a free state.

A new photochrome–fluorescence–spin method for the simultaneous quantitative analysis of the redox status and viscosity of a medium has been developed [20]. The method of the viscosity measurement is based on the use of double fluorescence-nitroxide molecules.

BFL1

In such hybrid compounds, the nitroxide moiety quenches the fluorescence of the fluorophore (stilbene moiety). The reduction of nitroxide segment by an antioxidant (ascorbic acid) caused a rise in fluorescence of the fluorophore. The rate constant of the stilbene fragment photoisomerization in such systems depended on the viscosity of the medium. The synthesized dual stilbene-nitroxide probe was covalently immobilized onto the surface of a quartz plate as an eventual sensor. The immobilization procedure included a cyanogen bromide surface activation followed by smoothing with a protein tether. The rate of fluorescence change was monitored in aqueous glycerol solution of different viscosities and content of ascorbic acid. The dependence of k_{app} on the reciprocal absolute viscosity $1/\eta$ of the bulk mixture glycerin–water and the dependence of the initial intensity of fluorescence (I_o) on solution viscosity were also studied.

These dependences may serve as calibration curves for microviscosity determination in the vicinity of the fluorescence–photochrome probe. Appropriate calibration would make possible the determination of the viscosity of a media (in the range 1–500 cP) as well as ascorbate content in the range $(1–9) \times 10^{-4}$ M.

Beside the estimation of medium microviscosity, quartz plates modified by BFL1 were used for the quantitative analysis of ascorbate. For this purpose, a number of solutions of ascorbic acid were prepared with different concentrations, and the kinetics of change in steady-state fluorescence was recorded. There are two parallel processes that influence fluorescence: (a) *trans–cis* photoisomerization and (b) reduction of the nitroxide moiety in *trans*-BFL1. After a correction taking into account the influence of *trans–cis* photoisomerization (curve "b"), the pseudo-first-order kinetics of BFL reduction appears as curve "c." The correction was done by performing a parallel measurement under identical conditions but without adding ascorbate.

The dependence of the reduction rate constant on bulk ascorbate concentration was found to be logarithmic in the range of investigated ascorbate concentration (0.005–2 mM) (Figure 10.9).

All the above-mentioned experiments were performed with the use of very sensitive fluorescence technique by means of commercial fluorimeter and of modified quartz plates in a single fast measurement. The covalent immobilization of the label enables multiple use of the same plate without the need to prepare and calibrate additional solutions – an important advantage for possible industrial use.

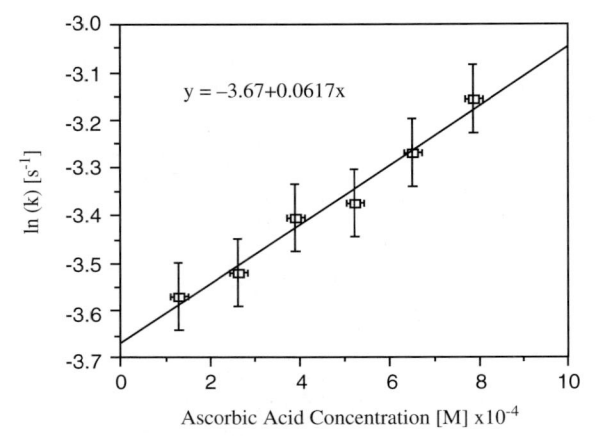

Figure 10.9 The dependence of the reduction rate constant on bulk ascorbate concentration was found to be logarithmic in the range of investigated ascorbate concentration (0.005 – 2 mM) [20].

Replacing the quartz plates by quartz optical fibers will facilitate the use of the given methods for the continuous monitoring of chemical and biological processes.

10.3.7.2 A Fluorescent–Photochrome Method for the Quantitative Characterization of Solid-Phase Antibody Orientation

A fluorescent–photochrome method of quantifying the orientation of solid-phase antibodies immobilized on a silica plate was proposed [9]. The method is based on the measurement of fluorescence quenching by a quencher in solution, rates of *trans–cis* photoisomerization and photodestruction of a stilbene-labeled hapten in an antibody binding site. These experimental parameters enable a quantitative description of the order of binding sites of antibodies immobilized on a surface and can be used to characterize the microviscosity and steric hindrance in the vicinity of the binding site. Furthermore, a theoretical method for the determination of the depth of immersion of the fluorescent label in a two-phase system was developed [9].

In the work [9], anti-dinitrophenyl (DNP) antibodies and stilbene-labeled DNP as a hapten were used. Possible arrangements of solid-phase antibodies bound to a fluorescent antigen are presented in Figure 10.10. Four different antibody immobilization techniques were examined: physical adsorption, covalent binding to a silane activated surface, Langmuir-Blodgett films, and the oriented method [9]. The fluorescence decay kinetics of a stilbene-DNP derivative (StDNP) bound to anti-DNP immobilized on quartz slides was investigated. The kinetics was shown to depend on local microviscosity in the vicinity of bound StDNP, while a decrease in initial intensity in the presence of different concentrations of potassium iodide indicated accessibility of the antibody biding. It was found that there was a well-ordered surface of monolayer antibodies when Langmuir–Blodgett films were used for immobilization.

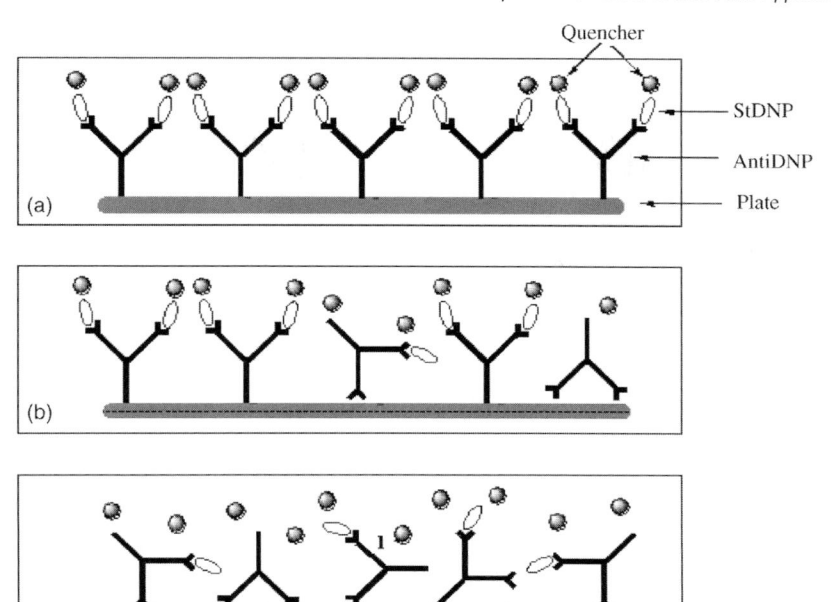

Figure 10.10 Possible arrangements of solid-phase antibodies bound to a fluorescent antigen. (a) A totally ordered system, (b) a partially ordered system, and (c) a completely disordered system [9]. (Reproduced with permission from Elsevier.)

The fluorescence–photochrome method is also suitable for characterizing the physical parameters of a wide range of immobilized systems, such as proteins, enzymes, lipids, and polymer films.

10.3.8
Triplet-Photochrome Method

The traditional fluorescence and electron-spin resonance methods for recording molecular collisions do not allow the study of translational diffusion and rare encounters of molecules in a viscous medium because of the short characteristic times of these methods. To measure the rate constants of rare encounters between macromolecules and to investigate the translation diffusion of labeled proteins and probes in a medium of high viscosity (such as biomembranes), a new triplet-photochrome labeling technique has been developed [5, 8, 12, 17].

The stilbene photoisomerization through the triplet potential surface can be sensitized by a donor molecule excited to its triplet state, which is energetically close to the stilbene excited triplet level T_1 ([52, 53]. The sensitizers (donors) with triplet energies of at least 255 kJ/mol (in a case of unsubstituted stilbene) transfer their energies to both *trans* and *cis* isomers of the stilbene molecule in the ground state in a diffusion-controlled process. The reaction proceeds from an initial donor–acceptor encounter complex, which generates the stilbene excited triplet states without any change in spin. From the excited triplet states of stilbene molecule, a relaxation process takes place on the triplet potential energy surface, leading to the deactivation transition occurrence. Finally, the triplet–triplet energy transfer drives the stilbene photoisomerization through the triplet pathway.

The triplet-photochrome method is based on the above-mentioned scheme. Starting from *cis*-stilbene, which is not fluorescent at the steady-state conditions, and measuring the rate of increase of emitted fluorescence, it has been possible to monitor the process of the sensitized *cis–trans* photoisomerization. The *cis*-stilbene concentration, which is proportional to fluorescence intensity, approaches the photostationary level exponentially with the rate constant

$$k = \left(k_t^T \zeta_t + k_c^T \zeta_c\right) \times \sigma I_{ex} \tau_{ph} \Phi_{ph}, \tag{10.12}$$

where k_t^T and k_c^T are the rate constants for the triplet–triplet energy transfer from a sensitizer to *trans*- and *cis*-stilbenes, respectively, ξ_t and ξ_c are the fractions of the *trans*- and *cis*-stilbene molecules, respectively, that undergo photoisomerization after encounters with the triplet sensitizer, and τ_{ph} and Φ_{ph} are the sensitizer's triplet lifetime and phosphorescence quantum yield, respectively. Equation 10.12 permits the calculation of the experimental rate constant $k_{exp} = \left(k_t^T \xi_t + k_c^T \xi_c\right)$ with the use of regular fluorescence technique if all other constants from this equation are measured independently or calibrated in a model system with these known values.

Due to the relatively long lifetime of the sensitizer triplet state and the possibility of integrating data on the stilbene photoisomerization, the apparent characteristic time of the method can reach hundreds of seconds. This unique property of the triplet-photochrome technique allows the investigation of slow diffusion processes, including encounters of proteins in membranes using very low concentrations of both the triplet and photochrome probes.

The triplet-photochrome labeling method has been used to study very rare encounters in a system containing the erythrosin B sensitizer and SITC photochrome probe [12]. Both types of molecules were covalently bound to α-chymotrypsin. The photoisomerization kinetics was monitored by fluorescence decay of the *trans*-SITS. The rate constants of the triplet–triplet energy transfer between erythrosin B and SITS (at room temperature and pH 7) were found to be $k_t^T = 10^7\ M^{-1}\ s^{-1}$. It should be emphasized that the concentration of the triplet sensitizer attached to the protein did not exceed 10^{-7} M in those experiments, and the collision frequencies were close to $10\ s^{-1}$ that are eight to nine orders of magnitude less than those measured with the regular fluorescence or ESR techniques.

The triplet-photochrome labeling technique was first applied to follow the protein–protein dynamic contacts in biomembranes [17]. SITS and Erythrosin-NCS (ERITC) were bound covalently to Na^+, K^+ ATPase. Triplet–triplet energy transfer from the light-excited triplet ERITC to *cis*-SITS initiated the *cis–trans* photoisomerization of *cis*-SITS. The photoisomerization kinetics of SITS was recorded with a regular spectrofluorimeter. The apparent rate constant of triplet–triplet energy transfer from ERITC to *cis*-SITS was found to be $k_{app} = 0.4 \times 10^3 \, M^{-1} s^{-1}$ (at 25 °C). The k_{app} value of the triplet–triplet energy transfer between unbound ERITC and SITS was measured in solution to be $7 \times 10^7 \, M^{-1} s^{-1}$. The drop of k_{app} in the case of labels bound to ATPase was a result of the increased media viscosity and steric factors.

10.3.9
Cascade Spin-Triplet-Photochrome Methods

Measurements of active encounters between molecules in native membranes containing ingredients including proteins are of prime importance. To estimate the encounters in high range of rate constants and distances between interacting molecules in membranes, a cascade of photochemical reactions for molecules diffusing in membranes has been designed and developed [14, 16]. A cascade of photochemical and photophysical reactions between a triplet sensitizer and a fluorescence photochrome probe exhibits the phenomena of *cis–trans* photoisomerization, triplet–triplet energy transfer, and triplet excited state quenching by a stable radical [7, 8, 14, 16].

The sensitized cascade reported in Ref. [16] consisting of triplet *cis–trans* photoisomerization of the excited stilbene includes the triplet sensitizer (erythrosin B), the photochrome stilbene derivative probe (4-dimethylamino-4′-aminostilbene), and nitroxide radicals (5-doxyl stearic acid) quenching the excited triplet state of the sensitizer (Figure 10.11).

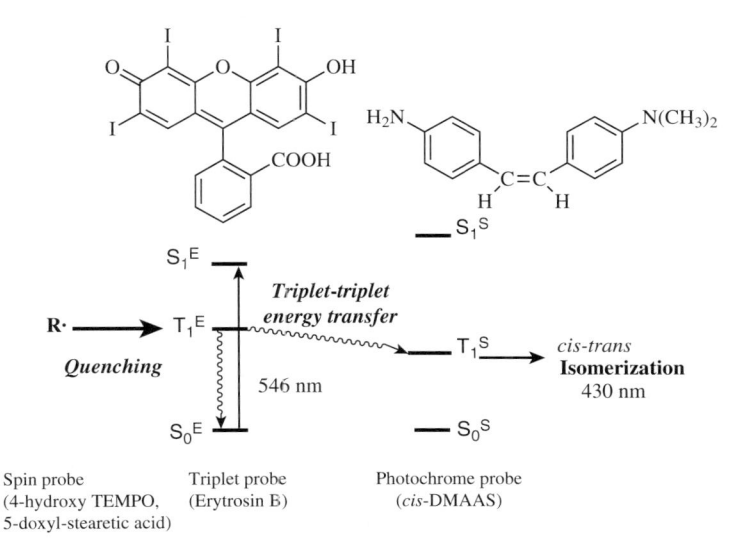

Figure 10.11 Representation of energy levels of cascade reactants and competition between the $T_1^E \rightarrow T_1^S$ and $T_1^E \rightarrow S_0^S$ processes [16]. (Reproduced with permission.)

The effective rate constant of the triplet–triplet energy transfer (k) in membrane can be calculated from the relation between apparent (experimental) rate constants k_{app} for probes in solution and in a membrane

$$k_{mem}^T = \frac{k_{app}^{mem} k_{sol}^T [E]_{sol} \tau_{ph}^{sol}}{k_{app}^{sol} [E]_{mem} \tau_{ph}^{mem}}, \tag{10.13}$$

where indices "sol" and "mem" are attributed to the encounter process in solution and membrane, respectively.

Experimental results gave the evidence of an efficient nitroxide radical inhibitor effect upon the sensitized cis–trans photoisomerization of DMAAS by quenching the sensitizer triplet state. The phosphorescence triplet lifetime of erythrosin B in PPDC membranes was measured to be 3×10^{-4} s. The experimental quenching rate constant of the cascade reaction k_q and the rate constant of the triplet–triplet energy transfer k_T evaluated in 2D terms were obtained as $k_q = (1.05 \pm 0.08) \times 10^{15}$ cm^2/(mol s) and $k_T = (1.26 \pm 0.21) \times 10^{12}$ cm^2/(mol s). Taking into consideration the efficiency of the triplet–triplet energy transfer $\gamma = 3.5$ %, the rate constant of encounters between the sensitizer and the photochrome was found as $k_{en} = 3.4 \times 10^{13}$ cm^2/(mol s).

Values of k_q and k_T in multilamellar liposomes were examined in frames of 2D model describing steady-state diffusion-controlled reactions including the dependence of the diffusion-controlled rate constant on the lifetime of the excited species [57]. The value of k_{en} obtained in this work together with the data on diffusion-controlled rate constants obtained by other methods of various characteristic times of eight orders of magnitude were found to be in good agreement with the above-mentioned advanced theory of diffusion-controlled reactions (Figure 10.12). Therefore, the Razi Naqvi and collaborators' theory [57] can be recommended for the analysis of experimental quenching data under long-irradiation steady-state conditions in two-dimensional geometry.

The expanded cascade method includes quenching of the excited triplet state of (E_T) by nitroxides at local concentrations starting from 1 µM. Such a combination maintains some facilities of the above-mentioned probes and has an essential advantage in the study of molecular dynamics and measurements of the local concentration of radicals. Due to the relatively long lifetime of the sensitizer triplet state, which represents the timescale of the method, and due to the opportunity to integrate the data on stilbene isomerization, the apparent characteristic time of the cascade method may reach hundreds of seconds. This property of the cascade system allows investigation of low-diffusion processes using very low concentration of probes.

In the frame of CSTPM, the following dynamics parameters of the cascade system components can be experimentally measured: the spin label rotation correlation time and spin relaxation parameters, the fluorescence and phosphorescence polarization correlation times, the singlet and triplet state quenching rate constants, the rate constant of photoisomerization, and the rate constant of the triplet–triplet energy transfer. This set of parameters is a cumulative characteristic of the dynamic state of biomembranes in the wide range of the probes' amplitude and characteristic time.

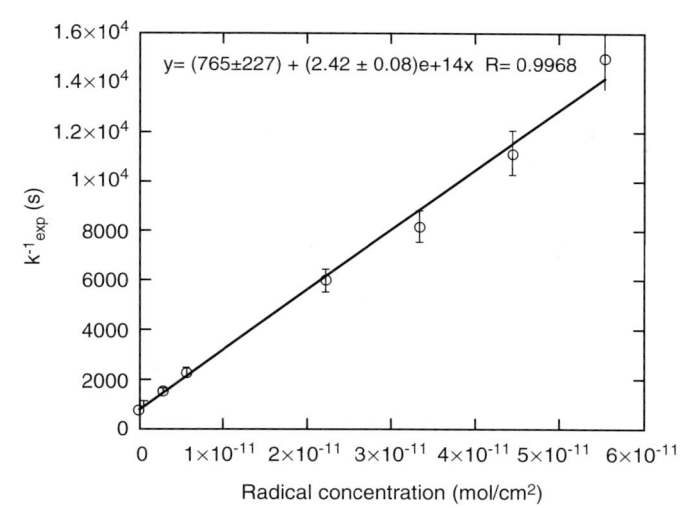

Figure 10.12 Theoretical (○) and experimental (●) dependences of logarithms of the diffusion-controlled rate constants (k_{diff}) on logarithms of the unimolecular decay of the excited species ($k_M = 1/\tau_M$), which characterizes the timescale of different methods. Experimental diffusion-controlled rate constants are marked as (●) [16]. (Reproduced with permission.)

Proficiency of the method can be expanded by a choice of the cascade participants with the higher efficiency of triplet–triplet energy transfer, higher sensitizer lifetime, and by an increase of the time of integration of experimental data on a photochrome photoisomerization. It should be noted that diffusion-controlled rate constants depending on k_M and on concentration more adequately depict the actual molecular dynamics in complex systems such as biological membranes. Eventually, for applications in biological system, it is also important to use photochrome probes with excitation and emission wavelengths in the near-infrared region.

10.3.10
Fluorescence–Photochrome Immunoassay

A rapid, sensitive, and quantitative novel immunoassay technique was developed that auspiciously combines the high sensitivity of fluorescence measurements with the high specificity of an antibody [10]. Fluorescence–photochrome immunoassay (FCIA) is based on the hypothesis that an appropriately designed stilbene–antigen analogue probe will face a considerable steric hindrance to *trans–cis* photoisomerization when bound within the combined constraints of both an antibody binding site and a second globular protein. Since *trans*-stilbenes exhibit fluorescence of high intensity while *cis*-derivatives are fluorescently silent, the competition process can be monitored by a very sensitive single steady-state fluorescence technique. FPIA is performed without separating the antibody-bound haptens from those that are free

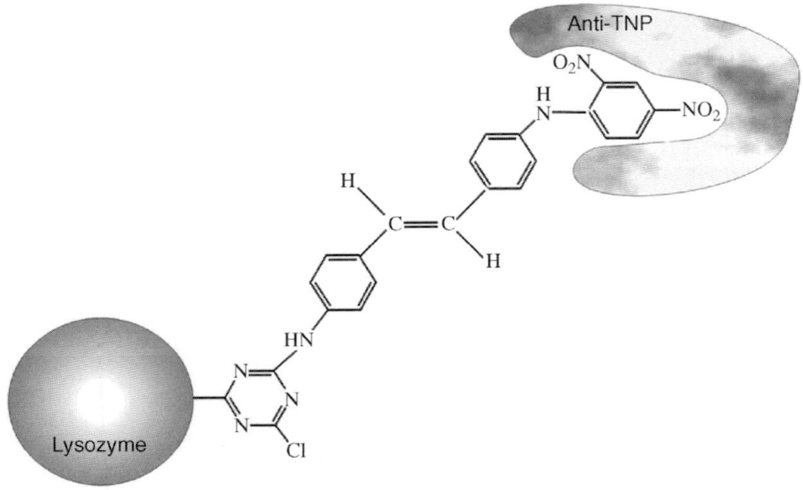

Figure 10.13 Scheme of the anti-TNP–DNP–DAS–Lys complex. The DNP–DAS label bound between two large protein molecules by means of noncovalent binding to an anti-TNP antibody on one side and on the other side via covalent linkage to lysozyme using cyanuric chloride as the cross-linker [10]. (Reproduced with permission from Elsevier.)

and utilizes fluorescence measurements from commonly available standard commercial fluorimeters.

Specifically, an appropriately designed 2,4-dinitrophenyl-hapten derivative of fluorescent *trans*-4,4′-diaminostilbene (DAS) was squeezed between two large globular proteins: lysozyme (Lys) from one side and anti-2,4,6-trinitrophenyl antibody (anti-TNP) from the other side, in order to provide the desired constricted environment to restrict *trans*/*cis*-stilbene isomerization within the anti-TNP–DNP–DAS–Lys adduct (Figure 10.13).

As was theoretically predicted by molecular computation and then experimentally verified, the *trans*–*cis* photoisomerization rate for the bound probe was found to be markedly inhibited, compared to that expected for the free probe in solution. The fluorescence–photochrome labeled probe was competitively displaced from the anti-TNP binding site in the presence of the picric acid hapten, and photoisomerization then commenced to produce the fluorescence-silent *cis*-stilbene diastereomer (Figure 10.14). The process of association and dissociation of a hapten–antibody complex was readily monitored by the fluorescence technique in the presence of both antibody-bound and free haptens.

The FPIA method can be potentially used either independently or as a complementary method to enzyme-linked immunosorbent assay (ELISA) and fluorescence polarization immunoassays (FPIA) technique. FCIA does not need polarization equipment and is not markedly influenced by light scattering effects. The FCIA technique can be expanded for analysis of enzymes and receptors including adaptation to fibro-optic techniques.

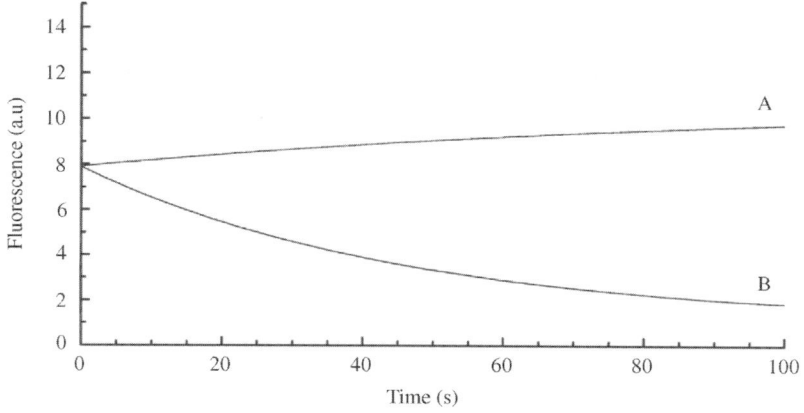

Figure 10.14 Time trace of anti-TNP–DNP–DAS–Lys solution before (a) and after (b) the addition of picric acid. Experimental conditions: excitation 360 nm, emission 410 nm, [DNP–DAS–Lys] = 2 µM, [anti-TNP] = 1.5 µM, [picric acid] = 20 µM; solution composition: DMF (20%), glycerol (30%), PBS buffer B (50%); $T = 25\,^\circ$C [10]. (Reproduced with permission from Elsevier.)

10.3.11
Suppermolecules Containing Stilbene and Fluorescent Quenching Groups

Suppermolecules containing a fluorescent group and fluorescent quenching segments, molecules liganding complexes of metals, can be used for the analysis of molecules and reactive radicals. In such supermolecules, the quenching of the fluorophore fluorescence can occur under certain conditions by a singlet–singlet energy transfer mechanism. The quenching efficiency strongly depends on the overlap integral between the donor fluorescence and the acceptor absorption spectra and the distance between the donor and the acceptor segments (Equation 10.4). Therefore, any change in chemical structure of the quenching segment can be monitored by a sensitive fluorescence technique. This approach was illustrated by examples of analysis of nitric oxide [11].

The proposed method was based on fluorescence, using a fluorophore-heme dual-functionality probe (FHP). The heme group can serve as an effective NO-trap due to its very fast reaction with NO and the high stability of the resulting complex. Since the heme was connected to a fluorophore as part of the FHP, the heme quenches the fluorophore fluorescence, under certain conditions, by a singlet–singlet energy transfer mechanism. This method was tested using myoglobin covalently modified by stilbene label 4-acetamido-4′-isothiocyanatostilbene-2,2′-disulfonic acid. The change in emission intensity of the stilbene fragment, versus an increasing concentration of NO precursors, clearly demonstrated the spectral sensitivity required to monitor the formation of a heme–NO complex in a concentration range of 10 nM – 2 µM (Figure 10.15) [11].

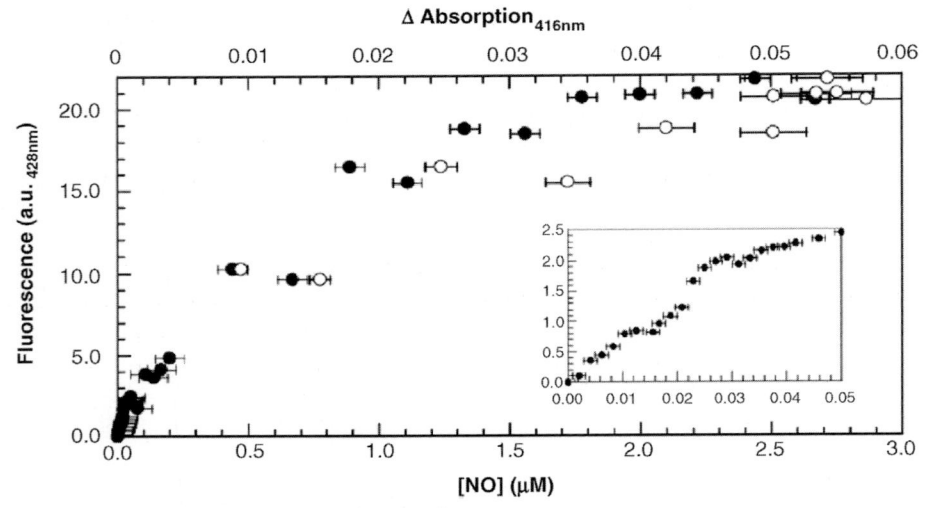

Figure 10.15 Calibration curve drawn as fluorescence intensity versus absorption (upper x-axis), which is proportional to NO concentration (lower x-axis). The graph insert is an extended version of the low concentration part of the calibration curve [11]. (Reproduced with permission from Elsevier.)

Furthermore, the new methodology for NO measurement was also found to be an effective assay using tissues from rabbit and porcine trachea epithelium in the presence of ATP and a specific inhibitor of nitric oxide synthase (NOS), L-NG-nitroarginine methyl ester hydrochloride (L-NAME) (Figure 10.16). The measured

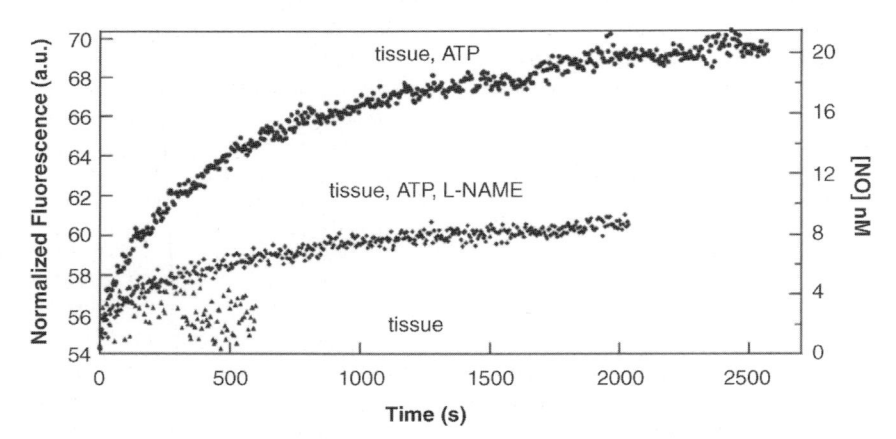

Figure 10.16 Time trace of fluorescence (left scale) and NO concentration (right scale) of Mb (Fe^{2+})-SITS solution (0.05 μM) with pieces of tissue (from rabbit trachea epithelium, 0.0192 g) and ATP (upper circle points), pieces of tissue (0.0203 g) with ATP including L-NAME (lower triangle points), and with piece of tissue without ATP. ATP was added at $t = 0$ s, L-NAME was added 10 min before time trace [11] (Reproduced with permission from Elsevier.)

NO flux (in an initial time interval) in tissue sample from rabbit trachea epithelia and porcine trachea epithelia was found to be as $\sim 7.9 \times 10^{-12}$ mol/s $\times$ g and $\sim 3.0 \times 10^{-12}$ mol/s $\times$ g, respectively.

After a modification, it would be possible to monitor nitric oxide in animal organs and, in future, in human body. The high ability of the tiny NO molecule to rapidly diffuse and penetrate would make it possible to monitor NO penetration through the skin.

In assessing the significance of theoretical and experimental results on stilbene probes presented in this chapter, we may conclude that these molecules possess a potential for solving miscellaneous problems in chemistry, biochemistry, and biophysics on molecular level. Owning to high sensitivity, simplicity, availability of fluorescence techniques, these stilbene probe methods can be widely employed. Some of these techniques may be adapted to fibro-optic sensing.

References

1 Lakowicz, J. (1983) *Principles of Fluorescence Spectroscopy*, Plenum Press, New York.
2 Lacowicz, J. (ed.) (1997) *Topics in Fluorescence Spectroscopy*, Plenum Press, New York.
3 Likhtenshtein, G.I. (1976) *Spin Labeling Method in Molecular Biology*, Wiley–Interscience, New York.
4 Likhtenshtein, G.I. (1993) *Biophysical Labeling Methods in Molecular Biology*, Cambridge University Press, New York.
5 Likhtenshtein, G.I. (2003) *New Trends In Enzyme Catalysis and Mimicking Chemical Reactions*, Kluwer Academic/Plenum Publishers, New-York.
6 Likhtenshtein, G.I., Yamauchi, J., Nakatuji, S., Smirnov, A., and Tamura, R. (2008) *Nitroxides: Application in Chemistry, Biomedicine, and Materials Science*, Wiley-VCH Verlag GmbH, Weinhem.
7 Likhtenshtein, G.I. (2009) *Applied Biochemistry and Biotechnology*, **152**, 135–155.
8 Papper, V. and Likhtenshtein, G.I. (2001) *Journal of Photochemistry and Photobiology A: Chemistry*, **140**, 39–52.
9 Ahluwalia, A., Papper, V., Chen, O., Likhtenshtein, G.I., and De Rossi, D. (2002) *Analytical Biochemistry*, **305**, 121–134.
10 Chen, O., Glaser, R., and Likhtenshtein, G.I. (2008) *Journal of Biochemical and Biophysical Methods*, **70**, 1073–1079.
11 Chen, O., Urlander, N., Likhtenshtein, G.I., and Priel, Z. (2008) *Journal of Biochemical and Biophysical Methods*, **70**, 1006–1013.
12 Mekler, V.M. and Likhtenshtein, G.I. (1986) *Biofizika*, **31**, 568–571.
13 Likhtenshtein, G.I., Bishara, R., Papper, V., Uzan, B., Fishov, I., Gill, D., and Parola, A.H. (1996) *Journal of Biochemical and Biophysical Methods*, **33**, 117–133.
14 Papper, V., Likhtenshtein, G.I., Medvedeva, N., and Khoudyakov, D.V. (1999) *Journal of Photochemistry and Photobiology A: Chemistry*, **122**, 79–85.
15 Papper, V., Medvedeva, N., Fishov, I., and Likhtenshtein, G.I. (2000) *Applied Biochemistry and Biotechnology*, **89** (2–3), 231–248.
16 Medvedeva, N., Papper, V., and Likhtenshten, G.I. (2005) *Physical Chemistry Chemical Physics*, **7**, 3368–3374.
17 Mekler, V.M. and Umarova, F.T. (1988) *Biofizika*, **33**, 720–722.
18 Likhtenshtein, G.I., Khudjakov, D.V., and Vogel, V.R. (1992) *Journal of Biochemical and Biophysical Methods*, **25** (4), 219–229.

19 Strashnikova, N.V., Papper, V., Parhomyuk, P., Ratner, V., Likhtenshtein, G.I., and Marks, R. (1999) *Journal of Photochemistry and Photobiology A: Chemistry*, **122**, 133–142.

20 Parkhomyuk-Ben Arye, P., Strashnikova, N., and Likhtenshtein, G.I. (2002) *Journal of Biochemical and Biophysical Methods*, **51**, 1–15.

21 Strashnikova, N.V., Medvedeva, N., and Likhtenshtein, G.I. (2001) *Journal of Biochemical and Biophysical Methods*, **48**, 43–60.

22 Haugland, R.P. (1996) *Handbook of Fluorescent Probes and Research Chemicals*, 6th edn, Molecular Probes, Leiden, p. 677.

23 Web Edition of The Handbook, Tenth Edition (http://probes.invitrogen.com/handbook/sections/0017.html).

24 Parker, C. (1968) *Photoluminescence in Solutions with Application to Photochemistry and Analytical Chemistry*, Elsevier, Amsterdam.

25 Bakhshhiev, N.A. (ed). (1989) *Solvatochromism, Problems and Methods*, Leningrad University Press, Leningrad.

26 Changenet, P., Choma, C.T., Gooding, E.F., DeGrado, E.F., DeGrado, W.F., and Hochstrasser, R.M. (2000) *Journal of Physical Chemistry. B*, **104**, 9322–9329.

27 Forster, T. (1948) *Annalen der Physik*, **2**, 55–75.

28 Zamaraev, K.I., Molin, Yu.N., and Salikhov, K.M. (1981) *Spin Exchange. Theory and Physicochemical Application*, Springer-Verlag, Heidelberg.

29 Ermolaev, V.L., Bodunov, E.N., Sveshnikova, E.B., and Shakhverdov, T.A. (1977) *Radiationless Transfer of Electronic Excitation Energy*, Nauka.

30 Likhtenshtein, G.I. (1996) *Journal of Photochemistry and Photobiology A: Chemistry*, **96**, 79–92.

31 Letard, J.F., Lapouyade, R., and Rettig, W. (1993) *Molecular Crystals and Liquid Crystals*, **236**, 41–46.

32 Jager, W.F., Volkers, A.A., and Neckers, D.C. (1995) *Macromolecules*, **28**, 8153–8158.

33 Huang, C., Fan, J., Lin, Z., Guo, B., and Peng, X. (2008) Abstracts of Papers, 235th ACS National Meeting, New Orleans, LA, April 6–10, 2008, ANYL-09.

34 Kabatc, J., Jedrzejewska, B., Bajorek, A., and Paczkowski, J. (2006) *Journal of Fluorescence*, **16**, 525–534.

35 Pucci, A., Bertoldo, M., and Bronco, S. (2005) *Macromolecular Rapid Communications*, **26**, 1043–1048.

36 Ono, M., Wilson, A., Nobrega, J., Westaway, D., Verhoeff, P., Zhuang, Z.-P., Kung, M.-P., and Kung, H.F. (2003) *Nuclear Medicine and Biology*, **30**, 565–571.

37 Parhi, A.K., Kung, M.-P., Ploessl, K., and Kung, H.F. (2008) *Tetrahedron Letters*, **49**, 3395–3399.

38 Wu, C., Wei, J., Tian, D., Feng, Y., Miller, R.H., and Wang, Y. (2008) *Journal of Medicinal Chemistry*, **51**, 6682–6688.

39 Kung, H.F., Kung, M.-P., and Zhuang, Z.-P. (2007) US Patent 7297820, November 20, 2007.

40 Xu, H., Zheng, M., Dai, D.-R., Xie, P., Bai, F.-L., and Zhang, R.-B. (1999) *Chinese Journal of Polymer Science*, **17**, 383–389.

41 Bova, M.P., Ding, L.L., Horwitz, J., and Fung, B.K. (1997) *The Journal of Biological Chemistry*, **272**, 29511–29517.

42 Ramamurthy, V. and Oliver, D. (1997) *Journal of Biological Chemistry*, **272**, 23239–23246.

43 Verkman, A.S., Skorecki, K.L., Jung, C.Y., and Ausiello, D.A. (1986) *American Journal of Physiology. Cell Physiology*, **251**, C541–C548.

44 Futamura, S. and Ohkawa, K. (1993) *Nippon Enerugi Gakkaishi*, **72**, 951–957.

45 Chang, Y.Y. and Cronan, J.E., Jr. (1997) *Biochemistry*, **36**, 11564–11573.

46 Krishnasastry, M., Walker, B., Braha, O., and Bayley, H. (1994) *FEBS Letters*, **356**, 66–71.

47 Speth, M., Baake, N., and Schulze, H.U. (1989) *Archives of Biochemistry and Biophysics*, **275**, 202–214.

48 Ignatova, M., Manolova, N., and Rashkov, I. (2007) *European Polymer Journal*, **43**, 1609–1623.

49 Legrum, B. and Passow, H. (1989) *Biochimica et Biophysica Acta*, **979**, 193–207.

50 Cardin, A.D., Smith, P.L., Hyde, L., Blankenship, D.T., Bowlin, T.L., Schroeder, K., Stauderman, K.A., Taylor, D.L., and Tyms, A.S. (1991) *Journal of Biological Chemistry*, **266**, 13355–13363.

51 Likhtenshtein, G.I. (2000) Depth of immersion of paramagnetic centers, in *Magnetic Resonance in Biology* (eds L. Berliner, S. Eaton, and G. Eaton), Kluwer Academic Publishers, Dordrecht, pp. 1–36.

52 Hammond, G.S. and Saltiel, J. (1962) *Journal of the American Chemical Society*, **84**, 4983–4984.

53 Saltiel, J. and Sun, Y.-P. (1989) *Journal of Physical Chemistry*, **93**, 6246–6250.

54 Nibbering, E.T.J., Fidder, H., and Pines, E. (2005) *Annual Review of Physical Chemistry*, **56**, 337–367.

55 Dugave, C. and Demange, L. (2003) *Chemical Reviews*, **103**, 2475–2532.

56 Chen, O., Glaser, R., and Likhtenshtein, G.I. (2003) *Biophysical Chemistry*, **103**, 139–156.

57 Razi Naqvi, K., Martins, J., and Melo, E. (2000) *The Journal of Physical Chemistry B*, **104**, 12035–12038.

11
Modern Methods of Stilbene Investigations

11.1
General

Practically, all modern methods of investigating the stilbene photodynamics are based on the use of laser techniques. Nevertheless, traditional chemical, photophysical, and photochemical methods have also widely used. A laser is an electrooptical device that emits coherent light radiation.

The term "laser" is an acronym for light amplification by stimulated emission of radiation. A typical laser emits light in a narrow, low-divergence beam, with a narrow wavelength spectrum ("monochromatic" light) [1–6]. The first working laser was demonstrated by Theodore Maiman on May 16, 1960 at Hughes Research Laboratories.

A laser consists of a gain medium inside a highly reflective optical cavity and a means to supply energy to the gain medium. The gain medium is a material with properties that allow it to amplify light by stimulated emission. The process of supplying the energy required for the amplification is called pumping. The energy is typically supplied as an electrical current or as light at a different wavelength. Such light may be provided by a flash lamp or perhaps another laser. Light of a specific wavelength that passes through the gain medium is optically and repeatedly amplified. The gain medium of a laser is a material that amplifies the beam by the process of stimulated emission. The gain medium absorbs pump energy, which raises some electrons into higher-energy quantum states. When the number of particles in one excited state exceeds the number of particles in some lower energy state, population inversion is achieved, and the amount of stimulated emission due to light that passes through is larger than the amount of absorption and the light is amplified.

Modern lasers cover radio, microwave, infrared, visible, ultraviolet, and X-ray spectra. There are following types of lasers: gas, chemical, excimer, solid-state, fiber-hosted, photonic crystal, free electron, and dye lasers. Taken in combination, the laser techniques allow one to produce a coherent beam of different frequency, power, and duration up to very short-duration pulses on the order of a few femtoseconds. Laser research has produced a variety of improved and specialized

Stilbenes. Applications in Chemistry, Life Sciences and Materials Science. Gertz Likhtenshtein
Copyright © 2010 WILEY-VCH Verlag GmbH & Co. KGaA, Weinheim
ISBN: 978-3-527-32388-3

laser types, optimized for different performance goals, including wavelength bands, maximum average output power, maximum peak output power, minimum output pulse duration, maximum power efficiency, maximum charging, and maximum firing.

Since the development of short-pulsed laser systems, ultrafast structurally resolving techniques allow a detailed insight into the dynamics of molecular structures [6, 7]. After the ultrashort optical trigger pulse induces a transition to a higher electronic state, the evolution of the molecular system is followed either by time-resolved fluorescence emission or by absorbance changes of a probe pulse tuned to electronic resonances in the near-UV or visible regions of the electromagnetic spectrum. A profound insight into the dynamics of molecular structures may be achieved using ultrafast structurally resolving techniques.

Recent developments in ultrafast laser technology have enabled the efficient generation of tunable femtosecond laser pulses from the UV to the far-infrared regions of the electromagnetic spectrum making femtosecond vibrational spectroscopy a versatile tool [8–10].

Time-resolved femtosecond absorption, fluorescence, IR, and Raman spectroscopy elucidate the molecular structure evolution during ultrafast chemical reactions [7–11]. The technique provides in real time direct insight into the structural dynamics of various systems including photoisomerization. In this chapter, we briefly describe modern methods of studying photochemical and photophysical processes that have been employed or can be employed in the stilbene photophysics and photochemistry.

11.2
Nanosecond Transient Absorption Spectroscopy

In the work [11], a system containing stilbene linkers, donor–acceptor-capped hairpins, and stilbene-linked hairpins (Figure 11.1) was investigated. The transient absorption spectra (Figure 11.2) and kinetics of the radical ions were monitored directly using a fully computerized kinetic spectrometer system ($\sim$7 ns response time). The excitation source was a nanosecond Nd:YAG laser (Continuum Surelite II, about 6 ns pulse width, 10 Hz repetition rate, $\sim$10 mJ/pulse/cm^2, 355 nm excitation). The sample excitation frequency was reduced to 1 Hz by an electromechanical shutter and used to excite sample solutions (0.25 ml) of the oligonucleotides saturated with argon. The transient absorbance was probed along a 1-cm optical path by light from a pulsed 75 W xenon arc lamp with its light beam oriented perpendicular to the laser beam. The signal was recorded by a Tektronix TDS 5052 oscilloscope operating in its high-resolution mode. Satisfactory signal/noise ratios were obtained even after a single laser shot.

Kinetics and thermodynamics of electron injection and charge recombination DNA hairpins possessing a stilbenediether electron donor linker observed with neighboring cytosine or thymine bases obtained by femtosecond transient absorption spectroscopy were investigated [12].

11.2 Nanosecond Transient Absorption Spectroscopy | 311

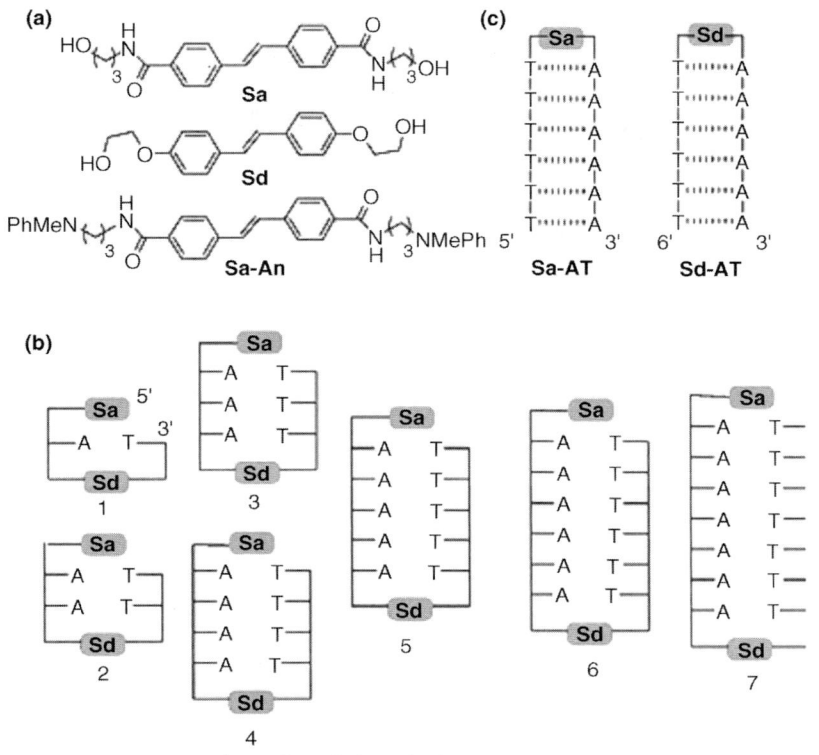

Figure 11.1 Structures of (a) stilbene linkers, (b) donor–acceptor capped hairpins, and (c) stilbene-linked hairpins used in Ref. [11]. (Reproduced with permission.)

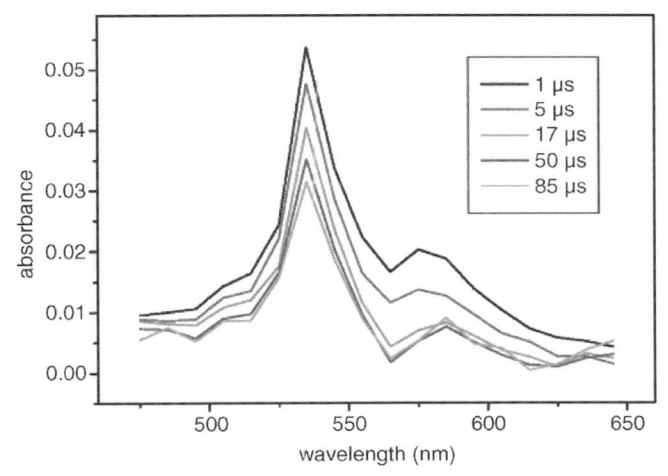

Figure 11.2 Nanosecond time-resolved spectra of stilbene. The 525 nm band is assigned to overlapping absorption of Sd^+ and Sa^-, and the 575 nm band is assigned to absorption of Sa^- [11]. (Reproduced with permission.)

Nanosecond transient absorption behavior of $Ln^{3+}[TTA]_3$(terpy) complexes at room temperature was explored [13]. The spectra were obtained by nanosecond flash photolysis technique. An excitation pulse of 355 nm was generated from the third harmonic output of a Q-switched Nd:YAG laser (EXSPLA NT342). The time duration of the excitation pulse was 5 ns pulse width at 10 Hz repetition rate. The probe light was obtained from a 50 W tungsten lamp. The probe light was collimated on the sample cell and then spectrally resolved by using a monochromator (Acton SpectraPro 2300i) equipped with a 1200 grooves/mm grating after passing the sample. The spectral resolution was about 5 nm for transient absorption experiment. The light signal was detected via a PMT (Hamamatsu, R928). The output signal from the PMT was recorded with a 500 MHz digital storage oscilloscope (Agilent Infiniium 54832B DSO) for the temporal profile measurement. Because the triplet-state dynamics of molecules in solution strongly depends on the concentration of oxygen molecules dissolved in solution, oxygen was removed rigorously by repeated freeze pump thaw cycles.

11.3
Femtosecond Broadband Pump-Probe Spectroscopy

The femtosecond broadband pump-probe spectroscopy was used to study the electronic energy relaxation and coherent vibrational dynamics [14]. A broadband white light continuum (WLC) was generated by amplified pulses from a laser. A fast optical multichannel analyzer combined with a noncollinear optical parametric amplifier allowed simultaneous acquisition of the differential transmission dynamics on the 500–700 nm wavelength range with a sub-10-fs temporal resolution. The broad spectral coverage enabled, on the one hand, a detailed study of the ultrafast bright-to-dark-state internal conversion process. On the other hand, the tracking of the motion of the vibrational wave packet launched on the ground-state multi-dimensional potential energy surface enables the straightforward acquisition and analysis of coherent vibrational dynamics, highlighting time – frequency domain features with extreme resolution.

A theory for the ultrafast pump-probe spectroscopy of large polyatomic molecules in condensed phases was developed in the work [15]. A multimode Brownian oscillator model was used to account for high-frequency molecular vibrations and local intermolecular modes as well as collective solvent motions. A semiclassical picture was provided using the density matrix in Liouville space. Conditions for the observation of quantum beats, spectral diffusion, and solvation dynamics (dynamic Stokes shift) are specified.

A detailed description of experimental setup of the femtosecond broadband pump-probe spectroscopy was given in Ref. [11, 16]. Schematic representation of the experimental setup is shown in Figure 11.3. The combination of femtosecond broadband pump-probe spectroscopy, with an expanded spectral range, and picosecond fluorescence decay measurements permitted more detailed analysis of electron transfer dynamics [17–19]. The authors disentangled the various kinetic

11.3 Femtosecond Broadband Pump-Probe Spectroscopy | 313

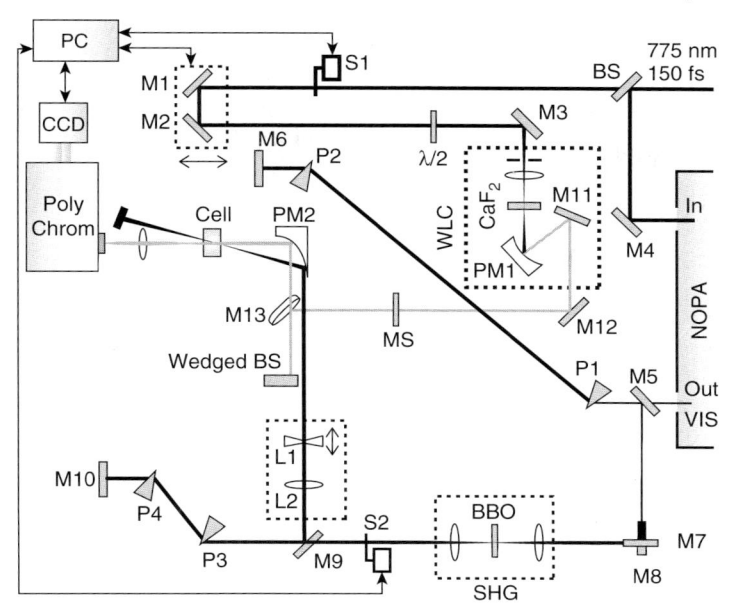

Figure 11.3 Schematic representation of the experimental setup. The following notations have been used: BS, beam splitter of the fundamental 775 nm beam; shutter 1 and shutter 2, two motorized shutters that block the probe and pump beams, respectively; M1, M2, M3, and M4, high reflection dielectric mirrors at 775 nm, 45°; λ/2, a λ/2 plate; CaF$_2$, calcium fluoride plate, used for white light continuum generation; M5, M6, M7, M8, M9, M10, M11, M12, and M13, aluminum mirrors; PM1, 30° off-axis parabolic mirror; PM2, 90° off-axis parabolic mirror; MS, dielectric mirror used to cut off the fundamental; wedged BS, wedged beam splitter, used to split the WLC into probe and reference beams; P1, P2, P3, and P4, fused silica prisms; L1 and L2, lenses with focal lengths −50 and 75 mm, respectively [16]. (Reproduced with permission.)

processes that are involved in photoinduced electron transfer in DNA modified by stilbene derivative: hole injection, hole arrival, and charge recombination. It was shown that each of these processes exhibits a characteristic distance dependence, resulting in the observation of a crossover from superexchange to interbase hopping as the mechanism for charge separation at a donor–acceptor plane-to-plane distance of about 10 Å (two base pairs) and at longer distances for charge recombination. Experimental studies on DNA charge separation and charge recombination dynamics were analyzed in terms of these two limiting mechanisms.

In Ref. [11], the changes in optical density after pumping with wavelength of 355–333 nm were probed by a femtosecond white light continuum generated by tight focusing of a small fraction of the output of a commercial Ti:Sp-based pump laser (CPA-2010, Clark-MXR) into a 3-mm calcium fluoride (CaF$_2$) plate. The WLC provides a usable probe source between 300 and 750 nm. The WLC was split into two beams (probe and reference) and focused on the sample using reflective optics. After passing through the sample, both probe and reference beams were spectrally dispersed and simultaneously detected on a CCD sensor. The pump pulse (1 kHz, 400 nJ) was generated by frequency doubling of the compressed output of a home-built NOPA system (from 666 to 708 nm, 7 J, 40 fs). To compensate for group velocity dispersion in the UV-pulse, an additional prism compressor was used. The overall time resolution of the setup is determined by the cross-correlation function between pump and probe pulses, which was typically 120–150 fs (fwhm, assuming a Gaussian line shape). A spectral resolution of 7–10 nm was obtained. All measurements were performed with magic angle (54.7°) setting for the polarization of pump with respect to the polarization of the probe pulse. A sample cell with 1.25 mm fused silica windows and an optical path of 1 mm was used for all measurements. A wire stirrer was used to ensure fresh sample volume was continuously used during the measurement.

The mechanism and dynamics of photoinduced charge separation and charge recombination have been investigated in synthetic DNA hairpins possessing donor and acceptor stilbenes (stilbene-4,4′-dicarboxylic acid, bis(3-hydroxypropyl)amide of stilbene-4,4′-dicarboxylic acid, bis(2-hydroxyethyl)stilbene 4,4′-diether) (Figure 11.1) using femtosecond broadband pump-probe spectroscopy, nanosecond transient absorption spectroscopy, and picosecond fluorescence decay measurements [11]. Nanosecond time-resolved spectra of stilbenes attached to DNA are shown in Figure 11.4.

These techniques permitted a detailed analysis of the formation and decay of the stilbene acceptor singlet state and of the charge-separated intermediates. It was found that when the donor and acceptor were separated by a single A:T base pair, charge-separated occurs via a single-step superexchange mechanism. However, when the donor and acceptor were separated by two or more A:T base pairs, charge separation occurs via a multistep process consisting of hole injection, hole transport, and hole trapping.

The femtosecond pump-probe absorption spectroscopy was used for the investigation of the S1-photoisomerization of *cis*-stilbene in compressed solvents [20]. The authors of the work [21] demonstrated a technique for femtosecond time-resolved optical pump-probe spectroscopy that allowed to scan over a nanosecond time delay at a kilohertz scan rate without mechanical delay line. Two mode-locked femtosecond lasers with 1 GHz repetition rate were linked at a fixed difference frequency of $\Delta f_R = 11$ kHz. One laser delivers the pump pulses, the other provides the probe pulses. The techniques enabled high-speed scanning over a 1-ns time delay with a time resolution of 230 fs.

11.4
Fluorescence Picosecond Time-Resolved Single Photon Counting

This technique was described in detail in Ref. [22–24].

The typical setup for the measurement of the fluorescent lifetime (τ_f) (Figure 11.5) described in Ref. [25] consisted of a Ti:sapphire laser (Spectra-Physics, Tsunami laser pumped by a 10 W Beamlok Ar-ion laser) that was operated in its picosecond lasing mode (1 ps pulses at 82 MHz). The fundamental train of pulses was pulse selected (Spectra-Physics, model 3980) to reduce its repetition rate to typically 0.8–4. MHz

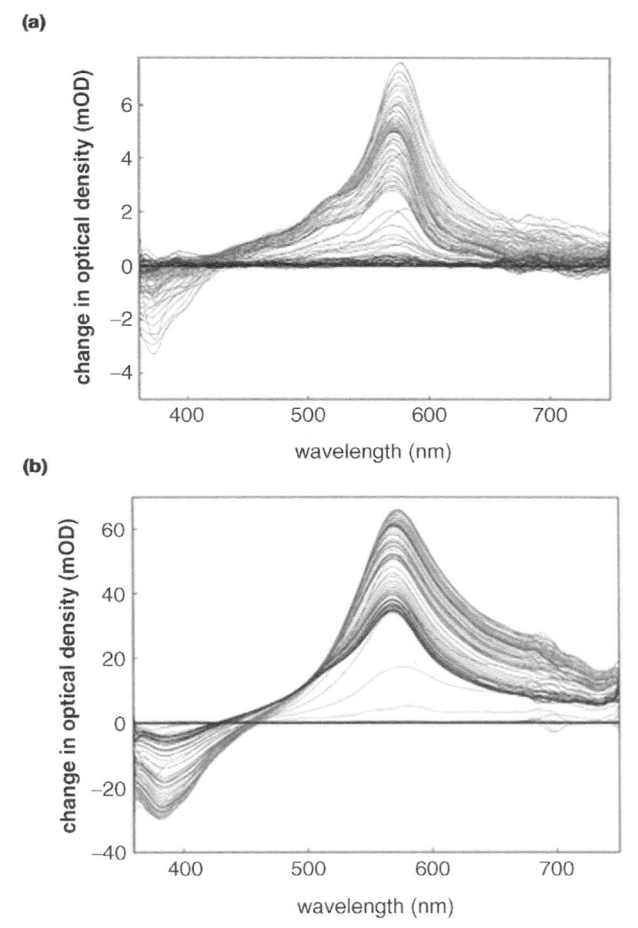

Figure 11.4 Temporal evolution of the pump-probe spectra of stilbenes conjugated with DNA (a) Sa-An and (b) Sa-AT (Figure 11.1) in the time range of −0.1–150 ps after excitation at 333 nm. Early spectra are shown in blue/green and late spectra are shown in orange/red colors (see details in Ref. [11]).

and then passed through a doubling LBO crystal. The laser was tuned between 690 and 800 nm using the Spectra-Physics blue optics set, and the double frequency used for the excitation of the stilbenes was between 345 and 400 nm. The detection system consisted of a Hamamatsu 3809U 6 m multichannel plate (MCP). The fluorescence light was focused on the entrance slit of the MCP after passing through a 1/8 m double monochromator. The electronic processing of signal was done by a combination of modular nim-bin units manufactured by Ortec, Tennelec, and Phillips Scientific. The instrument function was typically 25 ps and was reduced to below 17 ps when 0.1 mm slits were put in front of the sample. The time resolution of the single photon counting setup after data processing was below 3 ps in the 25 ns full-

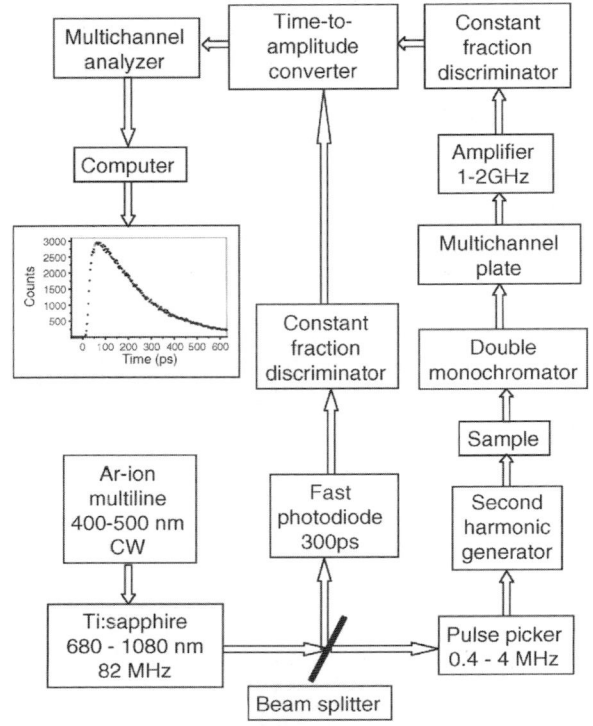

Figure 11.5 Time-correlated single-photon counting system [25].

scale range of the time-to-amplitude converter. Typical counting rates were kept below 5 kHz. The number of counts was between 4 and 10 k at the peak channel and these were collected by the Tennelec PCA3Card. Further signal processing and data analysis were done by personal computers.

The lifetimes of the first excited state of substituted stilbenes in solvents of different polarities were detected. An example of the time-resolved fluorescence decay profile is shown in Figure 11.6.

The time-resolved fluorescence behavior of two derivatives of 4-(dimethylamino)-4′-cyanostilbene (DCS) bearing a more voluminous (JCS) and less voluminous anilino group (ACS) was studied in ethanol by reconstructing the emission spectra using picosecond time-resolved single-photon counting technique [23]. For determining the time-resolved fluorescence spectra, the method described by Maroncelli and Fleming [22] was used. The time- and wavelength-dependent kinetics, $D(t, \lambda)$, was measured at fixed wavelengths covering the steady-state fluorescence spectrum. Kinetics of the stilbene JCS and change of time-resolved emission spectra of JCS is shown in Figure 11.7. These spectra exhibited a temporary isosbestic point, a clear indication of level dynamics between two emitting excited singlet states (LE and CT).

Another example of the fluorescence decay measurements was given in Ref. [11]. In the picosecond single-photon counting system, the sample was excited by a

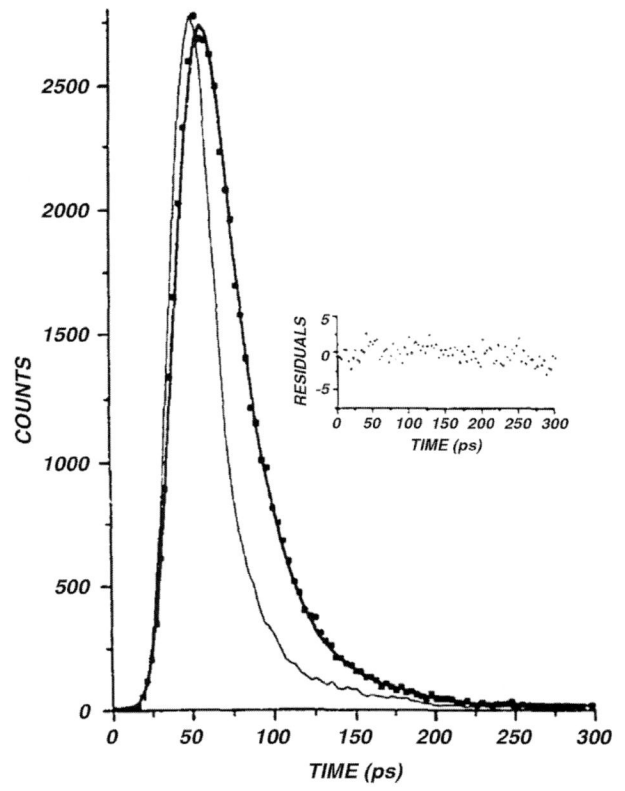

Figure 11.6 Time-resolved decay profile of 4-methoxy-4'-cyanostilbene in methyl ketone (8 mM). The instrument function (solid line) is 20 ps, and the experimental data (dots) were fitted after convolution by tf = 15 ps with $x = 0.8$ [24]. (Reproduced with permission of Elsevier.)

Coherent Mira 900 fs Ti:Sp laser that was pumped by an Innova 310 argon ion laser. The output of the Ti:Sp laser (700 nm) was passed through a Conoptics electro-optic light modulator system consisting of a model 350-160 Modulator, a model 25D Digital Amplifier, and a M305 Synchronous countdown device, to reduce the laser pulse frequency from 76 to 12 MHz, which was then doubled to provide excitation at 350 nm. Fluorescence emission was registered at 390 nm using an Aries FF250 monochromator. A Time Harp 100 PC card (PicoQuant, Germany) controlled by an IBM PC computer provided registration of the counts with rates up to 80 MHz. After deconvolution (PicoQuant FluoFit software), the time resolution of this apparatus was 35 ps. All experiments, including data collection and analysis, were controlled by an IBM PC computer using PicoQuant software.

Symmetrically substituted stilbenes with large multiphoton absorption cross section and strong two-photon-induced blue fluorescence were investigated [26]. These dye solutions exhibited linear transmission of 90% at wavelengths ≥

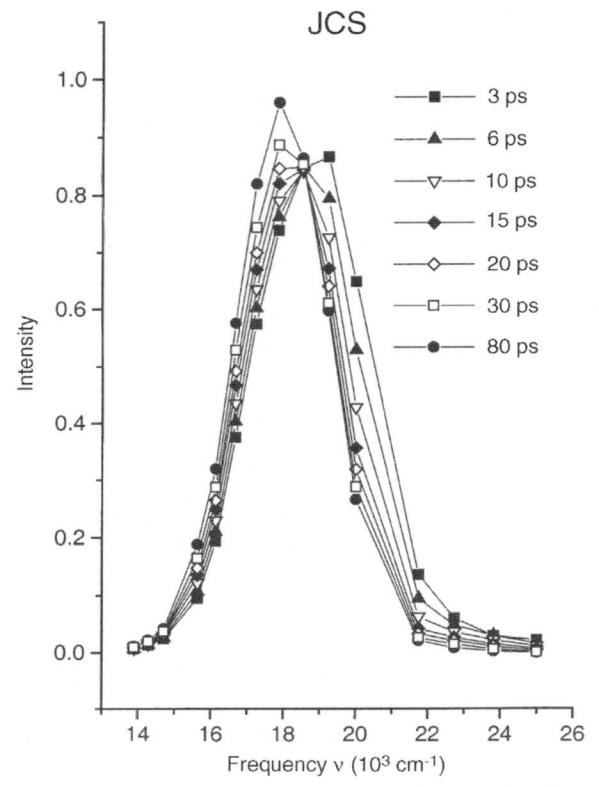

Figure 11.7 Time-resolved emission spectra of JCS in ethanol at 298 K reconstructed from emission decay curves. An isoemissive point is observed at 540 nm (18 520 cm^{-1}) [23]. (Reproduced with permission.)

500 nm at concentration of 0.005–0.0005 mol/dm^3. Pumped by 700 nm laser irradiation, they possess large two-photon absorption (TPA) cross sections of 44.5×40^{-48}–62.0×10^{-48} cm^4 s/photon, and strong upconversion blue fluorescence occurring at 437–452 nm. A large three-photon absorption cross section of 27.3×10^{-76} cm^6 s^2/photon has also been observed for one of these dyes under 990 nm laser irradiation.

11.5
The Fluorescence Upconversion Spectroscopy

Upconversion is a process by which light can be emitted with photon energies higher than those of the light generating the excitation. When a laser gain medium emits fluorescence as a consequence of being excited with incident light, the wavelength of the fluorescence is usually longer than that of the exciting light. This

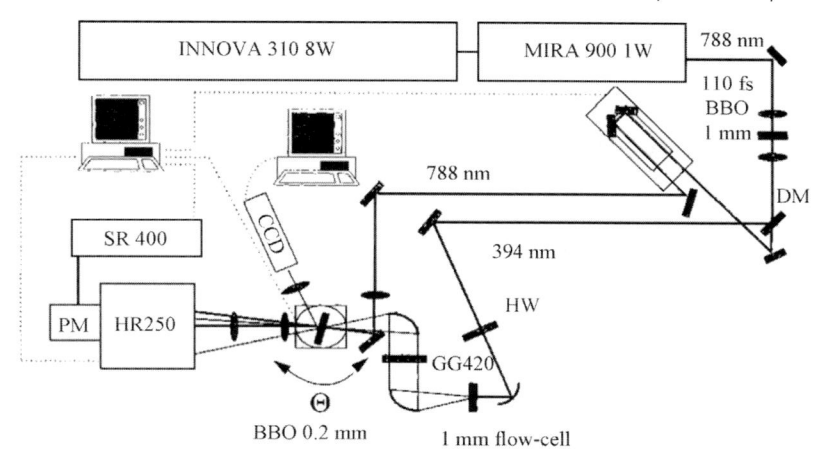

11.5 The Fluorescence Upconversion Spectroscopy

Figure 11.8 Schematic view of the experimental setup for the femtosecond fluorescence upconversion spectrometer. DM, dichroic mirror used to separate the second harmonic (394 nm) from the fundamental (788 nm); HW, half-wave plate used to control the polarization of the excitation pulse; GG420 is a Schott high-pass filter (in λ) used to eliminate scattered SHG light; PM, photomultiplier tube; CCD, video camera equipped with a charge-coupled device [27]. (Reproduced with permission.)

means that the photon energy is reduced. However, under some circumstances *upconversion fluorescence* can occur, where the wavelength of the emitted light is shorter. This is possible via excitation mechanisms that involve more than one absorbed photon per emitted photon as outlined below.

A setup for the femtosecond-resolved fluorescence upconversion experiments (Figure 11.8) has been described in the work [27]. A femtosecond laser system, consisting of a Ti:sapphire laser (Coherent MIRA 900), pumped by a 10 W cw solid-state laser (Coherent VERDI V10), produced 125 fs pulses at 800 nm with a repetition rate of 76 MHz and a 1.8 W average output power. The second harmonic (SH) was generated in a 0.5-mm thick BBO crystal and was separated from the fundamental beam by a dichroic splitter. The SH was used as pump pulse for the fluorescence excitation of the sample, whereas the fundamental pulse serves as gating pulse for the sum frequency ("upconversion") generation. The fluorescence light was focused into the upconversion crystal, and the upconverted light was spectrally analyzed using a 220 mm focal length double grating monochromator (SPEX 1680). The detection and amplification of the dispersed upconversion light was achieved by using a photomultiplier (Hamamatsu R1527P) in combination with a photon counter (Stanford SR400). Fluorescence upconversion experiments were performed at 0° and 90° angles between the polarization axes of excitation and detection. The cross-correlation trace between the laser fundamental (800 nm) and the SH (400 nm) gave a fwhm value of 175 fs for the apparatus function. Fluorescence upconversion spectra were corrected first by subtracting the background from the fluorescence upconversion spectra. The background was recorded by positioning the delay at "negative" time, so that the gating pulse arrives at the upconversion crystal well before the fluorescence signal. The current spectral correction curve, $R(\lambda)$, was determined experimentally by comparing the normalized fluorescence upconversion spectrum at long times (100 ps: $I^{100\,\text{ps}}(\lambda)$) with the normalized steady-state spectrum, $I^{SS}(\lambda)$, recorded for the same sample solution, so that the relationship $R(\lambda) = I^{SS}(\lambda)/I^{100\,\text{ps}}(\lambda)$ was obtained. All fluorescence upconversion spectra were subsequently multiplied by $R(\lambda)$.

In the fluorescence upconversion spectrometer used in Ref. [11], near-infrared femtosecond pulses of 60 fs from a mode-locked 82 MHz Ti:sapphire laser (Spectra-Physics Tsunami), pumped by an argon ion laser, were frequency doubled to produce light with wavelengths between 420 and 480 nm. After passing through a dichroic filter, the fundamental wavelength beam was sent to a delay line and the frequency-doubled pulses were used to excite the sample in a rotating cell. Detuning of the doubling crystal decreased frequency-doubled pulse energies to typically 0.1 nJ, reducing bleaching of the sample. The fluorescence was collected and focused on a rotating BBO crystal, together with the delayed near-infrared gate beam. The upconverted light was dispersed by a prism, sent through an iris and a UV filter, and focused onto the slits of a monochromator (ISA H10 UV, f/3.5). A Peltier element-cooled, low-noise photon counting photomultiplier (Hamamatsu R4220P) detected and a gated photon counter (Stanford Research Systems SR400) counted the signal. For isotropic decay measurements, the polarization of the pump beam was set to the magic angle with a variable wave plate (Berek's Polarization Compensator, New Focus). The sample (200–1500 l) was held in a rotating sample cell of 1–2 mm path length. In several solvents, bleaching occurred, and the solution was changed regularly. In a typical measurement, 5–15 scans were accumulated and averaged. Comparison of the first and last scans showed only slight deviations.

Frequency upconversion of 800 nm ultrashort 175 fs optical pulses by two-photon absorption in a stilbenoid compound-doped polymer (PMMA) optical fiber was reported [28]. By the intensity-dependent transmission method, the two-photon absorption cross section was deduced. The combination of a well-designed organic chromophore incorporated into a fiber geometry is appealing for the development of an upconversion blue polymer laser. Upconversion fluorescence and optical power limiting effects based on the two- and three-photon absorption process of a *trans*-4,4'bis(pyrrolidinyl)stilbene were investigated [29]. The molecular TPA cross section $\sigma 2'$ at 550–670 nm and the three-photon absorption (3PA) cross section $\sigma 3'$ at 720–1000 nm were measured. The 3PA-induced optical power-limiting properties were also illustrated at 980 nm.

The isomerization dynamics of *cis*-stilbene in the first excited singlet state was studied by fluorescence upconversion [30]. Lifetime measurements were made with subpicosecond resolution in 2-propanol, decanol, hexane, and tetradecane. The *cis*-stilbene fluorescence decay curves were single exponential in all solvents except for decanol, where they were adequately described by a double exponential. A picosecond component in the fluorescence anisotropy decay measurements made in 2-propanol suggests that motion along the reaction coordinate was directly measured.

11.6
Femtosecond Time-Resolved Fluorescence Depletion Spectroscopy

A method, femtosecond time-resolved stimulated emission pumping (SEP) fluorescence depletion (FS TR SEP FD), has been developed to study the vibrational relaxation of electronic excited states of molecules (Figures 11.9 and 11.10) [31]. The mechanism of FS TR SEP FD method is as follows: (i) a first ultrashort laser pulse is used to excite a molecule to the electronically excited states leading to a fluorescent emission with a lifetime of a few nanoseconds and (ii) with a specific delay time, a second ultrashort laser is introduced to SEP from upper state to electronic ground state. The intensity of initial fluorescence then should be reduced. The variation of the decrease of fluorescence intensity with the delay time between the two laser beams reflects the vibrational relaxation behavior of the electronic vibration state.

Two femtosecond laser pulses were used in the experiment of FS TR SEP FD. One pulse was used as a pump source to excite the molecule to the electronically excited state, to generate fluorescence. The other, with a specific delay time, was used to perform SEP from the electronic excited states to the ground state. The variation of the decrease in the fluorescence induced by the SEP with delay time reflects vibrational relaxation in the excited states. A homemade regenerative amplified self-mode-locking Ti:sapphire femtosecond laser, whose oscillator and amplifier

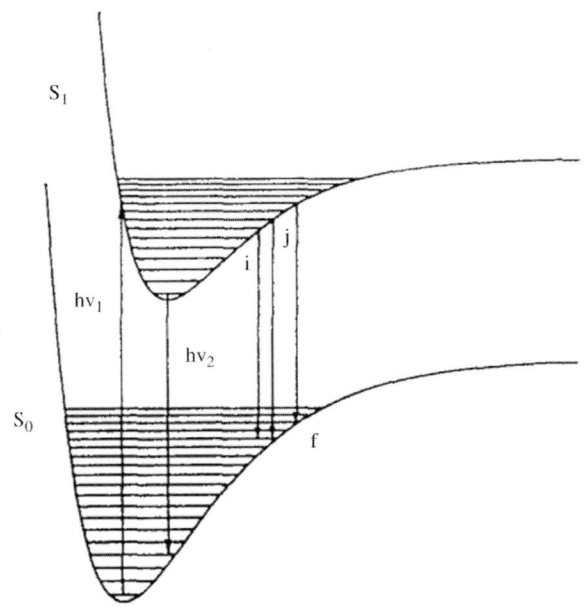

Figure 11.9 One-dimensional potential energy curve for the explanation of vibrational relaxation process observed by TR SEP FD [31]. (Reproduced with permission from Elsevier.)

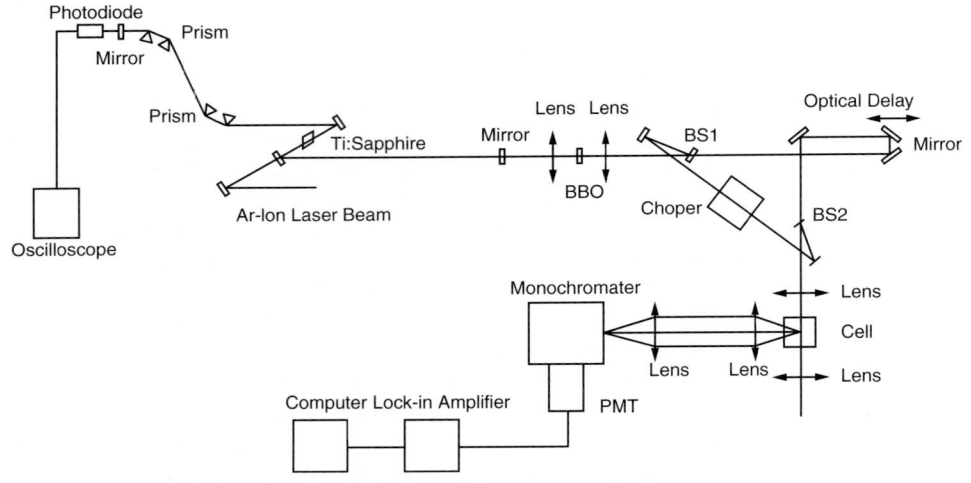

Figure 11.10 The schematic diagram of the experimental setup used in Ref. [31]. (Reproduced with permission from Elsevier.)

were pumped by a multilane argon ion laser (Innova 300, Coherent Corp.), and an Nd:YAG laser (SL903, Spectron Laser Systems, UK, power = 20 W, repetition rate 3 kHz), respectively, were used. A schematic diagram of the experimental setup is shown in Figure 11.10. The output power was about 700 mW at 3 kHz repetition rate at the wavelength of 800 nm with a pulse width of 60 fs (fwhm). A BBO crystal (0.3 mm BaB_2O_4, Fujian Castech Crystals Inc.) was used for frequency doubling to generate UV laser pulses (center wavelength = 400 nm, power = 170 mW, pulse width = 120 fs). The 400-nm pulse was used as an excitation pulse to generate fluorescence and the 800 nm pulse as a probe that was delayed by a translation stage (Unidex-100, Aerotech Inc.) with computer control, providing 2 m × 1 m path difference increment, equivalent to 6.67 fs. The probe beam was collinear with the pump beam and both of them were focused. The fused quartz sample cell was placed in a spot behind the focus where the beam diameter was 2 mm to avoid the thermal effect of the sample due to the laser heating. The probe polarization was at the magic angle with respect to the pump polarization. Fluorescence perpendicular to the incident beams was focused on a monochromator (Jarrel-Ash, Division of Fisher Scientific Company) and detected by a PMT (R456, Hamamatsu Corp.). The PMT signal was processed by a Boxcar (SR250, Stanford Corp.) and recorded by a computer. A frequency-tripling BBO crystal (0.5 mm BaB_2O_4, Fujian Castech Crystals Inc.) was used to generate a 266 nm pulse by both the pump and the probe pulses. The zero-time point and the time resolution were evaluated by detecting the correlation of 266 nm pulse power and the time delay between the probe pulse and the pump pulse. The sample cell and the frequency-tripling BBO crystal were fixed on a translation stage perpendicular to the incident beams that was used to ensure that the cell and the crystal had the same *positioning*.

11.7
High-Speed Asynchronous Optical Sampling

High-speed asynchronous optical sampling (ASOPS), a novel technique for ultrafast time-domain spectroscopy (TDS), was reported [32]. The technique employed two mode-locked femtosecond oscillators operating at a fixed repetition frequency difference as sources of pump and probe pulses. A system where the 1-GHz pulse repetition frequencies of two Ti:sapphire oscillators were linked at an offset of $\Delta f_R = 10\,\text{kHz}$ was developed (Figure 11.11). The relative time delay was repetitively ramped from zero to 1 ns within a scan time of $100\,\mu\text{s}$. Within only 1 s of data acquisition, a signal resolution of 6×10^{-7} was achieved for optical pump-probe spectroscopy over a time-delay window of 1 ns. When applied to terahertz TDS, the same acquisition time yielded high-resolution terahertz spectra with 37 dB signal-to-noise ratio under nitrogen purging of the spectrometer. This approach permits an unprecedented time-delay resolution of better than 160 fs. High-speed ASOPS provides the functionality of an all-optical oscilloscope with a bandwidth in excess of 3000 GHz and with a 1-GHz frequency resolution.

11.8
Multiphoton Excitation

Two-photon absorption is a nonlinear optical process caused by the simultaneous absorption of two photons of identical or different frequencies in order to excite a molecule from one state (usually the ground state) to a higher-energy electronic state [33]. The energy difference between the involved lower and upper states of the molecule is equal to the sum of the energies of the two photons. Two-photon absorption is many orders of magnitude weaker than linear absorption. TPA is a third-order nonlinear optical process. In particular, the imaginary part of the third-order nonlinear susceptibility is related to the extent of TPA in a given molecule. The selection rules for TPA are therefore different from those for one-photon absorption (OPA).

Two-photon absorption can be measured by several techniques. Two of them are two-photon excited fluorescence (TPEF) and nonlinear transmission (NLT). Pulsed lasers are most often used because TPA is a third-order nonlinear optical process and therefore is most efficient at very high intensities. In the nonresonant TPA, two photons combine to bridge an energy gap larger than the energies of each photon individually, and the transition occurs without the presence of the intermediate state. This can be viewed as being due to a "virtual" state created by the interaction of the photons with the molecule.

Advantages of the two-photon laser spectroscopy are as follows: high resolution, tolerance of infrared light by objects under investigation, different selection rules, and vibronic coupling. The last feature allows simultaneous accomplishment of two-photon and one-photon excitations.

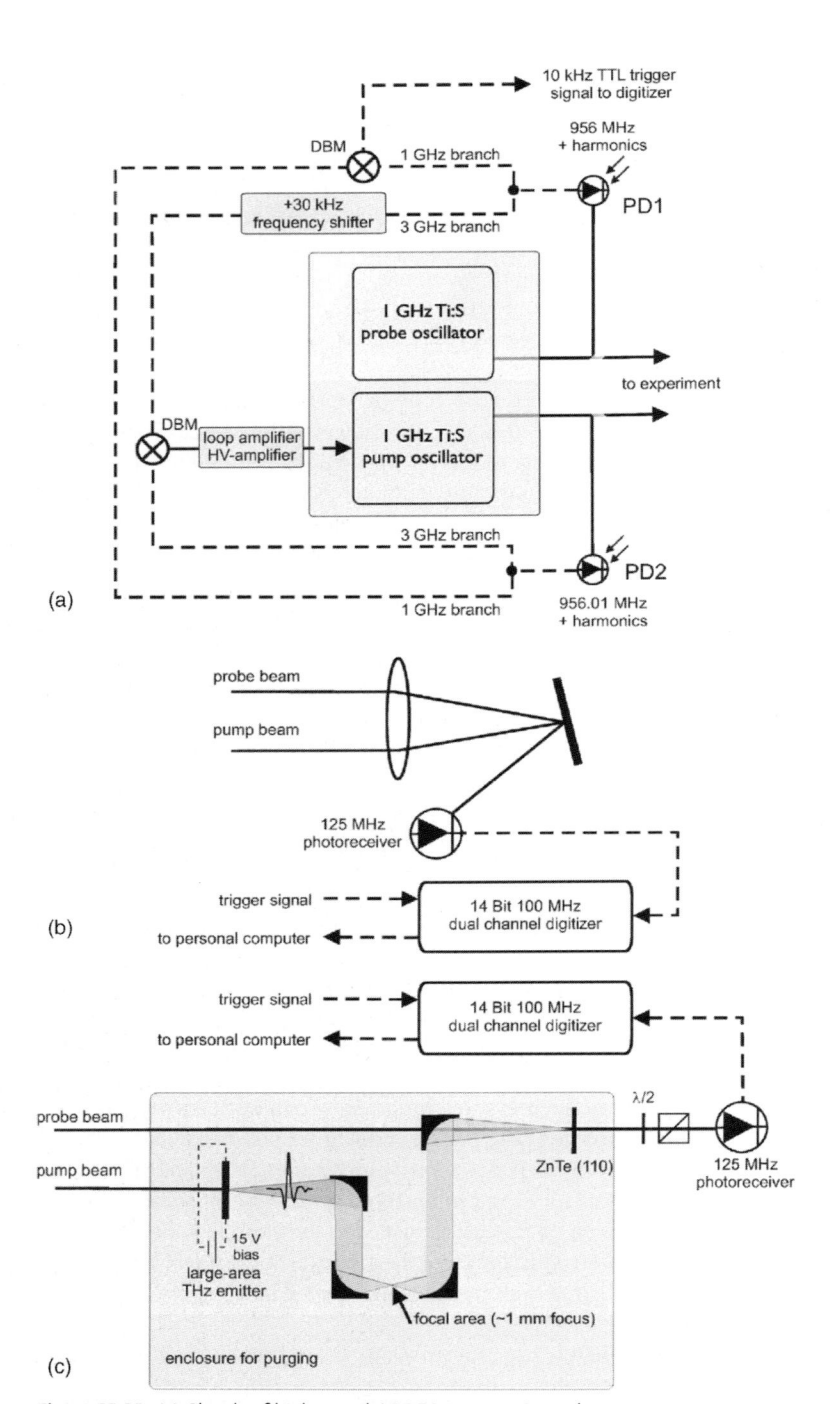

Figure 11.11 (a) Sketch of high-speed ASOPS system. Optical paths are represented by solid lines and electrical signal by dashed lines. (b) Setup for the optical pump-probe measurement. (c) High-speed ASOPS spectrometer setup [32]. (Reproduced with permission.)

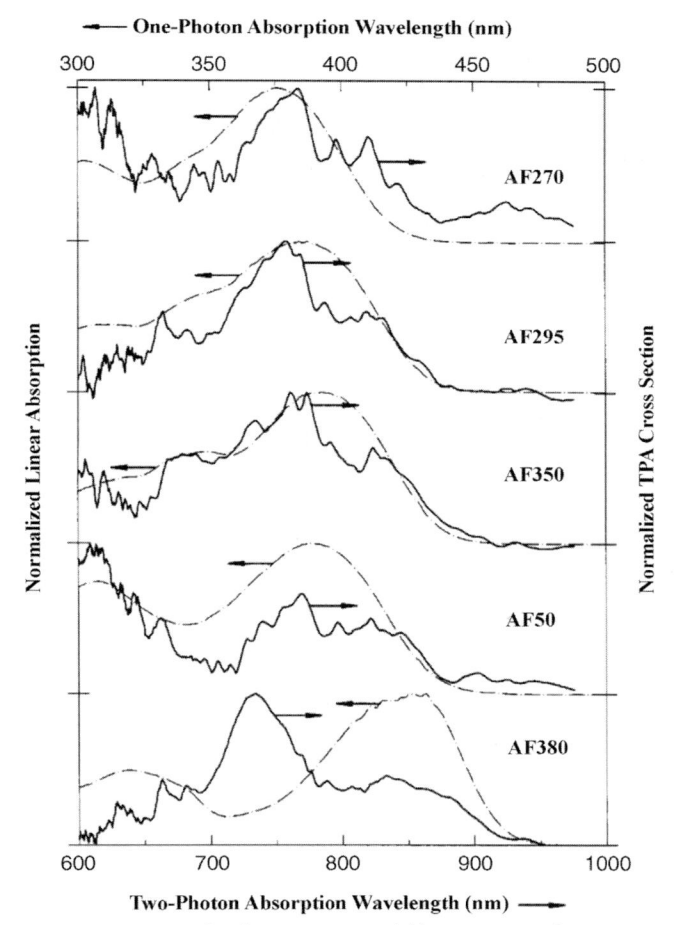

Figure 11.12 Normalized TPA spectra (solid line curves) and linear absorption spectra (dot-dashed line curves) for five chromophores solution in THF [34]. (Reproduced with permission.)

Degenerate two-photon absorption spectral properties of five chromophore solutions have been studied using a single and spectrally dispersed subpicosecond white light continuum beam (Figure 11.12) [34]. The nondegenerate TPA processes coming from different spectral components can be eliminated, and the direct nonlinear absorption spectrum attributed to degenerate TPA processes was obtained. Using this technique, the complete TPA spectra for five highly two-photon-active compounds were obtained in the spectral range of 600–950 nm on an absorption scale of TPA cross section.

Two-photon absorption spectra (650–1000 nm) of a series of stilbene chromophores were measured via a newly developed nonlinear absorption spectral technique based on a single and powerful femtosecond white light continuum

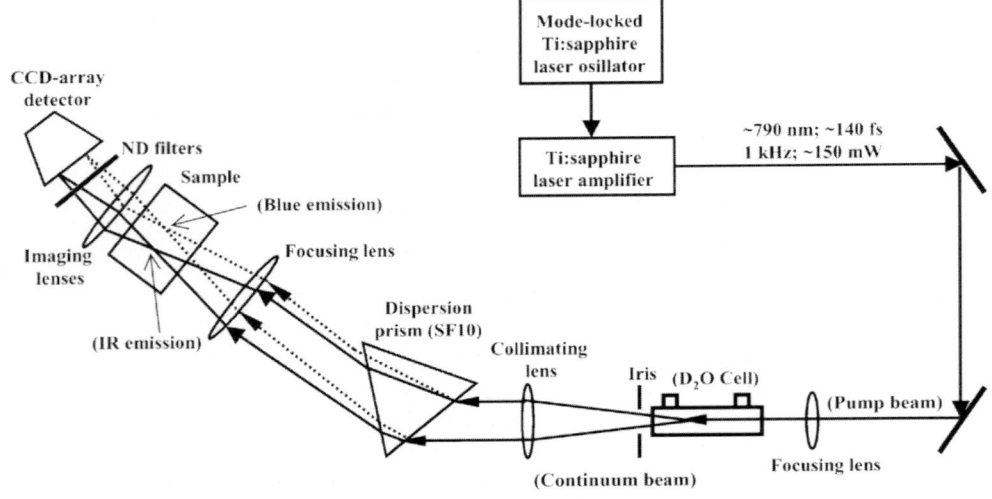

Figure 11.13 Optical setup for single femtosecond continuum beam-based degenerate two-photon absorption spectral measurement [35]. (Reproduced with permission.)

beam (Figure 11.13) [35]. Experimental results suggested that when either an electron donor or an electron acceptor was attached to a *trans*-stilbene at a *para*-position, an enhancement in molecular two-photon absorptivity was observed in both cases, particularly in the 650–800 nm region. However, the push–pull chromophores with both the donor and the acceptor groups showed larger overall two-photon absorption cross sections within the studied spectral region compared to their monosubstituted analogues. The combined results of the solvent effect and the ^{1}H-NMR studies indicated that stronger acceptors produce a more efficient intramolecular charge transfer character upon excitation, leading to increased molecular two-photon responses in this model compound set. A fairly good TPA-based optical power limiting behavior from one of the model chromophores was demonstrated.

A series of new (*E*)-4,4′-bis(diphenylamino)stilbene derivatives have been synthesized to investigate nonlinear absorptivities with attention paid to the peripheral substituent effect and multibranched modification effect by the open aperture femtosecond Z-scan technique and the nanosecond nonlinear optical transmission (NLT), respectively [36]. Two-photon fluorescence for (*E*)-4,4′-bis(diphenylamino)stilbene in THF was detected. It was found that substituent group attached to the periphery of BDPAS has no obvious contribution to TPA enhancement and that the dramatic increase in effective TPA cross sections of multibranched samples in nanosecond regime strongly suggested their larger excited-state absorption.

Two-photon molecular excitation was performed by very high local intensity provided by tight focusing in a laser scanning microscopy (LSM) [37]. This technique was combined with the temporal concentration of femtosecond pulsed lasers that produce a stream of pulses with a pulse duration of about 100 fs at a repetition rate of about 80 MHz. An average incident laser power that can saturate the fluorescence output has been estimated at about 50 mV (about 10^{31} photons/cm^2). The two-photon absorption technique was applied for investigating cross-linked polyurethane containing modified stilbene as potential frequency doubling or electro-optical controlling devices [38], properties of some donor–acceptor molecules with stilbene and azobenzene molecules as backbone [39], 4-fluorophenylethynyl- and 4-nitro-(E)-stilbenylethynylruthenium complexes [40], series donor–acceptor stilbene analogues [41], and so on.

Computations of nonlinear optical polymer polyphosphazenes with pendant nitro phenyl azo phenylamine and amino nitro stilbene groups were performed by using Gaussian '98 program to optimize the molecular structure and predict the hyperpolarizability of polyphosphazenes [6]. The authors used the nitro group as the acceptor and azaphosphane (R3P=N−) as the donor group. To study the effect of variation of nonlinear optic properties, the substituents (Rs) connected to the phosphorus atom were replaced by Me, amine, and Ph groups. It was found that both first-order polarizability and hyperpolarizabilities are larger for stilbene derivatives and are maximum for the Ph substitution. Second-order polarizability was higher for Me substitution. The two-photon absorption cross section for these molecules was also obtained. It was shown that both one-photon and two-photon absorption cross sections are maximum for the same excited state (first excited state in the case of stilbene and second excited state in the case of azobenzene derivatives. Two-photon absorption cross sections of the dominant TPA state for stilbene and azobenzene derivatives were measured.

A large second harmonic generation (SHG) excited by an Nd:YAG laser ($\lambda = 1.32$ μm) has been observed during the investigation of the optical poling process in 2-(stilbene-4-yl)benzoxazole derivative chromophores incorporated within oligoether acrylate photopolymer matrices [42]. It was revealed that the maximum output SHG was observed for the chromophore derivative molecules possessing the highest second-order hyperpolarizabilities and corresponding dipole moments. Linear absorption, single-photon-induced fluorescence, and two-photon-induced fluorescence of trans-4-(dimethylamino)-4'-[N-ethyl-N-(2-hydroxyethyl)amino]stilbene (DMAHAS) were experimentally studied [43]. The experiments showed a strong two-photon-induced blue fluorescence of 432 nm when pumped with 800 nm laser irradiation. Investigation of cross-linked polyurethane containing modified stilbenes revealed high second-order optical nonlinearity and would have potential application in frequency doubling or electro-optical controlling devices [44]. In the work [9], evidence was presented for three-photon parametric scattering in both uninfiltrated and infiltrated opal globular photonic crystals under pulsed laser excitation. The authors suggested that synthetic opal crystals can be used as photon traps for studying the emission spectra of organic and inorganic materials infiltrated in opal pores. The dynamic behavior of ultrashort laser pulse in a cascade three-level molecular system (the one-dimensional symmetry π-conjugated molecular material [4,4'-bis(dimethylamino)stilbene) was analyzed by solving Maxwell-Bloch equations and using density functional theory on *ab initio* level [45].

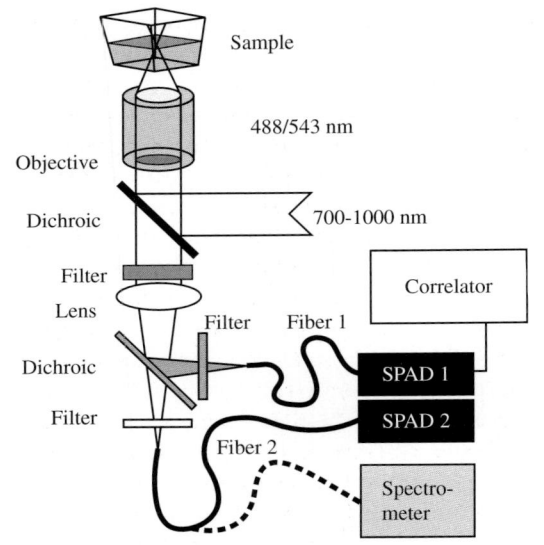

Figure 11.14 Optical setup as realized in an inverted microscope Olympus IX70/A single titanium:sapphire laser line with tunable wavelength is used for illumination, and the detected lines are split by a dichroic splitter after the collimating lens and imaged onto two different optical fibers. Both fibers can be coupled alternatively to a spectrometer [47]. (Reproduced with permission.)

In a dual-color cross-correlation fluorescence spectroscopy (DCCFS) experiment [46], a sample containing two fluorophores with different emissions in each molecule was irradiated with two lasers (or with one laser) to perform simultaneous excitation of the fluorophores. The DCCFS in combination with the confocal laser microscopy allows the separation of microscopic volume with two different fluorophores from volume with only one of them and, therefore, the monitoring of dissociation of the dual-labeled molecules or association of two single-labeled molecules. Optical setup as realized in an inverted microscope to perform simultaneous excitation of the fluorophores (Figure 11.14).

11.9
Time-Resolved Vibrational Spectroscopy

An important class of photoinduced chemistry of organic molecules is *cis–trans* isomerization [48]. A common feature of these *cis–trans* isomerization reactions is the ultrafast nature of the reaction dynamics taking place in a few picoseconds or less. Often, optical excitation leads to the formation of the isomerization product in its electronic ground state. Therefore, a large amount of internal vibrational energy is present immediately after isomerization. As a result, the vibrational fingerprint transitions initially appear often redshifted because of off-diagonal anharmonic coupling with highly populated low-frequency modes, before vibrational cooling sets in on a timescale of several tens of picoseconds.

Vibrational spectroscopy is a powerful tool for the study of molecular structure and dynamics. The typical vibrational frequency range of this spectroscopy is 100–4000 cm^{-1}, which corresponds to the energy range 0.3–12 kcal/mol. Because the resolution of vibrational spectroscopy is on the order of 5 cm^{-1}, the band shift on this order corresponds to a 0.02 kcal/mol. Vibrational transitions are correlated with specific vibrational motions by inspection of the transition frequencies. From identification of these fingerprint vibrational modes, conclusions can be drawn on specific structural motifs in the molecules. Vibrational transitions have bandwidths typically smaller (10–20 cm^{-1}) than those from electronic transitions (typically 200–2000 cm^{-1}), and it is thus less probable that different transition bands overlap in vibrational spectroscopy than in electronic spectroscopy. In addition, small molecular species may always be probed through their vibrations, and electronic transitions. Major disadvantages of vibrational spectroscopy, on the other hand, are the inherent lower cross sections of vibrational transitions and the frequent overlap of the absorption bands with those of the solvent [10].

Infrared spectra are related to changes of nuclear vibrational energy under absorption of electromagnetic radiation. In polyatomic molecules, the complex vibrational process may be resolved into a combination of n collective normal harmonic vibrations. If the parallel vibrations differ substantially in frequency, they may be regarded as independent. If the frequencies of two normal modes with frequency v_n are equal and a sufficiently strong dipolar interaction occurs between vibration modes, then, as a result of the resonance quantum mechanical effect (the Fermi resonance), this degenerate vibration splits into two modes with frequencies less and greater than v_n. The stronger the interaction, the higher the magnitude of splitting.

The vibrational processes in molecules are also reflected in the Raman spectra [49]. When the substance is irradiated at a frequency far from the frequency of its absorption, additional (satellite) lines may appear in the scattering light. The origin of such lines is accounted for by the fact that during the interaction of electromagnetic radiation, the molecular part of the radiant energy is transferred to the excited vibrational levels and part of the energy is released from the excited levels. A considerably more selective method is resonance Raman scattering (RRS). The selectivity of the method is due to the fact that the spectra display only vibrations associated with the electronic excitation of the chromophore being studied. After irradiation of the substance with monochromatic light of frequency v_0 in its absorption band, in the scattering light narrow RRS bands are observed with frequencies shifted relative to v_0. The origin of these bands is attributed to the electronic transition from the excited level to the first vibrational level of the ground state. In Raman differential spectroscopy, a conventional Raman spectrometer was adapted to measure small differences in the Raman spectra [50].

Experimentally, transient IR spectroscopy was performed in a spectrally resolved configuration. Femtosecond IR parametric devices deliver pulses with bandwidths of 150 cm^{-1} or more [51]. In order to observe shifts as small as the linewidths of

IR-active vibrations, the IR absorbance change was measured with a detector after spectral dispersion with a monochromator. As a side effect of this spectral dispersion, ground-state bleach signals often appear to grow at negative pulse delay with the dephasing time of the transition [52–54]. The time resolution of the experiment was given by the cross-correlation between the UV–vis-pump and IR-probe pulse (about 100–200 fs), and is typically dominated by group velocity mismatch in samples with thicknesses of about 100 μm. In case of probing Raman-active vibrations, the spectral resolution is determined by the monochromator, through which the spontaneous Raman emission is dispersed, and by the bandwidth of the gating pulse by which the Raman effect is induced. As a result, UV–vis-pump Raman probe spectroscopy has a temporal resolution of around 1 ps [55]. Resonance enhancement was often used by tuning the gating pulse close to or resonant with an electronic transition of the state that is probed. This has the advantage of isolating Raman bands of the state under inspection for observation. By comparing the intensities of anti-Stokes and Stokes lines of a particular vibration, it is possible to derive time-dependent changes in the excitation level of this vibration.

Among approaches in vibrational spectroscopy are differential and time-resolved IR and Raman spectroscopy, coherent anti-Stokes Raman scattering (CARS), Fourier transform infrared spectroscopy (FT-IR), multidimensional IR and RR spectroscopy, two-dimensional infrared echo and Raman echo [56], and ultrafast time-resolved spontaneous and coherent Raman spectroscopy: the structure and dynamics of photogenerated transient species [50, 57].

Time-resolved anti-Stokes Raman spectroscopy was used for monitoring vibrational relaxation dynamics in solution and provides information about specific modes in molecules under investigation [58, 59]. The experimental setup of a picosecond time-resolved Raman spectrometer is schematically shown in Figure 11.15 [59]. Probe-wavelength dependence of picosecond time-resolved anti-Stokes Raman spectra of a molecule under study allowed determination of

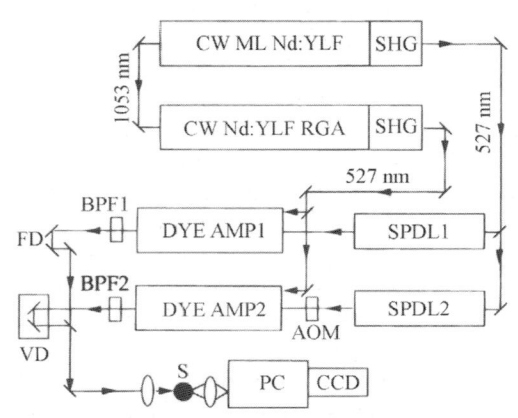

Figure 11.15 Block diagram of a time-resolved Raman spectrometer (see details in Ref. [59]. (Reproduced with permission.)

energy states of vibrationally excited molecules generated via internal conversion from the lowest excited single state. A pump pulse excites a molecule, and the anti-Stokes Raman spectrum of vibrationally excited S_0 state of the molecule was obtained by a probe pulse following the pump pulse after the delay time.

In coherent anti-Stokes Raman scattering, two ultrashort pulses of laser light (from femtoseconds to picoseconds in duration) arrive simultaneously at the sample of interest [57, 60]. The difference between the frequencies ($w_1 - w_2$) matches the frequency of a Raman active vibrational mode in the sample. A "probe" pulse (w_3) emits a signal pulse of frequency $w_1 - w_2 + w_3$ in a unique special direction. By scanning the delay time between the pump and the "probe" pulses, the delay in the vibrational coherence can be measured. The distinct advantage of CARS is that it is a background-free technique, since the signal propagates in a unique direction.

To overcome the problem of separating homogeneous and nonhomogeneous contributions to the line shape, a special technique, called the photon echo, has been developed [57]. The principal idea of this method is similar to spin-echo techniques in NMR and ESR. The photon spin echo technique generally involves five laser pulses of at least two different colors. Two time-coincident pulses of light create a Raman coherence at frequency w_g that is allowed to involve for longer time τ_1, after which the response to a single vibrational frequency occurs. At this point, another pulse pair is focused upon the sample. Each pulse in these pairs interacts with the system twice, reversing the coherence so that its frequency is w'_g. This coherence is allowed to involve for time τ_2, second delay, after which the response to a single vibrational frequency takes place. The ability to rephase inhomogeneity in Raman-active intermolecular vibrations increased with the use of five-order spectroscopic technique [57, 60]. Five-order spectroscopy relies on the existence of some sorts of nonlinearity, either in the coordinate dependence of polarizability or in the vibrational potential, and involves a three vibrational energy level. In this technique, the 2D response is obtained when the system is subjected to pairs of excitation pulses followed by the probe pulse. This technique provides the capacity for probing ultrafast intramolecular and intermolecular dynamic processes including charge transfer and chemical reactions.

A multidimensional nonlinear infrared spectroscopy was used to identify dynamic structures in liquids and conformational dynamics of molecules, peptides and, in principle, small proteins in solution [61]. This spectroscopy incorporates the ability to control the responses of particular vibrational transitions depending on their coupling to one another. Two-dimensional Raman echoes, femtosecond view of molecular structure and vibrational coherence, were investigated [62]. Two- and three-pulse IR photon echo techniques were used to eliminate the inhomogeneous broadening in the IR spectrum. In the third-order IR echo methods, three phase-locked IR pulses with wave vectors k_1, k_2, and k_3 are focused on the sample at time intervals. The IR photon echo eventually emitted and the complex 2D IR spectrum was obtained with the use of Fourier transformation. Geometry and time ordering of the incoming pulse sequence in fifth-order 2D spectroscopy are shown in Figure 11.16.

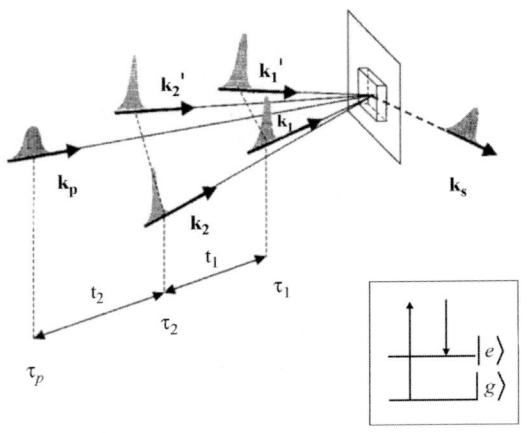

Figure 11.16 Geometry and time ordering of the incoming pulse sequence in fifth-order 2D spectroscopy. $\Delta k_1 \equiv k_1 - k'_1$ and $\Delta k_2 \equiv k_2 - k'_2$. Only one of the nine possible signal pulses in the direction $k_s = \Delta k_1 + k_p$ is shown. Inset shows the elementary Raman process [62]. (Reproduced with permission.)

Vibrational cooling rates in room temperature ionic liquids were measured with picosecond time-resolved Raman spectroscopy [63]. The 1570-cm^{-1} Raman band of the first excited singlet (S$_1$) state of *trans*-stilbene was used. The recorded vibrational cooling rates in 1-ethyl-3-methylimidazolium bis(trifluoromethylsulfonyl)imide (emimTf$_2$N) and 1-butyl-3-methylimidazolium bis(trifluoromethylsulfonyl)imide (bmimTf$_2$N) were close to those in ordinary molecular solvents despite a large difference in thermal diffusivity.

All these recent developments demonstrate that the fast and ultrafast absorption, fluorescence, and vibrational spectroscopies continue to evolve synergetically and at a rather rapid pace. It is our hope that even more progress in the instrumentation and its application in stilbene photophysics and photochemistry would follow in the coming years.

References

1 Silfvast, W.T. (1996) *Laser Fundamentals*, Cambridge University Press, Cambridge.
2 Csele, M. (2004) *Fundamentals of Light Sources and Lasers*, John Wiley & Sons, Inc., New York.
3 Young, M. (1992) *Optics and Lasers: Including Fibers and Optical Waveguides*, Springer-Verlag, Berlin.
4 Hecht, J. (1994) *Understanding Lasers: An Entry-level Guide*, IEEE Press, New York.
5 Laufer, G. (1996) *Introduction to Optics and Lasers in Engineering*, Cambridge University Press, Cambridge.
6 Pines, E., Pines, D., Barak, T., Magnes, B.-Z., Tolbert, L.M., and Haubrich, J.E. (1998) *Berichte Bunsengesellschaft für Physikalische Chemie*, **102**, 511–517.
7 Fleming, G.R. and Cho, M. (1996) *Annual Review of Physical Chemistry*, **47**, 109–134.

8 Zhong, Q., Wang, Z., Sun, Y., Zhu, Q., and Kong, F. (1996) *Chemical Physics Letters*, **248** (3–4), 277–282.

9 Goncharov, A.P. and Gorelik, V.S. (2007) *Inorganic Materials*, **43**, 386–391.

10 Nibbering, E.T.J., Fidder, H., and Pines, E. (2005) *Annual Review of Physical Chemistry*, **56**, 337–367.

11 Lewis, F.D., Zhu, H., Daublain, P., Torsten, F., Raytchev, M., Wang, Q., and Shafirovich, V. (2006) *Journal of the American Chemical Society*, **128**, 791–800.

12 Lewis, F.D., Liu, X., Miller, S.E., Hayes, R.T., and Wasielewski, M.R. (2002) *Journal of the American Chemical Society*, **124** (38), 11280–11281.

13 Baek, N.S., Nah, M.K., Kim, Y.H., and Kim, H.K. (2007) *Journal of Luminescence*, **127**, 707–712.

14 Polli, D., Antognazza, M.R., Brida, D., Lanzani, G., Cerullo, G., and Silvestri, S. (2008) *Chemical Physics*, **350**, 45–55.

15 Yan, Y.J. and Mukamel, S. (1990) *Physical Review A*, **41**, 6485–6504.

16 Raytchev, M., Pandurski, E., Buchvarov, I., Modrakowski, C., and Fiebig, T. (2003) *Journal of Physical Chemistry A*, **107**, 4592–4600.

17 Raytchev, M., Mayer, E., Amann, N., Wagenknecht, H.A., and Fiebig, T. (2004) *ChemPhysChem*, **5**, 706–712.

18 Kaden, P., Mayer-Enthart, E., Trifonov, A., Fiebig, T., and Wagenknecht, H.-A. (2005) *Angewandte Chemie – International Edition*, **44**, 1636–1639.

19 Shafirovich, V.Y., Courtney, S.H., Ya, N., and Geacintov, N.E. (1995) *Journal of the American Chemical Society*, **117**, 4920–4929.

20 Nikowa, L., Schwarzer, D., Troe, J., and Schroeder, J. (1993) *Springer Series in Chemical Physics*, **55** (Ultrafast Phenomena VIII), 603–605.

21 Bartels, A., Hudert, F., Janke, C., Dekorsy, T., and Kohler, K. (2006) *Applied Physics Letters*, **88**, 041117/1–041117/3.

22 Maroncelli, M. and Fleming, G.R. (1987) *Journal of Chemical Physics*, **86**, 6221.

23 Pines, D., Pines, E., and Rettig, W. (2003) *Journal of Physical Chemistry A*, **107**, 236–242.

24 Papper, V., Pines, D., Likhtenshtein, G.I., and Pines, E. (1997) *Journal of Photochemistry and Photobiology A: Chemistry*, **111**, 87–96.

25 Pines, E., Pines, D., Barak, T., Magnes, B.-Z., Tolbert, L.M., and Haubrich, J.E. (1998) *Berichte Bunsengesellschaft fur Physikalische Chemie*, **102**, 511–517.

26 Wang, X., Zhou, G., Wang, D., Wang, C., Fang, Q., and Jiang, M. (2001) *Bulletin of the Chemical Society of Japan*, **74**, 1977–1982.

27 Gustavsson, T., Cassara, L., Gulbinas, V., Gurzadyan, G., Mialocq, J.-C., Pommeret, S., Sorgius, M., and van der Meulen, P. (1998) *Journal of Physical Chemistry A*, **102**, 4229–4245.

28 Jordan, G., Kobayashi, T., Blau, W.J., Pfeiffer, S., and Hoerhold, H.-H. (2003) *Advanced Functional Materials*, **13**, 751–754.

29 Zhou, G., Wang, X., Wang, D., Shao, Z., and Jiang, M. (2002) *Applied Optics*, **41**, 1120–1123.

30 Todd, D.C., Jean, J.M., Rosenthal, S.J., Ruggiero, A.J., Yang, D., and Fleming, G.R. (1990) *Journal of Chemical Physics*, **93**, 8658–8668.

31 Qinghua, Z., Zhaohui, W., Ya, S., Qihe, Z., and Fanao, K. (1996) *Chemical Physics Letters*, **248**, 277–282.

32 Bartels, A., Cerna, R., Kistner, C., Thoma, A., Hudert, F., Janke, C., and Dekorsy, T. (2007) *Review of Scientific Instruments*, **78**, 035107/1–035107/8.

33 (a) Göppert-Mayer, M. (1931) *Annals of Physics*, **9**, 273–295; (b) Abella, I.D. (1962) *Physical Review Letters*, **9**, 453–456.

34 He, G.S., Lin, T.-C., Dai, J., Prasad, P.N., Kannan, R., Dombroskie, A.G., Vaia, R.A., and Tan, L.-S. (2004) *Journal of Chemical Physics*, **120**, 5275–5284.

35 Lin, T.-C., He, G.S., Prasad, P.N., and Tan, L.-S. (2004) *Journal of Materials Chemistry*, **14**, 982–991.

36 Huang, Z., Wang, X., Li, B., Lv, C., Xu, J., Jiang, W., Tao, X., Qian, S., Chui, Y., and Yang, P. (2007) *Optical Materials*, **29**, 1084–1090.

37 Denk, W., Strickler, J.H., and Webb, W.W. (1900) *Science*, **248**, 73–76.

38 Ye, M.X., Lei, J.L., Liu, L., and Wang, W. (2006) *Journal of Nonlinear Optical Physics & Materials*, **15**, 275–285.

39 Jha, P.C., Anusooya, P.Y., and Ramasesha, S. (2005) *Molecular Physics*, **103**, 1859–1873.

40 Morrall, J.P.L., Cifuentes, M.P., Humphrey, M.G., Kellens, R., Robijns, E., Asselberghs, I., Clays, K., Persoons, A., Samoc, M., and Willis, A.C. (2006) *Inorganica Chimica Acta*, **359**, 998–1005.

41 Park, G.J., Woo, S., and Ra, C.S. (2004) *Bulletin of the Korean Chemical Society*, **25**, 1427–1429.

42 Kityk, I.V., Makowska-Janusik, M., Gondek, E., Krzeminska, L., Danel, A., Plucinski, K.J., Benet, S., and Sahraoui, B. (2004) *Journal of Physics: Condensed Matter*, **16**, 231–239.

43 Yan, Y.-X., Wang, D., Zhao, X., Tao, X.-T., and Jiang, M.-H. (2003) *Chinese Journal of Chemistry*, **21**, 626–629.

44 Ye, Mingxin, Xu, L., Ji, L., Liu, L., and Wang, W. (2006) *Journal of Nonlinear Optical Physics & Materials*, **15**, 275–285.

45 Liu, J.-C., Ke, Z., Song, Y.-Z., and Wang, C.-K. (2006) *Wuli Xuebao*, **55**, 1803–1808.

46 Winkler, T., Kettling, U., Koltermann, A., and Eigen, M. (1999) *Proceedings of the National Academy of Sciences of the United States of America*, **96**, 1375–1378.

47 Heinze, M.G., Kolterman, A., and Schille, P. (2000) *Proceedings of the National Academy of Sciences of the United States of America*, **97**, 10377–110382.

48 Dugave, C. and Demange, L. (2003) *Chemical Reviews*, **103**, 2475–2532.

49 Gardiner, D.J. (1989) *Practical Raman spectroscopy*, Springer-Verlag.

50 Deng, H. and Callender, R. (1999) Raman spectroscopic studies of the structure, energies, and bond distortions of substrates bound to enzymes, in *Enzyme Kinetics and Mechanism, Part E* (eds V.L. Schramm and D.L. Purich), *Methods in Enzymology*, vol. 308, Academic Press, San Diego, CA, pp. 176–215.

51 Kaindl, R.A., Wurm, M., Reimann, K., Hamm, P., Weiner, A.M., and Woerner, M. (2000) *Journal of the Optical Society of America B: Optical Physics*, **17**, 2086–2094.

52 Wynne, K. and Hochstrasser, R.M. (1995) *Chemical Physics*, **193**, 211–236.

53 Hamm, P. (1995) *Chemical Physics*, **200**, 415–429.

54 Chachisvilis, M., Fidder, H., and Sundström, V. (1995) *Chemical Physics Letters*, **234**, 141–144.

55 Towrie, M., Parker, A.W., Shaikh, W., and Matousek, P. (1998) *Measurement Science & Technology*, **9**, 816–823.

56 Hamaguchi, H. and Gustafson, T.L. (1994) *Annual Review of Physical Chemistry*, **45**, 593–662.

57 Fourkas, J.T. (2001) *Multidimensional Raman spectroscopy*, in *Advances in Chemical Physics*, vol. 117 (eds I. Prigogine and S.A. Rice) John Wiley & Sons, Inc., New York, pp. 235–274.

58 Uchida, T., Ishikawa, H., Ishimori, K., Morishima, I., Nakajima, H., Aono, S., Mizutani, Y., and Kitagawa, T. (2000) *Biochemistry*, **39**, 12747–12752.

59 Nakabayashi, T., Okamoto, H., and Tasumi, M. (1997) *Journal of Physical Chemistry A*, **101**, 3494–3500.

60 Mukamel, S. (2000) *Annual Review of Physical Chemistry*, **51**, 691–729.

61 Asplund, M.C., Zanni, M.T., and Hochstrasser, R.M. (2000) *Proceedings of the National Academy of Sciences of the United States of America*, **97**, 8219–8224.

62 Mukamel, S., Pirytinski, A., and Chrnyak, V. (1999) *Accounts of Chemical Research*, **32**, 145–154.

63 Koichi, I., Kyousuke, Y., Yuta, T., and Hiro-o, H. (2007) *Chemistry Letters*, **36**, 504–505.

12
Conclusions

Both the theoretical materials and the experimental data presented in this book clearly demonstrate the significant progress that has recently been made in the application of stilbenes in chemistry, photochemistry, photophysics, materials science, biochemistry, biomedicine, and clinical research. This progress resulted to a great extent from interdisciplinary cooperation. Advances in synthetic chemistry that provided researchers in these areas with a wide assortment of stilbenes paved the way for their multiple applications in basic and applied research. Modifications of traditional physical techniques and advanced methods such as nano-, pico-, femto-second absorption, fluorescence and vibrational time-resolved spectroscopy, and theoretical approaches to the analysis of experimental data ensure profound photophysical and photochemical investigations in the area. Experts in biochemistry, biomedicine, and medicine effectively used natural and synthetic stilbenes in biochemical and preclinical studies and recently in clinical trials.

Let us summarize the main results, possibilities, and advantages of various fields of nitroxide application.

A combination of classical and modern synthesis methods allowed chemists to prepare thousands of new stilbenes, which possess a variety of chemical and physical properties, in solution, polymers, and on templates. The synthesized compounds show the rich chemistry involved in multiple reactions such as halogenation, oxidation, reduction, addition, substitution, polymerization, complexation, and so on.

Stilbenes have proved to be convenient models for detailed investigation into general mechanisms of biophysical processes such as the light absorption, fluorescence, phosphorescence, intersystem crossing, excimerization, and energy migration. These compounds undergo a series of photochemical reactions including isomerization, photocyclization, dimerization, addition, and charge transfer. These reactions run effectively in solutions, dendrimers, polymers, silica plates, and other matrices. Especially profound information has been gathered on direct and sensitized photoisomerization, a process that proved to be almost ideal "training area" in photochemistry and photophysics. Stilbenes and their derivatives, possessing above-mentioned properties, appear to be promising materials. These compounds form the basis for dyes and solid lasers, electro-optic, electrophotographic, light emitting, radioluminescence, and image-forming apparatuses.

Natural stilbenes such as resveratrol, combretastatin, pterostilbene, and its synthetic analogues demonstrate miscellaneous effects on the biological activity of cells, organs, and animals. These compounds possess antioxidant, anticancer, antiradiative, and antiaging activities, cause apoptosis and affect the signaling and genetic apparatuses. For these reasons, stilbenes have been finding massive application in clinical trials as therapeutic drugs.

Owning to their unique photophysical and photochemical properties, stilbenes are incorporated in biological and nonbiological systems using traditional and latest optical and luminescence methods, picosecond and femtosecond time-resolved techniques, and in particular allow to investigate micropolarity and macrostructure molecular dynamics of polymers, proteins, biomembranes, and other systems. Recently, a series of fluorescent methods of real-time biological analysis and investigation of molecular dynamics based on the use of stilbenes and potentially important for biochemistry, biophysics, biotechnology, and biomedicine were proposed and developed. In this approach, new types of stilbene molecular probes have been used: (i) fluorescent–photochrome molecules and (ii) supermolecules containing fluorescent and fluorescent quenching segments, and cascade system, containing spin, triplet, and fluorescence–photochrome probes.

We believe that research on stilbenes, which combines their fundamental importance for human welfare and intellectual fascination for investigation, will continue to solve exciting and complicated problems of chemistry, physics, materials science, biochemistry, and biomedicine.

Index

a

absorption excitation   82
absorption frequencies   67
absorption properties   1, 99
absorption spectra
– molecular probes   279
– photocyclization   138
– photophysics   71f.
acceptor groups   20, 106, 279
acetamido (iodoacetyl)amino stilbene-
    disulfonic acid (IASD)   287
acetonitrile (ACN)   86, 147
acetophenone   90
acetylcholine   215
acetyl-CoA carboxylase (ACC)   202
activation mechanism   112, 279
acute myeloid leukemias (AMLs)   233
addition reactions   53f., 143
additives   45
adipocytes   204
affinity enhancing modulators (DNA)   56
aging   209f.
aldol-type condensation   2f.
alicyclic group   161
aliphatic groups   9, 80, 161
alkenes reactions   45, 137, 143f.
alkenyl   7
alkoxy chains   23
alkynes   45
alpha-crystallin subunits   286
alpha-cyanostilbene fluorophores   176
alveolar epithelial cells   196
Alzheimer's disease   18, 205, 283
amine triphenylene   166
amines reactions   143f.
amino substituents   102
aminoalkyl phosphonate derivatives   227
aminostilbenes   82, 144, 281

AMP-activated protein kinase activity   202
analysis   1–42
angiogenesis inhibition   251
anhydrousliquidammonia   51
anilino groups   123, 316
animals   28, 32
– bioactive stilbenes   189, 204f.
– cancer protection   225f.
– preclinics   244f.
– pterostilbene   251f.
anionic polymerization   57f.
anthracene   125
anthraquinone   90
antibodies
– chemical reactions   60f.
– complexes   87
– donor–acceptor-substituted stilbene   28ff.
– fluorescent-photochrome method   296f.
anticancer effects   189, 262
anti-DNP antibody binding site   292f.
antihypertensive agents (AHA)   273
anti-inflammatory agents   250, 262
antimicrobial activity   16
antioxidant activity   193ff., 262
antiradiation damage   189
anti-TNP–DNP–DAS–Lys complex   302
apoptosis
– bioactivity   197f.
– combretastatin   233
– pterostilbene   251
– resveratrol   227
applications
– clinical   261–276
– molecular probes   280f.
– preclinical   225–260
Arachis hypogaea L.   32
arenes   45
aromatic aldehydes   2

aromatic amines 169
aromatic carbonyls 9
aromatic groups 161
artochamins 17
aryl hydrocarbon-induced CYP1A1 225
aryl iodides 19
aryl-/aroyl-substituted chalcone analogues 215
arylation 6
arylcoumarin 60
ascorbic acid 293f.
astilbin 191
asynchronous optical sampling 323f.
atmospheric pressure chemical ionization (APCI) 33
auristatin 268
Avogadro number 279
aza-stilbene derivatives 12
azobenzene 130
azodyes 18, 25, 55
azulene effect 120, 124

b
Ba(II) complexes 59
bacteria 189
Barton–Kellogg–Staudinger reaction 8f.
Beer–Lambert relationship 80
benzil derivatives 235
benzophenones 90ff.
benzothiazole 144
benzoxazolyl stilbene (BBS) 93, 283
benzyl ether type dendrons 19
beta-amyloid toxicity 18
BFLT immobilization 29
bichromophoric photochromes 13
biindanylidenes 113
bimolecular reactions 140f.
binder resins 165
binding sites 292f.
bioactive stilbenes 189–224
bioavailability 265
biochemical effect 202f.
biomembranes 289ff.
bis(oligonucleotide) conjugates 56
block copolymer formation 58
blue fluorescent antibodies 87
blue laser dyes 160
blue organic light-emitting device 167
bond formation 179
bond torsion 94, 102, 113
bonds *see* C–C bonds, double bonds
boron-containing stilbenes 161
Brassica napus 34
breast cancer 207

brighteners 2
broadband pump-probe spectroscopy 312f.
bromination 43, 58
bromostilbenes 90, 124
Bronsted acid sites 51f.
bryostatin 268
butanone 77

c
cadherin 249
caffeine 144
calixarene analogues 126
cancer cells 198, 225f., 233f.
capping groups 27
Caragana sinica 32
carbanion 2
carbazole groups 162ff.
carbazolyl groups 177
carbocation intermediates 144
carbolization 52f.
carbonyls 9
carbopalladation crosscoupling 11
carboplatin/combretastatin combinations 272
carcinoma 225
cardiac dysfunction 210
cardiopulmonary toxicity 271
cascade spin-triplet-photochrome methods 299f.
catalase 254
catalysts 46–51
catechin 192, 263
C–C bonds 2, 164
– image-forming apparatus 178
– one-photon absorption 73
– photocyclization 137
CC-chemokine receptor-5 208
cells
– bioactivity 204f.
– pterostilbene 250f.
– *see also* cancer cells
chalcone analogues 215
charge separation/recombination 314
charge transfer
– ionization 150f.
– photoisomerization 122
– photophysics 95f.
– precursors 105f.
– twisted intramolecular 80
chemical reactions 43–66
chemical structures 127
chemical synthesis 10ff.
chemically induced dynamic nuclear polarization (CIDNP) 152

chemopreventive activities   225f.
chemotherapeutic effects   201
chiral compounds   12
chloroacetate protecting groups   11
chlorobenzene   77
chloroform   44, 84
cholesterol   129, 216, 232
cholinesterase   202
chromic effects   279
chromophores   83
– aggregation   71, 130
– materials   162, 172
– molecular probes   279
– photoisomerization   101, 130
– photoreactions   146
– quencher pairs   289
– TPA spectra   325
cinnamic acid esters   4
cis-fixed stilbenophanes   141
cisplatin   242
cis-stilbene   1
– bromination   44
– fluorescence upconversion   319
– fluorescent-photochrome method   301
– molecular probes   290
– photocyclization   137
– photoisomerization   102
– polyphenol analogues   200
– potential energy surfaces   113
– triplet energy   69
– triplet-photochrome method   298
cis–trans isomerization   99
– dendrimers   126
– photocyclization   138
– triplet state   90
– vibrational spectroscopy   328
cis–trans photoisomerization   99
– molecular probes   293
– triplet-photochrome method   298
cleavage   49, 144, 193
clinics/clinical trials   261–276
coal liquefaction residues (CLR)   287
coherent anti-Stokes Raman scattering (CARS)   330
colchicine   60
colon carcinoma cells HT-29   227, 234
color-tunable light-emitting copolymers   177
combretastatin
– AC7700   246
– CA-4-P   233–248, 262, 267ff.
– bioactivity   189, 213f.
– chemical reactions   60
complexation   58f., 139
condensation   2f.

conductivity   105
conical intersection (COI)   105
conjugation   84, 102
coordination polymers   25
coronary arterial endothelial cells (CAECs)   213
coumarin copolymers   177
coupling   7ff., 75
– electronic   106
– HOMO–LUMO   104
– laser materials   165
cowpea mosaic virus (CPMV)   28, 88
cross-linking   293
cyanogen bromide activation   293
cyano-substituted stilbenes   106
cyanuric chloride   86, 293
cyclization   99
cycloaddition   17, 150
cyclodextrin/derivatives   23f., 44, 58
cyclohexane   77
cyclohexenone analogues   236
cyclooxygenase   218, 250
cyclophosphamide   242
cytochromes P   218
cytokine   195
cytometry analysis   230
cytotoxicity   198

d

D–A–D complexes   59
dansyl   84
DBASVP photopolymerization   149
deactivation   78f., 101
decay profiles   317
delocalization   106, 115
dendrimers   19f.
– charge transfer   150, 162
– fluorescence   84f.
– photoisomerization   126f.
– photoreactions   146f.
density functional theory (DFT)   68, 116
deoxyschweinfurthin   15
depletion spectroscopy   321f.
derivatives   1
– complexation   59
– DIDS, DNDS, SITS   210
– fluorescent-photochrome method   301
– molecular probes   285
– photoisomerization   121
– resveratrol   210, 227
– substituted groups   55
dermal fibroblasts   209
diabetic retinopathy   272
dialkylamino stilbene derivatives   280

diaminostilbene (DAS) 302
diarylamino group 177
diastereoisomer 143
diazetine dioxides decomposition 14
diazidostilbene-disulfonic (DASS) 288
diazo-thioketone coupling *see* Barton–Kellogg–Staudinger reaction
dibenzamido disulfonic stilbene (DBDS) 286
DIDS derivatives 210
dielectric polarization 72
Diels–Alder/Wittig olefination 5
dienes reactions 143f.
dienestrol (DIS) 28
diether linkers 56
diethylstilbestrol (DES) 28
diffusion coefficient 127
digital signal processing system (DSPS) 180
dihydrophenanthrene (DHP) 99
diimine Re(I) tricarbonyl complexes 80
dimethylamino stilbenes 280
dimethylamino-4′-aminostilbene (DMAAS) 291, 300
dimethylamino-4′-cyanostilbene (DCS) 80, 89
– photoisomerization 123
– single-photon counting 316
dimethylamino-4′-methoxy-stilbene(DMS) 83
dimethylamino-4′-nitrostilbene polymer (DANS) 163
dimethylformamide (DMF) 281
dimethylsilyl-4′-trifluoromethylstilbenes (HTS) 148
dimethylsulfoxide (DMSO) 77
dinitro-disulfonic stilbene (DNDS) 210, 288
dinitrophenyl (DNP) 296
dinitrophenylamino dimethylamino stilbene (StDNP) 61, 292
dioxygen 47, 91
diphenylamino stilbene derivatives 326
diphenylethylene 1
dipole moments 107, 279
direct photoisomerization 121f.
disilanyl-4′-trifluoromethylstilbene (DTS) 148
dismutase 254
disodium combretastatin 244ff.
distearoyl 216
disulfonate binding 288
disulfonic acids (DIDS/SITS) 210, 288ff.
DMAHAS 173, 327
DNA
– charge transfer ionization 154
– conjugates 27
– duplex stability 56

– hairpins 310
– polymerase 88
– resveratrol 195
dodecyloxy-substituted stilbenoid dendrimers 146
dolastatin 268
domino carbopalladation crosscoupling 11
donor–acceptor 76
– capped hairpins 310
– dye lasers 159
– photoisomerization 117
– substituted stilbenes 60, 89, 95
– *trans*-stilbene derivatives 170
donor emission spectrum 279
donor substituents 20, 106
double-bond cleavage 144, 193
double-bond torsions 102
double-bond twisting
– charge transfer 151
– chemical reactions 61
– linear quinoid structure 106f.
– photoisomerization 102
– Saltiel mechanism 102f.
dual color cross-correlation fluorescence spectroscopy (DCCFS) 327
dual path addition mechanism 45
dual thermal bond activation mechanism 112f.
duodenum 192, 233
dyes 2, 18
– BODIPY 285
– lasers 2, 159f.
– photoisomerization 130
– substituted groups reactions 55
dyspnea 271

e
Ehrenfest dynamics 150
eicosanoid synthesis 190
E-isomers 56
electrocyclic ring closure 245
electroluminescent material 164
electron–hole recombination 89
electron–hole transfer 153
electron transfer
– broadband pump-probe spectroscopy 312
– molecular probes 289
– photoisomerization 105
electronic coupling 106, 115
electronic polarization 117
electronic properties 105, 115
electronic transitions 329
electron-spin resonance methods 297
electrooptic materials 161f.

electrophilic fluorination 54
electrophotographic materials 165f.
electrospray ionization-tandem mass spectrometry (ESI-MS/MS) 31
electrostatic properties 279
Ellman method 202
emission energy 79
emission frequencies 67
emission spectra 82
– JCS 318
– molecular probes 279
– photocyclization 138
emodin 191
endothelial cells 267
energy migration 67, 84, 105
energy transfer
– molecular probes 278f.
– nonvertical 110f.
– photophysics 93f.
– triplet-photochrome method 300
enzyme activities 195, 202f., 265
enzyme phosphorylation 216
enzyme-linked immunosorbent assay (ELISA) 302
eosin 84
EP2-19G2 antibody 88
epithelial cells 196, 207
epoxidation 46f.
Erythrosin-NCS 298
Escherichia coli 212
E-stilbenes 1, 43–52
estrogens 207f.
ethiols 55
exchange integral 280
excimer emission 92f., 141
exciplexes 92f., 145
excited states 1
– molecular probes 279f.
– photophysics 67ff., 76f.
extracellular signal-regulated kinases (ERK1/2) 234
E–Z photointerconversion 138, 148

f

Fab 11G10 60
femtosecond broadband pump-probe spectroscopy 312f.
Fermi resonance 329
fibroblast growth 205, 209
fibro-optic biosensoring 293
fidelity-enhancing modulators (DNA) 56
Fischer carbene complexes 53
flavonoids 192, 212, 233
fluorene compounds 18

fluorescence 1, 67, 80f.
– excimers/exciplexes 92f.
– excitation 82
– Franck–Condon state 78
– molecular probes 278
– nitroxide molecules 294
– photoisomerization 101
– photophysics 72f., 79f.
fluorescence depletion spectroscopy 321f.
fluorescence picosecond time-resolved single photon counting 314f.
fluorescence resonance energy transfer (FRET) 279f.
fluorescence upconversion spectroscopy 318f.
fluorescence-photochrome immunoassay (FPHIA) 290, 301f.
fluorescence-photochrome method (FPM) 278, 290f., 296f.
fluorescent quenching groups 302f.
fluorescent whitening agent (FWA) 31, 183
fluorophore quenching 289
fluorophore-heme dualfunctionality probe (FHP) 302
fluorostilbenes 139
Förster distance 279
Fourier transform infrared spectroscopy (FT-IR) 330
Franck–Condon factor 110, 114
Franck–Condon state 75f., 105
free radical mechanism 47
frontier molecular orbitals (FMOs) 106
functional groups 2
fungi bioactive stilbenes 189

g

gamma interferon 208
gas chromatography–mass spectrometry (GC–MS) 28f.
gastric cancer 234
gelation 18, 128
generation number 84
globular proteins 60
gluconeogenic enzyme 216
glucuronide (RES(GLU)) 193
glutathione (GSH) 195
glutathione peroxidase 254
glyoxal 31, 274
gold complexes 173
green fluorescence protein (GFP) 235
Grignard reagents 10
Grubbs catalyst 13
guanidine 5
guest–host interactions 131, 162

h

halides 5
halocarbonium complexation 59
halogenation 43f.
Hamiltonian exchange interaction 280
Hammett relationships 44, 53, 75f., 117
hapten 30
heat-resistance stilbene derivatives 177
Heck reaction 5f., 83
heme quenches 302
hepatic sinusoidal endothelium (HSE) 126
Herzberg–Teller contribution 75
heterocyclic compounds 33
heteronuclear (HETCOR) spectra 84
hexafluoro-propanol (HFIP) 144
hexestrol 28
high-density lipoprotein (HDL) cholesterol 264
high-speed asynchronous optical sampling (ASOPS) 323f.
HIV activity 33
hole transporting agents 165
hole trapping 155
HOMO 71
– light-emitting materials 177
– photocyclization 137
– photoisomerization 104, 107
homocoupling stilbenes 16
homonuclear (COSY) spectra 84
Horner–Wadsworth–Emmons (HWE) reactions 13ff., 59
hula-twist (H-T) mechanism 102, 108
human breast (SKBR3) tumor 242
human liver 192
HUT78B1/3 cell lines 251
hybrid density functional theory 160
hydroboration 53
hydrogen abstraction 58
hydrogen transfer 138
hydrogenated soybean phosphatidylcholine (HSPC) 216
hydrogen-substituted stilbenes 106
hydrophilic sulfhydryl reagent 287
hydrophobic nanocapsules 141
hydroxystilbenes 9, 19, 49f.
hyperpolarizability materials 171, 183
hyper-Rayleigh scattering (HPS) 171
hypoxia 238, 271

i

image probing 283f.
image-forming materials 177f.
imidazoles 239
imines reactions 143f.
immersion depth 289f.
immobilized systems (quartz slides) 293f.
immunoassay 301
immunoconjugate 87
INDO quantum chemical method 71
indolic compounds 137
infrared spectra 329
initial area under the contrast agent concentration–time curve (IAUGC) 240
iNOS protein inhibition 255
interactions triplet states/phosphorescence 90f.
interferon 208
intersystem crossing 102, 289
intramolecular charge transfer 95f., 105f.
investigations methods 309–334
iodoacetamide 286
ion cyclotron resonance frequency (ICRF) 181
ionization 150f.
ischemia–reperfusion (I–R) 230
isomeric forms 1

j

Jablonski energy diagram 68
JCS anilino groups 316
Jin Que-gen 32

k

ketones 4
KHT tumor model 239
Knoevenagel condensation 183

l

labeling 278
laetevirenol 195
Langmuir–Blodgett films 27, 296
laser materials 159f.
laser scanning microscopy (LSM) 326
laser techniques 309
layers 167
leukemic cells 233
Leu[L89] antibody residues 60
levulinate protecting groups 11
lifetime
– exited state 67, 113
– fluorescence 78, 84
– molecular probes 279
ligand binding 60
light energy absorbance 102
light-emitting diode (LED) 166f.
light-emitting materials 175f.
linear electro-optic effect 162
linear free energy relationships (LFERs) 76

linear quinoid structure 106f.
linkers 56, 310
lipopolysaccharide (LPS) 250
liposomes 216
liquid chromatography 28f.
liquid crystals 137
lithium 51
living cationic polymerization (LCP) 184
local medium properties 278f.
loose-bolt effect 115
low-density lipoprotein (LDL)
  cholesterol 232, 252, 264
lucifer yellow iodoacetamide 286
luciferase 219
luminescence 67
– materials 161
– molecular probes 278
– stilbenoid chromophores 83
luminophores 84, 93
LUMO 71
– light-emitting materials 177
– photocyclization 137
– photoisomerization 104, 107
lung cancer H460 cells 238
luteolin 229
lymphoid cancers 225
lysozyme 302

m
Mach–Zehnder modulators 163
macrophages 208
magnetization 50
mammary tumors 207
materials 159–188
matrices reactions 147f.
matrix metalloproteinase 208
maximum-tolerated doses 271
Maxwell–Bloch equations 160, 170
McMurry reaction 8f.
media-melting mechanism 102, 109f.
medulloblastoma (MB) – Notch1/2 207
mefenamic acid 193, 232
melanoma 225, 251
melting mechanism 1, 110f.
meso-stilbene dibromide 58
metabolites 192f., 265
metal organic framework (MOF) 182
metalloproteinase 208
metalto-ligand charge transfer (MLCT) 25
methoxylated stilbenes 71
methoxyresorufin O-demethylation
  (MROD) 228
methoxy-substituted stilbenes 106

methyl methacrylate (MMA) 25, 56
methylarene 2
methylcyclohexane (MCH) 46, 103
methylethylketone (MEK) 77
methylglyoxal 31, 274
methyl-substituted stilbenes 106
Michael addition 53
Michaelis–Menten kinetics 192
microglia 205
microvascular density (MVD) 262
microwave-promoted Heck reaction 19
mitochondrial effects 198, 228, 233
mitogen-activated protein kinases (MAPKs) 234
Miyaura–Suzuki coupling 9, 15
Mizoroki–Heck reaction *see* Heck reaction
mobility 22, 146
modern investigations methods 309–334
molecular dynamics 278, 290ff.
molecular probes 277–308
monoclonal antibodies (Mabs) 60
monomers 92
multicatalytic processes 10
multiphoton excitation 323f.
multiple sclerosis 285
myelin 285
myeloid cancers/myeloma 225
myeloid leukemias 233
myoglobin 290

n
nanosecond transient absorption
  spectroscopy 310f.
naphthalene 234
naphthoquinones 137
Nazarov cyclization 215
Negishi–Stille coupling 7f.
neurofibrillary degeneration 209
neuronal excitability 232
neuroprotective effects 189
neutron–gamma pulse shape
  discrimination 180
NG-nitroarginine methyl ester hydrochloride
  (NAME) 302
nitric oxide 278
nitric oxide synthase (NOS) 218, 250, 303
nitroso oxide reactions 143f.
nitrostilbenes hydrolysis 54
nitrothienyl 238
nitroxide probe 129
nitroxide radicals 290, 300
nonlinear optics (NLO) 72
nonlinear optics materials 169f.
nonlinear transmission (NLT) 72, 323

nonresonant two-photon (NRTP) reaction 148
nonvertical triplet excitation transfer (NVET) 102–114f.
Notch1/2 – medulloblastoma 207
Nothobranchius furzeri 209
nuclear factor-kappaB (NF-κB) 250

o

Oil Yellow effect 120
OLED materials 163
olefination reactions 13ff.
olefinic C atoms pyramidalization 111
olefinic photoisomerization 1, 290
olefins methoxy-bromination 43
oligomers 195
one-bond flip mechanism 108
one-photon absorption (OPA) 72, 178, 323
Onsager model 279
ophthalmic combretastatin preparation 217, 272
optical properties 148
optical sampling 323f.
orbitals 67, 100, 160
organic optical materials 17
organozinc 7
ovarian (OW-1) tumor 242
oxalyl amide compounds 128
oxamide-based derivatives 18
oxidation 46ff.
oxygen-glucose deprivation (OGD) 232
oxygen-substituted acid 264
ozone 48

p

π-conjugated systems 105, 116, 150
π–π interactions 140
palladium-catalyzed reactions 6, 10f.
para-substituted stilbenes (PSS) 291
Parthenocissus laetevirens 195
pathogens 189
peanut resveratrol 192
perdeuterated *trans*-stilbene 130
Perkin reaction 9f.
peroxidase 245, 254
peroxisome proliferator-activated receptor (PPAR) 202, 232, 265
PGE2 inhibitory activity 273
pharmacokinetics 192f., 268f.
Phe^{L94} antibody residues 60
phenanthrene 147
phenanthroline rings 71
phenolic compounds 195
phenyl groups 2, 43, 102

phenyl–vinyl torsions 94, 112
phosphonate carbanions 4
phosphonium ylides 4
phosphorescence 67, 71
– triplet states interactions 90f.
– photoisomerization 101
phosphorylation 250
photoacoustic calorimetry (PAS) 121
photobleaching 163, 177
photochemical coupling 18
photochemical deactivation 102
photochemical reactions 106, 137–158
photochemical transformation 100
photochemistry 22
photochromes 13, 298
photoconductivity 162
photocyclization 137f.
photodestruction 296
photodimerization 140f.
photoisomerization 22, 70, 99–136
photolysis 55
photon counting 314f.
photon-excited fluorescence (TPEF) 72
photophysical/chemical properties 126, 278
photophysics 67–98
photoreactions 146f.
photoredox processes 67
photosensitizers 90
photoswitching processes 128f., 137
phthalocyanines 150
physically promoted reactions 19f.
phytoalexins 2, 189
piceatannol 262, 273
piceid 190
picosecond time-resolved single photon counting 314f.
pinosylvin 262, 273
piperidine 3, 9, 19
planar twisted intramolecular charge transfer (PICT) 102ff., 116
plants pterostilbene 254
plants resveratrol 190f.
plants stilbenes 189
plasma profiles 265, 270
Plasmopara viticola 34
platinum acetylide complexes 121
pleiadene 55
p–n junctions 166
polarity 117f., 278f.
polarization 67, 82
polar-substituted stilbenes 78
polyacetylenes 184
polyamidoamine dendrimers 21

polybutadiene 58
polyester stilbene 26
polyethylene glycol (PEG) 216
Polygonaceae 191
Polygonum cuspidatum 31f., 192, 204
polyhydroxylated ester analogues 10
polymer polyphosphazenes 327
polymer reactions 147f.
polymerase chain reaction (PCR) 250
polymer-dispersed liquid crystals (PDLC) 130, 148
polymerization 56f.
polymers reactions 147f.
poly(methyl methacrylate) (PMMA) 81, 130
polyphenols 192, 264
polyphenylene dendrimers 84, 126
polyphosphazenes 327
poly(propylene amine) dendrimers 85, 127, 147
polypropylene films 283
polystyrene films 155
polyurethane 169
positron emission therapy (PET) 273, 283
potassium 51
potential energy surfaces (PES) 102f., 112f., 155
power-limiting breakdown 160
preclinic effects 225–260
precursor mechanism 105f.
preparation 1–42
probing 280ff.
prostate cancer cell 199
prostatic adenocarcinoma 226
protein kinase C (PKC) 204ff.
proteins
– activities 196
– carbonylation/nitration 196
– complexation 60
– molecular probe/ dynamics 288f.
– photophysics 86f.
– polyphenol/resveratrol-bounded 265
protic solvents reactions 143f.
Pterocarpus marsupium Roxb. 273
pterostilbene 189ff.
– bioactivity 218f.
– clinics 273
– preclinics 232, 250f.
– resveratol 262
pulse radiolysis 57
pulse shape discrimination (PSD) 180
pump-damp-probe method (PDPM) 279
push–pull stilbenes 117
pyramidal neurons 232
pyrazoline analogues 235

pyridinium derivatives 114
pyrroloisoquinolines 139

q

quantitative solid-phase antibody orientation 296f.
quantitative structure–activity relationship (QSAR) 216
quantum yield 67
– fluorescence 82
– Franck–Condon state 78
– image-forming apparatus 180
– molecular probes 279
– phenanthrene 147
– photocyclization 139
– photoisomerization 99, 115, 120
quartz immobilized systems 293f.
quartz templates 86
quenching 91
– addition reactions 143
– fluorescence 88
– molecular probes 278, 289
– photoisomerization 124
quercetin 192, 233, 251, 263
quinones 53, 106, 245

r

radiative deactivation 78f.
radicals
– combretastatin 239
– formation 144
– polymerization 56f.
– two-photon excitation 149
radioluminescence scintillator materials 180f.
Raman spectroscopy 67, 329
reactions 143–148f.
reactive groups 287f.
reactive oxygen species (ROS) 190, 228
redshift 106
reduction 51f.
reductive coupling *see* McMurry reaction, coupling
reflective optical cavity 159
refraction index 279
regioselective oxidative coupling 49
relaxation shift 92, 279
reorganization 101, 108, 279
residues 60
resveratol
– aglycon (RES(AGL)) 193
– clinics 263
– bioactivity 189–213
– flavonoid pathway 31

– genetics 208f.
– polyhydroxylated ester analogues 10
– preclinics 225ff.
retinal neovascularization 250
retinoblastoma tumor cells 199
rhapontigenin 262, 273
rhubarb species 32
rigid surroundings 110f.
ring closure reactions 122
ring resonator 160
rotation mechanism 110
rotaxane 23
ruthenium complexes 173

s

salicylic acid 193
Saltiel mechanism 100ff.
Sandros plot 110
sarcoma 242
scintillator materials 180f.
SCN–ligands 26
SecA topology 286
second harmonic generation (SHG) 172
sensitivity 78, 124f.
Sicilian pistachio 31
Siegrist method 3f.
signaling pathways 207f.
silane-coupling agents 165
silanization 293
silica gel thin films 160
silyl chains 126
silylating agent 10
single-bond torsions 102
single-bond twisting 105f.
single-photon counting system 316
singlet states 1
– fluorescence 74f.
– molecular probes 290
– photoisomerization 1001
– photophysics 67f.
singlet–singlet absorption spectra 71
singlet–singlet energy transfer 93, 278, 302
SiO_2/SiO_2–TiO_2 surfaces 86
sirtuins 209, 265
sitosterol 190
SITS 210, 298
 see also derivatives, disulfonic acids
skin resveratrol deposition 193, 212
small molecules complexation 58f.
sodium 51
sodium tetrahydridoaluminate 57
sol–gel method 160
solid lasers materials 160f.

solid-phase antibody orientation 296f.
solid-phase extraction (SPE) 28
solutions 79f., 121f.
solvatochromism 278ff.
solvents 76, 106, 287f.
Sonogashira coupling 285
spin exchange 91, 110, 280
spin fluorescence-photochrome method 294
spin-echo techniques 330
spin-trapping 144
star-shaped stilbenoid phthalocyanine (SSS1Pc) 150
StDNP (dinitrophenylamino dimethylamino stilbene) 61, 292
stereoselectivity 47, 58
steric hindrance 151, 290
Stern–Volmer plots 80
stilbazoles 123
stilbene neutron detectors (SNDs) 180
stilbenoid chromophores 83
stilbenoid compounds 137
stilbenoid dendrimers 22
stilbenoid phenols 213
stilbenoids 2, 273f.
stilbenophanes 126f., 141
stilbenyl 84
stimulated emission pumping (SEP) fluorescence depletion 321
stimulated emission 159
Stokes shift 76, 110, 115f., 180
streptozotocin-nicotinamide-induced diabetes mellitus 254
structure probing 283f.
structure–reactivity correlation 76
styrenes 45, 56, 83
substituents 76, 117f.
substituted groups reactions 54f.
substituted stilbenes 106
substituted *trans*-stilbene derivatives 171
sulfation 192, 232
sulfotransferase 233
SUM(MA) fluorescence spectrum 25, 92
supermolecules 278, 302f.
supported liquid membranes (SLMs) 35
supported liquid-phase catalysts 51
supramolecular structure 142
surface microviscosity sensing 293f.
surfactant assemblies 137
susceptibility 50, 171
Suzuki–Miyaura coupling 9f., 15
synchronization factor 107
synchronous addition 48
syncope 271

synthesis 2ff., 10f.
– dendrimers 127
– resveratrol 197f., 229

t
TCNQ complexation 59
temperature dependence
– fluorescence quantum yields 82
– molecular probes 279
– photoisomerization 119, 124
templates 25f., 86f., 129f.
terphenyl 84
tetramethoxystilbene (TMST) 122
tetramethylethylene (TME) 143
theoreticals 101f., 113f., 278f.
thermal bond activation 112f.
thermal population 70
thermolysis 55
thiazolyl tetrazolium (MTT) assay 197
thienyl groups 119
thioflavin 283
thiol end-capped stilbenes 14
thiolate ions addition 53
third harmonic generation (THG) 160
three-photon absorption 170
time-dependent density functional theory (TD-DFT) 68, 105
time-domain spectroscopy (TDS) 322
time-resolved fluorescence depletion spectroscopy 321f.
time-resolved single photon counting 314f.
time-resolved spectra 311ff.
time-resolved vibrational spectroscopy 328f.
titanium nitride-catalyzed reduction 51
toluene 80, 146
toxicity 248, 271
trans–cis isomerization 95
trans–cis photoisomerization 76ff., 99
– fluorescence-photochrome method 302
– molecular probes 290ff.
– stilbenophanes 126
transferase 254
transient absorption spectroscopy 310f.
transient triplet–triplet (T–T) absorption spectra 71
transition metals catalysts 46
trans-methacryloyloxyethylcarbamoyloxy-methyl stilbenes (SUM/MA) 25
trans-pterostilbene *see* pterostilbene
trans-resveratrol 229
– clinics 262f.
– glucoside (trans-piceid) 190
  see also resveratrol

trans-stilbene 1
– azulene systems 80
– bromination 44
– complexation 59
– derivatives 293
– electronic states 68
– fluorescence spectra 91
– fluorescent-photochrome method 301
– Franck–Condon state 76
– molecular probes 286ff., 293ff.
– oxide materials 184
– photochemical behavior 119, 137
– photodimerization 142
– photoisomerization 102ff.
– polyphenol analogues 200
– pulse radiolysis 57
– radical cations 149
– vibrational spectroscopy 332
trans-ε,δ-viniferin 32
triacetate 227
triazoles 233
triflate 5
trifluoroethanol (TFE) 144
trinitrotoluene 278
triphenylamine 166
triphenylamino-substituted chromophores (TIOH) 175, 183
triplet states 1
– molecular probes 290
– photoisomerization 101
– photophysics 69f., 90f.
triplet-photochrome method 297f.
triplet–triplet (T–T) absorption spectra 71, 138
triplet–triplet energy transfer (TTET) 94, 110, 278
tryptophans 61
tubulins 60, 215f., 235
tumors 226, 239f., 262
twisted intramolecular charge transfer (TICT) 80
– molecular probes 281
– photoisomerization 102ff., 116
twisting 1
– charge transfer ionization 151
– chemical reactions 61
– photoisomerization 102–108, 117
two-photon absorption (TPA) 72f.
– image-forming apparatus 178
– laser materials 160, 170
– multiphoton excitation 323
two-photon excitation reactions 148f.
two-photon excited fluorescence (TPEF) 179, 323

two-photon fluorescence photophysics 72f.
Tyr^{H33} antibody residues 60
tyrosinase 245

u

ultrafast torsional isomerization 122
upconversion fluorescence spectroscopy 318
urethane acrylic monomer 25, 92
urethane-stilbene 87

v

V. amurensis 202, 208
V. angustifolium Aiton 191
V. thunbergi 192
Vaccinium myrtillus L. 191
Val^{H93} antibody residues 60
van der Waals nanocapsules 131, 140
vascular adhesion molecule 1 (VCAM-1) 126
vascular aging 213
vascular disrupting agents (VDA) 237, 262, 267f.
vascular endothelial growth factor (VEGF-Trap) 239–250, 273
vascular endothelial-cadherin 249
vascular targeting agents (VTA) 241, 273
ventricular tachycardia/fibrillation 231
very low-density lipoprotein (VLDL) 252
vibrational relaxation 117, 321
vibrational spectroscopy 328f.
vibronic two-photon absorption spectra 75
Vilsmeier–Haack reaction 183
vineatrol 228
viniferin 227
vinyl lithiation 52
virus–stilbene conjugate 28f.
viscosity effect 119f.
Vitis vinifera 262
Volta potential 148
volume-conserving mechanism 102, 108f.

w

Wadsworth–Emmons reaction 178
water–ethylene glycol catalyst 51
wavelength laser spectrum 159
Western immunoblotting 198, 204
white light continuum (WLC) 312
whitening agents 31, 183
wine resveratrol 190
Wittig reaction 4f.
Wittig–Horner reactions 13f., 176

x

xanthine oxidase 195
xenografts 239f.

y

yolk-sac membranes 205
Yucca schidigera 196

z

zeolites 153
Z-scan technique 169
Z-stilbenes 1
– bromination 44
– epoxidation 48
– photocyclization 138
– radical polymerization 56
zwitterionic states 78, 102